Advances in Intelligent Systems and Computing

Volume 434

Series editor

Janusz Kacprzyk, Polish Academy of Sciences, Warsaw, Poland
e-mail: kacprzyk@ibspan.waw.pl

About this Series

The series "Advances in Intelligent Systems and Computing" contains publications on theory, applications, and design methods of Intelligent Systems and Intelligent Computing. Virtually all disciplines such as engineering, natural sciences, computer and information science, ICT, economics, business, e-commerce, environment, healthcare, life science are covered. The list of topics spans all the areas of modern intelligent systems and computing.

The publications within "Advances in Intelligent Systems and Computing" are primarily textbooks and proceedings of important conferences, symposia and congresses. They cover significant recent developments in the field, both of a foundational and applicable character. An important characteristic feature of the series is the short publication time and world-wide distribution. This permits a rapid and broad dissemination of research results.

Advisory Board

Chairman

Nikhil R. Pal, Indian Statistical Institute, Kolkata, India
e-mail: nikhil@isical.ac.in

Members

Rafael Bello, Universidad Central "Marta Abreu" de Las Villas, Santa Clara, Cuba
e-mail: rbellop@uclv.edu.cu

Emilio S. Corchado, University of Salamanca, Salamanca, Spain
e-mail: escorchado@usal.es

Hani Hagras, University of Essex, Colchester, UK
e-mail: hani@essex.ac.uk

László T. Kóczy, Széchenyi István University, Győr, Hungary
e-mail: koczy@sze.hu

Vladik Kreinovich, University of Texas at El Paso, El Paso, USA
e-mail: vladik@utep.edu

Chin-Teng Lin, National Chiao Tung University, Hsinchu, Taiwan
e-mail: ctlin@mail.nctu.edu.tw

Jie Lu, University of Technology, Sydney, Australia
e-mail: Jie.Lu@uts.edu.au

Patricia Melin, Tijuana Institute of Technology, Tijuana, Mexico
e-mail: epmelin@hafsamx.org

Nadia Nedjah, State University of Rio de Janeiro, Rio de Janeiro, Brazil
e-mail: nadia@eng.uerj.br

Ngoc Thanh Nguyen, Wroclaw University of Technology, Wroclaw, Poland
e-mail: Ngoc-Thanh.Nguyen@pwr.edu.pl

Jun Wang, The Chinese University of Hong Kong, Shatin, Hong Kong
e-mail: jwang@mae.cuhk.edu.hk

More information about this series at http://www.springer.com/series/11156

Suresh Chandra Satapathy
Jyotsna Kumar Mandal · Siba K. Udgata
Vikrant Bhateja
Editors

Information Systems Design and Intelligent Applications

Proceedings of Third International
Conference INDIA 2016, Volume 2

 Springer

Editors
Suresh Chandra Satapathy
Department of Computer Science
  and Engineering
Anil Neerukonda Institute of Technology
  and Sciences
Visakhapatnam
India

Jyotsna Kumar Mandal
Kalyani University
Nadia, West Bengal
India

Siba K. Udgata
University of Hyderabad
Hyderabad
India

Vikrant Bhateja
Department of Electronics and
  Communication Engineering
Shri Ramswaroop Memorial Group
  of Professional Colleges
Lucknow, Uttar Pradesh
India

ISSN 2194-5357   ISSN 2194-5365   (electronic)
Advances in Intelligent Systems and Computing
ISBN 978-81-322-2750-2   ISBN 978-81-322-2752-6   (eBook)
DOI 10.1007/978-81-322-2752-6

Library of Congress Control Number: 2015960416

Preface

The papers in this volume were presented at the INDIA 2016: Third International Conference on Information System Design and Intelligent Applications. This conference was organized by the Department of CSE of Anil Neerukonda Institute of Technology and Sciences (ANITS) and ANITS CSI Student Branch with technical support of CSI, Division-V (Education and Research) during 8–9 January 2016. The conference was hosted in the ANITS campus. The objective of this international conference was to provide opportunities for researchers, academicians, industry personas and students to interact and exchange ideas, experience and expertise in the current trends and strategies for Information and Intelligent Techniques. Research submissions in various advanced technology areas were received and after a rigorous peer-review process with the help of programme committee members and external reviewers, 215 papers in three separate volumes (Volume I: 75, Volume II: 75, Volume III: 65) were accepted with an acceptance ratio of 0.38. The conference featured seven special sessions in various cutting edge technologies, which were conducted by eminent professors. Many distinguished personalities like Dr. Ashok Deshpande, Founding Chair: Berkeley Initiative in Soft Computing (BISC)—UC Berkeley CA; Guest Faculty, University of California Berkeley; Visiting Professor, University of New South Wales Canberra and Indian Institute of Technology Bombay, Mumbai, India, Dr. Parag Kulkarni, Pune; Dr. Aynur Ünal, Strategic Adviser and Visiting Full Professor, Department of Mechanical Engineering, IIT Guwahati; Dr. Goutam Sanyal, NIT, Durgapur; Dr. Naeem Hannoon, Universiti Teknologi MARA, Shah Alam, Malaysia; Dr. Rajib Mall, Indian Institute of Technology Kharagpur, India; Dr. B. Majhi, NIT-Rourkela; Dr. Vipin Tyagi, Jaypee University of Engineering and Technology, Guna; Prof. Bipin V. Mehta, President CSI; Dr. Durgesh Kumar Mishra, Chairman, Div-IV, CSI; Dr. Manas Kumar Sanyal, University of Kalyani; Prof. Amit Joshi, Sabar Institute, Gujarat; Dr. J.V.R. Murthy, JNTU, Kakinada; Dr. P.V.G.D. Prasad Reddy, CoE, Andhra University; Dr. K. Srujan Raju, CMR Technical Campus, Hyderabad; Dr. Swagatam Das, ISI Kolkata; Dr. B.K. Panigrahi, IIT Delhi; Dr. V. Suma, Dayananda Sagar Institute, Bangalore; Dr. P.S. Avadhani,

Vice-Principal, CoE(A), Andhra University and Chairman of CSI, Vizag Chapter, and many more graced the occasion as distinguished speaker, session chairs, panelist for panel discussions, etc., during the conference days.

Our sincere thanks to Dr. Neerukonda B.R. Prasad, Chairman, Shri V. Thapovardhan, Secretary and Correspondent, Dr. R. Govardhan Rao, Director (Admin) and Prof. V.S.R.K. Prasad, Principal of ANITS for their excellent support and encouragement to organize this conference of such magnitude.

Thanks are due to all special session chairs, track managers and distinguished reviewers for their timely technical support. Our entire organizing committee, staff of CSE department and student volunteers deserve a big pat for their tireless efforts to make the event a grand success. Special thanks to our Programme Chairs for carrying out an immaculate job. We place our special thanks here to our publication chairs, who did a great job to make the conference widely visible.

Lastly, our heartfelt thanks to all authors without whom the conference would never have happened. Their technical contributions made our proceedings rich and praiseworthy. We hope that readers will find the chapters useful and interesting.

Our sincere thanks to all sponsors, press, print and electronic media for their excellent coverage of the conference.

November 2015   Suresh Chandra Satapathy
                                                                 Jyotsna Kumar Mandal
                                                                        Siba K. Udgata
                                                                        Vikrant Bhateja

Organizing Committee

Chief Patrons

Dr. Neerukonda B.R. Prasad, Chairman, ANITS
Shri V. Thapovardhan, Secretary and Correspondent, ANITS, Visakhapatnam

Patrons

Prof. V.S.R.K. Prasad, Principal, ANITS, Visakhapatnam
Prof. R. Govardhan Rao, Director-Admin, ANITS, Visakhapatnam

Honorary Chairs

Dr. Bipin V. Mehta, President CSI, India
Dr. Anirban Basu, Vice-President, CSI, India

Advisory Committee

Prof. P.S. Avadhani, Chairman, CSI Vizag Chapter, Vice Principal, AU College of Engineering
Shri D.N. Rao, Vice Chairman and Chairman (Elect), CSI Vizag Chapter, Director (Operations), RINL, Vizag Steel Plant
Shri Y. Madhusudana Rao, Secretary, CSI Vizag Chapter, AGM (IT), Vizag Steel Plant
Shri Y. Satyanarayana, Treasurer, CSI Vizag Chapter, AGM (IT), Vizag Steel Plant

Organizing Chair

Dr. Suresh Chandra Satapathy, ANITS, Visakhapatnam

Organizing Members

All faculty and staff of Department of CSE, ANITS
Students Volunteers of ANITS CSI Student Branch

Program Chair

Dr. Manas Kumar Sanayal, University of Kalyani, West Bengal
Prof. Pritee Parwekar, ANITS

Publication Chair

Prof. Vikrant Bhateja, SRMGPC, Lucknow

Publication Co-chair

Mr. Amit Joshi, CSI Udaipur Chapter

Publicity Committee

Chair: Dr. K. Srujan Raju, CMR Technical Campus, Hyderabad
Co-chair: Dr. Venu Madhav Kuthadi,
Department of Applied Information Systems
Faculty of Management
University of Johannesburg
Auckland Park, Johannesburg, RSA

Special Session Chairs

Dr. Mahesh Chandra, BIT Mesra, India, Dr. Asutosh Kar, BITS, Hyderabad: "Modern Adaptive Filtering Algorithms and Applications for Biomedical Signal Processing Designs"
Dr. Vipin Tyagi, JIIT, Guna: "Cyber Security and Digital Forensics"
Dr. Anuja Arora, Dr. Parmeet, Dr. Shikha Mehta, JIIT, Noida-62: "Recent Trends in Data Intensive Computing and Applications"
Dr. Suma, Dayananda Sagar Institute, Bangalore: "Software Engineering and its Applications"
Hari Mohan Pandey, Ankit Chaudhary: "Patricia Ryser-Welch, Jagdish Raheja", "Hybrid Intelligence and Applications"
Hardeep Singh, Punjab: "ICT, IT Security & Prospective in Science, Engineering & Management"
Dr. Divakar Yadav, Dr. Vimal Kumar, JIIT, Noida-62: "Recent Trends in Information Retrieval"

Track Managers

Track #1: Image Processing, Machine Learning and Pattern Recognition—Dr. Steven L. Fernandez
Track #2: Data Engineering—Dr. Sireesha Rodda
Track #3: Software Engineering—Dr. Kavita Choudhary
Track #4: Intelligent Signal Processing and Soft Computing—Dr. Sayan Chakraborty

Technical Review Committee

Contents

About the Editors

Dr. Suresh Chandra Satapathy is currently working as Professor and Head, Department of Computer Science and Engineering, Anil Neerukonda Institute of Technology and Sciences (ANITS), Visakhapatnam, Andhra Pradesh, India. He obtained his Ph.D. in Computer Science Engineering from JNTUH, Hyderabad and his Master's degree in Computer Science and Engineering from National Institute of Technology (NIT), Rourkela, Odisha. He has more than 27 years of teaching and research experience. His research interests include machine learning, data mining, swarm intelligence studies and their applications to engineering. He has more than 98 publications to his credit in various reputed international journals and conference proceedings. He has edited many volumes from Springer AISC and LNCS in the past and he is also the editorial board member of a few international journals. He is a senior member of IEEE and Life Member of Computer society of India. Currently, he is the National Chairman of Division-V (Education and Research) of Computer Society of India.

Dr. Jyotsna Kumar Mandal has an M.Sc. in Physics from Jadavpur University in 1986, M.Tech. in Computer Science from University of Calcutta. He was awarded the Ph.D. in Computer Science & Engineering by Jadavpur University in 2000. Presently, he is working as Professor of Computer Science & Engineering and former Dean, Faculty of Engineering, Technology and Management, Kalyani University, Kalyani, Nadia, West Bengal for two consecutive terms. He started his career as lecturer at NERIST, Arunachal Pradesh in September, 1988. He has teaching and research experience of 28 years. His areas of research include coding theory, data and network security, remote sensing and GIS-based applications, data compression, error correction, visual cryptography, steganography, security in MANET, wireless networks and unify computing. He has produced 11 Ph.D. degrees of which three have been submitted (2015) and eight are ongoing. He has supervised 3 M.Phil. and 30 M.Tech. theses. He is life member of Computer Society of India since 1992, CRSI since 2009, ACM since 2012, IEEE since 2013 and Fellow member of IETE since 2012, Executive member of CSI Kolkata Chapter. He has delivered invited lectures and acted as programme chair of many international conferences and also edited nine volumes of proceedings from Springer AISC series, CSI 2012 from McGraw-Hill, CIMTA 2013 from Procedia Technology, Elsevier. He is reviewer of various international journals and conferences. He has over 355 articles and 5 books published to his credit.

Dr. Siba K. Udgata is a Professor of School of Computer and Information Sciences, University of Hyderabad, India. He is presently heading Centre for Modelling, Simulation and Design (CMSD), a high-performance computing facility at University of Hyderabad. He obtained his Master's followed by Ph.D. in Computer Science (mobile computing and wireless communication). His main research interests include wireless communication, mobile computing, wireless sensor networks and intelligent algorithms. He was a United Nations Fellow and worked in the United Nations University/International Institute for Software Technology (UNU/IIST), Macau, as research fellow in the year 2001. Dr. Udgata is working as principal investigator in many Government of India funded research projects, mainly for development of wireless sensor network applications and application of swarm intelligence techniques. He has published extensively in refereed international journals and conferences in India and abroad. He was also on the editorial board of many Springer LNCS/LNAI and Springer AISC Proceedings.

Prof. Vikrant Bhateja is Associate Professor, Department of Electronics and Communication Engineering, Shri Ramswaroop Memorial Group of Professional Colleges (SRMGPC), Lucknow, and also the Head (Academics & Quality Control) in the same college. His areas of research include digital image and video processing, computer vision, medical imaging, machine learning, pattern analysis and recognition, neural networks, soft computing and bio-inspired computing techniques. He has more than 90 quality publications in various international journals and conference proceedings. Professor Vikrant has been on TPC and chaired various sessions from the above domain in international conferences of IEEE and Springer. He has been the track chair and served in the core-technical/editorial teams for international conferences: FICTA 2014, CSI 2014 and INDIA 2015 under Springer-ASIC Series and INDIACom-2015, ICACCI-2015 under IEEE. He is associate editor in International Journal of Convergence Computing (IJConvC) and also serves on the editorial board of International Journal of Image Mining (IJIM) under Inderscience Publishers. At present, he is guest editor for two special issues floated in International Journal of Rough Sets and Data Analysis (IJRSDA) and International Journal of System Dynamics Applications (IJSDA) under IGI Global publications.

A Short Run Length Descriptor for Image Retrieval

Nishant Shrivastava and Vipin Tyagi

Abstract In this paper an image retrieval technique based on a novel Short Run Length Descriptor (SRLD) is proposed. SRLD can effectively represent image local and global information. It can be viewed as an integrated representation of both color and texture properties. HSV color space is quantized to 72 bins and SRLD is computed using short run lengths of size two and three for each color in different orientations. Short run lengths at all orientations are combined to get Short Run Length Histogram (SRLH) feature. SRLH can thoroughly describe the spatial correlation between color and texture and have the advantages of both statistical and structural approaches of texture representation. The experimental results clearly demonstrate the effectiveness of the proposed descriptor in image retrieval applications.

Keywords Run length histogram · Color descriptor · Texture descriptor · Texton detection · Color quantization

1 Introduction

In the recent years, growth of multimedia information from various sources has increased many folds. This has created the demand of an accurate Content Based Image Retrieval Systems (CBIR). Success of a general purpose CBIR largely depends upon the effectiveness of descriptors used to represent images. Low level features like color, texture and shape have been used for describing content of the images. The shape descriptors depends on the accuracy of the image segmentation

N. Shrivastava (✉) · V. Tyagi
Department of Computer Science and Engineering, Jaypee University of Engineering and Technology, Raghogarh, 473226 Guna, MP, India
e-mail: nishantuit@gmail.com

V. Tyagi
e-mail: dr.vipin.tyagi@gmail.com

© Springer India 2016
S.C. Satapathy et al. (eds.), *Information Systems Design and Intelligent Applications*, Advances in Intelligent Systems and Computing 434,
DOI 10.1007/978-81-322-2752-6_1

1

technique employed [1]. The accurate segmentation of images is still an open problem. Therefore a large number of CBIR techniques rely heavily on color and texture information of the image. Color and texture can provide the robust feature set for providing most discriminating information for natural images [2–4].

Color and texture integration using local binary patters have shown to obtain a good retrieval accuracy [5, 6]. Also many researchers have shown to integrate both the color and texture in a single descriptor using Textons [7–10]. A texton can be defined as the structural element used to capture the run lengths in an image at a particular orientation. However, the structure of textons is rigid. Moving Textons over the image may ignore some finer texture details which may be important for image discrimination. Each texton is moved separately for capturing texture information in a particular orientation. Individual texton maps are then combined to make a integrated descriptor for representing color and texture of the image. This consumes a lot of time.

To improve the texton representation, in this paper, a novel texture descriptor SRLD and its histogram SRLH is proposed. The images are quantized into 72 main colors in HSV color space. Short run length of each color of size 2 and 3 are extracted from the image. The size is chosen to overcome the limitation of texton based approaches which uses matrix of size 2×2 or 3×3 to extract texture features. Run lengths at each orientation are combined to make the final SRLH. The proposed SRLH can describe the correlation between color and texture in a detailed manner and have the advantages of both statistical and structural approaches of extracting texture. SRLH can be seen as an integrated representation of information gained from all type of textons together.

The rest of the paper is organized as follows: Sect. 2 describes the related work. Proposed SRLD and SRLH descriptors are discussed in Sect. 3. Experimental results are demonstrated in Sect. 4 and conclusions are summed up in Sect. 5.

2 Related Work

Various descriptors have been proposed for integrating color and texture of images [7–9, 11–13]. Micro-Structure Descriptors (MSD) [8] utilize underlying colors in microstructure with similar edge orientation to represent color, texture and orientation information of images. MSD uses 3×3 windows to detect microstructure in a quantized color image. Pixels having value similar to centre pixel in window of size 3×3 are only retained to define microstructures. MSD does not provide detailed correlation of color and texture as many textural patterns are left undetected as only centre pixel is considered.

Structure Element Descriptor (SED) [9] based scheme uses 2×2 matrix as shown in Fig. 1 to extract texture information at different orientation. The image is first quantized to 72 bins in HSV color space. Each of the five structuring elements are moved over the image for detecting textons of each of the 72 colors. A SED is detected when pixels having same value occur in colored part of template. After

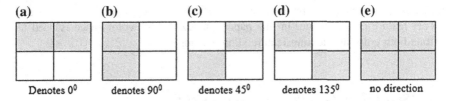

Fig. 1 Five texton type defined in SED

moving each of the 5 SED, the total count of SED of each type for each color form the SEH, with each bin representing the total number of SED found of a particular type for each color, is used to describe the image. SED is not capable of extracting all texture information as run lengths of odd size are not detected properly; also SED are redundant and overlap over each other as fifth SED is detected when all four are detected and vice versa. Moreover, the process takes a lot of time moving SED over the image.

Texton Co-occurrence Matrix (TCM) [10] based technique also uses textons to extract the texture. Textons are moved over the quantized image from left-to-right and top-to-bottom to detect textons with one pixel as the step-length. If the pixel values that fall in the texton template are the same, those pixels will form a texton, and their values are kept as the original values. Otherwise they will be set to zero. Each texton template can lead to a texton image, and five texton templates will lead to five texton images. Finally, all texton images are combined to get a single image.

Multi-Texton histogram (MTH) [7] based image retrieval integrates the advantage of co-occurrence matrix and histogram by representing the attribute of co-occurrence matrix by histogram. MTH uses first four texton out of 5 SED to extract texture from the image. Like SED, MTH also is not able to represent the full content of images. In [3], an adaptive color feature extraction scheme using the Binary Quaternion-Moment Preserving (BQMP) threshold technique is used to describe the distribution of image.

3 Proposed Method

3.1 Color Quantization in HSV Color Space

Color is the most commonly used feature in the CBIR since it is not affected by rotation, scaling and other transformation on the image. In this paper, we have selected HSV (Hue, Saturation, Vaue) color space since it is more perceptually uniform than other color spaces [14].

To obtain quantized image having 72 colors the images are converted from RGB to HSV color space and a non-uniform quantization technique [15] is used. One dimensional color feature vector P, is constructed using: $P = 9H + 3S + V$. Each image is quantized to 72 main colors and SRLD is computed to finally get the SRLH feature of the image. In this paper quantized HSV color space is used to extract both color and texture feature simultaneously.

3.2 Short Run Length Descriptor (SRLD)

The color, texture and shape features are extensively used in representation of images in content based image retrieval system. After quantization of image to 72 colors in HSV space, the texture information can be extracted using statistical and structural methods. SED has a limitation that only one type of SED can be used at a time, therefore it cannot describe all the repetitive structure in the image. Figure 4a shows an example portion of image having run length of 3 wrongly represented by 2×2 SED as of length 2. This confirms that the run lengths of odd size cannot be represented by SED. Figure 4b shows that the pair of 1 left undetected by moving SED of 2×2 with step length of 2 over the image. From Fig. 2, it is obvious that SED based methods can only represent the local characteristic of image and lacks in detail analysis of texture from the whole image. To integrate the color and texture information in a single descriptor including higher details of spatial correlation, we have proposed a more effective short run length descriptor (SRLD).

Capturing texture information using structuring elements is not flexible and may result in loss of some important discriminating texture patterns. The SRLD uses run lengths of size at most 3 to describe different texture structures hence is able to describe all repetitive texture patterns in the image. The size of run lengths is kept limited to 2 and 3 as the combination of 2 and 3 can describe any odd and even numbers. This is analogous to the texton based techniques using matrix of size 2×2 or 3×3 to extract texture. To capture orientation information the run lengths are extracted at $0°$, $45°$, $90°$ and $135°$ for each quantization level in the HSV color space. The process of extracting SRLD can be described in a simple 3-step strategy as follows:

1. Starting from (0, 0), scan each row of pixel from top to bottom. To avoid the extraction of wrong run length, counting of pixels terminates at the end of each row and start at the beginning of each row.

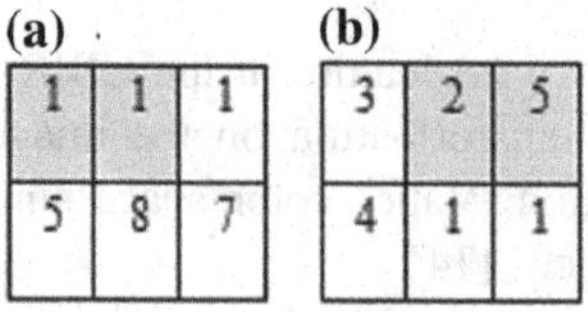

Fig. 2 An example showing a run length of 3 described by SED as of length 2 b undetected run length of pair of 1

2. Compute run lengths with size of at most 3 pixels excluding those of length 1. If the run length size is greater than 2 and 3 then break it into multiple smaller run length of size 2 and 3.
3. Count the number of run lengths of size 2 and 3 for each color for making final run length.

The above steps are used to extract SRLD at orientation of 0°. For other orientations the image is scanned in a column to column and diagonal to diagonal basis. The outcome of this process is a total of 4 run lengths one for each orientation. It can be easily observed that the run length representation is similar to texton based methods with more detail texture analysis.

3.3 Short Run Length Histogram

The run length computed above contains 2 entries for each color, first entry shows the number of run lengths of size 2 and other entry specify the total number of run lengths of size 3 in each orientation. All these run lengths are combined to form a single run length thereby having 8 entries for a single color. The first four entries represent the total number of run length of size 2 and the other 4 entries represent total run length of size 3 in each of the four possible orientations respectively. The final run length obtained is represented with the help of a histogram having 72 × 8 bins.

The method of SRLH computation is described in Fig. 3. For simplicity the quantized colors in HSV color space are denoted by alphabets a, b, c, d, e and f. The technique is illustrated using 6 colors, therefore the SRLH at each orientation contains 6 × 2 i.e. 12 bins. In real, the experiments are conducted with 72 colors and the histogram thus produced contains 72 × 2 i.e. 144 bins. Histograms at other three orientations are computed in a similar manner. All resulting histograms are merged to get a single histogram as shown in Fig. 4.

The histogram are combined starting from the first color (i.e. color a). For each color runs of size 2 in each of the 4 orientation are combined followed by combining the runs of size 3. The similar process is repeated for all the colors.

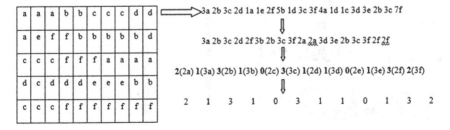

Fig. 3 The process of extraction of short run length histogram at an orientation of 0°

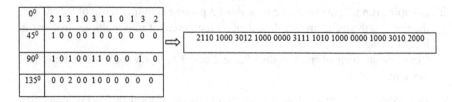

0^0	2 1 3 1 0 3 1 1 0 1 3 2
45^0	1 0 0 0 0 1 0 0 0 0 0 0
90^0	1 0 1 0 0 1 1 0 0 0 1 0
135^0	0 0 2 0 0 1 0 0 0 0 0 0

⇒ | 2110 1000 3012 1000 0000 3111 1010 1000 0000 1000 3010 2000 |

Fig. 4 The process of combining histograms into single histogram

The combined SRLH thus produced contains 6 × 8 bins (i.e. 48) for the case of 6 colors. In real experiments are performed using 72 colors in HSV color space therefore the final SRLH have a total of 72 × 8 (i.e. 576) bins. It may be easily noticed that the SRLH is similar to the texton histogram with higher texture details. For example in SED, each color is represented as 5 bins corresponding to 5 textons shown in Fig. 1, However, in the present method each color is represented as 8 bins corresponding to two different sizes of run length and 4 different orientation.

Figure 5 shows images and their corresponding SRLH. It can be easily depicted from the figure that the SRLH for similar images are similar. This confirms the effectiveness of SRLH in representing images. When the image is scaled, number of pixels in the images gets changed. SRLH may be different for original and scaled image. This problem can be solved by maintaining the proportion of pixels same in both images. To achieve this objective normalization is applied. Let $C_i (0 \leq i \leq 71)$

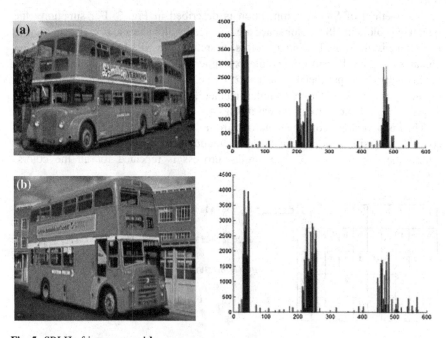

Fig. 5 SRLH of images **a** and **b**

denotes the quantized color in HSV color space. R_{1i}^n, R_{2i}^n, R_{3i}^n and R_{4i}^n denotes the number of run lengths of color i of size n at each of the four orientations respectively. The value of n can be either 2 or 3. The normalized value can be computed as:

$$r_{ji}^n = \frac{\left|R_{ji}^n\right|}{\sum_{j=1}^4 \left|R_{ji}^n\right|} \tag{1}$$

where, r_{ji}^n is the normalized bin value for orientation j. Similarly normalized bin value for n = 3 for color i is computed. Therefore each color is represented as 8 bins in the SRLH.

4 Experimental Results

To demonstrate the performance of the proposed descriptor, experiments are performed on MPEG-7 Common Color dataset (CCD) [16] (dataset-1) and Corel 11000 database (dataset-2) [17]. Dataset-1 (CCD) consists of 5000 images and a set of 50 Common Color Queries (CCQ) each with specified ground truth images. CCD consists of variety of still images produced from stock photo galleries, consecutive frames of news cast, sports channel and animations. In particular experiments we used as ground truth, the groups of images proposed in the MIRROR image retrieval system [16]. Chi-square distance is used as similarity measure for evaluating the performance of the system.

Figure 6 shows the retrieval performance comparison of the proposed SRLH with other three methods MTH, MSD and SED. It can be observed that SRLH has outperformed others on both Dataset-1 and Dataset-2. The reason being the MSD,

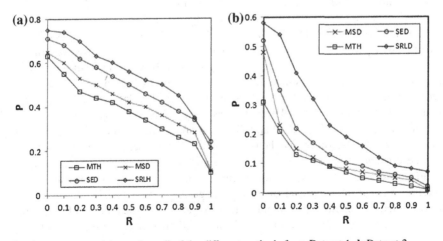

Fig. 6 Average precision and recall of the different methods for **a** Dataset-1. **b** Dataset-2

Fig. 7 Image retrieval for horses

MTH and SED based method does not represent various discriminating patterns of texture and hence has limited capability of describing color and texture of the image. The proposed SRLH is flexible and can represent detail texture information in the quantized color image. It can represent combined information gained from all type of textons of size 2×2 and 3×3 together in a single descriptor. Also in SRLD, orientation is captured without overlap. In Fig. 6a, for top 10 images average precision of SED, MSD and MTH based methods are 72, 65 and 61 % respectively. At this point SRLH outperforms others with average precision of 78 %. Similar conclusions can be drawn from Fig. 6b, using Dataset-2. An example of the sample retrieval result from our system using dataset-2 are shown in Fig. 7. Top 20 retrieved images are shown for query image of horse. Top left image is the query image and other images are relevant images retrieved from the system.

The extraction of MTH, MSD and SED descriptors require moving textons over the images multiple times and hence consume more time. Textons representing unique orientation of texture are moved for each color bin separately for making texton histogram. This takes a lot of time in high resolution images and is in feasible for real time retrieval of natural images. Proposed SRLH consumes less time as image is required to scan only 4 times regardless of the number of color bins used in the image. Each time run length in a particular orientation is extracted to make histogram.

5 Conclusions

In this paper a novel short run length descriptor for content based image retrieval is proposed which can represent color, texture and orientation information of the whole image in a compact and intuitive manner. The proposed SRLH can better represent the correlation between color and texture and can describe texture information extracted from all type of texton in a single descriptor. In addition texton based approaches like SED, EOAC [1], TCM and MTH consume more time in texton analysis and moving textons over the images. The proposed approach is faster as only the run lengths from the images are to be extracted in each orientation for the construction of feature vector. The experimental results on representative databases have shown that the proposed approach outperform other significantly and hence can be used in CBIR effectively.

References

1. F. Mahmoudi, J. Shanbehzadeh, Image retrieval based on shape similarity by edge orientation autocorrelogram, Pattern Recognition, 36 (8) (2003) 1725–1736.
2. N. Shrivastava, V. Tyagi, Content based image retrieval based on relative locations of multiple regions of interest using selective regions matching, Inform. Sci. Volume 259, (2014), 212–224.
3. C.H. Lin et al., A smart content-based image retrieval system based on color and texture feature, Image and Vision Computing, 27 (2009), 658–665.
4. N. Shrivastava, V. Tyagi, Multistage content- based image retrieval, CSI sixth international conference on software engineering (CONSEG) IEEE, (2012), 1–4.
5. Shrivastava N, Tyagi V (2013): An effective scheme for image texture classification based on binary local structure pattern, Visual Computer, Springer Berlin Verlag, pp. 1–10. doi:10.1007/s00371-013-0887-0.
6. Shrivastava N, Tyagi V (2015): An integrated approach for image retrieval using local binary pattern, Multimedia Tools and Applications, Springer. doi:10.1007/s11042-015-2589-2.
7. G.H. Liu, L. Zhang, Y.K. Hou, Z.Y. Li, J.Y. Yang, Image retrieval based on multitexton histogram, Pattern Recognition 43 (7) (2010) 2380–2389.
8. G.H. Liu, Z.Y. Li, L. Zhang, Y. Xu, Image retrieval based on micro-structure descriptor, Pattern Recognition 44 (9) (2011) 2123–2133.
9. W. Xingyuan, W. Zongyu, A novel method for image retrieval based on structure elements descriptor, Journal of Visual Communication and Image Representation, 24 (2013) 63–74.
10. G.H. Liu, J.Y. Yang, Image retrieval based on the texton co-occurrence matrix, Pattern Recognition 41 (12) (2008) 3521–3527.
11. G. Michèle, Z. Bertrand, Body color sets: a compact and reliable representation of images, Journal of Visual Communication and Image Representation, 22 (1) (2011) 48–60.
12. C.H. Lin, D.C. Huang, Y.K. Chan, K.H. Chen, Y.J. Chang, Fast color-spatial feature based image retrieval methods, Expert Systems with Applications, 38 (9) (2011) 11412–11420.
13. W.T. Chen, W.C. Liu, M.S. Chen, Adaptive color feature extraction base on image color distributions, IEEE Transactions on Image Processing 19 (8) (2010) 2005–2016.
14. Swain, M.J. and Ballard, D.H.: Color Indexing, Int. J. Computer Vision, 7(1) (1991) 11–32.

15. J.L. Liu, D.G. Kong, Image retrieval based on weighted blocks and color feature, International Conference on Mechatronic Science, Electric Engineering and Computer, Jilin, 2011, pp. 921–924.
16. Wong, K.-M., Cheung, K.-W., Po, L.-M.: MIRROR: an interactive content based image retrieval system. In: Proceedings of IEEE International Symposium on Circuit and Systems 2005, Japan, vol. 2, pp. 1541–1544 (2005).
17. Coral image database: http://wang.ist.psu.edu/docs/related/.

A New Curvelet Based Blind Semi-fragile Watermarking Scheme for Authentication and Tamper Detection of Digital Images

S. Nirmala and K.R. Chetan

Abstract A novel blind semi-fragile watermarking scheme for authentication and tamper detection of digital images is proposed in this paper. This watermarking scheme is based on Discrete Curvelet Transform (DCLT), which captures the information content of the image in few coefficients compared to other transforms. The novelty of the approach is that the first level coarse DCLT coefficients of the input image are quantized into 4 bits which is used as watermark and embedded into the pseudo randomly determined coefficients. At the receiver side, the extracted and generated first level coarse DCLT coefficients of the watermarked image are divided into blocks of uniform size. The difference in the energy between each block of extracted and generated coefficients is compared and if the difference exceeds threshold, the block is marked as tampered. This scheme exhibits higher Normalization Correlation Coefficient (NCC) values for various incidental attacks and is thus more robust than existing scheme [1]. The proposed scheme outperforms in localizing tampered regions compared to method [1].

Keywords Discrete curvelet transforms · Semi-fragile watermarking · Normalized correlation coefficient · Tamper detection · Incidental attacks · Intentional attacks

1 Introduction

Digital watermarking schemes protect the integrity and authenticity of the digital images. The watermarking schemes are broadly categorized as robust, fragile and semi-fragile. Robust watermarks are designed to resist attempts to remove or destroy the watermark. The fragile watermarks are designed to be easily destroyed

S. Nirmala (✉) · K.R. Chetan
Department of CSE, JNN College of Engineering, Shimoga, India
e-mail: nir_shiv_2002@yahoo.co.in

K.R. Chetan
e-mail: krc_555@yahoo.co.in

© Springer India 2016
S.C. Satapathy et al. (eds.), *Information Systems Design and Intelligent Applications*, Advances in Intelligent Systems and Computing 434,
DOI 10.1007/978-81-322-2752-6_2

for the minor manipulation on the watermarked image. A semi-fragile watermark combines the properties of fragile and robust watermarks [2]. The semi-fragile watermarks are robust against most of the incidental attacks and fragile to the intentional attacks. Many efforts on semi-fragile watermarking schemes were found in literature. Chang et al. [3] proposed a semi-fragile watermarking technique to improve the tamper detection sensitivity by analyzing and observing the impact of various image manipulations on the wavelet transformed coefficients. In [4–6], semi-fragile watermarking schemes based on the Computer Generated Hologram coding techniques were described. Maeno et al. [7] presented a semi-fragile watermarking scheme for extracting content-based image features from the approximation sub-band in the wavelet domain, to generate two complementary watermarks. A blind semi-fragile watermarking scheme for medical images was discussed in [8]. Wu et al. [9] proposed a semi-fragile watermarking scheme through parameterized Integer Wavelet transform.

A new watermarking approach based on Discrete Curvelet Transforms (DCLT) [10] was developed in [1]. In this method, a logo watermark is embedded into first level DCLT coefficients using an additive embedding scheme controlled by visibility parameter. The existing method [1] has many limitations. The Normalized Correlation Coefficient (NCC) values drops after a variance of 0.05 for the noise attacks. The existing method [1] is not completely blind as the first level coarse DCLT coefficients and visibility factor needs to be communicated to the receiver. In this paper, a blind semi-fragile watermarking is proposed that addresses all the afore-mentioned limitations. The authentication is performed based on the fact that incidental attacks do not change information content and the average energy in a block substantially. The rest of the paper is organized as follows: The proposed methodology is explained in Sect. 2. Section 3 presents the experimental results and comparative analysis. Discussions are carried out in Sect. 4. The conclusions are summarized in Sect. 5.

2 Proposed Semi-fragile Watermarking Scheme

Curvelet transform has been developed to overcome the limitations of wavelet and Gabor filters [11]. To achieve a complete coverage of the spectral domain and to capture more orientation details, curvelet transform has been developed [12]. The Digital Curvelet Transform (DCLT) is usually implemented in the frequency domain for higher efficiency reasons [13]. The implementation of DCLT is performed either using wrapping or unequally spaced fast Fourier transform (USFFT) algorithms [14]. In this paper, a novel method for semi-fragile watermarking using DCLT for authentication and tamper detection of digital images is presented.

2.1 Watermark Embedding

The process of generating and embedding watermark is depicted in Fig. 1. The input image is a gray scale image and is transformed using DCLT. The watermark is generated from the first level coarse DCLT coefficients.

These coefficients are quantized into four bits and embedded into the first level coarse DCLT coefficients at the locations determined using pseudo random method [15]. Subsequently, inverse DCLT is applied to get watermarked image. Suppose the size of the input image is $N \times N$, then the number of scales used in DCLT is varied from 2 to $\log_2(N)$. The perceptual quality of the watermarked image is measured using Peak Signal to Noise Ratio (PSNR) [16]. The average PSNR values of all the images in the corpus for different dimensions of the input image at different scales used in DCLT are shown in Table 1. It is observed from the values shown in Table 1 that, if the number of scales is less than or equal to $(\log_2(N) - 3)$, there is a significant increase in PSNR values.

The coarse coefficients of first level DCLT are extracted. The coefficients are quantized into 4 bits. The number of Least Significant Bits (LSBs) to be used is a tradeoff between accuracy of tamper detection and imperceptibility of the watermarked image. The accuracy of tamper detection is evaluated using the following equation:

$$Accuracy\ of\ Tamper\ Detection = \frac{Average\ Number\ of\ bits\ identified\ as\ tampered}{Average\ Number\ of\ bits\ actually\ tampered}$$

(1)

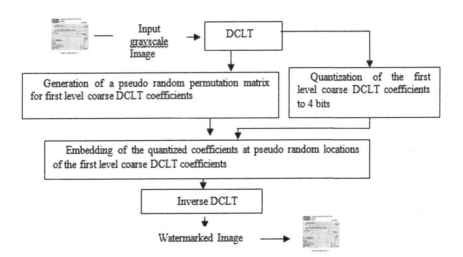

Fig. 1 Semi-fragile watermark embedding process

Table 1 PSNR values of the watermarked image at different scales of DCLT

Size of the input image $N \times N$	Range of the scale $(2 \text{ to } \log_2(N))$								
	2	3	4	5	6	7	8	9	10
128 × 128	28.45	35.56	44.61	44.78	44.89	45.01			
256 × 256	28.49	35.56	44.50	52.84	52.91	52.94	53.07		
512 × 512	29.02	36.61	45.01	52.91	53.70	53.91	54.01	54.1	
1024 × 1024	29.35	35.62	44.79	52.92	53.61	53.84	54.05	54.2	54.3

Table 2 Accuracy of the tamper detection and PSNR of the watermarked image

No. of least significant bits used	Accuracy of tamper detection (%)	PSNR values (dB)
2	78.65	66.62
3	87.50	59.61
4	94.27	53.15
5	95.57	36.24
6	96.88	26.58

It can be observed from the values shown in Table 2 that the accuracy of tamper detection values is more than 90 % for the number of bits used for quantization is above 3. It can also be inferred from the computed PSNR values that imperceptibility of the image and high accuracy of tamper detection decreases in large amount after 4 bits.

Each of the first level coarse DCLT coefficient is quantized to 4 bits and embedded into a coarse coefficient, whose location is decided by a pseudo random permutation [15]. The embedding of the watermark is done according to the Eq. (2) as follows:

$$D_{l_k}(m,n) = D_{l_k}^q(i,j)\, k = 1\ldots4 \qquad (2)$$

where, $D_{l_k}(m,n)$—kth LSB of the coarse first level DCLT coefficient at (m, n), $D_{l_k}^q(i,j)$—kth LSB of the quantized coarse first level DCLT coefficient at (i, j).

2.2 Watermark Extraction

The watermarked image may be subjected to incidental or intentional attacks during transmission. Checking the integrity and authenticity of the watermarked image are carried out during watermark extraction process. The process of watermark extraction is shown in Fig. 2. The watermarked image is transformed using first level DCLT.

The watermark is extracted from four LSBs of each first level DCLT coarse coefficient. It is inversely permuted and dequantized. At the receiver, tamper

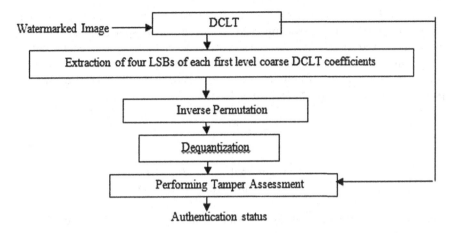

Fig. 2 Semi-fragile watermark extraction process

assessment is carried out by dividing the dequantized and generated first level coarse DCLT coefficients into blocks of uniform size. We have conducted experiments to determine the appropriate size of block for tamper assessment. The average accuracy and processing time required for tamper detection for all the images in the corpus is shown in Table 3. It can be observed from the values in Table 3 that, the tamper detection accuracy and processing time decreases with increase in the size of the block. We have arrived to the decision that the size of the block is set to 3 × 3 which is enough to achieve better (94 %) tamper assessment with reasonable processing time.

The energy of the dequantized and first level coarse DCLT coefficients are computed using the Eqs. (3) and (4) as follows:

$$E_1 = \frac{1}{N \times N} \sum_{i=1}^{N} \sum_{j=1}^{N} |D(i,j)| \qquad (3)$$

$$E_2 = \frac{1}{N \times N} \sum_{i=1}^{N} \sum_{j=1}^{N} |D_e(i,j)| \qquad (4)$$

Table 3 Accuracy of tamper detection and processing time for varying size of the blocks

Block size	Tamper detection accuracy (%)	Processing time (s)
2 × 2	96.35	5.1892
3 × 3	94.27	3.1121
4 × 4	83.85	2.1602
5 × 5	75.26	1.5123
6 × 6	60.16	1.0992

where, $N = 3$, size of each block, E_1—energy of each block of the first level DCLT coarse coefficients of the watermarked image, E_2—corresponding energy from the watermarked image, $D(i, j)$—generated first level DCLT coarse coefficient at (i, j) from the watermarked image, and $D_e(i, j)$—extracted first level DCLT coarse coefficient at location (i, j) from the watermarked image. The tamper assessment is performed using the Eqs. (5) and (6) as below:

$$T = \frac{|E_1 - E_2|}{\text{MAX}_E} \tag{5}$$

$$a(i) = \begin{cases} \text{tampered,} & T > 0.4 \\ \text{``not tampered''}, & otherwise \end{cases} \tag{6}$$

where, MAX_E—Maximum energy possible in first level coarse DCLT coefficient and T- tamper value and $a(i)$—Authentication status of each block i.

3 Experimental Results

We have created a corpus of different types of images for testing the proposed semi-fragile watermarking system. The corpus consist of various categories of images like Cheque, ID cards, Bills, Certificates, Marks cards and few images taken from the standard image database USC-SIPI [17]. The existing method [1] and the proposed method has been tested for tamper detection of all the images in the corpus. The results of tamper detection for the two sample cases which involves tampering of text and image respectively is shown in Fig. 3. It is evident from the

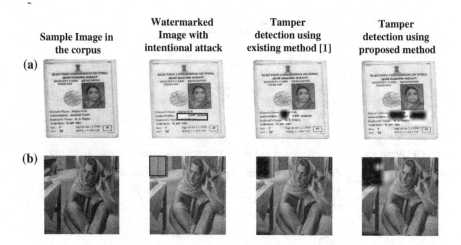

| Sample Image in the corpus | Watermarked Image with intentional attack | Tamper detection using existing method [1] | Tamper detection using proposed method |

Fig. 3 Results of tamper detection. **a** Text tampering. **b** Image tampering

visual inspection of results shown in Fig. 3 that the detection of tampered locations is more accurate in the case of proposed method.

4 Discussions

A comparative analysis of fragility and robustness of the existing [1] and proposed method are performed and are discussed in detail in the following subsections.

4.1 Robustness Analysis

The robustness is evaluated using the parameter Normalization Correlation Coefficient (NCC) [15] between the extracted and received first level DCLT coarse coefficients. The NCC values are computed for both existing method [1] and proposed method by varying the variance parameter of the noise. The graph is plotted for NCC values for each type of noise and is shown in Fig. 4. It is evident from Fig. 4 that the NCC values drop substantially for the existing method [1] for variance greater than 5. The watermarked image is also subjected to JPEG compression by varying the amount of compression. A plot depicting the robustness performance of the existing [1] and proposed schemes in terms of NCC values is shown in Fig. 5. It is observed from Fig. 5 that for quality factors from 15 to 5 %, NCC values are better for proposed scheme than the existing scheme [1].

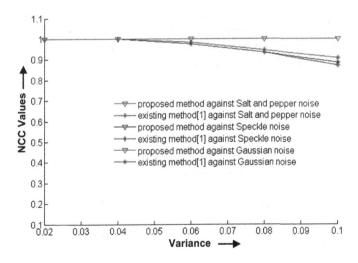

Fig. 4 Robustness under noise attacks

Fig. 5 Robustness against
JPEG compression

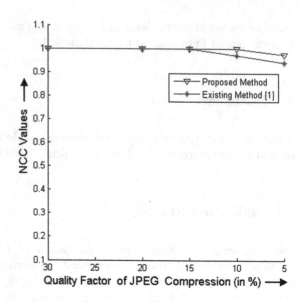

4.2 Fragility Analysis

The fragility is decided based on the tamper assessment capabilities of a
semi-fragile watermarking scheme. We have tested all the images in the corpus for
different types of tampering namely (i) inserting a new content (ii) deleting the
existing content (iii) modifying an existing content and (iv) performing multiple
attacks in the same image.

The performance of detection of tampered regions for both existing method [1]
and proposed method is measured in terms of parameter accuracy of tamper
detection using Eq. (3). The tamper detection values in Table 4 reveals that, the
proposed method results in better accuracy than the existing method [1]. The
average accuracy of tamper detection is around 94 % and at least 10 % more than
existing method [1]. However, there is a failure of around 6 % in accurate detection
and localization of tamper.

Table 4 Analysis of tamper detection against various intentional attacks

Attack	Accuracy of tamper detection	
	Existing method [1] (%)	Proposed method (%)
Insertion	82.69	93.27
Deletion	79.17	92.19
Modification	80.99	94.37
Multiple attacks	77.88	91.35

5 Discussions

A blind semi-fragile watermarking scheme based on DCLT has been proposed in this paper. From the results, it is inferred that this method exhibits higher NCC values and thus, more robust than the existing method [1]. The proposed approach leads to significant improvement in detection of tampered regions in the water-marked image compared to the existing method [1]. The proposed work can be further enhanced by using more sophisticated measures for evaluating tamper detection in an image. By embedding into selected curvelet coefficients, it could be possible to improve the accuracy of tamper detection and at the same time the embedding process could be made inexpensive. The selection of curvelet coefficients is considered as future work of the current study.

References

1. Ghofrani, S. et.al., Image content authentication and tamper localization based on semi fragile watermarking by using the Curvelet transform, TENCON 2012—2012 IEEE Region 10 Conference, Cebu, pp. 1–6, (2012).
2. Prabhishek Singh et.al., A Survey of Digital Watermarking Techniques, Applications and Attacks, International Journal of Engineering and Innovative Technology (IJEIT), Vol. 2, No. 9, pp. 165–175 (2013).
3. W.H. Chang and L.W. Chang, Semi-Fragile Watermarking for Image Authentication, Localization, and Recovery Using Tchebichef Moments, International Symposium on Communications and Information Technologies (ISCIT), pp. 749–754. (2010).
4. G. Schirripa, C. Simonetti and L. Cozzella, Fragile Digital Watermarking by Synthetic Holograms, Proc. of European Symposium on Optics/Fotonics in security & Defence, London, UK, pp. 173–182. (2004).
5. J. Dittmann, L. Croce Ferri and C. Vielhauer, Hologram Watermarks for Document Authentications, Proceedings of IEEE International Conference on Information Technology, Las Vegas, pp. 60–64 (2001).
6. Y. Aoki, Watermarking Technique Using Computer-Generated Holograms, Electronics and Communications in Japan, Part 3, Vol. 84, No. 1, pp. 21–31. (2001).
7. K. Maeno, Q. Sun, S. Chang and M. Suto, New semi-fragile image authentication watermarking techniques using random bias and nonuniform quantization, IEEE Trans. Multimedia, Vol. 8, No. 1, pp. 32–45. (2006).
8. L. Xin and L. Xiaoqi and W. Wing, A semi-fragile digital watermarking algorithm based on integer Wavelet matrix norm quantization for medical images, IEEE International conference on bioinformatics and biomedical engineering, pp. 776–779, (2008).
9. X. Wu and J. Huang and J. Hu and Y. Shi, Secure semi-fragile watermarking for Image authentication based on parameterized integer Wavelet, Journal of Computers, Vol. 17, No. 2, pp. 27–36, (2006).
10. D. L. Donoho and M. R. Duncan, Digital Curvelet transform: Strategy, implementation and experiments, Proc. Society optics and photonics, Vol. 4056, pp. 12–29, (2000).
11. L. Chen, G. Lu, and D. S. Zhang, Effects of Different Gabor Filter Parameters on Image Retrieval by Texture, in Proc. of IEEE 10th International Conference on Multi-Media Modelling, Australia, pp. 273–278. (2004).
12. E. J. Candès, L. Demanet, D. L. Donoho, and L. Ying, Fast Discrete Curvelet Transforms, Multiscale Modeling and Simulation, Vol. 5, pp. 861–899 (2005).

13. J.-L. Starck and M.J. Fadili, Numerical Issues When using Wavelets, in Encyclopedia of Complexity and Systems Science, Meyers, Robert (Ed.), Springer New York, Vol 14, pp 6352–6368, (2009).
14. E. Candes and D. Donoho, New tight frames of Curvelets and optimal representations of objects with C2 singularities Comm. Pure Appl. Mathematics, Vol. 57, No. 2, pp. 219–266, (2004).
15. Jaejin Lee, Chee Sun Won, A Watermarking Sequence Using Parities of Error Control Coding For Image Authentication And Correction, Consumer Electronics, IEEE Transactions, Vol. 46, No. 2, pp. 313–317 (2000).
16. Yevgeniy Dodis et.al., Threshold and proactive pseudo-random permutations, TCC'06 Proceedings of the Third conference on Theory of Cryptography, Verlag Berlin, Heidelberg, pp. 542–560, (2006).
17. http://sipi.usc.edu/database/.

Indexing Video by the Content

Mohammed Amin Belarbi, Saïd Mahmoudi and Ghalem Belalem

Abstract Indexing video by content represents an important research area that one can find in the field of intelligent search of videos. Visual characteristics such as color are one of the most relevant components used to achieve this task. We are proposing in this paper the basics of indexing by content, the various symbolic features and our approach. Our project is composed of a system based on two phases: an indexing process, which can take long time, and a search engine, which is done in real time because features are already extracted at the indexing phase.

Keywords Indexing · Indexing image by content · Indexing videos by content · Image processing · Video segmentation · Similarity measure

1 Introduction

Nowadays, with the rapid growth of cloud computing use, a lot of applications such as video on demand, social networking, video surveillance and secure cloud storage are widely used [1]. Multimedia database can stores a huge amount images, videos and sounds. These multimedia items can even been used in a professional field (computer aided diagnosis, tourism, education, museum, etc.) or just been used for

M.A. Belarbi (✉)
Faculty of Exact Sciences and Computer Science, Abdelhamid Ibn Badiss University, Mostaganem, Algeria
e-mail: belarbi_mohammed_amin@yahoo.fr

S. Mahmoudi
Faculty of Engineering, University of Mons, 20 Place du Parc, Mons, Belgium
e-mail: said.mahmoudi@umons.ac.be

G. Belalem
Department of Computer Science, Faculty of Exact and Applied Science, Ahmed Ben Bella University, Oran1, Algeria
e-mail: ghalem1dz@gmail.com

© Springer India 2016
S.C. Satapathy et al. (eds.), *Information Systems Design and Intelligent Applications*, Advances in Intelligent Systems and Computing 434,
DOI 10.1007/978-81-322-2752-6_3

personal use (memories, travel, family, events, movie collection, etc.). However, with the exponential development of existing multimedia database, we can point to the fact that standard applications designed to exploit these databases do not provide any good satisfactory. Indeed, the applications generally used to achieve both segmentation and video sequence retrieval are in the most of the cases enable to use and extract efficient characteristics. Multimedia indexing is an essential process used to allow fast and easy search in existing databases. The indexing process allows to minimize the response time and to improve the performances of any information retrieval system. In last years, content-based images retrieval methods ware generally combining images processing methods and database management systems. The latest developments in databases are aiming to accelerate research in multidimensional index, and to improve the automatic recognition capabilities. However, modern techniques in this field have to fundamentally change the way they achieve recognition. Indeed, it is now necessary to repeat the query several times and then synthesize the results before returning an answer, instead of just a single query as before. The rest of this paper is organized as follows: In Sect. 2, we present related researches about the previous standards of image retrieval systems. Section 3 explains the indexing video concepts. Section 4 reports the experimental results. Conclusions are drawn in Sect. 5.

2 Indexing and Research Images by the Visual Content

Image and videos storage and archiving for both TV channels, newspapers, museums, and also for Internet search engines has for a long time been done at the cost of manual annotation step using keywords. This indexing method represents a long and repetitive task for humans. Moreover, this task is very subjective to culture, knowledge and feelings of each person. That is why; there is a real need for research and indexing methods that are directly based on the content of the image [2]. The first prototype system that have attacked the attention of the research community was proposed in 1970. The first image indexing systems by the contents were created in the mid-90s and were generally designed for specialized and mostly closed databases. CBIR (Content-Based Image Retrieval Systems) systems are based on a search process allowing to retrieve an images from an image database by using visual characteristics. These characteristics, which are also called low-level characteristics are color, texture, shape [2].

2.1 Related Works

A CBIR system is a computer system allowing fast navigation and image retrieval in large database of digital images. Much works have already been done in this area. In many fields like commerce, government, universities, and hospitals, where large digital image collections are daily created. Many of these collections are the product of digitization of existing collections of analogue photography, diagrams, drawings, paintings, and prints. Usually, the only way to search these collections was by indexing keyword, or just by browsing. However, the databases of digital images open the way for research based on content [3]. In the following we will present a list of CBIR based systems.

Hirata and Kato [4] proposed an image retrieval system that can facilitates access to the images by query example. In their system, board extraction is carried on user requests. These edges are compared against the database images in a relatively complex process in which edges are displaced or deformed with respect to the corresponding other [5]. They do not provide an image indexing mechanism based on the contents [5].

The QBIC system [6], is one of the most noticeable systems for interrogation by image content, and was developed by IBM. It lets you compose a query based on various different visual properties such as shape, texture and color, that are semi-automatically extracted from images. QBIC use the R^*-tree like as image indexing method [5, 7].

In [8], authors propose an image retrieval system mainly based on visual characteristics and spatial information. For this purpose, they used dominant wavelet coefficients as feature description. Actually, they focused on effective features extraction from wavelet transform rather than the index structure to achieve fast recovery [5].

On the other side, VisualSEEk [9] is a content-based image retrieval system that allows querying by colored regions and spatial arrangement. The authors of this system developed an image similarity function based on color characteristic and space components.

VIPER is an image retrieval system that uses both color and spatial information of images to facilitate the recovery process. They first extract a set of dominant colors in the image and then derive the spatial information of the regions defined by these dominant colors. Thus, in their system, two images are similar in terms of color and spatial information if they have some major groups of the same color falling within the same image space [5].

2.2 Architecture of an Indexing and Search System Images

An image search system by content includes two phases: offline phase for indexing images databases and online phase for research as shown in Fig. 1.

These systems are running in two steps: (i) Indexing: where characteristics are automatically extracted from the image and stored in a digital vector called visual descriptor [10] looks good. (ii) Research: in this phase the system takes one or more requests from user and returns the result, which is a list of images ordered, based on their visual similarity between the descriptor and the query image using a distance measure [10].

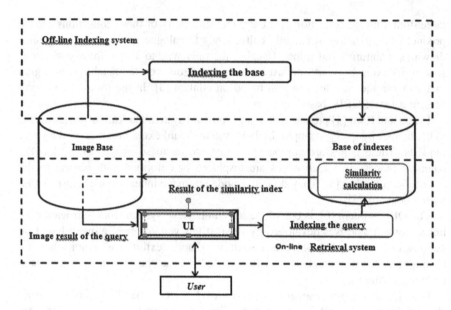

Fig. 1 The architecture of a system of indexing and image search

2.3 Symbolic Features

Colors. Color is the first characteristic that is used for image search. It is the most accessible characteristic because it is directly encoded in each pixel.

The conventional technique for calculating the areas of color is that of the histogram. The idea is to determine the color present in the image and the proportion of the area of the image it fills. The resulting output is called color histogram.

The color histograms can be built in many color ranges, RGB (Red, Green, Blue), the color histogram is produced by cutting the colors of the image into a number of boxes after that counting the number of pixels. Obviously, such a description of the process is simplistic but it is sufficient to search for images of sunset in a collection of maritime or cities images [2, 11].

This technique is often used; Histograms are quick and easy to calculate [12], robust to image manipulation such as rotation and translation [12, 13].

The use of histograms for indexing and image search leads to some problems. (i) Indeed, they have a large sizes, so it is difficult to create a fast and effective indexing using as in [14]. (ii) On the other side, they have no spatial information on colors positions [14]. (iii) They are sensitive for small changes in brightness, which is a problem to compare similar images, but acquired under different conditions [14].

Texture. There is no appropriate definition of texture. However, common sense definition is as follows: the Texture is the repetition of basic elements built from pixels that meet a certain order. Sand, water, the grass, the skin are all examples of textures [2]. It is a field of the image, which appears as a coherent and homogeneous [2].

Various methods are available that are used to describe texture characteristic such as gray-level co-occurrence matrix, Ranklet texture feature, Haar discrete wavelet transform, Gabor filter texture feature, etc. [15].

Shape. The shape is another important visual characteristic, which is however considered a difficult task to be fully automated; the form of an object refers to its profile and physical structure [2].

Shape characteristics are fundamental to systems such as databases of medical image, where the texture and color of objects are similar. In many cases, especially when the accurate detection is necessary, human intervention is necessary [2]. Various methods to extract shape feature are edge histogram descriptor (EHD), Sobel descriptor, SIFT, Fourier descriptor of PFT, etc. [15]. However, in storage applications and retrieving image, shape characteristics can be classified into local and global functions [2].

3   Video Indexing

The objective of a video indexing system is to allow a user to perform a research in a set of videos. Research is conducted on easy criteria to achieve such as the type of emission, actors or theme. But it is extremely complicated and long to get an accurate description of the contents of the videos. Video indexing by content is shown schematically as Fig. 2.

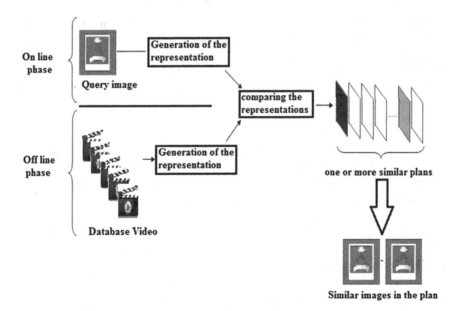

Fig. 2 Architecture of an indexing and retrieval system for video content

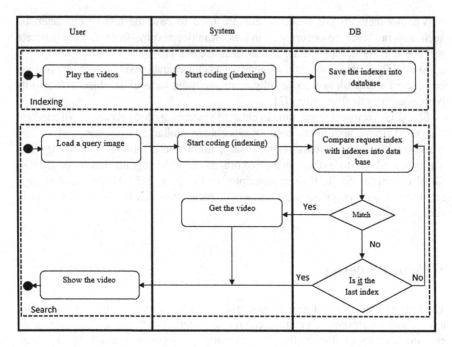

Fig. 3 Activity diagram of our approach

The objective of the proposed video indexing system is to allow a user to search through a set of videos. Figure 3 show the activity diagram of our approach.

- **Images cut**. In this step, we cut a video into a set of individual images. These images allow us to define the following image planes as shown in Fig. 4. After that the following step are processed.
- Application of a visual descriptor images (defined above).
- Calculating a measure of a similarity corresponding to the descriptor applied in the previous step.
- Segmenting a video into several basic units called "plans"; a plan is the shortest unit after the image that defines two shots as shown in Fig. 5.
- Select the key frame. We extract in this step the visual characteristics of each plan, these features are defined in one or more images called "key frames". The key frame of our approach is the first image of each plan because it contains images similar to this image as shown in Fig. 6.

4 Implementation

The general architecture of our application named Videoindexer is shown schematically in Fig. 7, which we developed in C++.

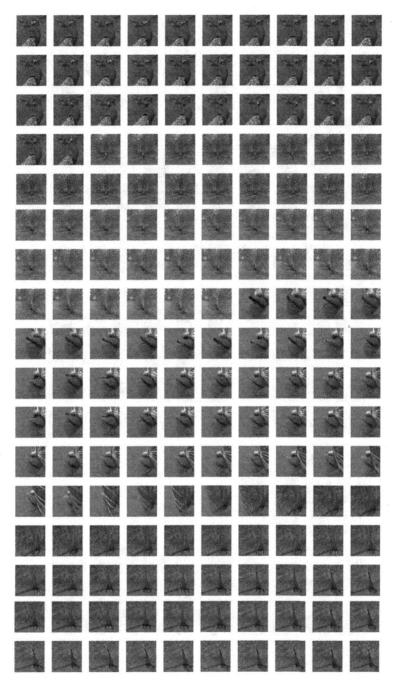

Fig. 4 Cutting into a sequence of images

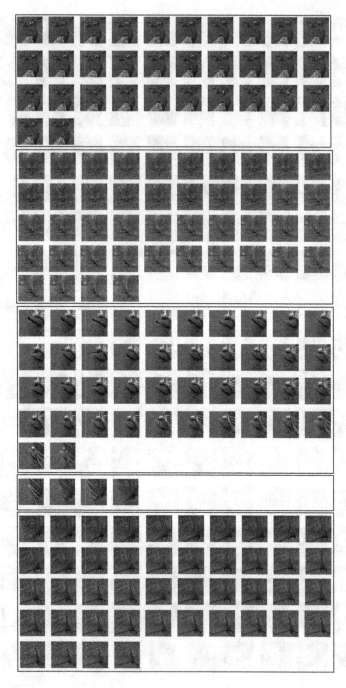

Fig. 5 Temporal segmentation (plan)

Fig. 6 Highlights (key frame of each plan)

Fig. 7 A general architecture of Videoindexer

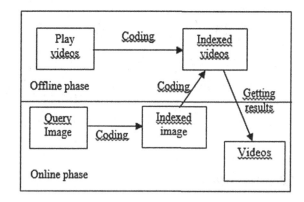

4.1 Indexing by the Content

Indexing by the Color. Color is one of the most component used in image retrieval and indexing [16, 17]. It is independent from the size of image and its orientation. In our system, we use the color space: RGB. We chose this space because it is commonly used.

Histogram and distance. Histograms are resistant to a number of changes on the image. They are invariant to rotations, translations [13], and scale changes [14]. But despite that, we must say that the histograms are sensitive to changes in illumination and lighting conditions [11, 14]. We used the histogram correlation technique, Batcharrya [18] and chi-square. With these methods, a distance of similarity is calculated to measure if two histograms are "close" to each other. H_1, H_2 are two histograms.

- Chi-square:

$$d(H_1, H_2) = \sum_I \frac{(H_1(I) - H_2(I))^2}{H_{12}(I)} \tag{1}$$

- Correlation:

$$d(H_1, H_2) = \frac{\sum_{i=1}^{N} \bar{H}_1(i)\bar{H}_2(i)}{\sqrt{\sum_{i=1}^{N} \bar{H}_1(i)^2 \sum_{i=1}^{N} \bar{H}_2(i)^2}} \tag{2}$$

$$\bar{H}(i) = H(i) - \frac{1}{N}\sum_{i=1}^{N} H(i) \tag{3}$$

- Bhattacharya:

$$d(H_1, H_2) = \sqrt{1 - \frac{1}{\sqrt{\bar{H}_1 \bar{H}_2 N^2}} \sum_{i=1}^{N} \sqrt{H_1(i)H_2(i)}} \tag{4}$$

4.2 Algorithm

Offline phase (Indexing). This phase is summarized in three steps: segmenting videos into individual images, then calculating their histograms and saving these extracted data in a database.

Algorithm 1: (*Indexing the videos*)

Begin:
1. Read Nbvideos;// number of videos
2. **For** i=1 **to** Nbvideos **do** {
3.   Read video (i);
4.   Read Nbimage; // frame number of the video
5.   **for** j=1 **to** Nbimage **do** {
6.   Read image (j);
7.   Calculate their histogram (Hr, Hg, Hb);
8.   Normalize the histogram;
9.   Indexing the histogram as XML format and store it in a file;
10.   Store the image (j) in a jpeg file;
11.   Storing in a DB the image path (j) and the path of XML file; } }
End

In line 4 of Algorithm 1, we read the frame number of the video, the number of images depends on the length of the video. Also, we calculate the histogram for each image then we must normalized it as in line 7. Then we store the histogram as XML format as in line 9.

Online phase. This is the second part of our approach

Algorithm 2: (Research the videos)

Begin:
1. Read the query image;
2. Read the threshold;
3. Choose a distance;
4. Calculate the histogram of the query image (H1R, H1G, H1B);
5. Normalized the histogram of the query image (H1R, H1G, H1B);
6. Read the content of the database;
7. **While** there are records in the database **do** {
8.   **for** i=1 **to** the last record into the database **do** {
9.   Read ith histogram recording (H2R, H2G, H2B)
10.   Retrieve all videos, which distance (h1, h2) < threshold } }
End

In line 3 of Algorithm 2, we choose a distance between the distances defined above such as chi-square. Also in line 10, we retrieve all the videos to display them to the user.

The research phase is quick, which due to the sequential search, and also because in this phase, we do not process any treatment of video and the image. We calculate just the histogram of the query image, and after that the system returns all the similar videos containing the image query.

We notice that our approach allows to save research time because we compare the query image with the key frames as shown in Fig. 8.

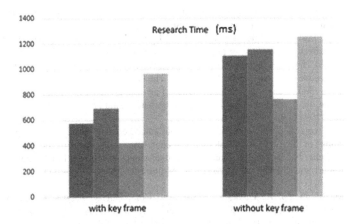

Fig. 8 The research time with key frame and without it

5 Conclusion

In this Paper, we have shown and comment experimental results obtained by the different strategies for calculating similarity. We found that these results positively guarantee the effectiveness of our approach and that a good video indexing process is the one that allows you to find the most relevant correspondence with the least possible number of calculations, and less time. The results obtained from the application allows us to judge and say that the histogram is effective for the overall comparison of the content of the image, simple, response time is fast.

In the future, we propose to improve our system to apply this attribute to image regions (to apply a space research) and combine other descriptors such as SURF, SIFT to improve the relevance of our system.

In our next works, we plan to discuss indexing video when data increase (Big Data), and we will try to reduce the dimensionality to accelerate research phase (response time).

References

1. Cheng, B., Zhuo, L., Zhang, J.: Comparative Study on Dimensionality Reduction in Large-Scale Image Retrieval. In: 2013 IEEE International Symposium on Multimedia. pp. 445–450. IEEE (2013).
2. Idris, F., Panchanathan, S.: Review of Image and Video Indexing Techniques. J. Vis. Commun. Image Represent. 8, 146–166 (1997).
3. Dubey, R.S., Student, M.T.C.S.E., Bhattacharjee, J.: Multi Feature Content Based Image Retrieval. 02, 2145–2149 (2010).
4. Hirata, K., Kato, T.: Query by visual example. In: Advances in Database Technology EDBT'92. pp. 56–71 (1992).
5. Lee, D.-H., Kim, H.-J.: A fast content-based indexing and retrieval technique by the shape information in large image database. J. Syst. Softw. 56, 165–182 (2001).
6. Faloutsos, C., Barber, R., Flickner, M., Hafner, J., Niblack, W., Petkovic, D., Equitz, W.: Efficient and effective querying by image content. J. Intell. Inf. Syst. 3, 231–262 (1994).
7. Flickner, M., Sawhney, H., Niblack, W., Ashley, J., Huang, Q., Dom, B., Gorkani, M., Hafher, J., Lee, D., Petkovie, D., Steele, D., Yanker, P.: The QBIC System. (1995).
8. Jacobs, C.E., Finkelstein, A., Salesin, D.H.: Fast multiresolution image querying. In: Proceedings of the 22nd annual conference on Computer graphics and interactive techniques. pp. 277–286 (1995).
9. Smith, J.R., Chang, S.-F.: VisualSEEk: a fully automated content-based image query system. In: Proceedings of the fourth ACM international conference on Multimedia. pp. 87–98 (1997).
10. Lew, M.S., Sebe, N., Djeraba, C., Jain, R.: Content-based multimedia information retrieval: State of the art and challenges. ACM Trans. Multimed. Comput. Commun. Appl. 2, 1–19 (2006).
11. Rasli, R.M., Muda, T.Z.T., Yusof, Y., Bakar, J.A.: Comparative Analysis of Content Based Image Retrieval Techniques Using Color Histogram: A Case Study of GLCM and K-Means Clustering. Intell. Syst. Model. Simulation, Int. Conf. 0, 283–286 (2012).
12. Zhang, Z., Li, W., Li, B.: An improving technique of color histogram in segmentation-based image retrieval. 5th Int. Conf. Inf. Assur. Secur. IAS 2009. 2, 381–384 (2009).

13. Krishnan, N., Sheerin Banu, M., Callins Christiyana, C.: Content based image retrieval using dominant color identification based on foreground objects. Proc. - Int. Conf. Comput. Intell. Multimed. Appl. ICCIMA 2007. 3, 190–194 (2008).

14. Marinov, O., Deen, M.J., Iniguez, B.: Charge transport in organic and polymer thin-film transistors : recent issues. Comput. Eng. 152, 189–209 (2005).

15. Choudhary, R., Raina, N., Chaudhary, N., Chauhan, R., Goudar, R.H.: An Integrated Approach to Content Based Image Retrieval. 2404–2410 (2014).

16. Takumi Kobayashi and Nobuyuki Otsu: Color Image Feature Extraction Using Color Index Local Auto-CorrElations. Icassp. 1057–1060 (2009).

17. Xue, B.X.B., Wanjun, L.W.L.: Research of Image Retrieval Based on Color. 2009 Int. Forum Comput. Sci. Appl. 1, 283–286 (2009).

18. Aherne, F.J., Thacker, N. a., Rockett, P.I.: The Bhattacharyya metric as an absolute similarity measure for frequency coded data. Kybernetika. 34, 363–368 (1998).

EIDPS: An Efficient Approach to Protect the Network and Intrusion Prevention

Rajalakshmi Selvaraj, Venu Madhav Kuthadi and Tshilidzi Marwala

Abstract Nowadays, Network Security is growing rapidly because no user specifically wants his/her computer system to be intruded by a malicious user or an attacker. The growing usage of cloud computing provides a different type of services, which leads users to face security issues. There are different types of security issues such as hacking intrusions worms and viruses, DoS etc. Since the entire needed resources are associated with everyone and are centrally monitored by main controller in cloud computing area it creates a simplest way for intruders. In addition, an experienced or knowledgeable attacker can get to make out the system's weakness and can hack the sensible information or any resource and so, it is essential to provide protection against attack or intrusion. Additionally, to handle poor performance or low latency for the clients, filtering malicious accesses becomes the main concern of an administrator. Some of the traditional Intrusion Detection and Prevention Systems fail to overcome the abovementioned problems. As a result, this research proposes a novel approach named Enhanced Intrusion Detection Prevention System to prevent, protect and respond the various network intrusions in the internet. Our proposed system use client-server architecture, which contains main server and several clients. Clients record the doubtful actions taking place in the Internet and record the suspicious information. Then, this recorded information is forwarded to the main server for further usage. After that, the main server analyses the received data and then make decision whether to provide a security alert or not; then the received information is displayed via an interface.

R. Selvaraj (✉)
Faculty of Engineering and the Built Environment, University of Johannesburg,
Johannesburg, South Africa
e-mail: selvarajr@biust.ac.bw

R. Selvaraj
Department of Computer Science, BIUST, Gaborone, Botswana

V.M. Kuthadi · T. Marwala
Department of AIS, University of Johannesburg, Johannesburg, South Africa
e-mail: vkuthadi@uj.ac.za

T. Marwala
e-mail: tmarwala@uj.ac.za

© Springer India 2016
S.C. Satapathy et al. (eds.), *Information Systems Design and Intelligent
Applications*, Advances in Intelligent Systems and Computing 434,
DOI 10.1007/978-81-322-2752-6_4

35

In addition, server verifies the data packets using IDPS and classifies the attacks using Support Vector Machine. Finally, as per the attack type Honeypot system sends irrelevant data to the attacker. The experimentally deployed proposed system results are shown in our framework which validates the authorized users and prevents the intrusions effectively rather than other frameworks or tools.

Keywords Network security · Intrusion detection prevention system · Honeypot system · Support vector machine

1 Introduction

Network security is one of the major domains of IT. Over the last decade, IT environment achieves the largest development because no one wants his/her system to be intruded and to get an intruded data from the attacker [1]. Regular internet usage in our daily life is getting increased and making a protection over the intrusion and intruders in the networks for providing the availability and reliable services to users is very essential. Various techniques such as DMZ (Demilitarized Zone) Firewall's are used but they are not effective [2]. Intrusion Detection System appeared as a solution to present best network security when these IDS are compared with other techniques. Intrusion Detection System constantly runs in a background of a system, watches network traffic and computer systems, and then it analyzes network traffic for potential network intrusions originating outside the firm in addition for system attacks or misuse created within the firm.

The administrator of a system relies on several devices to protect and monitor their systems and network. The administrator identify the irrelevant activity or use of a computer system or a network by watching activities and sending alerts by certain activities occur like scanning network traffic to resolve linked computer systems. An ID is a software application or device that watches system or network activities for policy violations or malicious activities and generates reports to an administrator. Several techniques have made an effort to stop an unauthorized access; IDPS (Intrusion Detection Prevention System) are mainly focused on detecting possible events, creating log files and reporting about intruder [3]. Additionally, firms use Intrusion Detection Prevention System for some other functions, like identifying issues with network safety policies, traditional threats, documenting and prevent individual users from breaking the safety policies. Intrusion Detection Prevention System has become an essential for security communications in every firm. The Honeypot system technology is a powerful complement for Intrusion Detection System that can really reduce the IDS burden [4]. Traditional IDS is proposed as per Honeypot technology and IDS should be mark out back via the movable user to attain inter-operability among different components, hence the Honeypot technique is used for obtaining the most promising intrusion for further attack tracking and data signature to attain automatic information [5].

IDS are classified into II different kinds such as anomaly detection system and misuse detection system, these categories are based on different intrusion detection methods. MDS (Misuse Detection Systems) are able to detect previous classified patterns of known attacks, and unknown attacks pattern couldn't be noticed. ADS utilize the systems at present processes and previous activity components to evaluate the new techniques and detect the un-known attacks effectively. Attack method is more difficult, single pattern matching or statistical analysis methods are difficult to locate the amount of network attacks. The traditional IDS can't exchange the data; it is difficult to locate the attack from attack source and even the attacker has produced vulnerabilities. Traditional IDS and other security products can't inter-operate. Information encryption, false alarms, traditional IDS are facing huge challenges. As a result, our proposed system implements a novel mechanism named Enhanced Intrusion Detection Prevention System to overcome the abovementioned issues. Our proposed model contains the combination of some tools such as Sebek, Dionaea, Snort IDS, Intrusion Detection Prevention System and Honeypot System and classification mechanism like Support Vector Machine. It also uses a client-server system architecture which has one main server and several clients. The client notifies the suspicious records and malicious codes. Then, the notified data is forward to main server for further process. After that, the server analyzes the received data packet and then the main server decides whether the data is in security warning or not. For analyzing packets and attacks we use Intrusion Detection Prevention System and Support Vector Machine. If the data is considered as attack then it forwards the data into Honeypot system and it will sends fake information.

2 Related Work

Research work [6] discusses in details about the IDS and its roles, where they focused the view point of intrusion such as victims and attackers. At present, a security standard is utilizing configured firewalls in mixture with the IDS. In research work [7], the author utilizes NB classifiers for detecting the intrusion. Furthermore, the authors build strict self-determination account between the aspects in a study results the accuracy of lower intrusion detection when the aspects are co-related and which are regular in case of intrusion detection. The research work [8] used Bayesian network for detecting intrusions. Moreover, they have a tendency to attack exact and construct decision network as per individual attacks characteristics. Research work [9] focus on detecting anomalous system call traces in privileged procedures, HMMs (Hidden Markov Models) was presented in research works [10–12]. Moreover, the modeling systems call by you may not provide exact classification for all time and as in several cases different connection level aspects are ignored. Additionally, hidden Markov models are generative systems and fail to form wide-range of needs between the key observations [13].

To defeat the single IDS weakness, a more amount of techniques have been presented that describes the collaborative usage of host based and network-based systems [14]. The author in [15, 16] proposed the system which utilizes both behavior-based and signature based techniques. Research work [17] describes a knowledge discovery framework for generating adaptive IDS schemes. Research work [18] proposed a framework for distributed intrusion detection as per the mobile agents. The research study [19] proposed a Honeyd with simulating networks. In network, Honeyd produce virtual hosts, and it is also utilized in the research of Honey net development. Honeyd is a slim daemon with several remarkable aspects. It may presume any OS's (Operating System) personality and should configure to present IP/TCP "services" such as SMTP, SSH, HTTP and etc. Research work [20] presents an Nmap Security Scanner, which is open source and free utility. Most of the people utilize Security Scanner for administration, security auditing, network discovery, and inventory. It uses raw Internet Protocol packets in new way to establish the hosts accessible on a system.

Research study [21] proposed an HIDS, which detects the intrusion and forward the information to the network administrator about possible attack occurrence in a secure network. The structural design uses a novel data evaluation method, which permits detection as per the facts of authorized user behavior variation in the system from the learned profile on behalf of the activity of authorized user. The author [22] proposed the values and definitions of Honeypot system; it provides several definitions and actual idea of Honeypot system. The author states that, the honeypot system as a resource and the attacker or malicious access has been compromised. The abovementioned quotes denotes a design of Honeypot system and it is our goal and prospect to have the scheme potentially exploited, attacked and probed and the honeypot systems are not a solution; the honeypot don't fix everything. The research work [23] provides a use of honeypot system in network security; they demonstrating a project set to support the Honeypot system network.

The author [24] proposes a secured maximum–communication honeypot for II different host types. The 1st type protects the system from corruption. The 2nd type assumes a corruption of the system but it can be easily re-installed. Different types of security devices are used for solving effortless analysis. Furthermore, network and host information permit a complete analysis for difficult attack scenarios and this solution fully depends on OSS (Open Source Software). The main focus of research work is to freeze confidential services from malicious sources which deal with spoofing Distributed Denial of Services attacks. This can be done via controlling network attack traffic to the DDoS source as per pushback technique to trace a specific source; roaming honeypot system was used to define the attacker capacity. In research study [25], the author proposed a honeypot technique as per the distributed intrusion tracking, which is totally different from existing honeypot, and this proposed honeypot utilize distributed exploitation to increase the total ability of the system.

Research work [26] developed IDS by combining two mechanisms like genetic algorithm and hybrid honeypot where minimum interaction honeypot system is used to interact with already classified attacks. Here, high interaction honeypot system is used for interacting unknown network attacks. Research work presents a novel Intrusion Detection System using C4.5 DT. In the intrusion rules construction process, instead of information gain, information gain ratio is utilized. The testing result shows that the proposed mechanism is feasible and has a maximum correctness rate. Research work [27], presented a Honeypot system in a network under severe surveillance that attract the intruders using virtual or original network and services over a network are used to evaluate the black hat events. This honeypot system are precious for new Intrusion Detection System signatures development, analyzing novel network attack tools, identifying new modes of hiding communication or DDoS tools.

3 Proposed Work

3.1 Proposed Work Overview

Our proposed system presents a novel Intrusion detection and prevention system named Enhanced IDPS System for protecting, preventing and responding the network from intrusion and attack. The Fig. 1 illustrates the system architecture of Enhanced Intrusion Detection and Prevention System. In order to protect the authorized user from attack our proposed system uses various mechanisms like Snort, Dionaea, Sebek, Honeypot server, IDPS and SVM. In our proposed system, there are two main parts like Server part and Client part. In client part, Snort IDS, Dionaea client, Sebek client side processes were used to validate the authorized users. In server part, Verifier, Dionaea server, Sebek server processes were performed to validate the user and analyze his/her packets. Then, normalization process is performed to normalize the client and server processes and validate the attacker details. After that it forwards the details into main server. In main server, two operations were performed such as: attack classification system and IDPS. IDPS is used to detect and prevent the packets which were sent by malicious user. Attack classification is used to identify the attacker based on support vector machine classification. Our classification system also adds new type of attack. After finding the attacker, our Honeypot system sends fake data to the attacker as per the attack type. Finally, original server sends valid data to the authorized users.

3.2 System Architecture

See Fig. 1.

3.3 Components of Proposed System

Our Enhanced Intrusion Detection Prevention System has two different parts. Such as Client and server.

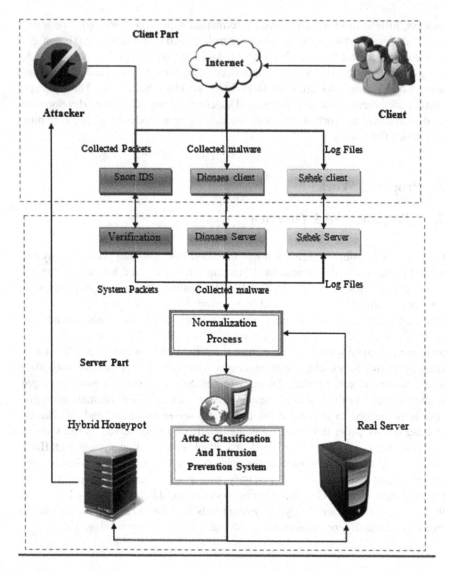

Fig. 1 System architecture

3.4 Client

In order to collect the information about the behavior of an intruder or attacker, clients are positioned in the same context. The client parts are self-governing each other and are independently activated. The information that has been acquired is distributed back to the main server to carry out further system analysis and its security updation. The client part is having three important tools:

- Snort IDS
- Dionaea client
- Sebek client

A. **Snort IDS**

Snort Intrusion Detection System focuses and filters data packets during intrusion detection process. It also identifies the separate attack pattern, warning messages and information.

B. **Dionaea client**

It attracts intruders and detains the malwares via simulating vulnerabilities and fundamental services.

C. **Sebek client**

Sebek Client records the suspicious activity of intruder or attacker for the period of communication with the Honeypot system in logs.

3.5 Server

Actually the main server is linked to several clients and centralizes the main server. The entire received records are stored in a DB (Database). The server part contains six important parts:

- Verifier
- Dionaea Server
- Sebek Server
- Attack Classification System and Intrusion Prevention System
- Honeypot System

A. **Verifier**

Verifier is used to check the packets that are received by Snort IDS. Additionally, it also verifies all the client's packets.

B. **Dionaea server**

Dionaea server allows malicious codes, which are sending to the Dionaea client part.

C. **Sebek server**

Sebek Server obtains and filters numerous sources representing directions or a link to received information storing process.

D. **Intrusion Prevention System**

The Fig. 2 illustrates the architecture of Attack Classification and Intrusion Prevention System.

The IDPS is the security appliance, which monitor system and network activities for unauthorized access. There are II types of IPS such as HIDPS and NIDPS. The key functions of IDPS are to find out malicious activity, and take action to stop/block it. It can able to perform two operations such as detection and prevention. Initially, IPS watches doubtful activity and network traffic then

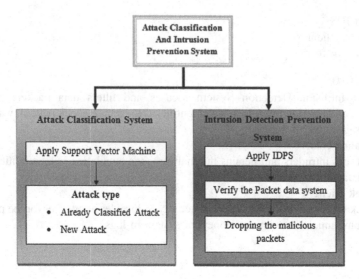

Fig. 2 Attack classification and IDPS architecture

it reacts to the malicious network traffic by stopping the user access or blocking Internet Protocol address in the network. IPS works as per the assumption, which states that the attacker's behavior differs from the valid user. There are three kinds of IDS such as Misuse detection, Anomaly detection and Hybrid mode detection system. Intrusion Prevention System should perform some actions like blocking the network traffic, drop the malicious data packets, send an alarm and stop the connection from the Intruding Internet Protocol address. An IPS can also correct un-fragment data packet streams, CRC errors, prevent Transmission Control Protocol issues, and clean up un-necessary network and transport layer options.

E. **Support Vector Machine (SVM)**

Support Vector Machine performs regression and classification tasks by creating boundaries of non-linear decision. The feature space nature is only helping to find these boundaries; this classification algorithm can show a large flexible degree in handling regression tasks and classification of varied difficulties. This includes various types of support vector models like Radial Basis Function, polynomial, linear, and sigmoid. Generally, the attack classification process consists of training and testing with several data examples. Each example in the SVM training dataset contains several features ("attributes") and one class label ("target values"). The main goal of Support Vector Machine is to generate a model which forecasts class label of data examples in the testing dataset which are specified in the features only. Four types of kernel functions are used to reach the objective as given below.

Radial Basis Function: The RBF is also called as the Gaussian kernel, it is of the form

$K_f(p_x, p_y) = \exp\left(-\frac{\|p_x - p_y\|}{2\sigma^2}\right)$, where σ denotes the width of the window.

Polynomial: the polynomial degree kernel dk and it is of the form
$K_f(p_x, p_y) = (p_x, p_y)^{dk}$

Linear: $K_f(p_x, p_y) = p_x^T p_y$

Sigmoid: The sigmoid kernel is $K_f(p_x, p_y) = \tanh(K_f(p_x, p_y) + r)$

To achieve more classification accuracy our proposed system utilizes Radial Basis Kernel function for attack classification. This classifier classifies the attacks based on the ranges of previously described attacks. If the received attributes are not matched with already classified attacks then it creates a new attack category.

F. **Honeypot System**

Honeypot system is a server, which sends irrelevant data to the attacker based on the attack type. It works as per the concept, "the entire network traffic coming to Honeypot is suspicious". Honeypot system is located somewhere in Demilitarized Zone. This denotes the original server is invisible or hidden to the intruders. Generally, Honeypot system are developed to watches the attackers activity, store logs files, and records the activities like processes started, file adds, compiles, changes and deletes. By collecting such details, Honeypot system increases the total security system of the organization. If enough information is collected it can be used to act against in severe condition. This information is used to calculate the attackers' skill level and identity their intention. Therefore, based on the attacker category Honeypot system sends irrelevant or fake information. A honeypot system is a system designed to learn how "black-hats" probe for and exploit weakness in the IT systems. It can be defined as an information system resource whose value lies in unauthorized or illicit use of that resource it can be used for production or research system.

4 Results and Discussion

To measure the performance of our proposed approach, a sequence of experiments on extracted dataset were conducted. Based on the following configuration our proposed method should be implemented (1) Windows 7, (2) Intel Pentium(R), (3) CPU G2020 and (4) processer speed 2.90 GHz.

Table 1 Attack categories

S. no.	Attacker category	Attack types
1	R2L	15
2	DOS	10
3	Probing	6
4	U2R	7

4.1   Dataset and Attack Categories

To check the effectiveness of Enhanced Intrusion Detection Prevention System, we have used KDD dataset. The KDD data set is used to find out the attacker categories and prevent those attackers or compromise the attackers. Table 1 shows several attack types and its categories that are utilized in this research work.

The attack categories and its types are described below:

R2L: imap, named, ftp-write, guess password, spy, send mail, snmp guess, snmp get attack, worm, xsnoop, xlock, warezmaster, warezclient, phf, multi-hop.

DOS: land, back, apacha2, pod, tear drop, process table, udpstorm, smurf, netpune, mailbomb.

Probing: saint, satan, namp, lpsweep, portsweep, msscan,

U2R: ps, perl, xtreme, rootkit, sqlattack, buffer over flow, load module, httprunnel.

4.2   Performance Evaluation

4.2.1   Computation Overhead

The Fig. 3 illustrates the computation overhead of our Enhanced Intrusion Detection Prevention System. Here, our system utilize minimum amount of resources from CPU (Control Processing Unit) to perform more number of request and response operations.

Fig. 3 CPU utilization of our proposed system

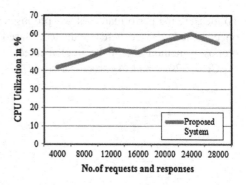

4.2.2 Total Number of Attackers Versus Total Number of Compromised Users

The Fig. 4 illustrates the attack detection and compromising (identified as attacker or malicious network traffic then send response to the attacker or drop the malicious network traffic) performance of our Enhanced Intrusion Detection Prevention System. For attack detection process, our system utilizes Support Vector Machine classification algorithm; Snort IDS and Intrusion Detection Prevention System for packet analysis and we use Honeypot system for compromising attack. The above mentioned algorithms improve the performance of our proposed system.

4.2.3 Overall Performance of Our Proposed System

The Fig. 5 compares the performance of the Enhanced Intrusion Detection Prevention System and the traditional network security techniques like Naive Bayes

Fig. 4 Attack detection and compromising process

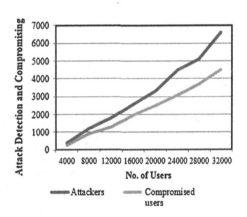

Fig. 5 Performance comparison of EIDS versus Naive Bayes classification system versus intrusion detection system

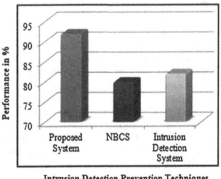

Attack Classification System and Intrusion Detection System. In comparison process, our EIDS achieves maximum performance because our system utilizes attack classification system, IPS and Honeypot System.

5   Conclusion

The computer network field is rapidly growing and making NIPS more complex gradually. Nowadays, new intrusion detection mechanisms are required to avoid intrusions from private data accessing. For that purpose, we presented a novel network security mechanism named Enhanced Intrusion Detection Prevention System. Our system provides real-time protection to authorized users with the help of Snort, Sebek and Dionaea. It also provides attack classification by using Support Vector Machine Classification and to prevent the attackers, malicious code and packet analysis we utilized Intrusion Detection Prevention System. Then, for compromising the attackers Honeypot system is utilized. Finally our system provides an effective protection to the authorized users as well as prevents the network intrusion in the network.

References

1. Verword T, Hunt R.: Intrusion detection techniques and approaches. Transaction on Computer Communication 25(15), 1356–1365 (2002).
2. Wang Yu, Cheng Xiaohui, Wang Sheng.: Anomaly Network Detection Model Based on Mobile Agent. In: 3rd International Conference on Measuring Technology and Mechatronics Automation. Shangshai, pp 504–507 (2011).
3. McHugh, John, Christie, Alan,; Allen, Julia.: Defending Yourself: The Role of Intrusion Detection System, IEEE Software 17(5), 42–51 (2000).
4. Jigiang Zhai, Yining Xie.: Research on network intrusion prevention system based on Snort. In: International Conference on Computer, Mechatronics, Control and Electronic Engineering . Changchun, pp 251–253 (2010).
5. Satish, L.K., Shobha, L.: A Survey on Website Attacks Detection And Prevention. International Journal of Advanced Engineering and Global Technology. 3(1), 238–244 (2015).
6. Amor, N.B, Benferhat, S, Elouedi, Z.: Naive Bayes vs. Decision Trees in Intrusion Detection Systems. In: Proc. ACM Symposium on Applied Computing, New York, pp. 420–424 (2004).
7. Kruegel, C, Mutz, D, Robertson, W, Valeur, F.: Bayesian Event Classification for Intrusion Detection. In: Proc. 19th Ann. Computer Security Applications Conference, NY, pp. 14–23 (2003).
8. Forrest, S.A. Hofmeyr, S.A, Somayaji, A, T.A. Longstaff.: A Sense of Self for Unix Processes," In: IEEE Symposium on Security and Privacy, Oakland, pp. 120–128 (1996).
9. Ye Du, Wang, H, Pang, Y.: A Hidden Markov Models-Based Anomaly Intrusion Detection Method. In: Fifth World Congress on Intelligent Control and Automation, Vol 5, pp. 4348–4351 (2004).
10. Warrender, C, Forrest, S, Pearlmutter.B.: Detecting Intrusions Using System Calls: Alternative Data Models. In: Proc. IEEE Symposium on Security and Privacy, Oakland, CA, pp. 133–145 (1999).

11. Wang, W, Guan, X.H, Zhang, X.L.: Modeling Program Behaviors by Hidden Markov Models for Intrusion Detection. In: Proc. Int'l Conf. Machine Learning and Cybernetics, Vol 5, pp. 2830–2835 (2004).
12. Selvaraj, R., Kuthadi, V.M., Marwala, T.: An Effective ODAIDS-HPs approach for Preventing, Detecting and Responding to DDoS Attacks. British Journal of Applied Science & Technology 5(5): 500–509 (2015).
13. Lafferty, J., McCallum, A., Pereira, F.: Conditional Random Fields: Probabilistic Models for Segmenting and Labeling Sequence Data. In: Proc. 18th Int'l Conf. on Machine Learning, pp. 282–289 (2001).
14. Wu, YS, Foo, B, Y. Mei, Y, Bagchi, S.: Collaborative Intrusion Detection System (CIDS): A Framework for Accurate and Efficient IDS. In: Proceedings 19th Annual Computer Security Applications Conference. pp. 234–244 (2003).
15. Ertoz, L, A. Lazarevic, A, Eilertson, E, P.-N. Tan, Dokas, Kumar, V, Srivastava, J.: Protecting against Cyber Threats in Networked Information Systems. In: Proc. SPIE Battlespace Digitization and Network Centric Systems III, pp. 51–56, 2003.
16. Tombini, E., Debar, E., L. Me, Ducasse, M.: A Serial Combination of Anomaly and Misuse IDSes Applied to HTTP Traffic. In: Proc. 20th Ann. Computer Security Applications Conference. pp. 428–437, 2004.
17. Lee, W., Stolfo, S.J., Mok, K.: A Data Mining Framework for Building Intrusion Detection Model. In: Proceedings of IEEE Symposium on Security and Privacy, Oakland, pp. 120–132, 1999.
18. Kuthadi, V.M., Rajendra, C., & Selvaraj, R.: A study of security challenges in wireless sensor networks. JATIT 20 (1), 39–44 (2010)
19. Vokorokos, L., A. Baláž, A.: Host-based Intrusion Detection System. In: 14th International Conference on Intelligent Engineering Systems, Budapest, pp. 43–47 (2010).
20. D. Boughaci, D, Drias, H, Bendib, A, Y. Bouznit, Y, Benhamou, B.: Distributed Intrusion Detection Framework Based on Mobile Agents. In: Proc. Int'l Conf. Dependability of Computer Systems, pp. 248–255, 2006.
21. Divya, Amit, C.: GHIDS: A Hybrid Honeypot System Using Genetic Algorithm, International Journal of Computer Technology & Applications 3(1),187–191 (2012).
22. Jeremy, B., Jean, F.L, Christian, T.: Security and Results of a Large-Scale High Interaction Honeypot. Journal of Computers 4(5), 395–404 (2009).
23. Jiqiang Zhai, Keqi Wang.: Design and Implementation of Dynamic Virtual Network. In: Proceedings of 2011 International Conference on Electronic & amp; Mechanical Engineering and Information Technology, Vol 4, pp 2131–2134 (2011).
24. Das, V.V.: Honeypot Scheme for Distributed Denial-of-Service Attack. In: Proceedings of the International conference on Advanced Computer Control, Singapore, pp 497–501 (2009).
25. Jabez J, Muthukumar, B.: Intrusion Detection System: Time Probability Method and Hyperbolic Hopfield Neural Network. Journal of Theoretical and Applied Information Technology 67(1), 65–77 (2014).
26. Juan Wang, Guiyang, Qiren Yang, Dasen Ren.: An intrusion detection algorithm based on decision tree technology. In: Proceedings of Asia-Pacific Conference on Information Processing, Shenzhen, pp. 333–335 (2009).
27. Yun, Y., Hongli, Y.: Design of Distributed Honeypot System Based on Intrusion Tracking. In: IEEE 3rd International Conference on Communication Software and Networks, Xian, pp. 196–198 (2011).

Slot Utilization and Performance Improvement in Hadoop Cluster

K. Radha and B. Thirumala Rao

Abstract In Recent Years, Map Reduce is utilized by the fruitful associations (Yahoo, Face book). Map Reduce is a prominent High-Performance figuring to prepare huge information in extensive groups. Different algorithms are proposed to address Data Locality, Straggler Problem and Slot under usage because of pre-arrangement of particular map and reduce phases which are not interchangeable. Straggler issue will happen because of unavoidable runtime dispute for memory, processor and system transmission capacity. Speculative Execution Performance Balancing is proposed to adjust the usage for single jobs and cluster of jobs. Slot Prescheduling accomplishes better data locality. Delay Scheduling is viable methodologies for enhancing the data locality in map reduce. For Map Reduce workloads job order optimization is challenging issue. MROrder1 is proposed to perform the job ordering automatically consequently the jobs which are arrived in Hadoop FIFO Buffer. Reducing the expense of Map Reduce Cluster and to build the usage of Map Reduce Clusters is a key testing issue. Restricted of accomplishing this objective is to streamline the Map Reduce jobs execution on clusters. This paper exhibits the key difficulties for performance improvement and utilization of Hadoop Cluster.

Keywords Data locality · Map Reduce · Delay scheduling · MROrder1 · Speculative execution performance balancing

K. Radha (✉) · B.T. Rao
KL University, Guntur, Andhra Pradesh, India
e-mail: radha.saitej@gmail.com

B.T. Rao
e-mail: thirumail@yahoo.com

© Springer India 2016   49
S.C. Satapathy et al. (eds.), *Information Systems Design and Intelligent Applications*, Advances in Intelligent Systems and Computing 434,
DOI 10.1007/978-81-322-2752-6_5

1 Introduction

Presently a Days, Map Reduce is utilized by the effective associations (Yahoo, Face book) for huge information escalated applications. Apache Hadoop is an open source execution of Map Reduce to bolster bunch preparing for Map Reduce workloads. Hadoop Distributed File System is a circulated stockpiling region which compasses crosswise over a large number of product equipment hubs. HDFS gives productive throughput, adaptation to internal failure, spilling information access, dependability. To store enormous volume of information and to prepare the monstrous volume of information HDFS is utilized. The Apache Hadoop structure contains the accompanying modules:

Hadoop Common—this module comprise of utilities and libraries that are needed by different modules of Hadoop.
Hadoop Distributed File System (HDFS)—It is a conveyed document framework which stores the information on the merchandise machines. It is additionally gives a high total data transmission over the bunch.
Hadoop YARN—It asset administration stage. It is dependable to oversee he process assets over the groups and utilizing them to plan the clients' applications.
Hadoop Map Reduce—It is a programming model utilized for vast scale information handling

Every one of these modules is naturally taken care of in the product application by the Hadoop system. Apache Hadoop's HDFS parts are gotten from Google's Map Reduce and Google File System (GFS) individually. Hadoop Distributed File System contains Name Node and Data Node. Name Node deals with the record framework metadata and Data Nodes are utilized to store the first information. HDFS has remarkable components, for example, Rack awareness, File consents, protected mode, fsck, fetchdt, rebalance, Upgrade and rollback, Secondary Name node, Checkpoint node and reinforcement node. HDFS objective is to utilize regularly accessible servers in a huge group where every single server has an arrangement of reasonable interior plate drives. For the effective execution, Map Reduce Application Programming Interface doles out the workloads on these servers where the information is put away and to be prepared. This idea is called as Data Locality. For Instance, consider a case in which there are 1000 machines in a group. Every single machine has three inside circle drives. In the event that disappointment rate of the bunch is made out of 3000 economical drives and 1000 cheap servers. Interim to disappointments rates is connected with reasonable equipment. Hadoop has constructed in issue remuneration and adaptation to non-critical failure abilities. It is same as with HDFS, as the information is divided into squares and lumps and duplicates of these pieces/squares are put away on different servers over the Hadoop bunch. Give us a chance to take one contextual investigation, expecting that a record comprises the telephone quantities of the considerable number of inhabitants in the North America. Who have their last name begins with R cloud been put away on server 1 and who are having the last name start with An are on server 2 and child. In the Hadoop environment, cuts of this phonebook would be put away and circulated on the whole group. To remake the information of the entire telephone directory, my project would need get to the squares from each server in the group. To accomplish higher accessibility, HDFS reproduces littler bits of information onto to extra servers as a matter of course. There is a possibility of excess in the names; this repetition urged to keep away from the disappointment condition and to give an adaptation to non-critical failure arrangement. Information excess is permitting the Hadoop bunch to separation that telephone directory passages into little pieces and run those littler occupations on every one of the servers in the group for effective versatility. Thus, we accomplish the benefit of information area; it is extremely critical while working with the gigantic datasets.

Typical workflow of Hadoop

- Load the data into the cluster (HDFS writes)
- Analysis of data (Map Reduce)
- Save the results into the cluster (HDFS writes)
- Read the results from the cluster (HDFS reads)

Example: How frequently did the clients sort "Discount" into messages sent to client administration (Fig. 1).

An expansive information record comprises of messages sent to the client administration office. Administrator of Customer Relationship or client adminis-tration division needs a depiction to see how frequently "Discount" was written by his clients. This would help the Manager to accept the interest on organization

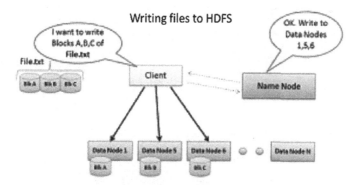

- Client consults Name Node
- Client writes block directly to one Data Node
- Data Nodes replicates block
- Cycle repeats for next block

Fig. 1 Workflow of Hadoop

returns and trades division and staff. It is a straightforward word check issue. The customer will stack the information into the bunch (File1.txt), present an occupation depicting how to dissect that information (word tally), the group will store the outcomes in another document called Results1.txt and the customer will read the outcomes record. To enhance the execution of guide decrease, when to begin the lessen assignments is key issue. At the point when the guide assignments yield turned out to be expansive, the execution of the guide Reduce planning calculation effected genuinely. Taking into account the past works, framework spaces assets waster will bring about decrease undertakings sticking around. To diminish undertakings begin time in Hadoop stage, (SARS) Self-Adaptive Scheduling cal-culation is proposed [1]. It chooses the begin time purposes of each decrease assignments powerfully as indicated by every employment setting which incorpo-rates size of guide yield, undertaking fruition time. At the point when contrasted with FIFO, Fair Scheduler and Capacity Scheduler diminish fulfillment time and normal reaction time is diminished by assessing the occupation finishing time, framework normal reaction time and decreases consummation time. Existing guide diminish system regard employments in general procedure. Dynamic corresponding scheduler gives more employment sharing and prioritization ability in planning furthermore brings about expanding offer of bunch assets and has more separation in administration levels of diverse occupations (Fig. 2).

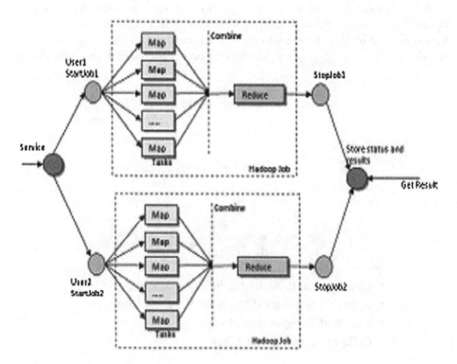

Fig. 2 Map Reduce structure

Delay scheduling is proposed to address the distinction between data locality and fairness. Time estimation and enhancement for HADOOP jobs investigated by Orchestrating an Ensemble of Map Reduce Jobs for Minimizing Their Make compass to minimize the aggregate consummation time of an arrangement of (cluster) of map reduce jobs [2]. The examines on Map Reduce planning calculation concentrate on advancing the bunch workloads, processing time and information correspondence. Extensional guide diminish assignment booking calculation is proposed for due date limitations in Hadoop Platform: MTSD. It permits client to indicate a vocation due date and to make the job be done before the due date. Scheduling method is proposed for multi-job map reduce workloads has the capacity alertly assemble execution models of the executing workloads then utilize these models for planning purposes. Decrease Task Scheduling is one of the key issues which impact the execution of the framework. Shared bunch environment contains unmistakable occupations which are running from different clients in the meantime. In the event that it shows up among the unmistakable clients in the same time, submitted employments from different clients won't be pushed ahead until the openings are discharged, lessen space assets are involved for quite a while, asset use rate is lower and it will broaden the normal reaction time of Hadoop framework which influences the throughput of Hadoop group [1]. In the previous decade, to create information serious applications World Wide Web has been has received as a perfect stage. Agent information serious web applications, for example, information mining are important to get to bigger datasets going from a couple of gigabytes to petabyte. Google influences the Map Reduce worldview to prepare give or take 20 petabyte of information for each day in parallel way. Map Reduce framework is isolated into different little undertakings running on numerous machines in a vast scale bunch. Contrasted with web information serious applications exploratory information concentrated application are profiting from the Hadoop framework. Map Reduce structure make straightforward without particular information of dispersed programming to make guide decrease capacities over different hubs in a group, in light of the fact that permits a with no programming learning. Map Reduce assembles the outcomes over the different hubs and return a solitary result or set of results. Map Reduce offers adaptation to non-critical failure.

Section 2 Discusses about the Challenges of Performance Improvement of Hadoop Cluster, Sect. 3 Discuss the Proposed Approach, Sect. 4 Gives the Details about Performance Optimization for Map Reduce Jobs and Finally Concludes the Paper.

2 Challenges of Performance Improvement of Hadoop Cluster

In Map Reduce, there are three difficulties, such as Straggler Problem, Data Locality and Slot under Utilization. Map Reduce is a popular computing paradigm for expansive information handling in distributed computing. Slot based Map

Reduce framework (Hadoop MRv1) is experiencing poor performance because of its unoptimized resource allocation. Resource allocation can be optimized from three distinct circumstances. Slots can be underutilized because of pre-setup of map and reduce slots which are not replaceable, because map slots are completely used while reduce slots are unfilled and vice versa. 1. Computational Resources (CPU Cores) are preoccupied into Map slots and Reduce slots; these are essential computational units which are statically pre-designed. Map Reduce has Slot Allocation constraint assumption (map slots can be allotted to map tasks and reduce slots can be allocated to reduce tasks) and General Execution limitation (map tasks are executed before the reduce tasks). It is watched that, (A). There is essentially distinctive framework usage and executions for map reduce workloads under diverse slot configurations. (B) Even under the optimal map reduce slot configuration there can be numerous reduce slots are idle which affects system performance and utilization. 2. Because of unavoidable runtime conflict of processor, memory system transmission capacity and different resources, there can be straggled (unpredictable gathering of different resources) map or reduce slots are idle which affects the system performance and utilization cause noteworthy deferral of the whole job.3. Information area augmentation is essential for execution change of guide decrease workloads and effective space usage and [3]. The aforementioned are the three difficulties for the use and execution change of Hadoop Clusters. To address the above difficulties, Dynamic MR can be proposed which concentrates on HFS (Hadoop Fair Scheduler) than FIFO Scheduler. Group usage and execution for Map Reduce occupations under HFS are poorer than FIFO Scheduler; DynamicMR can be utilized for FIFO Scheduler moreover. Dynamic MR contains three streamlining methods, such as Speculative execution Balancing, Dynamic Hadoop Slot Allocation (DHSA) and Slot Pre Scheduling.

3 Approaches for Slot Under Utilization and Performance Improvement

3.1 Dynamic Hadoop Slot Allocation (DHSA)

To beat the issue of slot under utilization, Dynamic Hadoop Slot Allocation (DHSA) is proposed [4]. It Allows reallocation of slots to either map or reduce slots relying upon their need. In Slot Allocation Constraint, If there are lack of map slots, map tasks will go through all the guide openings furthermore it utilizes unused reduce slots, likewise, if the quantity of reduce tasks is more prominent than the quantity of reduce slots then reduce slots can utilize unused map slots. Map Tasks uses map slots and Reduce tasks uses reduce slots. Contrasted with YARN, to control the proportion of running map and reduce tasks during runtime, pre-design of map and reduce slots per slave node still work. Without control system, it effortlessly happens that there are numerous reduce tasks running for data shuffling, which causes the system blockage genuinely. At the point when there are pending tasks, Dynamic Hadoop Slot Allocation expands the slot utilization and keeps up the fairness. There are two difficulties in Dynamic Hadoop Slot Allocation, for example, Fairness and the resource prerequisites among the map and reduce slots are distinctive. Fairness is accomplished when every one of the pools are dispensed with the same number of resources. In Hadoop Fair Scheduler, Fairness is a vital parameter. In Hadoop Fair Scheduler, first task slots are apportioned over the pools then inside of the pool slots are allotted to the jobs [5]. It shows diverse execution designs. Reduce task expends more resources, for example, system transmission capacity and memory. To conquer the fairness challenge, Dynamic Hadoop Slot Allocation is proposed. It comprises of two sorts of opening distribution strategies, for example, Pool Independent Dynamic Hadoop Slot Allocation and Pool Dependent Dynamic Hadoop Slot Allocation. To assign slots crosswise over pools with least ensure at map phase and reduce phase Hadoop Fair Scheduler embraces max-min fairness. Pool-Independent Dynamic Hadoop Slot Allocation expands the Hadoop Fair Scheduler through the distribution of slots from the cluster at the Global Level, autonomous of pools. As demonstrated in Fig. 3. When the quantity of typed slots via typed pools inside of the map and reduce phase are same it consider fair. There are two sorts of Slot designation process, for example, Intra Phase dynamic slot allocation and Inter-phase dynamic slot allocation. In the Intra Phase dynamic slot allocation, each pool is separated into two sub pools, such as map phase pool and reduce phase pool. At each phase each pool gets its offer of pools. Whose slot exceeds its share can take the unused slots from the other pools of the same phase. Construct up in light of the maximum min fairness approach an over-burden map phase pool A can take map slots from map phase pools B or C when pools B or C is underutilized. In Inter-phase element dynamic slot allocation, to boost the cluster utilization number of map and reduce slots at the map phase and reduce phase is lacking for the map tasks it will acquire some unmoving reduce slots for map tasks and the other way around. There are four situations, Let X_M and X_R be the aggregate number of map and reduce tasks, while S_M and S_R be the aggregate number of map and reduce slots configured by the client.

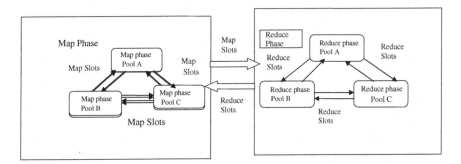

Fig. 3 Slot allocation flow for PI-DHSA

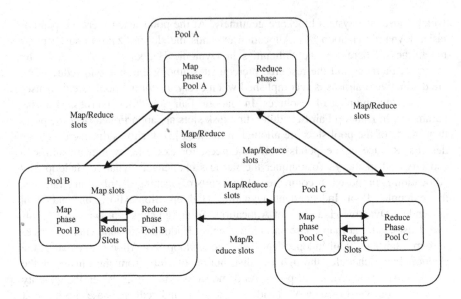

Fig. 4 Slot allocation flow for pool dependent-dynamic Hadoop slot allocation

$X_M <= S_M$ and $X_R \leq S_R$ no borrowing of map and reduce slots, the map tasks are run on map slots and reduce tasks run on reduce slots.

Case B. When $X_M > S_M$ and $X_R < S_R$ reduce tasks are allotted to reduce slots first and then use those idle reduce slots for running map tasks.

Case C. $X_M < S_M$ and $X_R > S_R$ to run the reduce tasks unused map slots can be scheduled.

Case D. $X_M > S_M$ and $X_R > S_R$, the system should be completely in busy state and similar to Case A, there will be no movement of map and reduce slots.

Pool-Dependent Hadoop Slot Allocation (PD-DHSA)

As shown in Fig. 4. Every pool contains map-phase pool and reduce-phase pool. Pool dependent Dynamic Hadoop Slot Allocation consists of two processes such as Inter pool dynamic slot allocation, Intra pool dynamic slot allocation.

A. **Intra pool dynamic slot allocation**

Based on the max-min fairness at every phase, every typed phase pool receive its regarding its demand (map slots demand, reduce slots demand) between two phase there are four relationships.

Case I. map slots demand < map share and reduce slots demand < reduce share: unused map slots can be borrowed for its overloaded reduce tasks from its reduce-phase pool first before producing to other pools.

Case II. map slot demand > map share, reduce slots demand < reduce share: some unused reduce slots can be satisfied for its map tasks from its map-phase pool first before giving to other pools.

Case III. map slots demand ≤ map share, reduce slots demand ≤ reduce share. Map and reduce slots are enough for its own usage. It can borrow some unused map slots and reduce slots to other pool.

Case IV. map slot demand > mapshare, reduce slot demand > reduce share: both ap and reduce slots are insufficient. It might lend some of the unused map or reduce slots from other pools through inter-pool dynamic slot allocation given below.

B. **Inter-pool dynamic slot allocation**

When, map slots demand + reduceslotdemand ≤ mapshare + reduceshare, no need to borrow the map and reduce slots from other pools.

Slot allocation process in Pool-Dependent Dynamic Hadoop Slot Allocation

Whenever task trackers gets a heartbeat instead of assigning map and reduce slots individually, it regards them all in all amid the allotment process. It first computes the most extreme number of free slots than can be relegated at each round for the heartbeat task tracker. At that point it begins the space distribution for pools. As demonstrated in Fig. 5. For each pool there four slot allocations. Case (i). We will first attempt the map tasks assignment if there are any unmoving map slots for the task tracker and there are pending map tasks for the pool. Case (ii). on the off chance that case (i) comes up short following the condition does not hold or it can't discover a guide assignment fulfilling the information area level, then keep on attempting decrease errands allotments when there are pending lessen undertakings and unmoving diminish spaces. A space weight based methodology is proposed to address the issue of asset prerequisites challenge in DHSA. For Map and Reduce openings allot with distinctive weight values. For Example, task tracker with map reduce slot configuration of 12/6. As indicated by heterogeneous resource requirements, expect the weights for map reduce slots are 1 and 2, hence, the total

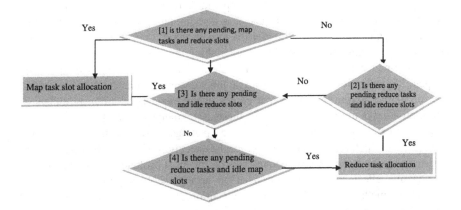

Fig. 5 Slot allocation flow for each pool under pool dependent dynamic Hadoop slot allocation

resource weight is $12 \times 1 + 6 \times 2 = 24$. Maximum number of running map tasks is 24 in the compute node with slot weight based methodology for dynamic borrowing where as number of running reduce tasks is at most $12/2 + 6 = 12$ instead of 24.

3.2 Speculative Execution Performance Balancing (SEPB)

This system is utilized to address the straggler issue, that is, moderate running of task's impact on a single job's execution time by running a backup task on another machine. To beat the straggler issue, Speculative Execution. Performance Balancing is proposed to adjust the performance tradeoff between a single job and a batch of jobs. Speculative execution can handle the straggler issue to enhance the performance for a single job at the cost of cluster efficiency. For moderate running the tasks Map Reduce work's execution time is touchy. There are numerous motivations to bring about a straggler which incorporates misconfiguration and defective equipment [6]. Stragglers are characterized into two sorts, for example, delicate straggler and hard straggler. In Hard Straggler task went into a halt status as a result of its endless waiting for the resources. If we kill the task then only then only we can stop and complete the task. It contains the backup task. In delicate straggler, computation task should be possible effectively yet it will take additional time contrast with common tasks. There are two cases for delicate straggler and backup task. Delicate straggler no needs to run the backup task, it finishes the task first (C1) and Soft straggler completes later the backup task (C2). Speculative Execution Performance Balancing in Hadoop is proposed to manage the straggler. Utilizing the Heuristic calculation called LATE (Longest Approximation Time Extended) SEPB distinguishes the straggling task rapidly [6]. When the straggling task is alertly distinguished, straggler can't be killed instantly because of the accompanying reasons, for example, to separate hard straggler and delicate straggler Hadoop does not have system and it is discriminating to choose, whether the delicate straggler the delicate straggler fits in with the (C1) or (C2) before running. Between the straggler and move down undertaking there is a reckoning overlay. At the point when both of two tasks is finished the assignment killing operation will happen. Regardless of the fact that a single job's execution time can be lessened by the speculative execution however it comes at the expense of group proficiency. Speculative tasks are financially savvy; these assignments will seek some different assets, for example, decrease spaces, guide openings, system with other running tasks of jobs which have negative effect for the execution of set of jobs. Henceforth, it raises a testing issue for speculative tasks on the most proficient method to lessen its negative effect for the execution of an arrangement of jobs. Element Slot assignment method called Speculative Execution Performance balancing is proposed for speculative task. It adjusts the tradeoff among the single work's and a bunch of work's execution time by powerfully deciding when the time has come to designate spaces and timetable for theoretical errands. Opening asset in-effectiveness issue for Hadoop group is recognized by the Speculative Execution Performance Balancing. Deferring the planning of Speculative undertaking is a testing issue.

3.3 Slot Prescheduling

To beat the issue of data locality, Slot Pre-Scheduling is proposed with no effect on fairness. Slot Pre-Scheduling enhances the Data Locality and expense of fairness. Delay Scheduling is successful methodologies for the change of Data Locality in Map Reduce [7]. Slot Prescheduling accomplishes better Data Locality when there are no presently nearby tasks are accessible, it delays the slot assignments in jobs. It achieves the load balance between slave nodes. To augment the data locality we can pre-allocate those slots of the node to jobs. Execution and improving so as to open use effectiveness is enhanced data locality for map tasks with information on that node with no reasonable unmoving map slots. To retain the task computation at the computing node with the local data (Data Locality) is an effective methodology. This methodology enhances the productivity of the slot utilization and execution. To enhance the information area in Map Reduce Delay Scheduler is proposed [3]. There are two area issues with Naïve Fair Sharing, for example, sticky slots and head-of-line-scheduling. In two cases, without local data on a node to keep up the fairness a scheduler is compelled to force to launch a task from a job. Delay Scheduling is proposed to enhance the data locality to wait for a scheduling opportunity on a node with local data. Naïve Fair Sharing Algorithm is utilized to share a cluster fairly among jobs is to always assign free slots to the job has fewer running tasks. Locality expands the throughput because of system data transmission of the cluster's disks [8]. Node locality is generally effective. Running on the same (rack region is speedier than the running off-rack. Head-of-line scheduling happened among the jobs which have small input files. At whatever point an job achieves the least running tasks one of its tasks dispatched on the following slot turns out to be free if the head of the line job is little, it is, it is unexpectedly to have data on the node that is given to it. This is the issue in Head-of-line scheduling. For Example, a job with data on 30 % of the nodes will accomplish just 30 % locality. Second issue is sticky slots, if fair sharing is utilized, sticky slots issue happens with large jobs. Issue with stick slots is that a job is allocated to the slot more than once. For instance, in a 100-node cluster with one slot for each node if there are 20 employments, each occupation has 20 running assignments. In the event that occupation j finishes an undertaking on nodes n and Node n asks for another assignment. At this stage, j has 19 running tasks while all the other jobs have 20. Naïve Fair sharing algorithm is again allotting the slot on node n to the job. Subsequently jobs never leave their original slots in consistent state, which prompts poor data locality because of data documents are striped by means of cluster, such that every job needs to run a few tasks on every machine. Sticky slots won't happen in Hadoop because of a mistake in how Hadoop tallies running assignments. In the wake of completing their work Hadoop undertakings enter a submit pending state, where demand authorization to rename their yields to its last filename. It has less effect on reaction time and throughput. Contrasted with deferral scheduler, burden parity is accomplished to the detriment of burden equalization crosswise over slave hubs. Frequently there are a percentage of the unmoving openings can't be designated on account of burden adjusting limitation in runtime. To boost the information territory spaces of the hub can be pre-assigned [4]. On the off chance that information region is enhanced burden adjusting will influence truly. Conversely, load parity will contrarily influence the information territory. Dynamic Hadoop Slot Allocation (DHSA), Speculative execution Performance adjusting (SEPB) and Slot Prescheduling procedures are joined together to frame; Dynamic MR. Dynamic MR is an orderly Slot Allocation framework which enhances the execution of Map Reduce workloads.

4 Performance Optimization for Map Reduce Jobs

Execution streamlining for Map Reduce occupations is ordered into Resource Allocation enhancement and Scheduling, Data locality improvement Speculative Execution advancement and Map Reduce enhancement on Cloud Computing. Past Research says that, for the productivity change and execution change of the cluster utilization can be accomplished through the Data locality enhancement [9, 10]. In speculative execution enhancement, task scheduling technique is imperative in Map Reduce to manage straggler issue for single job which incorporates BASE, LATE and Maximum Cost Performance (MCP) [11]. Longest Approximate Time to End (LATE) is utilized as a part of heterogeneous situations, LATE is a speculative execution algorithm which organizes the assignments to theorize, chooses the quick hubs to keep running on, topping the speculative tasks. Advantage Aware Speculative Execution (BASE) calculation [11] is proposed to enhance the execution of LATE algorithm. BASE algorithm ends the unneeded runs and assesses the conceivable advantage of speculative tasks. In Map Reduce enhancement on distributed computing, there are different advancement meets expectations for Map Reduce on distributed computing, for example, improvement of Budget and Deadline Constraints to deal with the asset portion for Map Reduce workloads for each parameter and expense models for streamlining the errand planning [12, 13]. For Map Reduce jobs on an arrangement of provisioned heterogeneous machines in cloud platforms, consider task scheduling algorithm regarding deadline and budget constraints [14].

For a given budget, global optimal scheduling algorithm is proposed by combining the Dynamic Programming techniques and in-stage local greedy algorithm. Global optimal scheduling algorithm achieves the minimum scheduling length of workflow O (jB)2. Further, Gradual Refinement (GR) and Global Greedy Budget (GGB) are created by taking the particular covetous techniques to effectively disperse the global reduce spending plan to refine the arrangements continuously. Various Choice Knapsack Problem through a parallel change minimizes the expense when the due date of the reckoning is altered. To convey the monetary allowance for execution enhancements of the Map Reduce work processes, Greedy and Multiple-Choice Knapsack issue ideal calculations demonstrates the proficiency in expense viability [14]. In Resource Allocation advancement and Scheduling for Map Reduce jobs there is some asset distribution enhancement work and calculation planning. To enhance the framework execution, work execution request in a Map Reduce workload in fundamental. For Map Reduce workloads, job order optimization is testing issue for the accompanying reasons after the map reduce task can be performed, due to pre-arrangement, both map slots and reduce slots are having constrained computing resources and map slots are designated to map tasks and reduce slots are distributed to reduce tasks [15–17]. MROrder1 is proposed to perform the job ordering consequently the jobs which are arrived in Hadoop FIFO Buffer. MROrder contains two sorts of parts, for example, ordering engine and policy module. Ordering engine contains two sorts of methodologies such as algorithmic based methodology and simulation-based ordering approach which gives job ordering. Policy module task is to choose how to perform job ordering and when to perform job ordering. MROrder comprises of completion time and make span. Rent the suitable size Map Reduce cluster consumes resources for what they have used on pay per pattern. Map Reduce cluster devours assets for what they have utilized on pay per design. Decreasing expense of Map Reduce Cluster and to expand the usage of Map Reduce Clusters is a key testing issue. Restricted of accomplishing this objective is to upgrade the Map Reduce occupations execution on groups. Work execution request has huge effect on group asset usage and general finish time [2, 18]. Dynamic MR enhances the Map Reduce workloads execution and use essentially with 46–115 % for single occupations and for cluster of employments 49–112 %. Contrasted with YARN Dynamic MR reliably beats for group of jobs around 2–9 % [4].

5   Conclusion

To enhance the performance and slot utilization in Map Reduce, different systems are proposed, such as Slot Pre scheduling, Dynamic Hadoop Slot Allocation (DHSA) and Speculative Execution Performance Balancing (SEPB). MROrder1 performs the job ordering consequently the jobs which are arrived in Hadoop FIFO Buffer. For the system performance, job execution order in a Map Reduce workload in essential. In Map Reduce improvement on distributed computing, there are different optimization works for Map Reduce on distributed computing, such as enhancement of Budget and Deadline Constraints to deal with the resource allocation for Map Reduce workloads and optimize the task scheduling.

References

1. Zhuo Tang, Lingang Jiang, Junqin, Zhou, Kenli Li, Keqin Li, "A self-adaptive scheduling algorithm for reduce start time", Future Generation Computer Systems, Elsevier, Volumes 43–44, February 2015, Pages 51–60.
2. A. Verma, L. Cherkasova, and R. H. Campbell, "Orchestrating an ensemble of MapReduce jobs for minimizing their makespan,"IEEE Trans. Dependency Secure Comput., vol. 10, no. 5, pp. 314–327, Sep./Oct. 2013.
3. M. Zaharia, D. Borthakur, J. Sarma, K. Elmeleegy, S. Schenker, and I. Stoica, "Delay scheduling: A simple technique for achieving locality and fairness in cluster scheduling," in Proc. 5th Eur. Conf. Comput. Syst., 2010, pp. 265–278.
4. Shanjiang Tang, Bu-Sung Lee, Bingsheng He, "DynamicMR: A Dynamic Slot Allocation Optimization Framework for MapReduceClusters", IEEE Transactions on Cloud Computing, Vol. 2, No. 3, pp. 333–347, July-September 2014.
5. M. Zaharia, D. Borthakur, J. Sarma, K. Elmeleegy, S. Schenker, and I. Stoica, "Job scheduling for multi-user Mapreduce clusters,"Univ. California, Berkeley, CA, USA, Tech. Rep. EECS-2009-55 2009.
6. M. Zaharia, A. Konwinski, A. D. Joseph, R. Katz, and I. Stoica, "Improving MapReduce performance in heterogeneous environments," in Proc. 8th USENIX Conf. Operating Syst. Des. Implementation, 2008, pp. 29–42.
7. S. J. Tang, B. S. Lee, B. S. He, and H. K. Liu, "Long-term resource fairness: Towards economic fairness on pay-as-you-use computing systems," in Proc. 28th ACM Int. Conf. Supercomput., 2014, pp. 251–260.
8. J. Dean and S. Ghemawat. MapReduce: Simplified Data Processing on Large Clusters. Commun. ACM, 51(1):107–113, 2008.
9. M. Zaharia, D. Borthakur, J. Sarma, K. Elmeleegy, S. Schenker, and I. Stoica, "Delay scheduling: A simple technique for achieving locality and fairness in cluster scheduling," in Proc. 5th Eur. Conf. Comput. Syst., 2010, pp. 265–278.
10. M. Hammoud, and M. F. Sakr, "Locality-aware reduce task scheduling for mapreduce," in Proc. IEEE 3rd Int. Conf. Cl ud Comput. Technol. Sci., 2011, pp. 570–576.
11. Z. H. Guo, G. Fox, M. Zhou, and Y. Ruan, "Improving resource utilization in mapreduce," in Proc. IEEE Int. Conf. Cluster Comput.,2012, pp. 402–410.
12. J. Polo, and Y. Becerra, D. Carrera, "Deadline-based MapReduce workload management," IEEE Trans. Netw. Serv. Manage., vol. 10, no. 2, pp. 231–244, Jun. 2013.
13. M. A. Rodriguez, and R. Buyya, "Deadline based resource provisioning and scheduling algorithm for scientific workflows clouds," IEEE Trans. Cloud Comput., vol. 99, 2014.
14. Y. Wang and W. Shi, "Budget-driven scheduling algorithms for batches of MapReduce jobs in heterogeneous clouds," IEEE Trans. Cloud Comput., 2014.
15. B. Moseley, A. Dasgupta, R. Kumar, and T. Sarl, "On scheduling in map-reduce and flow-shops," in Proc. Annu. ACM Symp. Parallelism Algorithms Archit., 2011. pp. 289–298.
16. S. J. Tang, B. S. Lee, and B. S. He, "MROrder: Flexible job ordering optimization for online MapReduce workloads," in Proc. 19th Int. Conf. Parallel Process., 2013, pp. 291–304.
17. S. J. Tang, B. S. Lee, R. Fan, and B. S. He, "Dynamic job ordering and slot configurations for MapReduce workloads," Tech. Rep. NTU/SCE-2014-4, SCE, NTU, Singapore, April, 2014.
18. A. Verma, L. Cherkasova, and R. Campbell, "Two sides of a coin Optimizing the schedule of MapReduce jobs to minimize their makespan and improve cluster performance," in Proc. IEEE 20th Int. Symp. Model., Anal. Simul. Comput. Telecommun. Syst., 2012, pp. 11–18.

An Efficient Educational Data Mining Approach to Support E-learning

Padmaja Appalla, Venu Madhav Kuthadi and Tshilidzi Marwala

Abstract Currently, data mining technique has become popular in online learning environment for learning as this technique works on huge amount of dataset. There are several e-learning systems which are based on the classroom, providing the natural interface of human-computer and the communication among to multi-modality, and the computing technology of integrated pervasive in the classroom. The pervasive computing for the development of some of the requirements is being raised for system openness, scalability and extensibility. Everyone could access the materials of learning from anywhere with the help of internet. Increased access of users also leads to security requirements and higher efficiency in data retrieval. To address these concerns, we propose an educational data mining approach and OTP system for improving the efficiency of data retrieval and security. This system provides more accuracy in data mining and secure data transmission.

Keywords JSM · OTP · RSA

1 Introduction

Currently, more number of researchers are focusing on e-learning system and data mining technique. The e-learning system works on a large database [1]. In educational system, the data mining approaches are utilized to find the student activities

P. Appalla (✉) · T. Marwala
Faculty of Engineering and the Built Environment, University of Johannesburg,
Johannesburg, South Africa
e-mail: padmaja_app@yahoo.com

T. Marwala
e-mail: tmarwala@uj.ac.za

V.M. Kuthadi
Department of AIS, University of Johannesburg, Johannesburg, South Africa
e-mail: vkuthadi@uj.ac.za

© Springer India 2016
S.C. Satapathy et al. (eds.), *Information Systems Design and Intelligent Applications*, Advances in Intelligent Systems and Computing 434,
DOI 10.1007/978-81-322-2752-6_6

63

and status of learning. The importance of E-learning is increasing and so is the role of data mining in e-learning for retrieving the data. Online multimedia education plays an important role for distance education. Due to this, more number of video assets are being produced by various organizations. This system is hard to connect, share, explore and reuse. Many videos are mainly annotated through the semantic links, and also the services create semantic links amidst the video datasets and allow their data to move internally.

The related issues are in the construction for the learning environment being tied within main features of the systems: hypertextuality, decentralizing versus centralizing, synchronicity, multimedia and interactivity. Furthermore, distance learning with the existing techniques could reformulate on the basis of advancement of the technology, making it more reliable and effective. The architecture based on the security is needed for the improvement of learning environment for getting the requirement of authorization and authentication [2]. The activity of online learning is required for assessing the personalization improvement and decision making. The activity of e-learning are not accurately described.

Our proposed technique uses JSM + Dictionary algorithms to get the semantic links between user query and video assets. Also our proposed system finds the weightage of every document. There are so many systems for providing the e-learning facility based on the classroom that making an interaction between the computer-human and communication through multi-modality, computing technology for integrating pervasive in the classrooms, the development over pervasive computing, system openness for raising the novel requirement, scalability, and extensibility. Everyone could access the learning materials from any place and route by using internet connection.

This research model is providing solution for the activity of e-learning. This approach is containing (1) the identification of the real user or authorized user and providing the access abilities based on given rights, (2) Providing the learning content in an easy-to-access manner, (3) making classroom interesting and interactive between tutor and students, (4) demonstration of the scalability, (5) Tracking the activity of student by web log mining algorithm, (6) Generating OTP to improve security while receiving the requested data. (7) Domain and sub-domain based searching system reduces the searching and provides accurate results. There are more requirements over the e-learning materials that have been optimized. Hopefully, the results have targeted to give the experience for convenient searching within the users.

2 Related Work

In this research [3], the author introduced navigation and semantic search methods for the purpose of semantic visualization in electronic domain. This concept contains navigation and querying steps and these steps are more. In the step of querying, the relations between the concepts are used and this concept contains properties to determine the relevant resources for the given keywords; whenever the new keyword is searched it mainly focus on its resources. In navigation the retrieved content from search are formed into seven various categories as per the sources. Also these retrieved results are ranked with help of distance and similarity measures. To test and adopt the steps of semantic visualization, we have created and implemented Semantic web portal. This portal provides good services in semantic navigation. In this research [4], to control and manage huge distributed systems, the policy based management approaches is been utilized. In most management systems, the policies are mainly utilized to alter the system's activities. Policies are mainly derived as per the responsibilities and authorization imperatives on subject as well as object entities. The responsibility policy mainly specifies the negative and positive responsibilities of subject. Like other system, this e-learning system also utilizes the policy based system to control and manage privacy and security aspects in the system. In this responsibility, policy could be utilized to state: finding purpose, restricting use, security, providing certificates for limiting the set and openness. The e-learning system is mainly based on the policy, the administrator of system can state some fundamental rules for the purpose of general process of system, and the additional policies may be merged as per the entities preferences. For every entity of system and communication among the entities the collections of policy will be available. In addition, regulatory bodies and governments might have some regulations and privacy laws [5]. These rules might be added to common policies and converted into electronic policies. Some kind of conflicts may occur on these rules.

An efficient methodology must be implemented to streamline the online activities, to identify and resolve the conflicts of policy. So the facility for the policy negotiation and specification will be more efficient for the e-learning system. Here user and online learning provider can find the conflicts of policy as well as negotiate the solution.

The specification of WS-Security is mainly based on the draft which is proposed by Microsoft, VeriSign and Microsoft in 2002, and it was a very good feature extension and flexible to Symbol Object Access Protocol to implement security in web services. This web service utilizes HTTP as a Transport protocol and Symbol Object Access Protocol as message protocol for the purpose of enabling business to transfer Symbol Object Access Protocol message in very good secure surroundings. The W3C defines the specifications of SML encryption; SAML and XML signature are used for signing digitally and encrypting messages which is based on XML. The Security of Web Service is a set of block and it mainly used to facilitate the requirements of a web service collection of WS-Security specifications.

This is a new e-signature technology and is mainly utilized for XML data exchange. The specification of XML signature [6] introduced by the W3C/IETF explains the formats of electronic signature with help of XML, the formation of conditions and electronic signature for the purpose of verification. It provides solutions for more number of security issues like integrity, non-reputation and authentication. The real-time advantage of this feature are multiple and partial signature. The old system permits electronic signature on particular tags in XML data, after that it enables more number of electronic signatures to be added in an Extensible Markup language document.

In digital signature, the elements are <Signed Info>. This element indicates the information is signed and illustrates what kind of techniques is utilized. The signature contains two steps.

1. Finds the object's digest.
2. Encrypt the digest and private key.

The XML signature of object is signed. This signed object contains reference object, Signature Method element and Canonicalization Method element. This element is a sign for all Signed Info elements.

<Signature Method> states the algorithm is mainly utilized to create the signature. This is a group of key dependent and digests algorithm and probably other algorithm like padding.

<Canonicalization Method> states that the method is mainly utilized for canonicalization.

This technique is permitting the different type of characters and attributes for applying in the file of XML, where the process of canonicalization is much important [2]. Because the format of XML and the way of parsing by various intermediaries and processors which can modify the information as same as the signature. But the data which is signed is still equivalent logically.

<Reference> finds the connection among the files being digested. At any place, more number of <Reference> elements mainly belongs to a particular segment of XML data. This contains digest mechanism and output of digest value is determined on the value. After that the sign is verified through reference and validation of signature.

<Transforms> tag explains the transformation algorithm utilized to convert the data prior to sign. This is not a main element and it contains the collections of many transform elements. For instance, whenever the signature object is in binary then it converted into the scheme of Base-64 for eliminating the unauthorized object format of XML. There are other mechanisms like XSLT and XPATH transformation.

The extension of current web is called as semantic web [7] and this permits the description of data to be more accurately explained by using very well-known words mostly understood by computers and people. In this semantic web, the data are very explained by novel W3C standard called RDF. This semantic web is one of the search engines. The present web sites are utilized by computers as well as people to accurately place and collect data which is being published on the semantic web.

The concept of the ontology [8] is a best approach for searching of semantic web. RDF and OWL is the model of data representation, which is recommended by W3C and it is utilized to indicate Ontologies. The Web which is semantic will always support automation, discovery, reuse and integration of information and provides solutions for interoperability issue. Now a day, research on semantic web search engine is in starting stage like existing search engine such as Bing, Yahoo, and Google.

As per the community of educational data mining, the educations data mining explained an important technique and this technique offers best method for finding the data type of which is from educational resources utilized to analyze regarding the educational resources. The data mining techniques helps to improve the educational fields. This EDM considers current emerging fields with growing method by gathering knowledge from various educational systems.

The existing data mining algorithms differs like different level utilization of educational data mining. But the process of EDM mainly focuses on the gathering of data, understanding the data, data retrieval regarding student learning and work. The analysis of data is done on EDM like a process of research and are more sequentially regarding to statistics of educational, machine learning, visualization and psychometrics.

Videos are main resources in EDM [9], students can get more information from videos spontaneously and effectively better than text resources. And video resources play key role in distance learning. Therefore, to enhance the e-learning outputs resources of video must be clearly annotated using approaches of data mining, domain experts and course creator. Also if every annotation part in video resources has full details of advanced information it will help the user to examine learning topic in full representation. As well as if annotation is connected to extra learning data from internal and external resources so that will helps students to observe more clearly about subjects from lots of view [10].

Semantic web is an important for WWW creation where information representation on internet would be explained. So it has possibility to handle various learning applications. The basic creation of SW is to put vocabularies and ontological concepts to explain the object in machine readable format. These vocabularies and concepts are published and gathered from various web data. In this, each part of definition is mainly supported by explained logic. The existing research on semantic web mainly concentrated on explaining Ontologies based on different technology of learning examination and the domain. From the technical point aspect the connected data represents the allocated information on web and this can be entirely readout using any machines and its descriptions are explained externally. The system of e-learning is helping the students and tutors in learning process based on the classroom technique. Anyway, the pervasive computing creation, the novel requirement is increasing the system scalability, openness and extensibility.

Whenever the collections are not referring it, it offers structure for investigation of web services. The specification of UDDI mainly uses the IETE and W3C standards like HTTP, XML, DNS protocols. It also has capability to accept old version. The registration of UDDI is free to E-learning worldwide rather than its length.

In [11] UDDI entered one phase called public beta-testing. The founders are IBM, Ariba and Microsoft. Now process the registry server that works as an interoperable server with other servers as same as the data enters into the registry server, then it is exchanged through servers over different business. The beta version of the UDDI has been scheduled for the end of first quarter; some of the company will behave like operators for the business registry of UDDI.

The Simple Object Access Protocol (SOAP) is providing the lightweight and simple mechanism for swapping the typed and structured information amid of the distributed and decentralized environment by using XML. The Simple Object Access Protocol (SOAP) is not defining any semantic application like particular semantic implementation or programming model, instead of defining the general mechanism over the expression of semantic application which provides an encoding mechanism and model of modular packaging for data encoding with the modules. This allows the Simple Object Access Protocol (SOAP) to be implemented in the ranking system of large variety from the system of messaging to RPC.

There are three parts in the SOAP:

- The SOAP is covering the construction that defines the framework to express the message like what is in a message; whether it is compulsory or optional; and who should deal with it.
- The rules of Simple Object Access Protocol (SOAP) encoding are defining a mechanism of serialization that could be used in the exchanging of the application instances that defines the data types.
- The presentation of the Simple Object Access Protocol (SOAP) RPC is defining the convention that should be used for representing the call and responses of remote procedure.

These are being described together like Simple Object Access Protocol (SOAP) parts, they seems to be orthogonal. Particularly, the rules of encoding and envelope are being defined by using distinct namespace for promoting the ease by the process of modularity. The author in [12], explains about various cryptography algorithms like AES, DES, and etc. This technique mainly involves secure data sharing with help of one key based data conversion and two key based data conversion. Vector Space model is one of the best models for finding the similarity between the two documents or file and query. This uses two approaches. One is IDF, TF. This is used to find total number of terms occurred in one document and number of terms occurred in other documents. This acts as Main role in data mining concepts [13]. This research work [14] proposed a good mechanism for improving the security and maintaining the connection between user and tutors. By using image processing they are identifying the hand signal and faces based on histogram feature.

2.1 Problem Statement

More number of E-learning Systems is based on classrooms which offer natural interface within human-computer together and communication over the multi-modality, connecting technology of the pervasive computing through classroom. The new requirements are raised with the pervasive computing development for the scalability, extensibility and System openness. With help of internet connection any one can access learning materials in any location and any route.

The important issues reading the creation of online learning environments are integrated to the important features of system: decentralizing versus centralizing, hypertextuality, interactivity, synchronicity and multimedia. In addition, the exact steps which are applied in distance learning will be created as per rendering it much effectively, advantages of technology and the security architecture which is reliable and most necessary to increase the learning performance to satisfy the requirements of authorization and authentication. The activity of e-learning requires to be assessed for the purpose of decision making and personalization improvement.

In the activity of online learning, the video resources are not much efficiently described. So the efficiency of data mining is totally reduced. Also the security in data transmission is not much efficient.

3 Proposed Work

3.1 Overall Architecture

Our proposed system mainly concentrates on improving the security, interactions between users and tutors and to improve the efficiency of data retrieval. For this purpose we have proposed JSM + Dictionary and RSA + OTP. These two algorithms improve the data transmission and efficiency of data retrieval.

3.2 Tutors and User Authentication

Initially the user makes registration by entering required field like username, name and class time, branch and etc. After entering the required field the server will check whether these fields are valid or not and generates one public key and private key and send response with private key. After registering, the user can login and utilize the service (Fig. 1).

3.3 Tutors Upload Files and Videos

Initially the tutors make login by entering username and password, and then uploads the files and videos. In this process the tutors select the file (video or text file), select domain and enter the short description about the files and upload the files into server. Then the server stores the files in that particular tutor name. Likewise every tutor can upload the file into server.

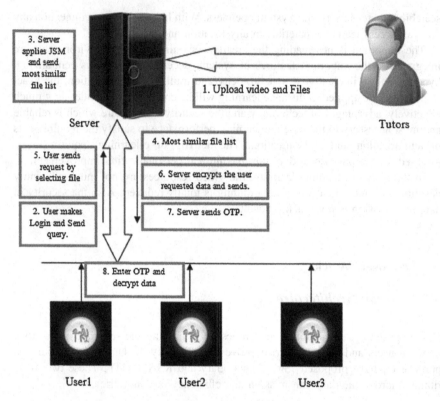

Fig. 1 Architecture for data transmission in e-learning system

3.4 User Data Request and Secure Data Transmission

First the user enters the query and selects the domain and file type (text or video). In this all Pdf document comes under text and all video files comes under video. After selecting, the user submits to server side. Then the server finds the domain and files type and apply JSM algorithm for mining the relevant document.

JSM: This is a best technique for finding the similarity between user query and available data. This model is derived from hypothesis technique and probabilistic approach. Jensen-Shannon-Model represents every document by probability distribution. A documents probability distribution is

$$Q = s(m, n)/An$$

s(m,n) represents how many times a word is repeated in document n and the sum of appearing word in n. Jensen-Shannon Model calculates the distance among two documents and returns ranked list. Jensen-Shannon Model ranks objective document by the distance of their probability distributions to source documents (Query):

$$JSM(x, n) = W(QP + Qn)/2 - W(Qp) + Q(Qn),$$
$$W(Q) = w(q(m)),$$
$$W(b) = -b \log b,$$

W(Q) represents probability entropy distribution q, qp and qn are probability distribution of two documents (a "document" and a "query"), respectively. In definition, w(q) = 0 we calculate distance among the two documents by 1 − JSM(p,n).

To improve the accuracy of similarity we are using Wikipedia dictionary. This will improve the semantic relations between two keywords. Then sort the document based on similarity and send the list of document with the name. Then user selects the document and send request to user. Then the server generates one OTP to encrypt the data using user's public key and send file to requested user and send OTP to user's mail id, and then the user will get the encrypted data after submitting the OTP and decrypt the data using his private key (Fig. 2).

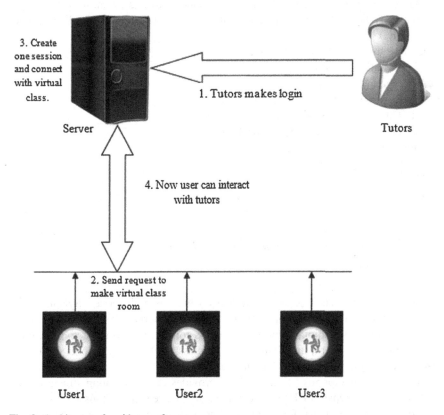

Fig. 2 Architecture for video conference

RSA: This is an Asymmetric key encryption algorithm and provides good security for data.

This algorithm generates two key in 2048 bits format. That key is known as private and public key, these private and public key is decrypting and encrypting the message, respectively.

The RSA includes some significant steps for solving a problem, which are specified below:

1. Let us consider a and b is the two prime numbers.
2. 2 Calculate M = a*b
   Where, M is the characteristic of two big prime numbers.
3. Choose an E_n (Encryption Key) like that it is not a circumstance of $(a - 1)*(b - 1)$.
   i.e. $\emptyset(m)) = (a - 1)*(b - 1)$
   For computing encryption examples E_n must be $1 < E_n < \emptyset(m)$ given that gcd $(E_n, \emptyset(m)) = 1$
   The major important of computing gcd is that $\emptyset(m)$ and E_n must be Prime of relative. Where $\emptyset(m)$ is the Function of Euler Totient and E_n is the Encryption Key (E_n).
4. Choose the D_n (Decryption Key),
   Which fulfill the Eqn
   $D_n^*E_n \bmod (a - 1)^*(b - 1) = 1$
5. For E_n (Encryption):
   CT (Cipher Text) = PT (Plain Text)^{En} mod M
   Cipher Text = (Plain Text)^{En} mod M, or
   Cipher Text = S^{En} mod M
6. For D_n (Decryption):
   PT = (CT)^{En} mod M
   Plain Text = (Cipher Text)^{En} mod M

3.5 Creation of Virtual Class Room Between Users and Tutors

E-learning activity gets flexible solution from this research model. Synchronous connectivity is a combination of VOIP (Voice-Over Internet Protocol), video, text chat, real-time presentation, splitting the rooms for group of small activities, presentation of white board, and instruments of class pooling. These tools are organized into single interface. This interface is used to create an e-classroom environment. Synchronous online teaching provides careful learning structure and plan. Tutors can prepare PPT to share at the time of live synchronous session. This technology adds flexibility for better teacher-student interaction and meeting the increasing requirements of the learners. This flexibility structure is planned to reduce the transactional level of distance among the student and instructor.

4 Result and Discussion

Table 1 shows the performance of our proposed system with existing system. Our proposed technique is JSM + Dictionary. This technique provides good result in finding the similarity between user given query and file. Our proposed technique provides 95 % on data retrieval

The graph in Fig. 3 shows the comparisons of similarity between existing and proposed system. In this graph the percentage decreases when the number of dataset increases. But in proposed system the decreasing level is very less.

Table 2 shows that our proposed RSA (2048 bit) algorithm provides good performance in security than the existing algorithm like DES and AES. In this table,

Table 1 Comparison of VSM without dictionary and JSM with dictionary

S. no.	(Query, no of dataset)	Similarity accuracy of VSM without dictionary (%)	Similarity accuracy of JSM + dictionary (%)
1	(1, 200)	85	95
2	(1, 400)	83	94
3	(1, 600)	80	92
4	(1, 800)	76	90
5	(1, 1000)	70	88

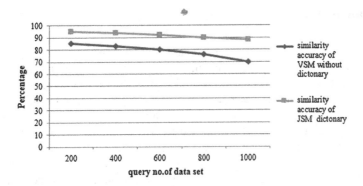

Fig. 3 Comparison graph for VSM and JSM

Table 2 Comparison of existing and proposed techniques

S. no.	No. of user's request (%)	DES security mechanism (%)	AES security mechanism (%)	RSA (2048 bit key) + OTP (%)
1	100	77	83	93
2	200	77	80	92
3	300	83	78	91
4	400	81	75	90

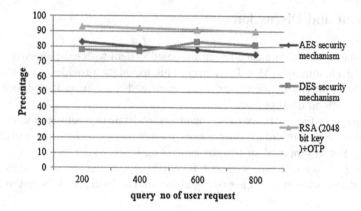

Fig. 4 Comparison graph for existing and proposed techniques

DES provides 78 % and AES provides 83 % but RSA (2048 bit) + OTP provides 93 % security.

The graph in Fig. 4 shows the comparison of AES, DES and RSA + OTP and illustrates that the proposed system provides better performance than the AES, DES security mechanism.

5 Conclusion

In this paper, we proposed an enhanced data mining approach for mining relevant results from the database and provide more security in the data transmission. This proposed system uses JSM + Wikipedia dictionary for finding similarity between user query and documents description and provides more accurate similarity and this is main thread to improve accuracy of data retrieval and improve the interactions among user and tutors. This proposed security mechanism provides high level key for decryption and encryption. The comparison graph (shown in Fig. 4) shows the performance of proposed system and existing system. In future, we are also planning to reduce the memory size and further improve the access performance.

References

1. Yue Suo, Naoki Miyata, Hiroki Morikawa, Toru Ishida, Yuanchun Shi.: Open Smart Classroom: Extensible and Scalable Learning System in Smart Space Using Web Service Technology. IEEE Transactions on Knowledge and Data Engineering 21(6), 814–828 (2008).
2. Akram, A., Samir, A. El-Seoud.: Web Services Based Authentication System for E-Learning. IJCIS 5(2), 74–78 (2007).

3. Myungjin, L., Wooju, K., June, S.H., Sangun P.: Semantic Association-Based Search And Visualization Method On The Semantic Web Portal. International Journal of Computer Networks & Communications 2(1), 140–152 (2010).
4. Khalil El-Khatib, Larry Korba, Yuefei Xu, George Yee.: Privacy and Security in E-Learning. International Journal of Distance Education Technologies 1(4), 321–334 (2003).
5. Shaik, G.S., Rajendra. C.: An Integration of clustering and of Adaptive E-Learning Database to Support The Analysis of Learning Processes. International Journal of Advanced Engineering and Global Technology, 1(01), 11–17 (2013).
6. Zhang Weiyan, Zhi-Jie Wu, Xia Tao. Web Services messages in Communication Research. Computer Engineering and Design, 26(10), 2621–2623 (2005).
7. Berners-Lee, T., Hendler, J. and Lassila, O.: The Semantic Web. Scientific American. pp. 29–37 (2001).
8. Dieter, F., Wahlster, W., Henry, L.,: Spinning the Semantic Web: Bringing the World Wide Web to Its Full Potential, MIT Press, Cambridge, (2002).
9. T.-D. Wu, Y.-Y. Yeh, and Y.-M. Chou.: Video Learning Object Extraction and Standardized Metadata. In: Proc. Int'l Conf. Computer Science and Software Engineering. Vol. 6, pp. 332–335 (2008).
10. Hausenblas, M., Karnstedt, M.: Understanding Linked Open Data as a Web-Scale Database", In: Proceedings of the Second Int'l Conf. Advances in Databases Knowledge and Data Applications, Menuires, pp. 56–61 (2010).
11. Ling Guo, Xin Xiang and YuanChun Shi.: Use Web Usage Mining to assist Online E-Learning Assessment. In: Proceedings of the IEEE International Conference on Advanced Learning Technologies, Finland, (2004).
12. Mitali, Vijay, K., Arvind, S.: Survey on Various Cryptography Techniques, International Journal of Emerging Trends & Technology in Computer Science 2(4), 307–312 (2014).
13. Vaibhav, S., Vinay, S.: Vector Space Model: An Information Retrieval System. International Journal of Advanced Engineering Research and Studies. 4(2), 141–143 (2015).
14. Bousaaid, M., Ayaou, T., Estraillier, P.: System Interactive Cyber Presence for E- Learning to Break Down Learner Isolation. International Journal of Computer Applications 3(16), pp. 35–40 (2015).

A Re-ranking Approach Personalized Web Search Results by Using Privacy Protection

Vijayalakshmi Kakulapati and Sunitha Devi Bigul

Abstract Various search services quality on the Internet can be improved by personalized web search. Users face sort of dissatisfaction when the results fetched by search engines are not related to the query they have asked for. This irrelevance result is retrieved huge based on the enormous variety of consumers' perspective and backgrounds, as well as the ambiguity of the contents. However, evidences show that the user's private information which they search has become public due to the proliferation of Personalized Web Search. The proposed framework RPS implement re-ranking technique, which adaptively make simpler user profiles by queries while respecting the consumer particular constraints of privacy. The great challenge in personalized web search is Privacy protection. To increase the efficiency and accuracy of web search privacy we use Greedy IL algorithm, i.e. GreedyDP and GreedyIL, for runtime generalization. Experiment assessment results show that the privacy-preserving personalized framework and re-ranking approach is highly effective and accurate enough for user profiling privacy personalization on the web search.

Keywords Web search · Privacy · User profile · Ranking · Query

1 Introduction

Now days, to acquire any useful data about anything on the internet, the very important gateway which help in achieving this is a web search engine. Sometimes these search engines retrieve results for users with moot info that don't fulfill user

V. Kakulapati (✉)
Department of Computer Science and Engineering, Guru Nanak Institutions
Technical Campus, Ibrahimpatnam, Hyderabad, India
e-mail: vldms@yahoo.com

S.D. Bigul
Department of Computer Science and Engineering, CMRIT, Hyderabad, India

© Springer India 2016
S.C. Satapathy et al. (eds.), *Information Systems Design and Intelligent Applications*, Advances in Intelligent Systems and Computing 434,
DOI 10.1007/978-81-322-2752-6_7

77

desires. This connectedness is predicated on the sizable amount of user contexts further maximum amount open retrieval info. The customized net search could be a common variety of look for techniques which give higher retrieved results and meet user desires. Gathering and analyzing user info offers user intention behind the sitting question. The customized net search is of two varieties, one is the click through knowledge and another is user identification technique. Within the click through knowledge users retrieve the tendency to clicked web content within the question record. Though this strategy has been with the efficiency question result [1], a Click through knowledge will work solely on continual queries. This is often the most disadvantage of this strategy.

Profile primarily based techniques are successful for a large vary of inquiries, yet are seen to be unsteady beneath some conditions. Profile primarily based strategies improve search results with difficult user models created from user identification methodology. There are a unit favorable circumstances and inconveniences for all styles of PWS techniques, the profile primarily based PWS is additional successful in enhancing the character of the internet hunt. By utilizing personal and behavior data of user profile, that area unit usually gathered from the internet logs [2, 3], consumer query history [4–6], bookmarks [7], click through knowledge [1, 8, 9], user documents [2, 10]. Unfortunately, individual personal data will simply relate a user's non-public data. Security problems build uneasiness in clients yet as reduce excitement in giving custom-built pursuit. For look after the user has to privacy in personalized internet search folks to believe two things in the search method. One is, they need to enhance the search quality victimization personal question of the user. Another issue is, they need to cover non-public data accessible within the user's profile so as to stay privacy risk in check. However, a few of previous history [10, 11] demonstrates that people are ready for agreement security on the off likelihood that they improve question things by provision user profile. In a perfect case, important measures of knowledge are often obtained at the value of solely tiny a part of user profile known as generalized profile. During this means privacy of the user are often protected with none negotiation. There's a balance earned between search quality and generalization gives privacy protection.

Privacy preserving PWS existing works don't seem to be satisfactory. The problems with the existing system are as follows:

1. The previous profile based mostly PWS do no generalize profile at runtime. A user profile is generalized one time which is in offline mode and queries accustomed modify square measure from a similar user. One issue reportable is [1] that typically profile based mostly personalization technique doesn't support unexpected queries. On-line identification is that the higher method, however, no earlier works have supported this.
2. The previous system doesn't contemplate modification in privacy needs. During this user privacy aren't protected properly.
3. While making customized search results several existing personalization techniques need continual user interactions.

All the above drawback of the system is resolved in our RPS (Re-ranking Privacy preserving search) structure. Construction works with the belief that the queries are with none sensitive data, progressing to shield not solely the privacy of individual users, however additionally retentive their quality for PWS. Framework usually uses two phases. One is the offline phase and other one is the online phase.

1. A consumer poses a keyword Q on the search engine, a generalized consumer profile G is creating by proxy gratifying confidentiality.
2. The keyword Q and the consumer profile G are then sent together to Personalized Web Server log file.
3. The Search results tailored according to profile is sent back P to proxy server log file.
4. At last, the web logs either present unrefined results P to the consumer or ordered them P' for the entire consumer profile.

Our main contributions are as listed below:

1. A Re-ranking privacy-preserving personalized architecture RPS, generalizes outlines the query based on the consumer privacy requirements.
2. The two metrics, web search personalization utility and privacy possibility are taken into consideration and originate the difficulty of the privacy-preserving personalized search.
3. Two simple and efficient algorithms GreedyDP and GreedyIL are developed to facilitate dynamic profiling.
4. Re-ranking algorithm applied on the generalized personalized web search.
5. Client can decide to personalize a query in RPS before each runtime profiling.

2 Related Work

2.1 Profile-Based Personalization

For the improved search results we tend to use profile primarily based personalization. To facilitate totally different personalization methods several profile representations are out there. Most of the class-conscious representations are made with weighted topic hierarchy. Our framework doesn't concentrate on the implementation of user profiles; it will with efficiency implement any class-conscious illustration supported information taxonomy.

In order to scale back human participation in performance mensuration, researchers have projected alternative metrics of customized internet search like an average preciseness [10, 12], level rating [13], and normal Rank [5, 9]. In this paper, we tend to the use typical preciseness measure projected by Dou et al. [1] that measures usefulness of personalization in cps. We tend to propose two

prognostic metrics, specifically measure of service and measure of confidentiality
on a profile while not demanding consumer response.

2.2 Privacy Protection System in Personalized Web Search

Privacy protection issues are classified into two categories [14] for PWS. The first
category contains of those treat privacy because the detection of a private. The
second category contains of those contemplate the kindliness of the info, notably
the consumer profiles, representation to the Personalized Web Search server.
Distinctive work within the study of protective consumer identifications attempt to
resolve the confidentiality downside completely dissimilar levels, together with the
simulated uniqueness, the cluster distinctiveness, no uniqueness, and no individual
information. Resolution of the major stage is confirmed breakable. The next two
levels are unreasonable as a result of the high price in message and cryptography.
The prevailing attempts specialize in the subsequent level. Each [2, 8] give on-line
namelessness on consumer profiles by make a bunch of profile for k consumers.

Exploitation this advance, the association between the question and the consumer is broken. Mix up queries among a bunch of consumers of United Nations
agency concern them [9] to plan as a worthless consumer profile protocol [3]. As a
consequence, any individual cannot profile an explicit entity. All of these efforts
assume the continuation of an expectation of third-party anonymizer, that isn't
promptly out there over the web at giant. Viejo and Castell_a-Roca [4] use
inheritance social networks rather than the moderator to supply an imprecise consumer profile to the online computer program. Within the theme, each consumer
acts as a groundwork activity of his or her neighbors. Consumers will attempt to
propose the question on behalf of the United Nations agency issued it, or promote it
to different neighbors.

2.3 User Profile Generalization

Removing topics with low sensitivity is reserve. Hence, merely forbidding sensitive
topics don't defend the consumer's confidentiality wants. To resolve this drawback
with forbidding, we have a tendency to propose a brand new technique. This
method identifies and removes set of topics from user profile specified the privacy
risk is in check. This method is named generalization, and also the output of this
method could be a generalized profile. Generalization is assessed into offline
generalization and on-line generalization. Offline generalization is performed while
not involving consumer queries. But it's unreasonable to perform offline simplification as a result of the output during this method might contain topic branches
tangential to a question. Online generalization [15] avoids reserve privacy revealing

and additionally removes topics tangential to the present question. Over generalization causes ambiguity in personalization, resulting in poor search results.

The dilemma of confidentiality maintaining generalization in the cycle is outlined supported utility and risk. Utility calculates the personalization service of the comprehensive profile, whereas risk measures the privacy possibility of exposing the profile.

2.4 The Re-ranking Approach

In this architecture, we present an absolutely unique method for fabricating ontological consumer profiles by allocating significant scores to existing suggestions in domain ontology. All of these profiles are continued and revised as explained interests of predecessor reference realm ontology. In this regard, we propose an extending commencement algorithmic program for maintaining interest scores within the consumer profile holding the user's current performance. The RPS experimental results show that supported the significant scores and the semantic proof for associate degree, ontological consumer profile with success provides the consumer with a customized read of the search results by the delivery results nearer to the uppermost after they are appropriate to the consumer.

2.5 Generalization Metrics

(1) **Utility Metric** This metric predicts the search quality of the query on a generalized profile. We have a tendency to remodel the utility prediction downside to the analysis of characteristic power of a given question on a generalized profile. Similar suggestion has been created in [11] to form of utility; however, this measure cannot be utilized for downside settings, as we've got a profile along with hierarchical data structure rather than flat one.

(2) **Privacy Metric** When a generalized profile is exposed the entire kindliness contained in normalized kind is outlined as privacy possibility. If the unique profile is uncovered the chance of exposing all insightful topics is the peaks.

2.6 Profile Generalization Algorithms

(1) **Brute Force Algorithm** *The m*ost favorable generalization is created by generating all rooted sub trees of our seed profile by using Brute Force algorithm and the associate tree and the best service is taken as the consequence.

(2) **GreedyDP Algorithm** We apply this algorithm on a generalized profile. We remove the leaf topic of this profile to generate optimal profile. Algorithm works [16] in a bottom up the manner. With the repeated iterations we generate profile with maximum distinguishing power and satisfying δ risk constraint. And this is the final output of GreedyDP algorithm.

(3) **GreedyIL Algorithm** GreedyIL algorithm [16] reduces the information loss. When δ risk is satisfied stop the iterative process and this reduces the computational cost. Then it simplifies the computation of information loss. It reduces the need of information loss recomputation.

3 Framework for Privacy Preserving and Personalization

The Framework describes our proposed key components of the framework and re-ranking approach for personalization.

3.1 The Proposed Framework

Our framework (Fig. 1) implements a re-ranking process and enables an effective personalization using the user query log and click through data. The framework consists of five components: Request Handler, Query Processor, Result Handler, Event Handler and Response.

3.2 Greedy DP Algorithm

In this planned the model of RPS, hand and glove with a greedy algorithmic rule Greedy DP named as Greedy Utility to sustain online recognition supported on prognosticative measures of personalization effectiveness and confidentiality problem. Greedy algorithmic rule Greedy DP works during a bottom up the manner. The most downside of Greedy DP is that it needs computation of all candidate profiles generated from tries of prune-leaf manner. Formally, we denote by $gi - t\ gi + 1$ the procedure of trimming leaf t from G_i to obtain Gi + 1. Visibly, the most favorable profile G * can be created with a finite-length transitive closure of trim-leaf. Greedy DP algorithm employed in a bottom up manner. This algorithm starts from G0, in every ith iteration, Greedy DP choose a leaf topic t∈TGi(q) for trimming, striving to maximize the effectiveness of the output of the recent iteration, namely Gi + 1. In these iterations, maintain a preeminent profile-so-far, which gives the Gi + 1 having the highest perceptive power while satisfying δ-problem restriction. The iterative procedure concludes when the profile is indiscriminate to a

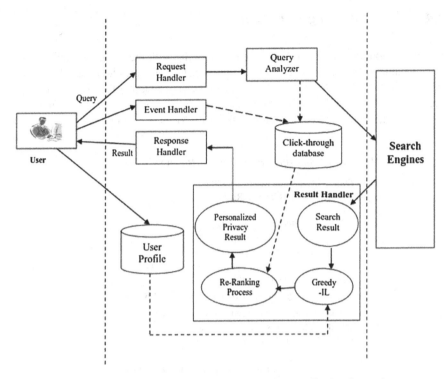

Fig. 1 Framework for re-ranking approach for privacy and personalized web search

root-topic. The final result (G*) of the algorithm will be the best-profile. The major obstacle of Greedy DP is that it requires recomputation of all candidate profiles (together with their perceptive power and privacy issue) produced from tries of trim-leaf on all tЄTGi (q). This will lead considerable memory requirements and computational cost.

3.3 Greedy IL Algorithm

In this planned a brand new profile generalization formula referred to as Greedy IL. The Greedy IL formula improves the potency of the generalization using heuristics supported varied conclusions. One of the results is that any trim-leaf operation shrinks the perceptive authority of the profile. In different statements, the refugee demonstrates monotonicity by trim-leaf. Greedy IL any diminishes this live with Heuristic. The less iterations the algorithm desires the bigger the isolation threshold.

3.4   The Re-ranking Approach

In this method, the ranking rule is about in stepping with the category attain of the item capable the quantity of selections of the constant user profile in past. From the primary ranking tend to plan replacement ranking to redefine the consumer preference record. We tend to utilize Probalistic Similarity measure and cosine function Similarity measure for Item cosine and Ranking for base Search.

1. *Algorithm: Ranking* (**Privacy Personalized Result** Set *PPRS)*

Input: Privacy Personalized Result Set PPRS.
Output: Arranged Result List with Ranking r.
    do
             if (PPRS i >PPRS j) then
                  Swap (Ii,Ij)
          else
                  Return PPRS I with ranking Order
        Until (no more Items in PPRS)

2. *Algorithm: Re-ranking (Ranked* **Privacy Personalized Result** Set RPPR *S)*

Input: Ranked Privacy Personalized Result Set RPPRS.
Output: Ordered Result List with Re-Ranking r.
        CTD<--GetClick_ThroughData (q, r, s);
        do
             if (CTD=True && RPPRS i > RPPRS j) then
                  Swap (Ii, Ij)
          else
                  Return RPPRS I with Re-ranking Order
        Until (no more Items in RPPRS)

4   Experimental Evaluation

Required datasets for experiment evaluation is collected through java based application. We created 50 user profiles (UP) with different interests and then perform query search for each user using different queries and create more than 1000 click-through database records for evaluation. We use the Yahoo search engine to retrieve the search results. To measure the effectiveness of the proposed framework approach we measure the personalized precision rate and recall of obtain results for existing (i.e., based on click-through data only) and proposed using profile-based with Greedy-IL. The measure of results is considered based on the number of similar and relevant results are re-ranked based on both existing and proposed approach.

Table 1 Precision and recall without user profile

Without user profile (WO-UP)

Query	Number of total search result	Relevant results to the query	Number of similar and relevant results	Precision	Recall
Movies	50	6	4	0.08	0.667
Sports	50	9	5	0.1	0.556
Music	50	11	5	0.1	0.455
Electronic	50	10	5	0.1	0.5
Travels	50	11	6	0.12	0.545

Table 2 Precision and recall without user profile

With user profile (W-UP)

Query	Number of total search result	Relevant results to the query	Number of similar and relevant results	Precision	Recall
Movies	50	31	8	0.16	0.258
Sports	50	18	7	0.14	0.389
Music	50	38	8	0.16	0.211
Electronic	50	38	7	0.14	0.184
Travels	50	35	8	0.16	0.229

In the first run we evaluate the framework without user-profile and Greedy-IL.
We run the query in five different domains as *Movies, Sports, Music, Electronic and
Travels* with click-through data and user-profiles. We observe that most of the
results which are similar and relevance to the query keywords, but not in relevance
to the user profile interest in case of existing click-through based, whereas high
relevancy is observed in case user-profile with Greedy-IL as shown below in
Tables 1 and 2.

4.1 Personalized Precision Measure

It is a measure of correctly predicted results by the system among all the predicted
results. It is defining as the number of relevant results retrieved by a search divided
by the total number of results retrieved by that search.

$$Personalize\ Precision\ (PP) = \frac{|number\ of\ simillar\ and\ relevant\ results|}{|No.\ of\ total\ Search\ Result|} \times 100$$

4.2 Personalize Recall Measure

Recall is a measurement of correctly predicted results by the system among the positive results. Recall is defined as the number of relevant results retrieved by a search divided by the total number of existing relevant results.

$$Recall\ Ratio = \frac{|The\ number\ of\ simillar\ and\ relevant\ results|}{|Relevant\ results\ to\ query|} \times 100$$

Figures 2 and 3 illustrates personalized precision and recall performance at different query categories of search results with the help of click-through data and user profile with Greedy-IL. The result shows an improvement in the personalized precision rate with different query categories in case of with-UP and the low recall rate in compare to without-UP. It's suggested more appropriate to meet the satisfactory level of motivation of the proposal. Improvisation is due to an online

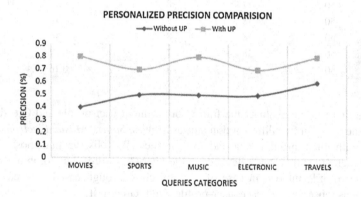

Fig. 2 Personalized precision comparison between with and without user-profile

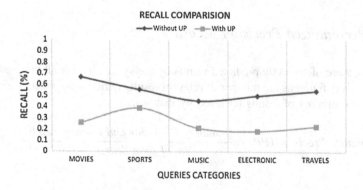

Fig. 3 Recall comparison between with and without user-profile

prediction method used to make a decision regarding relevancy using user profile for high query relevancy and make more beneficial for accurate in information retrieval. The results revealed that using user-profiles could achieve accurate and quality search results while preserving the user's tailored privacy requirements. The final results conforms the effectiveness and efficiency of privacy personalized search.

5 Conclusion

In this work, we proposed a re-ranking privacy protection framework called RPS for privatized website searching. RPS could be used by any Personalized Web Search that holds user profiles in a definite hierarchical categorization. The framework allowed consumers to specify customized privacy related requirements via the hierarchically categorized profiles. Along with this, RPS also had acted to generalize user profiles to ensure the personal privacy without undermining the search quality. User profiles are generalized using greedy IL. The result induces us to keep up a precedence queue of trim-leaf actors in a digressive arrangement of the data mislaying caused by the operators. The queue q permits quick retrieval of the most effective candidate operator. Filtering results is re ranked by using ranking rule supported RPS and results are shown to the user.

6 Future Enhancement

In our further work we will also make an attempt to resist opponents with a wider relationship among topics or capability to capture a more relevant personalized privacy preserving the result of queries posed by users based on the privacy of user profiles. We will also look into better-brushed methods to build the consumer profile, and more effective measures forecast the performance of RPS.

References

1. Silvia Quarteroni, Suresh Manandhar "User Modeling for Personalized Question Answering" AI*IA 2007: Artificial Intelligence and Human-Oriented Computing Lecture Notes in Computer Science Volume 4733, 2007, pp 386–397.
2. J. Teevan, S.T. Dumais, and E. Horvitz, "Personalizing Search via Automated Analysis of Interests and Activities," Proc. 28th Ann. Int'l ACM SIGIR Conf. Research and Development in Information Retrieval (SIGIR), pp. 449–456, 2005.
3. G. Chen, H. Bai, L. Shou, K. Chen, and Y. Gao, "Ups: Efficient Privacy Protection in Personalized Web Search," Proc. 34th Int'l ACM SIGIR Conf. Research and Development in Information, pp. 615– 624, 2011.

4. http://www.citeulike.org/user/abellogin/article/776870M. Spertta and S. Gach, "Personalizing Search Based on User Search Histories," Proc. IEEE/WIC/ACM Int'l Conf. Web Intelligence (WI), 2005.
5. F. Qiu and J. Cho, "Automatic Identification of User Interest for Personalized Search," Proc. 15th Int'l Conf. World Wide Web (WWW), pp. 727–736, 2006.
6. Y. Xu, K. Wang, B. Zhang, and Z. Chen, "Privacy-Enhancing Personalized Web Search," Proc. 16th Int'l Conf. World Wide Web (WWW), pp. 591–600, 2007.
7. A. Krause and E. Horvitz, "A Utility-Theoretic Approach to Privacy in Online Services," J. Artificial Intelligence Research, vol. 39, pp. 633–662, 2010.
8. F. Qiu and J. Cho, "Automatic Identification of User Interest for Personalized Search," Proc. 15th Int'l Conf. World Wide Web (WWW), pp. 727–736, 2006.
9. Archana Ukande, Nitin Shivale, "Supporting Privacy Protection in Personalized Web Search with Secured User Profile" International Journal of Science and Research (IJSR), Volume 3 Issue 12, December 2014.
10. X. Shen, S. Dumais, and E. Horvitz.: Analysis of topic dynamics in Web search". In Proceedings of the International Conference on World Wide Web, pages 1102–1103, 2005.
11. P. Sudhaselvanaya et al "Confidential User Query Profile Construction for Personalized Web Search", International Journal On Engineering Technology and Sciences–ISSN (P): 2349–3968, Volume 1 Issue 6, October 2014.
12. Ahu Sieg et al., "Web Search Personalization with Ontological User Profiles", CIKM'07, November 6–8, 2007, Lisboa, Portugal. Copyright 2007 ACM 978-1- 59593-803-9/07/0011.
13. V. Ramya, S. Gowthami, "Enhance privacy search in web search engine using greedy algorithm" International Journal of Scientific Research Engineering & Technology, ISSN 2278–0882 Volume 3, Issue 8, November 2014.
14. Y. Zhu, L. Xiong, and C. Verdery, "Anonymizing User Profiles for Personalized Web Search," Proc. 19th Int'l Conf. World Wide Web (WWW), pp. 1225–1226, 2010.
15. Yusuke Hosoi et al, "Generalization User Profiles to Context Profiles and Its Application to Context-aware Document Clustering", ISBN: 978-960-474-361-2, Recent Advances in Computer Engineering, Communications and Information Technology.
16. Lidan Shou, He Bai, Ke Chen, and Gang Chen, "Supporting Privacy Protection in PersonalizedWeb Search", IEEE Transactions on Knowledge and Data Engineering Vol. 26 No. 2, 2014 doi no. 10.1109/TKDE.2012.201.

An Efficient Virtualization Server Infrastructure for e-Schools of India

Gantasala Sai Hemanth and Sk. Noor Mahammad

Abstract According to 2011 census, the literacy rate in India is 74.04 %, a substantial increase when compared to that of 65.38 % in 2001 but it is a way below
than the average world literacy rate of 84 %. Out of 177 countries covered as per
Human Development Report, India ranks 126 among the in the literacy rate (http://
www.rediff.com/news/2007/nov/20illi.htm [1]). Since 90 % schools are located in
rural areas, Rural education in India is to given utmost importance. Access to
quality education is far behind in rural areas since there are fewer committed
teachers because the qualification and source of income for the teachers are very
less than what it is expected to be. Since there is an upward trend in growth of rural
Internet access, creating rural specific applications enable immense growth of rural
India. So using the World Wide Web as Education delivery medium would be the
legitimate solution as it can not only eliminate most of the issues in rural education
but also disseminate the best quality education. This paper provides a novel solution
through implementation of e-school by using concept of virtualization. Server
virtualization is the cost-effective implementation with least server consolidation.

Keywords Virtualization · Server infrastructure · Hypervisor · E-learning

1 Introduction

Bearing the largest illiterate population currently, literacy rate in India is increasing
very sluggishly from the past few years and it will take 40–50 years to achieve
universal literacy considering the current progress [2]. As per 2011 statistics, there

G. Sai Hemanth (✉) · Sk.Noor Mahammad
Indian Institute of Information Technology, Design and Manufacturing (IIITDM)
Kancheepuram, Chennai 600 127, India
e-mail: coe12b005@iiitdm.ac.in
URL: http://www.iiitdm.ac.in

Sk.Noor Mahammad
e-mail: noor@iiitdm.ac.in
URL: http://www.iiitdm.ac.in

© Springer India 2016   89
S.C. Satapathy et al. (eds.), *Information Systems Design and Intelligent
Applications*, Advances in Intelligent Systems and Computing 434,
DOI 10.1007/978-81-322-2752-6_8

is considerable deviation in literacy rate among urban population (84.1 %) and rural population (67.8 %). More than 50 % of the rural students in 5th grade are unable to grasp/read a second grade text book and moreover they are not able to solve simple arithmetic problems, according to a survey.

In the Government run educational sector, inadequacy/inefficiency of the teaching staff is one of the daunting factors contributing to the low literacy in rural India. According to a survey made on 3700 randomly selected schools from 20 states, it was concluded that teacher-absenteeism is 25 %, which means 25 % of the teachers were absent at any random time. The dearth in teaching staff is witnessed all over India bearing an average pupil teacher ratio with 1:42 and apparently Bihar being the worst with pupil teacher ratio of 1:83. Introduction of para-teacher scheme is expected to overcome the issue of teacher shortage and teacher absen-teeism. Para educators are generally members of same community, who teach as well as communicate their ideas, share many experiences and cultural practices with their students which includes primary languages. The para-teacher scheme has come out of flying colors when implemented in Rajasthan under the project named Shiksha karmi project by overcoming the short of teacher and teacher absenteeism. In some cases, even though the infrastructure of school is good there is scarcity in availability of teachers for each every class [3]. Even there are many schools with only two teachers teaching for the whole school. So this consequence led to the scenario of *single teacher teaching the same syllabus to all the students ranging from class I to class V seated in a single classroom. Rapid urbanization is also contributing factor for poor teacher count*, since low salary in rural areas are unable to meet the daily needs so the teaching staff are slowly migrating from rural to urban areas where salary is decent enough.

In both language and math, there is a substantial gap between what the text books expect from the children and what children can actually do. Although the children's level of understanding do show an improvement over the course of the year most of the children are at least two grades below the level of proficiency assumed by the text books. In the Indian public school system, kids are not able to grasp/attain knowledge and skills required at a commensurate pace with their age. The evaluations, if they happen, are focused on a child being able to vomit what they have learnt by heart. Study of history is reduced to dates and study of language to reciting poems. The only thing children learning about our great leaders are their birth, death and important program's launched dates but not about the message given by them to our society or about their philosophy. The curriculum is designed to leave the kids with no skills and perspectives to be able to build a constructive life. Marks have been given the prime importance in present educational system, that can make or break your career and are considered to be the best metric of judging one's knowledge on the subject. Lack of understanding of concepts leading to rote memorization to score good marks. Examinations are designed in such a way that all the questions appear from within the text book, so there is no point in reading out of syllabus and to actually understand the concept. Availability and Standard of text books is also a major concern.

So in order to address all the above requirements huge amount of money is to be invested which is not practically possible. *The cost effective solution to meet the above stated requirements are through E-Schools a Web based interactive learning. The information is disseminated at faster rates very effectively* via *web based approach. The best advantage of the web based learning is that any one can learn any thing, at any time irrespective of distance between tutor-student and medium of communication.* Since students are unable to reach the quality education, this is the best possible way to service the education at their door step. The Web based approach enhances context based learning for students, consistency of learning materials, flexibility, accessibility and convenience. This approach will also become a medium for teachers to discover and learn abundant knowledge available in the Internet. In this approach discussions on particular topics create opportunities to share the student perspective's beyond the traditional concept/idea. Study materials can be updated in real time which is not possible in case of traditional text books. In this paper Web based learning, e-learning and online education are used inter-changeably by ignoring the subtle differences between them.

2 Literature Review

2.1 E-learning

The only way to access E-learning has been restricted to using high tech tools that are either web distributed or web-based via Internet or intra net. E-learning can be a form of application, program, object or website but ultimately it's aim is to provide a quality learning opportunity to individuals [4]. Using e-learning different education philosophies can be implement with ease. This Strategy motivates the students to develop their own knowledge out of their own imagination and this approach incorporates highly structured ways which can be used to thrive the students to pre-determined conclusions. E-learning can be used in two major ways (1) presentation of educational content; (2) facilitation of education process. E-learning can make information available and also play a major role in student's self construction of knowledge. It is important to note that technology is not content; technology is not process; rather it can provide access to both. So we can conclude that e-learning is a form educational convergence [5]. The benefits of e-learning include cost effectiveness, consistency, timely content, flexible accessibility. More importantly this approach develops the knowledge of Internet and computer skills that will help the students through out their lives and careers. American psychological society did a survey on e-learning and inferred that: 'Learning outcome using computer-based instruction is far more fruitful/favourable than the conventional ways of teaching learning process', as per statistics. By leveraging e-learning for *online tests* and *quizzes*, the necessity of *paper-based assessment can be reduced*. With e-learning a teacher can host a guest lecture virtually, without being physically present in front of the students with the same level of interaction as that of previous scenario i.e., face-to-face education.

2.2 E-learning Abroad

In 1953, the first sensational history was created by offering first televised classes on KUHT (today called Houston PBS) by the University of House. After television, the personal computer and WEB were the next major inventions that made web-based education possible. Ultimately in 1989 the University of Phoenix became the first institution to launch a fully online collegiate institutions. In 2014, More than one million enrollments has been done for online courses in USA reported by American sources and apparently public schools offer more online courses than private schools. Statistics of enrollments for online education is as follows: it is 7 % in 2002, 30 % in 2012, 53 % in 2014–15. Market research already reveals that Africa is the growing leader in e-learning, with 15 % annual growth rate for the next 4 years, and individual countries such as Senegal and Zambia exhibiting up to 30 % growth in e-learning deployments. The Initiative made by UNESCO and NGO, like one laptop per one child urges the importance of web-based learning. In order to promote the well being of student-athletes, the National Collegiate Athletic Association came up with a initiative, a membership organisation dedicated to teach the right skill to succeed in playing field as well as help in building up their career via Online education providing a wide variety of courses so that a student can take any course to fulfill minimum credit requirement. According to a survey of Britain's Open University, e-learning consumes 90 % less energy than traditional courses, thereby divulging the fact that e-learning is eco-friendly. The amount of CO_2 emissions (per student) is also reduced up to 85 %. By 2019, 50 % of all classes will be delivered online world wide.

2.3 E-learning in India

Being one of the largest education systems in the world, India with a network of more than 1 million schools and 18,000 higher educational Institutions. The Department of Electronics and Information Technology, Ministry of Communications and IT, Government of India has been rigorously working on developing tools and technologies for setting up of e-learning education in India [6]. As a result the following are some of major outcomes (1) Training of trainers in E-learning at Aurangabad, Kolkata, Imphal, Calicut, Gorakhpur had trained 600 teachers altogether. (2) Data compression and decompression techniques and its applications to E-learning/education has implemented data compression techniques applicable to images, scanned documents and videos. (3) IIT Kanpur developed an open source free Learning management System (LMS) named as Brihaspati which included virtual classrooms. LMS is a software application for administration, tracking and reporting basically managing purpose for e-learning training or education programs. (4) Online practical science experiments on CBSE syllabus for the subjects of Physics, Biology, Chemistry for class 9 and 10 that included interactive simulations based on mathematical models, videos were conducted. (5) E-learning in 204 high schools in rural and tribal area of Srikakulam district of Andhra Pradesh has been set up. The Government of India formed an alliance with HCL technologies to launch *Sakshat*, the world's most inexpensive tablet (pricing around INR 1500/-). This effort is a part of NMEICT scheme, aiming to connect 25,000 colleges and 400 universities through the same Sakshat portal. Acharya Nagarjuna University, Dr. BR Ambedkar Open University, Annamalai University and Amity University have launched there distance degree programs in online mode. TechNavio's analysts forecast that online education in India to grow at a compound annual growth rate of 17.4 % over the period of 2014–19. Arunachal Pradesh became the first state in our country to implement web based education in all the schools. Meanwhile some of well known organizations which has implemented e-learning are classteacher.com, meritnation.com, extramarks.com, adhyapak.com, classontheweb.com etc., offering e-materials which includes textbooks to practice tests and lab exams for K-12 (primary and secondary education) students. IIT's and IISc has already initiated e-learning through the website nptel.ac.in in engineering, science and humanities streams for free of cost. In July 2014, Indian government has allocated INR 500 Crore for the implementation of the project digital India campaign that aims to setup broadband services in rural areas.

3 Proposed System Architecture

Since providing free and online education education is the only technical solution to boost the rate of literacy, the system architecture is proposed at the server side to handle millions of requests efficiently and effectively. In order to implement web based education covering all the schools especially rural schools a centralized server(s) is required. Typically four servers are required, namely (1) Exam server; (2) Administration server; (3) Digital library server; and (4) Video steaming server. Overall requirements and maintenance for managing these servers are simply too high. Virtualization technology provides the cost effective solution as it can lower the consumption of power, cooling, hardware as well as cost of facilities. It can be effectively used to simplify administration and maintenance without compromising on performance. The ability to simulate a hardware platform, such as a network resource, storage device or server, in software is called Virtualization. As just like the traditional hardware solution would, all the functionalities of virtualized system are separated from the hardware and simulated as a virtual instance with the ability to operate individually. Apparently, multiple virtual devices or machines can be supported with a single hardware platform, which can be easily scaled up or down as per the requirement. As a result, a virtualized solution is typically much more cost effective, portable and scalable than a traditional hardware-based solution. Two kinds of virtualizations are required to implemented in this architecture, (1) Server Virtualization (2) Network Virtualization. Server Virtualization is intended to mask server resources (both hardware and software), including the number of discrete physical servers, OS's and processors from the clients. Network virtualization is a method of splitting the available bandwidth into channels by combining available resources in a network. Each channel is independent of other, and each of which can be assigned (or reassigned) to a particular server or device in real time [7]. Network virtualization (NV) is intended to enhance the flexibility, reliability, scalability, security, optimize network speed, and to improve productivity, efficiency and require less man power by accomplishing many tasks automatically, there by disguising the true complexity of the implemented network. Virtualization of network is necessary because when a server attached to a network, all hardware and software networking resources (switches, routers etc.) are viewed as logical instances (soft-ware based view). The virtual network (software) is responsible for an intelligent abstraction that makes it easy to deploy and manage network services and underlying network resources whereas the physical networking devices are merely responsible to forward the packets. Hence, NV can guide the network to support better optimised virtualized environments by creating virtual networks within a virtualized infrastructure thereby enabling NV to support complex requirements in multi-tenancy environments. Type-1 Virtualization is used in this architecture. Type-1 Virtualization is also referred as 'native' or 'bare metal' hypervisor which run directly on system hardware, this gives better performance when compared to its counterpart Type-2 virtualization.

3.1 Hypervisor

The Hypervisor (Virtual Machine (VM) Monitor) is the key innovation behind server consolidation which is an extraction of the physical hardware to allow multiple operating systems and their applications to share the hardware resources among them. The underlying hardware is viewed as unshared and a complete machine by VM even though portions of it may not exist or may be shared by multiple VM's. An example of this is the virtual NIC (vNIC). Efficient communication among VM's within the hypervisor as well as efficient communication to the external network is achieved by the hypervisor as it permits the communication of VM's to the physical networking infrastructure by attaching the server's physical NIC's to the hypervisor's logical infrastructure. A single hardware host can be shared among multiple operating systems with the help of hypervisor AKA virtual machine manager. Each operating system appears to have the host's processor, memory, and other resources all to itself.

Fig. 1 Conventional architecture for sever virtualization

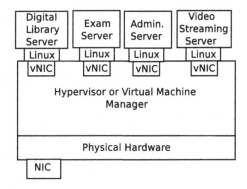

3.2 Virtual Networking

In order to make virtual networking infrastructure more productive, Virtual switch is required. The Virtual switch attaches physical NIC to vNIC's and also enable communication between vNIC's as shown in Fig. 1. A virtual switch does more than mere forwarding data packets, before passing the packets it can intelligently direct the communication on the network by inspecting them [8]. It reduces the complexity of network configuration. Virtual switch has a key advantage over physical switch as it is compatible with the new functionality.

3.3 Bridging

Bridges are used to connect separate network segments into a extended Ethernet. It is a layer 2 device that works at the data link layer and delivers packets based on the destination MAC address. Hence Bridges can be used by intelligently managing the network traffic [9]. When connected by a bridge, the VM acts as if it were a normal physical machine to the outside world. Since the VM's implemented on the same hypervisor gets the same MAC address as that of the hardware, we need to create software bridge interfaces so that, each node of VM connected to bridge gets unique port number.

Since the server consists of two NIC's. One NIC can be interfaced (Bridged) to three VM servers (Digital Library Server, Exam Server, Administration Server) and the other NIC can be bridged directly with video streaming server, since for video streaming a lot of bandwidth is required and henceforth it is allocated a separate channel (NIC). Now by bridging the three VM's bridged to single NIC appears as three different servers and apparently video stream server is separate server as it is allocated to other NIC as shown in Fig. 2. Hence by bridging performance is not all

Fig. 2 Proposed architecture
for sever virtualization

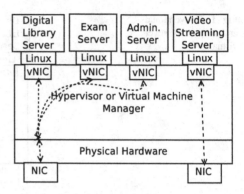

compromised since data packets are directly forwarded to destination VM from the physical NIC with the help of hypervisor and bridging but whereas if NAT is configured within the switch instead of bridging then it uses routing protocols as routing is implemented at IP layer (L3) which involves immense of Host CPU computation for delivering the packets to the destination which is a big performance overhead. The disadvantage of bridging is that it cannot be used for implementing in wireless drivers.

Virtual machine CPU architecture should be same as that of the hardware architecture. Otherwise we need to perform QEMU (quick emulator) emulation which is a performance overhead. QEMU is used to emulate another architecture (examples are emulator for ARM/IBM POWER PC architecture using x86 processor). It recompiles x86 processor code to give ARM Arch code. Suppose all VM's are of Ubuntu server, then KVM is the hypervisor used as it is of both type 1 and type 2 hypervisor. KVM is a fork of QEMU and when KVM is run on a machine without any hardware virtualization extension, it switches back to QEMU. Typically virtualization extensions for intel architecture is VT-X and AMD architecture is AMD-V [10]. Hence emulating VM's is two types (1) KVM with acceleration-type 'KVM', instructions meant for VM are directly executed in physical state CPU; (2) QEMU with acceleration-type 'TCG', Tiny Code Generator which will optimally translate and execute virtual CPU instructions. By this we can conclude that KVM requires QEMU to provide complete virtualization solution.

3.4 Description About VM Servers

The Exam server, digital library server, administration server (directory structure is shown in Fig. 3) running on a single physical NIC is to be implemented with traditional web servers, HTTP/HTTPS protocols. The directory structure implementation details are shown in Figs. 4 and 5 respectively.

Fig. 3 Administrative server directory structure

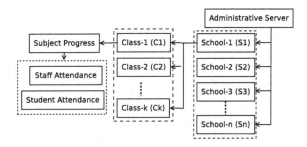

Fig. 4 Digital library directory structure

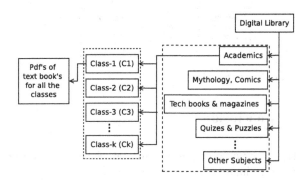

Fig. 5 Exam server directory structure

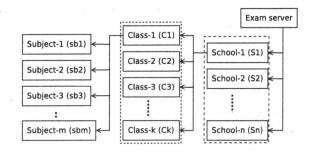

For video streaming server, the protocol to be implemented is real-time streaming protocol, which is effective in streaming media servers. Network media server will not be able to see videos on hard disk unless we have media server software (plex media server for Ubuntu). In Plex, trans-coder uses CPU so there is no need of GPU. Hence, there is no need of graphics card. The directory structure for video streaming server is shown in Fig. 6.

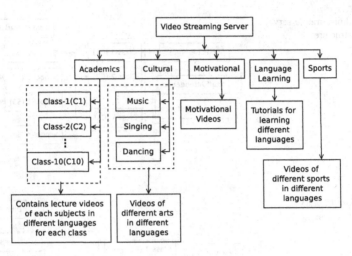

Fig. 6 Video streaming server directory structure

4 Conclusion

This paper proposed an importance of the e-learning requirements for Indian scenario, to improve the literacy rate, overcoming the inadequate/inefficient teacher strength and quality education. This paper also gives the detailed implementation of the e-school infrastructure for Indian schools. An efficient server virtualization infrastructure can be used to implement the e-schools. Implementing an effective e-learning strategy can be an invaluable investment that stands to generate greater learning outcomes in educational sector. The proposed solution is a cost effective, reliable and easy reach solution for rural India.

References

1. India has the largest number of illiterates in the world, November 2007, http://www.rediff.com/news/2007/nov/20illi.htm.
2. Modern Indian Education System, http://www.scribd.com/doc/37720377/Modern-Indian-Education-System#scribd.
3. Global campaign for education more-teachers needed, http://unicef.in/Story/746/Global-Campaign-for-Education-more-teachers-needed.
4. Moore, Joi L., Camille Dickson-Deane, and Krista Galyen. "e-Learning, online learning, and distance learning environments: Are they the same?." The Internet and Higher Education 14, no. 2, pp. 129–135, (2011).
5. Nichols, M, A theory for eLearning. Educational Technology & Society, 6(2), pp. 1–10, (2003).
6. Sunil Kumar Sharma, Javed Wasim, Dr. Jamshed Siddiqui, E-Learning in India, (2014).

7. Sherwood, Rob, Glen Gibb, Kok-Kiong Yap, Guido Appenzeller, Martin Casado, Nick McKeown, and Guru Parulkar, Flowvisor: A network virtualization layer, OpenFlow Switch Consortium, Tech. Rep (2009).
8. B. Pfaff, J. Pettit, T. Koponen, K. Amidon, M. Casado, S. Shenker, Extending Networking into the Virtualization Layer, (2009).
9. Bridged Networking, https://help.ubuntu.com/community/KVM/Networking.
10. Desai, Ankita, Rachana Oza, Pratik Sharma, and Bhautik Patel, Hypervisor: A survey on concepts and taxonomy, International Journal of Innovative Technology and Exploring Engineering (IJITEE) pp. 2278–3075, (2013).

A Novel Approach for Horizontal Privacy Preserving Data Mining

Hanumantha Rao Jalla and P.N. Girija

Abstract Many business applications use data mining techniques. Small organizations collaborate with each other to develop few applications to run their business smoothly in competitive world. While developing an application the organization wants to share data among themselves. So, it leads to the privacy issues of the individual customers, like personal information. This paper proposes a method which combines Walsh Hadamard Transformation (WHT) and existing data perturbation techniques to ensure privacy preservation for business applications. The proposed technique transforms original data into a new domain that achieves privacy related issues of individual customers of an organization. Experiments were conducted on two real data sets. From the observations it is concluded that the proposed technique gives acceptable accuracy with K-Nearest Neighbour (K-NN) classifier. Finally, the calculation of data distortion measures were done.

Keywords Horizontal privacy preserving · Walsh hadamard transformation · Data perturbation · Classification

1 Introduction

Explosive growth in data storing and data processing technologies has led to creation of huge databases that contains fruitful information. Data mining techniques are used to retrieve hidden patterns from the large databases. In distributed data

H.R. Jalla (✉)
Department of Information Technology, CBIT, Hyderabad, T.S, India
e-mail: hanucs2000@gmail.com

P.N. Girija
School of Computer and Information Sciences, UoH, Hyderabad, T.S, India
e-mail: pn_girija@yahoo.com

© Springer India 2016   101
S.C. Satapathy et al. (eds.), *Information Systems Design and Intelligent Applications*, Advances in Intelligent Systems and Computing 434,
DOI 10.1007/978-81-322-2752-6_9

mining a group of clients share their data with trusted third party. For example a health organization collects data about diseases of a particular geographical area that is nearby to the organization. To improve the quality of information and collaborate with other organizations, which benefits the participated clients to conduct their business smoothly. The Third party performs a knowledge based technique on the collected data from group of clients. In this scenario, clients can be grouped under three categories like honest, semi-honest and malicious. First category honest client always obey the protocol and will not alter the data or information. Second category, semi-honest client follows the protocol but tries to acquire the knowledge about other clients while executing. Third category, malicious client or unauthorized client always tries to access other's data and alter the information. To provide security of the data, this paper proposes a novel approach.

Many researchers address this problem by using cryptography, perturbation and reconstruction based techniques. This paper proposes an approach for Horizontal Privacy Preserving Data Mining (HPPDM), with combination of different transformations. Consider a Trusted Party (TP) among group of clients. Then TP communicate with other clients by using symmetric cryptography algorithm and assigns transformation techniques to each client. In this work, transformation techniques such as WHT, Simple Additive Noise (SAN), Multiplicative Noise (MN) and FIrst and Second order sum and Inner product Preservation (FISIP) are used. Proposed model is evaluated with data distortion and privacy measures such as Value Difference (VD) and Position Difference parameters like CP, RP, CK and RK.

This paper is organized as follows: Sect. 2 discusses about the Related Work. Section 3 focus on Transformation Techniques. Section 4 explains about Proposed Model. Section 5 discusses Experimental Results and in Sect. 6 Conclusion and Future Work discussed.

2 Related Work

Recently there has been lot of research addressing the issue of Privacy Preserving Data Mining (PPDM). PPDM techniques are mostly divided into two categories such as data perturbation and cryptographic techniques. In data perturbation methods, data owners can alter the data before sharing with data miner. Cryptography techniques are used in distributed data mining scenario to provide privacy to individual customers.

Let X be a numerical attribute and Y be the perturbed value of X. Traub [1] proposed SAN as $Y = X + e$ and MN as $Y = X * e$. Other perturbation methods such as Micro Aggregation (MA), Univariate Micro Aggregation (UMA) and

Multivariate Micro Aggregation (MMA) are proposed in [2–4]. Algorithm based Transformations are discussed in Sect. 3.

Distributed Data Mining is divided into two categories Horizontal and vertical in PPDM. Yao [5] introduced two-way Communication protocol using cryptography. Murat and Clifton proposed a secure K-NN classifier [6]. In [7] Yang et al. proposed a simple cryptographic approach i.e. many customers participate without loss of privacy and accuracy of a classifier. A frame work was proposed [8] which include general model as well as multi-round algorithms for HPPDM by using a privacy preserving K-NN classifier. Kantarcioglu and Vaidya proposed privacy preserving Naive Bayes classifier for Horizontally Partition data in [9]. Xu and Yi [10] discussed about classification of privacy preserving Distributed Data Mining protocols.

3 Transformation Techniques

3.1 Walsh Hadamard Transformation

Definition The Hadmard transform H_n is a $2^n \times 2^n$ matrix, the Hadamard matrix (scaled by normalization factor), that transforms 2^n real numbers X_n into 2^n real numbers X_k. The Walsh-Hadamard transform of a signal X of size $N = 2^n$, is the matrix vector product $X*H_n$. Where

$$H_N = \overset{n}{\underset{i=1}{}} \otimes H_2 = \underbrace{H_2 \otimes H_2 \otimes \ldots \otimes H_2}_{n}$$

The matrix $H_2 = \begin{bmatrix} 1 & 1 \\ 1 & -1 \end{bmatrix}$ and $\otimes$ denotes the tensor or kronecker product. The tensor product of two matrices is obtained by replacing each entry of first matrix by multiplying the first matrix with corresponding elements of second matrix. For example

$$H_4 = \begin{bmatrix} 1 & 1 \\ 1 & -1 \end{bmatrix} \otimes \begin{bmatrix} 1 & 1 \\ 1 & -1 \end{bmatrix} = \begin{bmatrix} 1 & 1 & 1 & 1 \\ 1 & -1 & 1 & -1 \\ 1 & 1 & -1 & -1 \\ 1 & -1 & -1 & 1 \end{bmatrix}$$

The Walsh-Hadamard transformation generates an orthogonal matrix, it preserves Euclidean distance between the data points. This can be used in Image Processing and Signal Processing.

3.2 First and Second Order Sum and Inner Product Preservation (FISIP)

FISIP is a distance and correlation preservation Transformation [11].

Definition The matrix representation of a linear transformation can be written as $A = [A_i] = [A_1 \ A_2 \ ... \ A_k]$, Additionally, A_i can be written as $A_i = [A_{im}]$. Then the transformation is called a FISIP transformation if A has following properties.

a. $\sum\limits_{m=1}^{k} A_{im} = 1.$

b. $\sum\limits_{m=1}^{k} A_{im}^2 = 1.$

c. $\sum\limits_{m=1}^{k} A_{im} A_{jm} = 0, \ for \ i \neq j$

$$A^{[k]} = [a_{ij}], 1 \leq j \leq k, a_{ii} = \frac{2-k}{k}, a_{ij} = \frac{2}{k}$$

$$A^2 = \begin{bmatrix} 0 & 1 \\ 1 & 0 \end{bmatrix}, A^3 = \begin{bmatrix} \frac{-1}{3} & \frac{2}{3} & \frac{2}{3} \\ \frac{2}{3} & \frac{-1}{3} & \frac{2}{3} \\ \frac{2}{3} & \frac{2}{3} & \frac{-1}{3} \end{bmatrix}$$

3.3 Discrete Cosine Transformation (DCT)

DCT works on real numbers and gives following real coefficients:

$$f_i = \left(\frac{2}{n}\right)^{\frac{1}{2}} \sum_{k=0}^{n-1} \Lambda_k x_k \cos[(2k+1)i\Pi/2n]$$

where for k = 0 and 1 otherwise. These transforms are unitary and Euclidean distance between two sequences is preserved [12].

3.4 Randomization

Randomization is one of the simple approaches to PPDM. Randomization involves perturbing numerical data. Let X is a confidential attribute Y be a perturbed data

[1]. SAN is defined as $Y = X + e$, where e is random value drawn from a distribution with mean value zero and variance 1.

MN is defined as $Y = X * e$, e is a random value.

4 Proposed Model

This paper proposes a new approach for HPPDM as shown in Fig. 1. In this approach a group of clients select a Trusted Party who has capability to retrieve information from large data. Symmetric cryptography algorithm is used for communication between clients. Suppose client1 wants to collaborate with other client i.e., client2. Client1 sends a request to client2 for approval of collaboration. If client2 sends acceptance response both the clients choose their own transformation/modification techniques to modify the data.

This work focuses on perturbs numeric data. Numerical attributes are considered as a confidential attribute. Different transformations techniques are used to modify original data, transformations techniques discussed in Sect. 3. Both clients modify their original data using transformation techniques then modified data will be sent to the Trusted Party. TP decrypts modified data which is collected from clients and performs knowledge based technique. K-Nearest Neighbor (K-NN) as knowledge based technique.

Theorem *Suppose that* $T: R^n \rightarrow R^n$ *is a linear transformation with matrix A, then the linear transformation T preserves scalar products and therefore distance between points/vectors if and only if the associated matrix A is orthogonal.*

Here $T(X) = T_1(X_1) + T_2(X_2)$, T_1 and T_2 are linear transformations subsequently; T is also a linear Transformation. Whereas X_1 and X_2 are the original data of the respective clients.

Fig. 1 Model for HPPDM

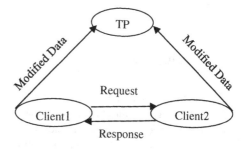

Proposed Algorithm

Client1:

```
Input:  D₁[M₁][N]
Output: D₁'[M₁][N]

Begin
        If (N<2ⁿ)   (n=0, 1, 2, 3.....)
                Append Columns with Zeros its nearest 2ⁿ
        Endif
        A=WHT (N) // generate WHT matrix
        MD₁= D₁ * A;
End
```

Client2:

```
Input:  D₂ [M₂][N]
Output: D'₂[M₂][N]

Begin
        Choose any one of the transformation technique discussed in section
3.
End
```

Trusted Party:

```
Input:  D₁ [M₁][N] ,D₂[M₂][N]
Output: D [M][N]

Begin
        Step1: Collect modified data from client1
        Step2: Collect modified data from client2
        Step3: Merge both modified data sets which are received from client1
        and client2
        Step4: Perform knowledge based technique.
        Step5: Share knowledge i.e., obtained from step4 to both the clients.

End
```

Any data like parameters, keys and modified data that needs to be securely shared between the clients and TP must be ensure using Symmetric Cryptography Algorithms.

5   Experimental Work

Experimental work conducted on two real datasets Iris and WDBC collected from [13]. Assume that datasets taken as matrix format, row indicates an object and column indicates an attribute. Divide entire dataset into two parts numerical and categorical attributes. Among that only numerical attributes are considered and it is shared between the clients.

The distribution of the data is done in two methods. Generate a random value using random function which is considered as the percentage of the total records. In method 1: For client1 the data records are sent from record 1 to the percentage of random value. And the remaining records are sent to client2. In method 2: first n records are leaving (n value change as per the dataset size) and then consider the random value to select next number of records to send for client 1. For client2 merge few records from first n records and remaining from the left out records of the whole data set and send them. If any client choose WHT as transformation technique data pre-processing is required if Number of Attributes are less than 2^n, $n = 0, 1, 2, 3,\ldots$ add number of columns to its nearest 2^n (Table 1).

K-NN is used as a classification technique from WEKA Tool [14]. While conducting experiments K value set to 3, 5, and 7. Tenfold cross validation is used when running the K-NN algorithm. In this paper, consider four combinations of linear transformations such as WHT-WHT, WHT-DCT, WHT-FISIP and WHT-SAN. Follow two methods in data distribution, select 4(2 values are below 50 and 2 values are above 50 to 100) random values in each method per a dataset. 35 and 135 records are skipped in IRIS and WDBC datasets respectively as per method 2 discussed above. Classifier results of IRIS data set shown in Tables 2, 3, 4 and 5. Classifier Accuracy of WDBC shown in Tables 6, 7, 8 and 9. IRIS original data gives 96.00 % using K-NN. Modified IRIS data gives acceptable accuracy on k = 7 using all methods. WDBC Original data set gives 97.18 and modified WDBC gives acceptable accuracy in all cases. Calculated different privacy measures VD, RP, CP, RK and CK from [15].

Calculated average values of all data distributions shown in Tables 10 and 11. The higher values of RP and CP and the lower value of RK and CK, and the analysis show more privacy is preserved [15].

Table 1 Data set description

Data set	No. of records	No. of attributes	No. of classes
IRIS	150	4	3
WDBC	569	31	2

Table 2 Accuracy on Iris data set using combination of WHT-WHT

%split (client1-client2)	Accuracy (%)					
	K = 3		K = 5		K = 7	
	Org	Mod	Org	Mod	Org	Mod
28-72	96.00	96.00	96.00	96.67	96.00	94.67
45-65	96.00	96.00	96.00	96.67	96.00	94.67
65-35	96.00	96.00	96.00	96.67	96.00	94.67
71-29	96.00	96.00	96.00	96.67	96.00	94.67
51-49*	95.33	95.33	96	94	96.67	95.33
55-45*	95.33	96	955.33	96	96.67	96
70-30*	95.33	95.33	96	95.33	94.67	95.33
76-24*	95.33	94.67	95.33	95.33	96	95.33

*Skip first 35 records for data distribution

Table 3 Accuracy on Iris data set using combination of WHT-DCT

%split (client1-client2)	Accuracy (%)					
	K = 3		K = 5		K = 7	
	Org	Mod	Org	Mod	Org	Mod
27-73	96.00	96.67	96.00	97.33	96.00	97.33
46-54	96.00	96.67	96.00	96.67	96.00	98
61-39	96.00	100	96.00	100	96.00	100
75-25	96.00	97.33	96.00	97.33	96.00	98
26-74*	96.00	98.67	96.00	98	96.67	98
45-55*	96.00	96.67	96.00	96.67	96.67	96.67
54-46*	94.00	98	95.33	98	95.33	98
58-42*	95.33	98	95.33	98	97.33	98

*Skip first 35 records for data distribution

Table 4 Accuracy on Iris data set using combination of WHT-FISIP

%split (client1-client2)	Accuracy (%)					
	K = 3		K = 5		K = 7	
	Org	Mod	Org	Mod	Org	Mod
29-71	96.00	96.00	96.00	96.00	96.00	96.00
39-61	96.00	96.00	96.00	94.67	96.00	94.67
55-45	96.00	97.33	96.00	97.33	96.00	97.33
76-24	96.00	98.67	96.00	98	96.00	98
34-66*	95.33	97.33	94.67	97.33	96	97.33
47-53*	95.33	99.33	96.00	99.33	96.00	99.33
53-47*	94.00	98.67	96.00	98	96.00	98
57-43*	95.33	98	95.33	98.00	96.00	98

*Skip first 35 records for data distribution

Table 5 Accuracy on Iris data set using combination of WHT-SAN

%split (client1-client2)	Accuracy (%)					
	K = 3		K = 5		K = 7	
	Org	Mod	Org	Mod	Org	Mod
25-75	96.00	96.00	96.00	95.33	96.00	94.67
34-66	96.00	94.00	96.00	94.00	96.00	94.67
58-42	96.00	99.33	96.00	99.33	96.00	99.33
79-31*	96.00	98.66	96.00	98.00	96.00	98.00
24-76*	95.33	96.00	96.67	96.00	96.67	96.00
50-50*	95.33	98.67	96.00	98.00	96.00	98.67
55-45*	95.33	98.00	96.00	98.00	97.33	97.33

*Skip first 35 records for data distribution

Table 6 Accuracy on WDBC data set using combination of WHT-WHT

%split (client1-client2)	Accuracy (%)					
	K = 3		K = 5		K = 7	
	Org	Mod	Org	Mod	Org	Mod
17-83	96.83	92.44	97.01	92.79	97.18	92.44
35-65	96.83	92.44	97.01	92.79	97.18	92.61
50-50	96.83	92.44	97.01	92.79	97.18	92.61
68-32	96.83	92.44	97.01	92.79	97.18	92.61
35-65**	97.12	92.26	97.36	93.84	97.36	93.32
48-52**	97.18	92.61	96.83	92.97	97.01	93.49
59-41**	97.01	92.09	96.48	92.07	97.18	93.14

**Skip 135 records for data distribution

Table 7 Accuracy on WDBC data set using combination of WHT-DCT

%split (client1-client2)	Accuracy (%)					
	K = 3		K = 5		K = 7	
	Org	Mod	Org	Mod	Org	Mod
31-69	96.83	57.82	97.01	61.68	97.18	62.21
49-51	96.83	93.67	97.01	94.02	97.18	94.20
63-37	96.83	93.49	97.01	93.32	97.18	92.44
79-21	96.83	92.79	97.01	92.97	97.18	92.79
19-81**	97.01	92.61	96.66	92.44	96.66	92.09
37-63**	96.66	92.97	97.01	92.79	97.01	92.26
69-31**	97.01	92.26	96.66	93.32	97.18	92.97
75-25**	97.01	92.26	97.01	93.32	96.83	92.79

**Skip 135 records for data distribution

Table 8 Accuracy on WDBC data set using combination of WHT-FISIP

%split (client1-client2)	Accuracy (%)					
	K = 3		K = 5		K = 7	
	Org	Mod	Org	Mod	Org	Mod
24-76	96.83	93.32	97.01	93.67	97.18	93.84
36-64	96.83	93.49	97.01	93.84	97.18	93.84
65-35	96.83	93.32	97.01	92.97	97.18	93.32
83-27	96.83	93.49	97.01	93.32	97.18	92.79
30-70**	96.83	92.44	96.66	93.32	97.18	91.56
46-54**	96.83	92.44	97.18	91.91	97.18	91.56
65-35**	97.18	92.97	96.66	92.44	96.83	92.44
74-26**	96.84	93.67	96.83	93.84	96.83	93.67

**Skip 135 records for data distribution

Table 9 Accuracy on WDBC data set using combination of WHT-SAN

%split (client1-client2)	Accuracy (%)					
	K = 3		K = 5		K = 7	
	Org	Mod	Org	Mod	Org	Mod
21-79	96.83	93.67	97.01	93.84	97.18	94.02
36-64	96.83	93.49	97.01	94.02	97.18	93.67
51-49	96.83	94.02	97.01	94.37	97.18	94.02
82-18	96.83	93.14	97.01	93.14	97.18	92.79
18-82[**]	96.83	92.44	96.83	92.79	97.18	92.44
44-56[**]	96.66	93.32	96.83	91.56	97.01	92.09
60-40[**]	97.01	75.08	96.49	75.78	97.19	74.73
74-36[**]	96.48	93.14	96.83	93.67	96.83	93.67

[**]Skip 135 records for data distribution

Table 10 Privacy measures on IRIS

Method	VD	CP	RP	CK	RK
WHT+WHT	1.2737	0.50	50.9770	0.50	0.7623
WHT+DCT	0.9345	0.5000	53.2433	0.500	0.0067
WHT+FISIP	0.6704	0.5	54.1233	0.5	0.0050
WHT+SAN	0.7944	0	48.1533	1	0.1033

Table 11 Privacy measures on WDBC

Method	VD	CP	RP	CK	RK
WHT+WHT	5.0548	9.8000	174.4946	0	0.0062
WHT+DCT	4.2033	9.7333	185.1981	0.0333	0.0019
WHT+FISIP	3.1417	10.8667	173.1258	0	0.0024
WHT+SAN	0.9609	9.3333	180.2974	0.0333	0.0021

6 Conclusion and Future Work

This paper proposes a new approach for Horizontal PPDM based on combination of linear transformations. It is a simple and efficient approach to protect privacy of individual customers by inference from the experimental results. This approach will be extended in future to more than two clients and different combinations of transformation techniques and will be applied to vertically partitioned data also.

References

1. J.F. Traub, Y. Yemini, and H. Wozniakowski, "The StatisticalSecurity of a Statistical Database," ACM Trans. Database Systems, vol. 9, no. 4, pp. 672–679, 1984.
2. C.C. Aggarwal and P.S. Yu, "A Condensation Approach to Privacy Preserving Data Mining," Proc. Ninth Int'l Conf. Extending Database Technology, pp. 183–199, 2004.
3. D. Defays and P. Nanopoulos, "Panels of Enterprises andConfidentiality: The Small Aggregates Method," Proc. Statistics Canada Symp. 92 Design and Analysis of Longitudinal Surveys, pp. 195–204, 1993.
4. J. Domingo-Ferrer and J.M. Mateo-Sanz, "Practical Data-Oriented Microaggregation for Statistical Disclosure Control," IEEE Trans. Knowledge and Data Eng., vol. 14, no. 1, pp. 189–201, 2002.
5. C.C. Yao, "How to generate and Exchange Secrets", IEEE, 1986.
6. M. Kantarcioglu and C. Clifton. "Privately computing a distributed k-nn classifier". PKDD, v. 3202, LNCS, pp. 279–290, 2004.
7. Z. Yang, S. Zhong, R. Wright, "Privacy-preserving Classification of Customer Data without Loss of Accuracy", In: Proceedings of the Fifth SIAM International Conference on Data Mining, pp. 92–102, NewportBeach, CA, April 21–23, 2005.
8. L. Xiong, S. Chitti and L. Liu. k Nearest Neighbor Classification across Multiple Private Databases. CIKM'06, pp. 840–841, Arlington, Virginia, USA, November 5–11, 2006.
9. M. Kantarcioglu and J. Vaidya. Privacy preserving naïve Bayes classifierfor horizontally partitioned data. In IEEE ICDM Workshop on Privacy Preserving Data Mining, Melbourne, FL, pp. 3–9, November 2003.
10. ZhuojiaXu, Xun Yi, "Classification of Privacy-preserving Distributed Data Mining Protocols", IEEE, 2011.
11. Jen-Wei Huang, Jun-Wei Su and Ming-Syan Chen, "FISIP: A Distance and Correlation Preserving Transformation for Privacy Preserving Data Mining" IEEE, 2011.
12. Shibnath Mukharjee, Zhiyuan Chen, Aryya Gangopadhyay, "A Privacy-preserving technique for Euclidean distance-based mining algorithms using Fourier-related transforms", the VLDB Journal, pp (293–315), 2006.
13. http://kdd.ics.uci.edu/.
14. http://www.wekaito.ac.nz/ml/weka.
15. ShutingXu, Jun Zhang, Dianwei Han, and Jie Wang, (2005) "Data distortion for privacy protection in a terrorist Analysis system", P. Kantor et al (Eds.): ISI 2005, LNCS 3495, pp. 459–464.

References

1.

2.

3.

4.

5.

6.

7.

8.

9.

10.

11.

12.

Randomized Cryptosystem Based on Linear Transformation

K. Adi Narayana Reddy, B. Vishnuvardhan
and G. Shyama Chandra Prasad

Abstract The secure transmission of any form of data over a communication medium is primary important across the globe or in research arena. Cryptography is a branch of cryptology and it provides security for data transmission between any communicating parties. The Hill cipher is one of the symmetric key substitution algorithms. Hill Cipher is vulnerable to known plaintext attack. This paper presents randomized cryptosystem based on linear transformation using variable length sub key groups. The proposed technique shares a prime circulant matrix as a secret key. The security analysis and performance of the method are studied and presented.

Keywords Circulant matrix · Determinant · Hill cipher · Sub key group · Substitution cipher

1 Introduction

Today, information is one of the most valuable assets. Information transmission across the network is of prime importance in the present age. Cryptography is the branch of cryptology and it provides security to the transmitted data between the communicating parties. There are various algorithms to provide security for the information. Traditional symmetric ciphers use substitution in which each character is replaced by other character. Lester S. Hill invented the Hill cipher in 1929. Hill cipher is a classical substitution technique that has been developed based on linear

K.A.N. Reddy (✉)
Department of CSE, ACE Engineering College, Hyderabad, India
e-mail: aadi.iitkgp@gmail.com

B. Vishnuvardhan
Department of IT, JNTU, Jagityala, Karimanagar, India
e-mail: mailvishnu@yahoo.co.in

G.S.C. Prasad
Department of CSE, Matrusri Engineering College, Saidabad, Hyderabad, India

© Springer India 2016
S.C. Satapathy et al. (eds.), *Information Systems Design and Intelligent Applications*, Advances in Intelligent Systems and Computing 434,
DOI 10.1007/978-81-322-2752-6_10

113

transformation. It has both advantages and disadvantages. The main advantages are disguising letter frequencies of the plaintext; high speed, high throughput, and the simplicity because of using matrix multiplication and inversion for enciphering and deciphering. The disadvantages are, it is vulnerable to known plaintext attack and the inverse of the shared key matrix may not exist always. To overcome the drawbacks of Hill cipher algorithm many modifications are presented. In our paper we present a modification to the Hill cipher by the utilization of special matrices called circulant matrices. A circulant matrix is a matrix where each row is rotated one element to the right relative to the preceding row vector. In literature circulant matrices are used in many of the cryptographic algorithms. Advanced Encryption Standard (AES) uses circulant matrices to provide diffusion at bit level in mix columns step. Circulant matrices can be used to improve the efficiency of Lattice-based cryptographic functions. Cryptographic hash function Whirlpool uses circulant matrices.

The paper is systematized accordingly: Sect. 2 presents an over view of Hill cipher modifications. Section 3 presents a proposed Hill cipher modification. Section 4 explains security analysis. Conclusion of the proposal is in the Sect. 5.

2 Literature Review on Hill Cipher Modifications

Many researchers improved the security of linear transformation based cryptosystem. Yeh et al. [14] presented an algorithm which thwarts the known-plaintext attack, but it is not efficient for dealing bulk data, because too many mathematical calculations. Saeednia [11] presented an improvement to the original Hill cipher, which prevents the known-plaintext attack on encrypted data but it is vulnerable to known-plaintext attack on permutated vector because the permutated vector is encrypted with the original key matrix. Ismail [4] tried a new scheme HillMRIV (Hill Multiplying Rows by Initial Vector) using IV (Initial Vector) but Rangel-Romeror et al. [8] proved that If IV is not chosen carefully, some of the new keys to be generated by the algorithm, may not be invertible over Z_m, this make encryption/decryption process useless and also vulnerable to known-plaintext attack and also proved that it is vulnerable to known-plaintext attack. Lin et al. [7] improved the security of Hill cipher by using several random numbers. It thwarts the known-plaintext attack but Toorani et al. [12, 13] proved that it is vulnerable to chosen ciphertext attack and he improved the security, which encrypts each block of plaintext using random number and are generated recursively using one-way hash function but Keliher et al. [6] proved that it is still vulnerable to chosen plaintext attack. Ahmed and Chefranov [1–3] improved the algorithm by using eigen values but it is not efficient because the time complexity is more and too many seeds are exchanged. Reddy et al. [9, 10] improved the security of the cryptosystem by using circulant matrices but the time complexity is more. Again Kaipa et al. [5] improved the security of the algorithm by adding nonlinearity using byte substitution over GF (2^8) and simple substitution using variable length sub key groups. It is efficient but the cryptanalyst can find the length of sub key groups by collecting pair of same ciphertext and plaintext blocks. In this paper randomness will be included to the linear transformation based cryptosystem to overcome chosen-plaintext and chosen-ciphertext attacks and to reduce the time complexity.

3 Proposed Cryptosystem

In this paper an attempt is made to propose a randomized encryption algorithm which produces more than one ciphertext for the same plaintext. The following sub sections explain the proposed method.

3.1 Algorithm

Let M be the message to be transmitted. The message is divided into 'm' blocks each of size 'n' where 'm' and 'n' are positive integers and pad the last block if necessary. Let M_i be the ith partitioned block (i = 1, 2, ... m) and size of each M_i is 'n'. Let C_i be ciphertext of the ith block corresponding to the ith of block plaintext. In this paper the randomness is added to the linear transformation based cryptosystem. Each element of the plaintext block is replaced by a randomly selected element from the corresponding indexed sub key group. The randomly selected element will not be exchanged with the receiver. In this method key generation and sub key group generation is similar to hybrid cryptosystem [3]. Choose a prime number 'p'. The following steps illustrate the algorithm.

1. **Step 1: Key Generation**. Select randomly 'n' numbers $(k_1, k_2, ... k_n)$ such that GCD $(k_1, k_2, ... k_n) = 1$. Assume $k_i \in Z_p$. Rotate each row vector relatively right to the preceding row vector to generate a shared key matrix $K_{n \times n}$. The generated key matrix is called prime circulant matrix.
2. **Step 2: Sub Key Group Generation**. Let $r = \sum_{i=1}^{n} k_i$ mod p. A sequence of 'p' pseudo random numbers S_i (i = 0, ..., p − 1) are generated with initial seed as r. The sub key groups are generated with following steps as

   Step 1: initialize i = 0
   Step 2: j = i + S[i] % b
   Step 3: $S_G[j] = \{i\}$
   Step 4: i ++
   Step 5: goto step 2

3. **Step 3: Encryption.** The encryption process encrypts each block of plaintext using the following steps.

   3.1. Initially the transformation is applied as Y = KM mod p.
   3.2. Convert each element of the block into base b number system
   3.3. Replace each digit of the element by a randomly chosen element from the corresponding sub key group.
   3.4. Transmit the ciphertext block to the other end user

4. **Step 4: Decryption.** The encryption process encrypts each block of plaintext using the following steps

   4.1. Replace each element by an index of the sub key group which it belongs
   4.2. Convert the base b number system into equivalent decimal number system
   4.3. The inverse linear transformation is applied as M = K^{-1}Y mod p
   4.4. This produces the plaintext corresponding to ciphertext

3.2 Example

Consider a prime number p as 53 and the set of relatively prime numbers as [4, 11]. Generate shared key matrix $K_{3 \times 3}$. Assume the plaintext block M = [3, 10, 12]. Generate a sequence of 'p' pseudo-random number with seed value as r = 45. Assume b = 5 and generate five sub-key groups (S_G) from the random number sequence. The sub key groups are random and of variable length.

$$S_G[0] = \{0, 6, 17, 21, 24, 25, 31, 38, 50\}$$
$$S_G[1] = \{1, 4, 9, 12, 16, 29, 30, 34, 39, 40, 43, 44, 46, 48, 49\}$$
$$S_G[2] = \{2, 3, 13, 22, 23, 26, 37, 45, 51, 52\}$$
$$S_G[3] = \{7, 10, 15, 19, 20, 27, 33, 42\}$$
$$S_G[4] = \{5, 8, 11, 14, 18, 28, 32, 35, 36, 41, 47\}$$
$$Y = KM \bmod p = KM \bmod 53 = [0, 42, 44]$$
$$0 \rightarrow 0.000 \left(0.5^2 + 0.5^1 + 0.5^0\right)$$
$$42 \rightarrow 132 \left(1.5^2 + 3.5^1 + 2.5^0\right)$$
$$44 \rightarrow 134 \left(1.5^2 + 3.5^1 + 4.5^0\right)$$

Each of the digits is replaced by an element from the corresponding sub key group

The possible ciphertext pairs are presented in Table 1.

The same plaintext is mapped to many ciphertext pairs

After communicating the ciphertext pair (C_1, C_2) to the receiver, the decryption process outputs the plaintext as [3, 10, 12].

Table 1 Ciphertext corresponding to plaintext

Plaintext	Base b number system	Ciphertext 1	Ciphertext 2	Ciphertext
12	000	6	17	50
14	132	4	15	23
3	134	30	42	5

4 Performance Analysis

The performance analysis is carried out by considering the computational cost and security analysis which are to show the efficiency of the algorithm.

4.1 Computational Cost

The time complexity measures the running time of the algorithm. The time complexity of the proposed algorithm to encrypt and to decrypt the text is O (mn^2) which is shown in the Eq. (2), where 'm' is number of blocks and 'n' is size of each block, which is same as that of original Hill cipher. In this process T_{Enc} and T_{Dec} denote the running time for encryption and decryption of 'm' block of plaintext respectively.

$$T_{Enc}(m) \cong m(n^2)T_{Mul} + m(n^2)T_{Add}$$
$$T_{Dec}(m) \cong m(n^2)T_{Mul} + m(n^2)T_{Add} + mnT_s \tag{1}$$

In which T_{Add}, T_{Mul}, and T_s are the time complexities for scalar modular addition, multiplication, and search for the index respectively.

$$T_{Enc}(m) \cong m(n^2)c_1 + m(n^2)c_2 \cong O(mn^2)$$
$$T_{Dec}(m) \cong m(n^2)c_1 + m(n^2)c_2 + mnc_3 \cong O(mn^2) \tag{2}$$

where c_1, c_2 and c_3 are the time constants for addition, multiplication and index search respectively. The running time of proposed randomized LTCM and other methods are analysed and presented in the Fig. 1. The running time of proposed randomized LTCM method is equal to the linear transformation based cipher. The proposed method is better than other methods.

Fig. 1 Encryption time

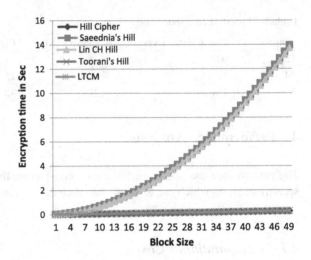

4.2 Security Analysis

The key matrix is shared secretly by the participants. The attacker tries to obtain the key by various attacks but it is difficult because the random selection of elements from sub key groups. It is difficult to know the elements of the sub key groups because each sub key group is of variable length and generated by modulo which is an one-way function.

The proposed cryptosystem overcomes all the drawbacks of linear transformation based cipher and symmetric key algorithms. This is secure against known-plaintext, chosen-plaintext and chosen-ciphertext attacks because one plaintext block is mapped to $(b*l_1*\ldots*l_n)^n$ ciphertext blocks where l_i is the length of the corresponding ith sub key group and these groups are variable length. This is due to the random selection of element from the corresponding sub key group. Therefore, the cryptanalyst can no longer encrypt a random plaintext looking for correct ciphertext. To illustrate this assume that the cryptanalyst has collected a ciphertext C_i and guessed the corresponding plaintext M_i correctly but when he/she encrypt the plaintext block M_i the corresponding ciphertext block C_j will be completely different. Now he/she cannot confirm M_i is correct plaintext for the ciphertext C_i.

5 Conclusion

The structure of the proposed cryptosystem is similar to substitution ciphers i.e. initially the linear transformation is applied on the original plaintext block then the result is replaced by a randomly selected element from the corresponding sub key group. The sub key groups are of variable length and each sub key group is generated randomly using one-way modulo function. The proposed randomized encryption algorithm produces more than one ciphertext for one plaintext because each element of the block is replaced by a randomly selected element from the corresponding sub key group. The proposed cryptosystem is free from all the security attacks and it has reduced the memory size from n^2 to n, because key matrix is generated from the first row of the matrix and is simple to implement and produces high throughput.

References

1. Ahmed, Y.M. and A.G. Chefranov, 2009. Hill cipher modification based on eigenvalues hcm-EE. Proceedings of the 2th International Conference on Security of Information and Networks, Oct. 6–10, ACM Press, New York, USA., pp: 164–167. DOI: 10.1145/1626195. 1626237.

2. Ahmed, Y.M. and Alexander Chefranov, 2011. Hill cipher modification based on pseudo-random eigen values HCM-PRE. Applied Mathematics and Information Sciences (SCI-E) 8(2), pp. 505–516.

3. Ahmed, Y.M. and Alexander Chefranov. Hill cipher modification based generalized permutation matrix SHC-GPM, Information Science letter, 1, pp. 91–102.

4. Ismail, I.A., M. Amin and H. Diab, 2006. How to repair the hill cipher. J. Zhej. Univ. Sci. A., 7: 2022–2030. DOI: 10.1631/jzus.2006.A2022.

5. Kaipa, A.N.R., V.V. Bulusu, R.R. Koduru and D.P. Kavati, 2014. A Hybrid Cryptosystem using Variable Length Sub Key Groups and Byte Substitution. J. Comput. Sci., 10:251–254.

6. Keliher, L. and A.Z. Delaney, 2013. Cryptanalysis of the toorani-falahati hill ciphers. Mount Allison University. http://eprint.iacr.org/2013/592.pdf.

7. Lin, C.H., C.Y. Lee and C.Y. Lee, 2004. Comments on Saeednia's improved scheme for the hill cipher. J. Chin. Instit. Eng., 27: 743–746. DOI: 10.1080/02533839.2004.9670922.

8. Rangel-Romeror, Y., R. Vega-Garcia, A. Menchaca-Mendez, D. Acoltzi-Cervantes and L. Martinez-Ramos et al., 2008. Comments on "How to repair the Hill cipher". J. Zhej. Univ. Sci. A., 9: 211–214. DOI: 10.1631/jzus.A072143.

9. Reddy, K.A., B. Vishnuvardhan, Madhuviswanath and A.V.N. Krishna, 2012. A modified hill cipher based on circulant matrices. Proceedings of the 2nd International Conference on Computer, Communication, Control and Information Technology, Feb. 25–26, Elsevier Ltd., pp: 114–118. DOI: 10.1016/j.protcy.2012.05.016.

10. Reddy, K. A., B. Vishnuvardhan, Durgaprasad, 2012. Generalized Affine Transformation Based on Circulant Matrices. Internatonal Journal of Distributed and Parallel Systems, Vol. 3, No. 5, pp. 159–166.

11. Saeednia, S., 2000. How to make the hill cipher secure. Cryptologia, 24: 353–360. DOI: 10. 1080/01611190008984253.

12. Toorani, M. and A. Falahati, 2009. A secure variant of the hill cipher. Proceedings of the IEEE Symposium on Computers and Communications, Jul. 5–8, IEEE Xplore Press, Sousse, pp: 313–316. DOI: 10.1109/ISCC.2009.5202241.

13. Toorani, M. and A. Falahati, 2011. A secure cryptosystem based on affine transformation. Sec. Commun. Netw., 4: 207–215. DOI: 10.1002/sec.137.

14. Yeh, Y.S., T.C. Wu, C.C. Chang and. W.C. Yang, 1991. A new cryptosystem using matrix transformation. Proceedings of the 25th IEEE International Carnahan Conference on Security Technology, Oct. 1–3, IEEE Xplore Press, Taipei, pp: 131–138. DOI: 10.1109/CCST.1991. 202204.

Literature Survey to Improve Image Retrieval Efficiency by Visual Attention Model

T. Venkata Ramana, K. Venugopal Rao
and G. Shyama Chandra Prasad

Abstract Now a day's CBIR is facing several performance issues because of the growth of digital world. To overcome the issues of CBIR, one challenging task is using the simulation of the visual attention model. To implement visual attention model several factors to be considered like similarity measures, Saliency model. Whereas the traditional CBIR focuses on image features. This paper presents analysis of different concepts which are used to improve the image retrieval efficiency. After analyzing it was understood that there exists some gap to concentrate in increasing the effectiveness of image retrievals. In accomplishing the gap we are presenting a kind of scope which improvises the performance issues in image retrievals.

Keywords Visual attention model · Saliency model · CBIR · Similarity measures · Image retrieval

1 Introduction

Digital libraries are growing in rapid pace in increasing the volumes of electronic data day by day, which is due to the electronic gadgets like cell phones, web cameras and cam carders etc. To satisfy the needs of the user an expert system is needed to have the effective retrieval of similar images for the given query image

T.V. Ramana (✉)
JNTU, Hyderabad, India
e-mail: meetramana_12@yahoo.co.in

K.V. Rao
CSE Department, Narayanamma Engineering College, Hyderabad, India
e-mail: kvgrao@gmail.com

G.S.C. Prasad
CSE Department, Matrusri Engineering College, Hyderabad, India
e-mail: gscprasad@gmail.com

© Springer India 2016
S.C. Satapathy et al. (eds.), *Information Systems Design and Intelligent Applications*, Advances in Intelligent Systems and Computing 434,
DOI 10.1007/978-81-322-2752-6_11

[1]. CBIR system is one of such experts systems that highly rely on appropriate extraction of features and similarity measures used for retrieval [2]. The area has gained wide range of attention from researchers to investigate various adopted methodologies, their drawbacks, research scope, etc. [3–11]. This domain became complex because of the diversification of the image contents and also made interesting [2].

The recent development ensures the importance of Visual information retrieval like images and videos, in the meantime so much of research work is going on in implementing in many real world applications such as Biological sciences, medical sciences environmental and health care, in digitally stored libraries and social media such as twitter, LinkedIn, etc. CBIR understands and analyzes the visual content of the images [12]. It represents an image using the renowned visual information like color, texture, shape, etc. [13, 14]. All these features are often referred as basic features of the image, which undergoes lot of variations according to the need and specifications of the image [15–17]. Since the image acquisition varies with respect to illumination, angle of acquisition, depth, etc., it is a challenging task to define a best limited set of features to describe the entire image library.

Similarity measure is another processing stage that defines the performance of the CBIR system [18]. In this stage, the similarity between the query image and the images in the database is determined using the distance between their feature vectors [19]. The database images that exhibit least distance with the input query are identified as required images and they are accessed from the database. Despite numerous distance measures are reported in the literature [4], Euclidean distance and other norm-based distance measures gain popularity due to its simplicity. Nevertheless, these norm-based distance measures often fail due to the global variances and practical dynamics of the image acquisition technology [18]. From the above Kullback divergence is identified as most promising one to rectify the aforesaid problem [20], still it is facing some issues like compatibility with frequent domain representations such as wavelets [18].

The first initiative step has taken by the IBM company towards image retrieval by proposing query-by image content (QBIC), the other name of it is content based image retrieval. Basic features considered in CBIR includes color (distribution of color intensity across image), texture (Homogeneity of visual patterns), shape (boundaries, or the interiors of objects depicted in the image), spatial relations (the relationship or arrangement of low level features in space) or combination of above features were used. In order to add to the image databases every image has to undergo the feature extraction process, so that all low level features are extracted before it gets added to the database. While performing the process of feature extraction, like extracting the image color, texture or shape are extracted in feature extraction stage. If the user wants to retrieve an image, he provides a sample image and applying extraction process and then the similarity measurement engine is accountable in estimating the similarities between the query image and the images present in the database. After completion of the similarity measure, ranking is given by considering the similar features of the given query image. Approaches for feature extraction include histogram and color movement (Niblack et al.), color histogram (Sahwney et al.), region histogram (Chad carson et al.), Fourier transforms, Gabor filter and statistical methods. There are many sophisticated algorithms are there to which will be used to describe low level image features like color, shape, texture and spatial features approaches. But these algorithms did not gave the satisfaction, clarity and comfort to visualize it.

This is because of not having the basic features of image in describing high level concepts in the users mind such as searching of a image of a baby boy who is laughing a lot. The only way a machine is able to perform automatic extraction is by extracting the low level features that represented by the color, texture, shape and spatial from images with a good degree of efficiency. Till now there are different CBIR systems which are implemented are available, still there exists the problem in retrieving the images by considering the pixel content. It is not solved till now. To have the Semantic image retrieval, to identify the higher level concepts it requires a feedback mechanism with the human intervention. Out of the different methods, the most common method for comparing two images in content-based image retrieval (typically an example image and an image from the database) is using an image distance measure. The functionality of distance measure is to have the comparison of two images to know the similarity of two images in different dimensions like shape, color, texture and others. After applying the distance measure method if the result is a distance of 0, then it means an exact match with the query, with respect to the dimensions that were considered. If the result of distance measure function is greater than 0, then it indicates various degrees of similarities between the images. Search results then can be sorted based on their distance to the queried image. Color histograms are used to know the color similarity in Distance measures. Apart from that segment color proportion by region and by spatial relationship among several color regions are also used.

In all the methods of retrieval two issues are ignored. They are Performance and usability. There is high variability of interpretation in judging the success of an algorithm. To judge the Performance of an algorithm, evaluation can be done through precision and recall but still it is not meeting the requirements of the user. Queries to incorporate complete context are critical to the success of CBIR.

2   Literature Review

(A)   Murala et al. [1] have used local tetra patterns (LTrPs) to construct an image indexing and retrieval system. In contrast to the local binary pattern (LBP) and local ternary pattern (LTP), LTrPs determines the first-order derivatives in both horizontal and vertical directions to identify the pixel direction and hence to represent the reference pixel and the neighbors. They have also introduced a methodology to determine the nth order LTrP using (n − 1)th order horizontal and vertical derivatives. The experimental results have demonstrated that LTrP based CBIR system outperforms LBP, LTP and local derivative patterns based CBIR systems. Though the methodology is proved for its computational efficiency and good performance for diverse texture patterns, the feature descriptor is falling under low level category. The extracted features consider only the pixel relationship and not the pixel relevance with the respective image category. As a result, these feature descriptors probably fail to handle different classes of images, but with similar visual patterns.

(B) Su et al. [2] have introduced a CBIR system in which retrieval accuracy is defined by the users feedback. They have addressed that the conventional relevance feedback based CBIR system has entertained iterative process to ensure the relevance between the retrieval results and the query input. To overcome this, they have proposed navigation—pattern based relevance feedback (NRPF) method that substantially minimizes the number of iterations incurred to ensure the retrieval precision. The method has also been facilitated by three query refinement strategies such as Query Point Movement (QPM), Query Reweighting (QR), and Query Expansion (QEX). Performance demonstration is given by them over conventional systems. The relevance feedback model has been entertained well in the recent days. But now a days this is not that much effective because of huge growth of data. However, these semi-automatic retrieval systems do not cope well with practical scenarios, where retrieval efficiency plays the major role rather than the retrieval precision. These methods are insensitive to similarity measures as the relevance feedback plays crucial role. Under the scenario of worst case performance, the methodology tends to move towards manual retrieval process.

(C) Guo et al. [21] have proposed an indexing method based on Error-Diffusion Block Truncation Coding (EDBTC). Here, the image features are extracted by vector quantization (VQ) on color quantized and bit mapped images obtained from EDBTC. To know the image similarities between the images of the query and the images present in the data base, authors had introduced bit pattern features and color histogram.

(D) Guo et al. [22] have exploited low complexity ordered dither block truncation coding (ODBTC) for CBIR system. ODBTC has been used to compress the image as quantizes and bitmap image. Here, the image is represented as color co-occurrence feature (CCF) and bit pattern features (BPF). ODBTC is shown high performance on BTC based image retrieval systems in CBIR. In [21], the adopted scheme (EDBTC) for extracting features such as color histogram feature (CHF) and bit pattern histogram feature (BHF) were promising in the image compression scheme. In the same way [22] has also adopted ODBTC to extract CCF and BPF features. Any way out of the extracted features, 4 are low level, since they consider only the pixel intensity and the structure of pixel neighborhood. Likewise [1], the retrieval efficiency will be reduced when trying to retrieve visually similar images of dissimilar classes. By this, the practical relevance is found to be less while working on with such feature descriptors.

(E) Liu et al. [23] had introduced a computational vision attention model, known as saliency structure model, to support CBIR system. This saliency model is implemented in three steps. Initially at the beginning stage, they have detected saliency regions using color volume and edge information rather than basic image features. After getting the output of the first step next step is global suppressing of maps has been performed using the energy feature of the gray level co-occurrence. Finally, the last step is used to construct the saliency structure histogram to represent the image. The saliency structure model has been experimentally proved for its performance over bag of visual words (BOW) model and micro-structure descriptor. Even after having precise information with the help of descriptor, similarity measure is considered which can improve the efficiency of image retrieval. However, they have not provided adequate significance to the similarity measure and hence it is quite complex to decide upon the performance accomplishment.

(F) Stella Vetova et al. [24], had proposed a novel algorithm which is used for extracting the image features by using Dual-Tree Complex Wavelet Transform (DT CWT) in the CBIR system. In their experiment they have shown the results of algorithm which satisfies the conditions of feature extraction rate, feature vector length and high information concentration is necessary for CBIR system. Authors has discussed Dual-Tree Complex Wavelet Transform as a filter bank (FB) structure, running process, conditions for shift-invariance and applications. This mechanism is suitable for edge and surface detection in image processing. By using this algorithm authors achieved high accuracy in comparison among the feature sets.

3 Scope of Research

To develop the CBIR system which has to cope up with the human visual system, a lot more has to done with visual attention model. To implement the human visual system a simulation model is needed which is termed as computational visual model. The high level challenges reside in this area has attracted the world wide researchers to work on it. The existing global features are not promising to build the visual attention model. For instance, the feature descriptors [1, 21, 22] are still under the category of basic features and hence they does not represent the image in the human perception. Semi-automatic methodologies can rectify the issue, but it may tend to be manual retrieval under worst case scenarios [2].

Recently, saliency model is found to be suitable for image representation [25, 26]. However, constructing visual attention models are still challenging [23]. Despite successful development has been made in [23] to construct the model as per the requirement, there is no adequate significance given for exploiting suitable similarity measure. These are the primary research gaps to be considered in my proposal for further development of an effective CBIR system.

Advantages of the Proposal: In the proposed CBIR system, three major components play key roles. They are Saliency model, higher order statistics and decision rules library. In Saliency model, we will be giving an input image which has to be subsampled for extracting the visual features of the image like color, shape, intensity respectively. These features are used in fine tuning of images by using some of the decision rules, so that visually similar images are extracted. While implementing this mechanism we can check which rules and image features are to be considered so that we can achieve highest relevant image for the given query. Upon this different other methods like similarity measures are applied to have the effective image retrieval.

4   Conclusions

In order to fulfill the gap between CBIR and Visual attention model, each should be implemented then the performance of both should be investigated. To know the performance of each comparison should be made for each of the methodology and applying some other techniques like similarity measures and even we can apply saliency model to know the efficiency. The performance of the methodology will be quantified using renowned performance metrics such as efficiency, precision, recall and F-scores. Even we can use MATLAB to implement these methodologies.

References

1. Murala, S.; Maheshwari, R.P.; Balasubramanian, R., "Local Tetra Patterns: A New Feature Descriptor for Content-Based Image Retrieval", IEEE Transactions on Image Processing, Vol. 21, No. 5, pp. 2874–2886, 2012.
2. Ja-Hwung Su; Wei-Jyun Huang; Yu, P.S.; Tseng, V.S., "Efficient Relevance Feedback for Content-Based Image Retrieval by Mining User Navigation Patterns" IEEE Transactions on Knowledge and Data Engineering, Vol. 23, No. 3, pp. 360–372, 2011.
3. Y. Ruiand, T.S. Huang, "Imageretrieval: Current techniques, promising directions and open issues," J. Visual Commun. Image Represent", vol. 10, no. 1, pp. 39–62, Mar. 1999.
4. A.W.M. Smeulders, M. Worring, S. Santini, A. Gupta, and R. Jain, "Content-based image retrieval at the end of the early years," IEEE Trans. Pattern Anal. Mach. Intell., vol. 22, no. 12, pp. 1349–1380, Dec. 2000.
5. M. Kokare, B.N. Chatterji, and P.K. Biswas, "A survey on current content based image retrieval methods," IETE J. Res., vol. 48, no. 3&4, pp. 261–271, 2002.
6. Y. Liu, D. Zhang, G. Lu, and W.-Y. Ma, "A survey of content-based image retrieval with high-level semantics," Pattern Recogn., vol. 40, no. 1, pp. 262–282, Jan. 2007.
7. G. Qiu, "Color Image Indexing Using BTC," IEEE Trans. Image Processing, Vol. 12, No. 1, Jan. 2003.
8. C. H. Lin, R. T. Chen, and Y. K. Chan, "A smart content-based image retrieval system based on color and texture feature," Image and Vision Computing, vol. 27, no. 6, pp. 658–665, May 2009.
9. N. Jhanwar, S. Chaudhuri, G. Seetharaman and B. Zavidovique, "Content based image retrieval using motif co-occurrence matrix," Image and Vision Computing, vol. 22, pp. 1211–1220, Dec. 2004.
10. P. W. Huang and S. K. Dai, "Image retrieval by texture similarity," Pattern Recognition, vol. 36, no. 3, pp. 665–679, Mar. 2003.

11. C. C. Lai, and Y. C. Chen, "A user-oriented image retrieval system based on interactive genetic algorithm," IEEE Trans. Inst. Meas., vol. 60, no. 10, October 2011.
12. K. Seetharaman, "Image retrieval based on micro-level spatial structure features and content analysis using Full Range Gaussian Markov Random Field model", Engineering Applications of Artificial Intelligence, Vol. 40, pp. 103–116, 2015.
13. K. Vu, K.A. Hua, and N. Jiang,"Improving Image Retrieval Effectiveness in Query-by-Example Environment," Proc. 2003 ACM Symp. Applied Computing, pp. 774–781, 2003.
14. W. Xingyuan and W. Zongyu,-"The method for image retrieval based on multi-factors correlation utilizing block truncation coding," Pattern Recog., vol. 47, no. 10, pp. 3293–3303, 2014.
15. M. Unser, "Texture classification by wavelet packet signatures," IEEE Trans. Pattern Anal. Mach. Intell., vol. 15, no. 11, pp. 1186–1191, Nov. 1993.
16. B. S. Manjunath and W. Y. Ma, "Texture features for browsing and retrieval of image data," IEEE Trans. Pattern Anal. Mach. Intell., vol. 18, no. 8, pp. 837–842, Aug. 1996.
17. M. Kokare, P. K. Biswas, and B. N. Chatterji, "Texture image retrieval using rotated wavelet filters," Pattern Recogn. Lett., vol. 28, no. 10, pp. 1240–1249, Jul. 2007.
18. Missaoui, R.; Sarifuddin, M.; Vaillancourt, J., "Similarity measures for efficient content-based image retrieval", IEE Proceedings—Vision, Image and Signal Processing, Vol. 152, No. 6, pp. 875–887, 2005.
19. Hua Yuan; Xiao-Ping Zhang, "Statistical Modeling in the Wavelet Domain for Compact Feature Extraction and Similarity Measure of Images", IEEE Transactions on Circuits and Systems for Video Technology, Vol. 20, No. 3, pp. 439–445, 2010.
20. M. N. Do and M. Vetterli, "Wavelet-based texture retrieval using generalized Gaussian density and Kullback–Leibler distance," IEEE Trans. Image Process., vol. 11, no. 2, pp. 146–158, Feb. 2002.
21. Jing-Ming Guo; Prasetyo, H.; Jen-Ho Chen "Content-Based Image Retrieval Using Error Diffusion Block Truncation Coding Features", IEEE Transactions on Circuits and Systems for Video Technology, vol. 25, No. 3, pp. 466–481, 2015.
22. Jing-Ming Guo; Prasetyo, H., "Content-Based Image Retrieval Using Features Extracted From Halftoning-Based Block Truncation Coding", IEEE Transactions on Image Processing, Vol. 24, No. 3, pp. 1010–1024, 2015.
23. Guang-Hai Liu, Jing-Yu Yang, ZuoYong Li, "Content-based image retrieval using computational visual attention model", Pattern Recognition, In Press, Corrected Proof, 2015.
24. Stella vetova, Ivan ivanov., " Image Features Extraction Using The Dual-Tree Complex Wavelet transform",- Advances in Applied and Pure Mathematics, 978-960-474-380-3.
25. A. Toet, "Computational versus psychophysical bottom-up image saliency: a comparative evaluation study", IEEE Trans. Pattern Anal. Mach. Intell. 33 (11) (2011) 2131–2146.
26. A. Borji, L. Itti., "State-of-the-art in visual attention modeling", IEEE Trans. Pattern Anal. Mach. Intell. Vol. 35, No. 1, pp. 185–207, 2013.

Substrate Integrated Waveguide Based 4-Port Crossover for Beam-Forming Applications

P. Rahul Lal, Prasanth M. Warrier and Sreedevi K. Menon

Abstract In this paper an effective crossover is designed using Substrate Integrated Waveguide (SIW) technology. Reflection, transmission and isolation of the crossover is studied along with the electric field to substantiate the effectiveness of the crossover. At the operating frequency, an isolation better than 35 dB is achieved between the decoupled ports with a transmission of ~ 1 dB through the coupled ports.

Keywords Substrate integrated waveguide (SIW) · Crossover · Beam-forming · Coupled ports · Decoupled ports

1 Introduction

Substrate Integrated Waveguide (SIW), the manipulator of electromagnetic waves is one of the most recent advancements in microwave communication and is replacing its counterpart, the conventional waveguide in almost all the fields. SIW finds its application in technologies like power dividers, couplers, microwave junctions etc. [1, 2]. By using SIW all these devices can now be used as planar structures, making it easier to interface with a PCB. Moreover these microwave passive devices are found to be much better in performance in comparison with their planar and waveguide counter-parts.

P. Rahul Lal (✉) · P.M. Warrier · S.K. Menon
Department of Electronics and Communication Engineering, Amrita School of Engineering,
Amrita Vishwa Vidyapeetham, Clappana P.O., Kollam, India
e-mail: rahullal@am.amrita.edu

P.M. Warrier
e-mail: prasanthmw@am.amrita.edu

S.K. Menon
e-mail: sreedevikmenon@am.amrita.edu

© Springer India 2016
S.C. Satapathy et al. (eds.), *Information Systems Design and Intelligent
Applications*, Advances in Intelligent Systems and Computing 434,
DOI 10.1007/978-81-322-2752-6_12

The term crossover finds its importance in a communicating network using more than one transmitter and receiver antennas [3]. In order to maintain the purity of the transmitted and received signals, network designer will have to keep the losses due to crossover to the minimum possible extent. For a crossover to be ideal at a particular frequency, the more the isolation between the decoupled ports the better it is. Usually crossovers are realised using air wedges, vias [4] etc. Further for compactness microstrip technology has also been used for crossover design [5, 6].

To the best knowledge, no work regarding SIW based crossover has been reported. In the proposed paper, a crossover is achieved by means of a 4-port SIW, which is first of its kind. A central region in the shape of a circle acts as the interface for the signals entering form the 4-ports. Out of the modifications which were made it was noticed that, a metallic via which is placed at the centre of the structure determines the isolation factor and coupling factor in this particular design. The detailed analysis is presented in the following sections.

2 Design

SIW is a special case of our conventional wave guide. To be specific, it can be considered as a Dielectric Filled wave guide. Guided wave length in such a case is given by [2],

$$\lambda_g = \frac{2\pi}{\sqrt{\frac{\varepsilon_r (2\pi f)^2}{c^2} - \left(\frac{\pi}{a}\right)^2}} \tag{1}$$

Coming to the SIW, the diameter and the pitch of the metallic vias are determined using the following conditions,

$$d < \frac{\lambda_g}{5} \quad \& \quad p < 2d \tag{2}$$

Cut off frequency for a conventional wave guide is given by,

$$f_c = \frac{c}{2\pi} \sqrt{\left(\frac{m\pi}{a}\right)^2 + \left(\frac{n\pi}{b}\right)^2} \tag{3}$$

where 'a' and 'b' are the length and breadth of the rectangular waveguide.

For TE_{10}, 'b' is less significant, so the equation reduces to,

$$f_c = \frac{c}{2a} \tag{4}$$

Length of the dielectric filled wave guide is given by,

$$a_d = \frac{a}{\sqrt{\varepsilon_r}} \tag{5}$$

Passing on to the design equations of SIW, separation between the metallic vias is given by,

$$a_s = a_d + \frac{d^2}{0.95p} \tag{6}$$

3 Simulation and Analysis

A four port network as crossover is as shown in Fig. 1. In the geometry, if it is to provide crossover characteristics, the adjacent ports should be decoupled ports. With respect to Port 1, Port 3 and Port 4 are the decoupled ports. This is because the wave entering through these ports will be orthogonal to each other and Port 2 is the coupled port. Similarly if Port 3 is the input port, Port 4 will be the coupled port and Port 1 and 2 will be the decoupled ports. This will give the ideal S-matrix for a matched crossover as follows,

$$[S] = \begin{bmatrix} 0 & 1 & 0 & 0 \\ 1 & 0 & 0 & 0 \\ 0 & 0 & 0 & 1 \\ 0 & 0 & 1 & 0 \end{bmatrix}$$

Substrate used for the design is Rogers RT Duroid having dielectric constant 10.2 and thickness 0.67 mm. For the crossover characteristics to be centered at 2.5 GHz, the pitch and diameter for SIW is calculated using the equations from (1)

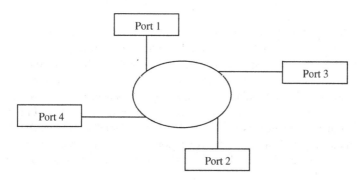

Fig. 1 Circular crossover

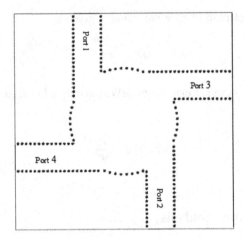

Fig. 2 Top view of the substrate integrated waveguide without the center metallic short

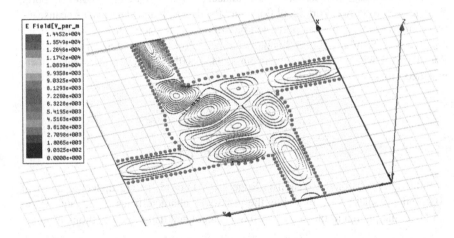

Fig. 3 Electric field pattern without the metallic short at the center

to (6). With respect to these dimensions, structure obtained is as shown in Fig. 2. The ports are equally spaced from each other with the adjacent ports perpendicular to each other. So the received or transmitted waves will be orthogonal polarizations, which make the adjacent ports decoupled, thus maintaining isolation. Since the opposite ports are of same polarization transmission is achieved between them.

The electric field flow for the SIW based crossover is as shown in Fig. 3 and it can be seen that power flow is there towards the decoupled ports too. As SIW is a metallic enclosure with array of shorting ports, the waves gets reflected at the boundary thus power flow is there to the decoupled ports giving a maximum isolation of 20 dB.

Fig. 4 Variation of
S-parameter with frequency

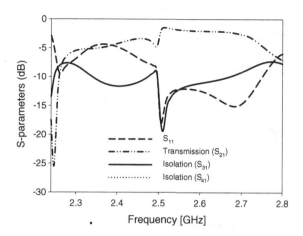

The variation of S-parameters with respect to the frequency is given in Fig. 4. From the graph it can be seen that at a frequency of 2.51 GHz, isolation between the decoupled ports is almost 19 dB and the power transmitted to the coupled port is about −1.4 dB.

For further improvement in isolation, one extra metallic short is placed at the centre of the crossover as shown in Fig. 5. As a result of this metallic short placed at the centre, waves bend around it and the entire power moves only in the forward direction, thus enhancing the isolation.

Electric Field pattern of the SIW based crossover when energized at Port 1 is depicted in Fig. 6.

From Fig. 6 it can be easily seen that the power entering the Port 1 is being coupled only to Port 2. Port 3 and Port 4 are completely isolated from Port 1. This confirms that the presented SIW based structure acts as an effective crossover.

Fig. 5 Top view of the
substrate integrated
waveguide with the center
metallic short

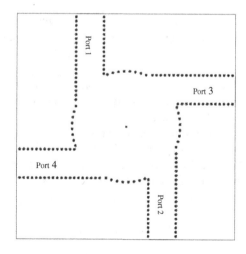

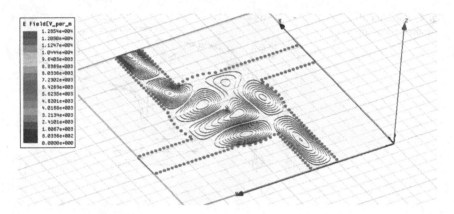

Fig. 6 Electric field pattern

Fig. 7 Variations of
S-parameters with frequency

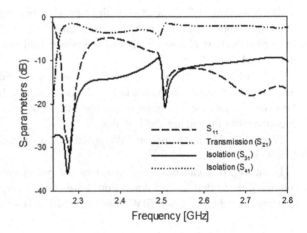

The variation of S-parameters with frequency for the modified crossover is depicted in Fig. 7. Transmission and isolation is obtained at the frequency 2.51 GHz as of the parent crossover with dual frequency characteristics. Crossover characteristics are achieved at a low frequency, 2.23 GHz in addition, due to the modification made with an isolation of almost 35 dB is achieved along with the transmission of −1.5 dB. As a result of this the modified crossover is a potential candidate for dual band applications.

The isolation and transmission of the proposed crossover is compared with some other recent papers dealing with various methods of crossovers and is presented in Table 1. From the comparisons made, the current design has better isolation, dual band of operation and is less complex.

Table 1 Comparison of crossover characteristics

Crossover	Transmission (dB)	Isolation (dB)
Ref. [3]	−0.4	20
Present work (without center conductor)	−1.4	19
Present work (with center conductor)	−1.5	35

4 Conclusion

A novel design of crossover using SIW is presented. The presented crossover is found to give an isolation of 37 dB between the ports which is better than the results in the reported list with dual band characteristics. This isolation is achieved with a transmission better than ∼1 dB, which enables the discussed crossover a potential candidate in beam forming networks.

Acknowledgments The project is partially funded by a grant from Information Technology Research Agency (ITRA), Department of Electronics and Information Technology (Deity), Government of India.

References

1. Bouchra Rahali and Mohammed Feham, Design of K-Band Substrate Integrated Waveguide Coupler, Circulator and Power Divider, International Journal of Information and Electronics Engineering, Vol. 4, No. 1, January 2014.
2. Zamzam Kordiboroujeni and Jens Bornemann, Department of Electrical and Computer Engineering, University of Victoria, Victoria, BC, Canada, Efficient Design of Substrate Integrated Waveguide Power Dividers for Antenna Feed Systems, 2013 7th European Conference on Antennas and Propagation (EuCAP).
3. Soon Young Eom,, Ariunzaya Batgerel, and Laxmikant Minz, Compact Broadband Microstrip Crossover with Isolation Improvement and Phase Compensation, IEEE Microwave and Wireless Components Letters, Vol. 24, No. 7, July 2014.
4. Bassem Henin, Amin Abbosh: Design of compact planar crossover using Sierpinski carpet microstrip patch: IET Microw, Antenna Propag., 2013, vol 7. Iss, 1, pp. 54–60, Dec. 2012.
5. A. Abbosh, S. Ibrahim, M. Karim: Ultra-Wideband Crossover Using Microstrip-to-Coplanar Waveguide Transition: Progressin Electromagnetics Research, (PIERS-C), 2012.
6. Heera P, Rahul Lal and Dr. Sreedevi K. Menon, Fully Planar Circular Microstrip Crossover, International Journal of Applied Engineering Research, ISSN 0973–4562 Vol. 10 No. 55 (2015).

Computational Model of Pixel Wise Temperature Prediction for Liver Tumor by High Intensity Focused Ultrasound Ablations

P. Revathy and V. Sadasivam

Abstract Medical imaging is a challenging research field. High-intensity focused ultrasound (HIFU) is a developing medical imaging method for non-invasive ablation of tumors. Based on the patients image of the tumor region, a computational model is proposed for planning and optimization of the HIFU treatment. A pixel wise temperature prediction based on the grey scale intensity values was done using MATLAB code making improvement in the Pennes Bio-Heat Transfer Equation (PBHTE). Also the defects in other heat equations like wulff and Klinger are considered in the proposed heat equation. As peak temperatures above 85–90 °C causes cavitation to the tissue exposed, the present study aims at maintaining the thermal dose applied to tumor tissue to be within the limit. Simulated temperature values lie in the patients safe limit avoiding preboiling and thus cavitation of tumor tissue is avoided.

Keywords MR guided HIFU · Temperature prediction · PBHTE · Pixel intensity based prediction · Thermal dose

1 Introduction

Due to the discovery of seminal physical phenomena such as X rays, ultrasound, radioactivity and magnetic resonance and the development of imaging instruments that harness them have provided some of the most effective diagnostic tools in medicine. Data sets in two, three or more dimensions convey increasingly vast and detailed information for clinical or research applications. The diagnostic information

P. Revathy (✉) · V. Sadasivam
Department of Computer Science and Engineering,
PSN College of Engineering and Technology, Tirunelveli, India
e-mail: kbrevathyseshu@gmail.com

V. Sadasivam
e-mail: principal@psncet.ac.in

© Springer India 2016
S.C. Satapathy et al. (eds.), *Information Systems Design and Intelligent Applications*, Advances in Intelligent Systems and Computing 434,
DOI 10.1007/978-81-322-2752-6_13

137

when sent in a timely, accurate manner will help the radiologist and the health care systems to give an accurate dosage of medicine without causing any harm to the adjacent organs or tissues. High Intensity Focused Ultrasound is a non-invasive method of treating tumor affected tissues. Magnetic Resonance Imaging (MRI) is a medical imaging technique used in radiology to investigate the anatomy and function of the body in both healthy and diseased environment. This technique is used for medical diagnosis, staging of disease and for follow-up without exposure to ionizing radiation. Hence it is recommended in reference to CT when either modality could yield the same information.

MR images are affected by intensity inhomogeneity, weak boundary, noise and the presence of similar objects close to each other. Applying HIFU to the organs like liver in the abdominal region is complicated due to the major blood vessels passing through. The target organ is below the ribs and so the left liver lobe is easily accessible than the right lobe. Other reasons like air bubbles, normal respiration of patients, bowel or fat as a hindrance to the ultrasonic rays are also reducing the accuracy of diagnosed heat to be applied. The large blood vessels carry away the heat applied to treat the affected area, and it becomes problematic to achieve complete necrosis of tumors. Magnetic Resonance guided High Intensity Focused Ultrasound (MRgHIFU) is a technique where HIFU beam heats and destroys the targeted tissue in non-invasive way and MRI is used to visualize patient's anatomy and controls the treatment by monitoring the tissue effect in real time. Thermal ablation is related to exposure time. To predict the temperature elevation in the tissue thermal models should be applied. Bio-Heat transfer Equation (BHTE) is the basic and often used model for prediction and the prediction is done based on applied acoustic pressure, absorption rate, Heat diffusion coefficient and Perfusion value [1].

Section 2 presents the literature review of segmentation, liver lesion calculation, Treatment considerations during HIFU and bioheat transfer. Section 3 discusses about the Heat equations and the thermal dose calculation. Results and discussion are made in Sect. 4. Finally, conclusions are made in Sect. 5.

2 Literature Review

2.1 Liver Tumor

Liver tumor is the abnormal growth in the liver [2]. Any tumor can be classified as primary or secondary tumor. Tumors are groupings of abnormal cells that cluster together to form a mass or lump. The unwanted growth of mass or lump disturbs the normal functionality of the human body. Tumors that originate in the liver may be benign and malignant known as primary tumors and the tumor that has spread to the liver from its original source of origin in another part of the body is a secondary tumor.

2.2 Segmentation of the Lesion Region

Automatic threshold based liver lesion segmentation method for 2D MRI images are proposed in this paper to determine the Region of Interest (ROI). Although Threshold based segmentation methods produce only rough results in liver tumor segmentation, they can be used to segment the lesion area as the contrast between the lesion and the liver is more significant. In [3, 4] threshold based segmentation was used as the main method to segment the tumor. The values for thresholding (i.e.) the higher and the lower limits are identified from intensity information of the grayscale input image. These values are used to isolate tumor directly. Morphological filter was used to mark the tumor from liver tissue. This is done as post-processing. As the output is sensitive to noise the neighborhood pixel values are used to reduce the false detection.

2.3 Lesion Value Calculation

The liver region is segmented and the morphological operators Dilate and Erode are used to filter the lesion values roughly. After creating morphological structuring element the dilate function returns the maximum value of all neighboring elements, whereas the erode function returns the minimum value. Now the connected components are identified in the binary image. Finally the label matrix is found for further processing. The canny edge detector is used for edge detection in gray scale images. Edge starts with the low sensitivity result and then grows it to include connected edge pixels from high sensitivity result. This helps to fill gaps in connecting edges.

2.4 Treatment with HIFU

The treatment process is classified as pretreatment preparation, treatment planning, ablation strategies and aftercare [5].

   During treatment the parameters to be considered are the heat at the focus and focal length of the transducer. The focusing of the target organ using a transducer is shown in the Fig. 1. The ideal focal length of transducer is calculated as the sum of the distance from the deepest layer of the tumor to the skin surface and 1 cm safety margin.

3 Heat Equations and Thermal Dose Calculation

The Kuznetsov, Zabolotskaya and Khokhlov (KZK) equation is solved numerically in frequency domain and so it is replaced by a coupled set of nonlinear partial differential equations [6]. The physical phenomena of the living tissues are studied

Fig. 1 HIFU Treatment
(focusing of target organ
using transducer)

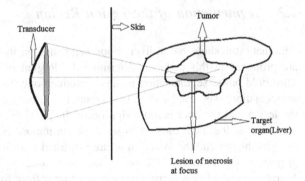

by the Bio-Heat Transfer Equations (BHTE). The traditional and the basic one is the Penne's bio-heat transfer equation. Other thermal models of bio heat transfer are the extended and modified versions of the original work of penne's [7]. Thus the temperature and thermal dose to be applied to the region of interest is calculated. Various methods are analyzed for studying the temperature of the tissue during treatment planning. The Penne's bio—heat transfer equation for blood perfused tissues is written as,

$$(\rho C_\rho)_t \frac{\partial T_t}{\partial t} = \nabla.(K_t \nabla T_t) + q_p + q_m \tag{1}$$

where q_p, q_m, ρ_t, C_ρ, T_t, K_t, t are the heat convention, metabolic heat transfer, tissue density, specific heat of blood, temperature of tissue, thermal conductivity and time respectively. The advantage of BHTE is that it predicts temperature fields and it is used in hyperthermia modeling. The limitation of BHTE is that it does not consider the effect of the direction of blood flow. The limitation of BHTE is overcome by Wulff Continuum model and Klinger continuum model [8]. In Wulff continuum model the heat transfer between blood and tissues is proportional to the temperature difference between these 2 media rather than between the two bloodstreams temperatures. Wulff's equation is given as,

$$(\rho C_\rho)_t \frac{\partial T_t}{\partial t} = \nabla.(k_t \nabla T_t) - \rho_b V_h C_b \nabla T_b - \nabla H_b \nabla \phi \tag{2}$$

where V_h and H_b are local mean blood velocity and specific enthalpy of blood respectively. The disadvantage of this method is that the local blood mass flux is hard to determine. The disadvantage of Penne's bio heat model is that it neglected the effect of blood flow within the tissue, to overcome this, in Klinger continuum model the convective heat caused by blood flow in the tissue was considered. Heat source and velocity of blood flow inside tissue was considered and the modified penne's model equation is as,

$$(\rho C_\rho)_t \frac{\partial \mathbf{Tt}}{\partial \mathbf{t}} + (\rho C)_b V_0 \nabla \mathbf{T_t} = k \nabla^2 \mathbf{T_t} + q_m \qquad (3)$$

where k_t, T_t, q_m are the thermal conductivity, tissue temperature (convective heat caused by blood flow inside the tissue) and metabolic heat transfer respectively. The blood and tissue parameters are listed with their values as discussed elsewhere [9, 10].

Comparing the Pennes BHTE, Wulff and Klinger the proposed model of heat equation is formed. The blood perfusion rate (W_{b1}) and the dynamic viscosity of blood (μ) values are considered in the calculation of thermal dose for the proposed model.

$$(\rho C)_t \frac{\partial Tt}{\partial t} = K_t \nabla^2 T_t - (\rho C)_b V \nabla T_t - (\rho C)_b (w_b + \mu) \nabla T_t + q_m \qquad (4)$$

4 Result and Discussion

4.1 Existing Results

When tested in a rabbit liver the rise in tissue temperature has reached 73 °C after 7 s during a continuous exposure of HIFU [11]. Left liver metastasis with a Volume of 10 cm × 7 cm × 10 cm was treated with HIFU with a sonication time of 1 h and 19 min and focal peak intensities ranged between 400 and 150 W. The 80 % of the lesion was ablated. Cell death occurred in temperature more than 60 °C [12]. The tests with gel phantom resulted in a temperature rise of 56 °C in about 4 s. The initial gel temperature was 24 °C, and reached 66.49 °C in about 5.7 s. That is in broad agreement with the study by Gail ter Haar (2001) [13], showed that the temperature will exceed 56 °C in 1–2 s at an intensity of 1 kW cm^2 [14]. The simulated temperature rises above 85–90 °C and causes preboiling or cavitation. And high temperatures in the range of 65–85 °C is used for ablation and this high temperature ensures ablation of tumor cells [15].

4.2 Simulation Results of the Proposed Equation

A cancer affected liver image is taken for analysis, the pixel values are analyzed for two set of pixel intensities set A and set B and the results are plotted in Tables 1 and 2. Both set of values are taken from the total pixel intensities of the gray scale input image. They are two sets of 3 × 3 values taken for analysis. Set A pixel intensities are in the range 114 to 196. The proposed model has reduced the over prediction of

Table 1 Set A-Elevation in temperature values for PBHTE versus proposed heat model

Pixel intensity (candela)	Elevation in temperature	
	PBHTE	Proposed heat model
114	84.70195	81.99984
126	78.13434	78.51421
127	77.34939	78.52047
146	69.18063	74.24557
148	68.39726	73.85524
148	68.39726	73.85524
149	68.00544	73.88179
176	60.17813	69.56939
196	55.85021	67.62164

Fig. 2 Temperature analysis for PBHTE and proposed method SET A values

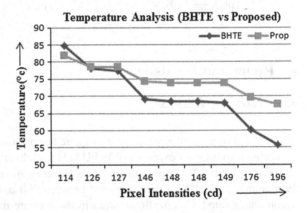

temperature for the intensity value 114 Table 1. Set A-Elevation in temperature values for PBHTE versus Proposed heat model.

So when the over prediction is corrected, the unwanted heat applied to the tissue is avoided. For the next intensity value 126, both the models predict the same temperature value. This temperature value is taken as breakpoint. Below this breakpoint value the temperature predicted is increasing, due to considering the parameters blood perfusion (W_{b1}) and blood viscosity (μ). And above the breakpoint temperature the temperature predicted has reduced. The graph for Table 1 is shown in the Fig. 2. It shows the close relationship between the PBHTE and the proposed heat prediction model. At a certain point the graph meets and then diverts. Tissue boiling is avoided in the proposed heat prediction model as it has not predicted heat above 85 °C.

Elevation in temperature values are plotted for PBHTE versus Proposed heat model for values Set B in Table 2.

Set B pixel intenities are in the range from 107 to 122. The temperature prediction for the proposed heat model has shown decresing values than the PBHTE model. The graph for Table 2 is plotted in Fig. 3.

Table 2 Set B-Elevation in temperature values for PBHTE versus proposed heat model

Pixel intensity (candela)	Elevation in temperature	
	PBHTE	Proposed heat model
107	89.32471	84.3177
108	88.94889	83.92999
110	87.40564	83.1576
111	86.6317	83.16753
114	84.70195	81.99984
116	83.54884	81.2749
117	82.7696	81.23283
122	80.06692	79.68127
122	80.06692	79.68127

Fig. 3 Temperature analysis for PBHTE and proposed method SET B values

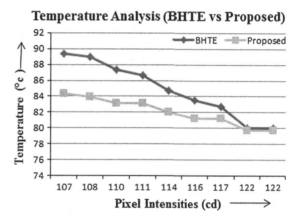

By combining both SET A and SET B values and eliminating the repeated values the Table 3 is obtained. The range of pixel intensities from 107 to 122 has predicted decreased temperature values avoiding overdosage when compared with PBHE model. Pixel intensity 126 has shown similar response for both the heat models PBHTE and proposed.

Remaining pixel intensities in the range 127 to 196 has shown increased heat prediction taking into consideration the convectional heat carried away by the blood vessels passing through the liver. The graph for Table 3 is plotted in Fig. 4.

In the above figure the temperature values are plotted for the intensity values of 3 * 3 lesion area chosen. In the graph the three points are noticed. First one is that the over prediction in the first value is corrected in the proposed method. Second the next value is overlapping means there is equal prediction. This point can be considered as the breakpoint. Then from the breakpoint value the graph is increasing. The increased values of temperature denote the improving accuracy. The number of iterations the temperature has to be applied to the focal point depends upon the depth of the tumor. The tumor is focused such that it is 8 cm from the transducer.

Table 3 Combined temperature values for SET A and B

Intensity	BHTE	Proposed
Dose value is decreased avoiding over dosage		
107	89.32471	84.3177
108	88.94889	83.92999
110	87.410564	83.1576
111	86.6317	83.16753
114	84.70195	81.99984
116	83.54884	81.2749
117	82.7696	81.23283
122	80.06692	79.68127
Similar response		
126	78.13434	78.51421
Dose value increases due to increased accuracy		
127	77.34939	78.52047
146	69.18063	74.24557
148	68.39726	73.85524
149	68.00544	73.88179
176	60.17813	69.56939
196	55.85021	67.62164

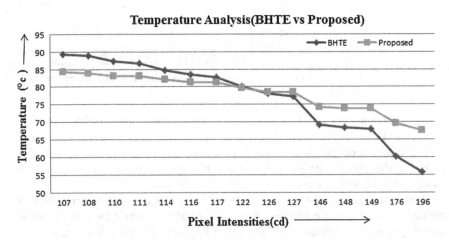

Fig. 4 Combined temperature analysis for SET A and B

Analysis of temperature rise with respect to time is done for the pixel value 156. Number of sonication considered is 10, and from the results obtained it is found that the present approach has decreased the number of iterations of dose. Thus the exposure time of a patient to the radiations is reduced.

5 Conclusion

The new prediction algorithm has reduced the over prediction of the temperature values and also improved the prediction of tissue temperature in the necessary pixels where the acoustic streaming due to larger blood vessels flowing through the liver is considered. As temperature greater than 85 °C causes tissue boiling and may lead to severe damage to the nearby region or pixels, the proposed heat prediction model has shown the maximum heat predicted for highly infected tissue intensity 107 as 84.3177 °C. Further this method can be improved by increasing the accuracy of tumor area prediction. In this research paper threshold based segmentation is used and in future other efficient segmentation techniques like active contours can be used for better segmentation.

References

1. A. Jolesz Kullervo H. Hynynen, "MRI-guided focused ultrasound surgery" book, 2008, ISBN: p13: 978-0-84 93 - 7370-1 2008, by Informa Health care USA.
2. Vinita Dixit and Jyotika Pruthi, Review of Image Processing Techniques for Automatic Detection of tumor in human liver, *International journal of computer science and mobile computing*, Vol. 3, No. 3, 2014, pp. 371–378.
3. Abdel-massieh, N.H. Inf. Technol. Dept., Menoufia Univ., Elkom, Egypt Hadhoud, M.M, Amin, K.M, Fully automatic liver tumor segmentation from abdominal CT scans, Computer Engineering and Systems (ICCES), 2010 International Conference, 2010, pp. 197–202.
4. Suhuai Luo, Xuechen Li and Jiaming Li, Review on the methods of Automatic liver Segmentation from abdominal images, *Journal of Computer and Communications*, Vol 2, No 2, 2014, 2, pp. 1–7.
5. Shehata IA, Treatment with high intensity focused ultrasound: Secrets revealed, *Eur J Radiol*, Vol. 81, No. 3, 2011, pp. 534–541.
6. Filonenko, E. A, and V. A. Khokhlova, Effect of acoustic non linearity on heating of biological tissue by high intensity focused ultrasound, *Acoustical Physics*, Vol. 47, No. 4, 2001, pp. 468–475.
7. H.H. Pennes, Analysis of tissue and arterial blood temperature in the resting human forearm, *Journal of Applied Physiology*, Vol. 1, No. 2, 1948, pp. 93–122.
8. K. Khanafer and K. Vafai, "Synthesis of mathematical models representing bioheat transport," chapter 1 in Advances in Numerical Heat Transfer, Volume 3, CRC Press, pp. 1–28.
9. Tzu-Ching Shih, Tzyy-Leng Horng, Huang-wen, Kuen Cheng, Tzung-chi Huang, Numerical analysis of coupled effects of pulsatile blood flow and thermal relaxation time during thermal therapy, *International journal of heat and mass transfer*, Vol 55, 2012, pp 3763–3773.
10. Shahnazari M, Aghanajafi C, Azimifar M & Jamali H, Investigation of bioheat transfer equation of pennes via a new method based on wrm & homotopy perturbation, *International Journal of Research and Reviews in Applied Sciences*, 2013, Vol. 17, No. 3, pp. 306–314.
11. Yu-Feng Zhou., "High intensity focused ultrasound in clinical tumor ablation," World Journal of Clincal Oncology, Vol 2, Issue 1, 2011, pp 8–27.
12. Michele Rossi1, Claudio Raspanti, Ernesto Mazza, Ilario Menchi, Angelo Raffaele De Gaudio and Riccardo Naspetti, "High-intensity focused ultrasound provides palliation for liver metastasis causing gastric outlet obstruction: case report," Journal of Therapeutic Ultrasound, Vol 1, Issue 9, 2013, pp 1–6.
13. Gail ter Haar., " Acoustic surgery" Physics Today, Vol 54, Issue 12, 2001, pp 29–34.

14. Qiu, Zhen, Chao Xie, S. Cochran, G. Corner, Z. Huang, C. Song, and S. Daglish. "The development of a robotic approach to therapeutic ultrasound." In Journal of Physics: Conference Series, Vol. 181, Issue 1, 2009, pp 1–8.
15. Solovchuk, Maxim A, San Chao Hwang, Hsu Chang, Marc Thiriet, Tony WH Shew. Temperature elevation by HIFU in ex vivo porcine muscle: MRI measurement and simulation study. Medical physics; Vol. 41, 2014, pp 1–13.

Implementation of Server Load Balancing in Software Defined Networking

Sukhveer Kaur and Japinder Singh

Abstract Network management is very painful and tedious process in large networks having hundreds of switches and routers. Software Defined Networking (SDN) is a new way for creating, designing and managing networks which aims to change this current undesirable situation. The main idea of SDN consists in logically centralizing network control in a SDN controller, which controls and monitors the behavior of the network. In this paper we developed an SDN application that performs server load balancing. The Main problem with traditional load balancer is that they use dedicated hardware. That hardware is very expensive and inflexible. Network Administrators cannot write their own algorithms since traditional load balancers are vendor locked. Therefore to solve these problems, we created SDN application which turned simple OpenFlow device into powerful load balancer. There are already certain existing load balancing algorithms in SDN but main problem with all these algorithms is that every request and return message has to pass through the load balancer. It introduces unnecessary latency. To solve these problems we implemented direct routing based load balancing algorithms. In Direct Routing, the load balancer is not involved in the return message from the web server to the client. It means server responds directly to client bypassing the load balancer thus improving performance.

Keywords SDN · OpenFlow · Load balancer · Load balancing algorithms

S. Kaur (✉) · J. Singh
Shaheed Bhagat Singh State Technical Campus, Ferozepur, India
e-mail: bhullarsukh96@gmail.com

J. Singh
e-mail: japitaneja@gmail.com

© Springer India 2016
S.C. Satapathy et al. (eds.), *Information Systems Design and Intelligent Applications*, Advances in Intelligent Systems and Computing 434,
DOI 10.1007/978-81-322-2752-6_14

147

1   Introduction

Currently networking traffic load is very heavy and growing very rapidly. Server overload and network congestion are major problems. Online services such as web sites, search engines and social networking sites require multiple servers to offer enhanced capacity and high reliability. Load balancer is used to distribute requests to multiple servers. Cardellini et al. [1] discusses about the traditional load balancers use dedicated hardware, are very costly and difficult to program. Ghaffarinejad et al. [2] proposed that the load balancing implementation does not require separate load balancer device and allows for flexibility of network topology. Our solution can be scaled as the number of switches and servers increases, while handling client requests at line rates.

Teo et al. [3] explain that the load balancing is a method used for distributing requests to multiple servers. This is done using highly expensive vendor specific hardware devices. Load balancer helps in enhancing performance of networks by properly utilizing available resources. The purpose is to minimize response time, latency and increasing throughput. The distribution of traffic can be based on random, round-robin, current server loads, and location based request strategies.

To solve the problems of traditional load balancing, we created SDN application which turned simple OpenFlow device into powerful load balancer. Our load balancing application runs on an SDN controller. To achieve our goal we used the POX controller which is implemented in Python.

Our main contribution in this work is the development and evaluation of a server load based load balancing solution for a Software Defined Networking. We have:

- Implemented 2 load balancing algorithms namely Direct Routing Based on Server Load and Direct Routing Based on Server Connections in SDN. In Direct Routing, the load balancer is not involved in the return message from the web server to the client.
- Tested our load balancing algorithms in real environment.
- Compared these algorithms with Round Robin strategy based on various parameters. The parameters are Response Time (sec), Throughput (mbps) and Transaction Rate (trans/sec).

The outline of our paper is as follow. Section 2 contains load balancing background and related work. Section 3 contains load balancing architecture. Section 4 covers Experimental Evaluation and Sect. 5 contains conclusion and future work.

2   Background and Related Work

Feamster et al. [4] discusses about SDN that decoupled the Control Plane from the Data Plane. SDN was developed to enable innovation and simple programmatic control of Data plane. The development of new applications and simplified protocol

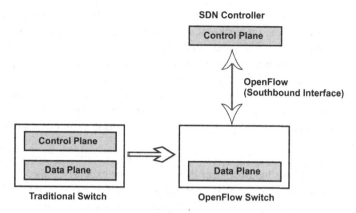

Fig. 1 Separation of control plane and data plane

management is much easier because the control plane is decoupled from the data plane. Software defined networking (SDN) is futuristic paradigm that is powerful, easy to manage, less expensive and flexible, making it suitable for today's dynamic, high bit rate applications. SDN allows control plane to be programmable. It abstracts underneath infrastructure. The OpenFlow protocol is basis for creating SDN solutions.

Data Plane, Control Plane and SDN applications are the three components of SDN architecture, as explained by Mendonca et al. [5]. The Data Plane consists of OpenFlow based physical or virtual switches. Decoupling has allowed the control plane to evolve independently resulting in short evolutionary cycle. This led to development of various control technologies. The forwarding plane evolution cycle aimed at faster packet delivery. OpenFlow switches separate the forwarding and control plane. The separated control plane is called SDN controller.

OpenFlow protocol is used for communication between the decoupled data plane and control plane as shown in Fig. 1. OpenFlow protocol standardizes information exchange between two planes, as discussed by Suzuki et al. [6]. In the SDN architecture, the OpenFlow switch maintain one flow table and secure channel is used for communication between OpenFlow switch and control plane using OpenFlow protocol. Each Flow table contains several flow rules, each flow rule decides how particular packet will be handled by the OpenFlow switch. Each Flow rule contains three components: Matching rules are used to match incoming packets; Match field can be ingress port, dst MAC address, dst IP. Counters are used to gather information such as number of incoming packets and outgoing packets or number of bytes for the specific flow. Actions decide how to handle matching packets such as forward a packet or drop a packet.

When an OpenFlow switch receives a packet, header fields in the packet are parsed and matched with the matching rules component of the flow table rules. If match is found, Open flow switch apply the number of actions that are specified in matched flow rule. If match is not found in flow table, the action specified in table-miss entry is performed by OpenFlow switch. Each flow table maintains a table-miss entry to handle the packet that is not matched. Actions in the table-miss entry can be drop the packet and send the packet to the controller using the OpenFlow protocol. For communication between control and data plane, OpenFlow protocol is used, as explained by Lara et al. [7]. OpenFlow protocol specifies different messages that can be exchanged over secure channel between control and data plane. Using the OpenFlow protocol a remote controller can update, delete, or add flow entries from the flow tables of switch. This process can occur proactively or reactively.

Load Balancing is a technique used to distribute large number of requests across multiple devices. These load balancer devices are very expensive, specialized and vendor-specific devices. Load balancer increases network performance by using the available resources properly and helps in improving response time, transactions per second and throughput.

Wang et al. [8] discusses about partitioning algorithm that divides client traffic in proportion to the load balancing weights. The limitation of this paper is it does not deliver performance measurements. Koerner et al. [9] discusses about multiple load balancing where one load balancer is used for balancing web servers while other controller is used for balancing e-mail servers. Shang et al. [10] discuss about load balancing technology in the OpenFlow environment. Uppal et al. [11] use the number of pending requests in the servers queue as Load. The NOX Controller listens for current load information from servers. When a new flow request is received, it selects the server with lowest current load and current load information gets updated by the NOX Controller.

The main problem with all these implementations is that every request and return message has to pass through the load balancer. This means that it introduces unnecessary latency. Therefore to solve these problems we implemented direct routing based load balancing algorithms (Direct Routing Based on Server Load, Direct Routing Based on Server Connections). In Direct Routing, the load balancer is not involved in the return message from the web server to the client. It means server responds directly to client bypassing the load balancer. Direct Routing improves performance because in this case load balancer handles packets in inbound direction only. Mostly traffic in outbound direction is larger than traffic in inbound direction.

3 Load Balancing Architecture

Our load balancing architecture contains POX controller, OpenFlow switch, multiple server machines and clients connected to the ports of the OpenFlow switch. Static IP address has been configured on each server. A list of currently connected servers to the OpenFlow switch is maintained by POX controller.

When a switch receives a packet from a client, POX controller determines how to handle this flow based on load balancing algorithm. The POX controller adds

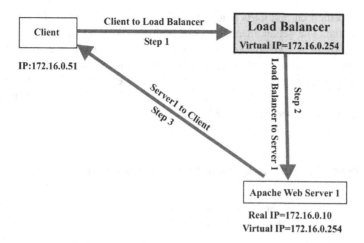

Fig. 2 Direct routing

new flow entries into the flow table of switch using the OpenFlow protocol. The flow entry contains the rule that changes the destination MAC of the request packet with the MAC address of the selected server. After packet header modification, the switch forwards the packet to the server. In return journey, the servers directly send the reply to the client bypassing the load balancer. It means load balancer is not included in the return message from the web server to web client as shown in Fig. 2.

We will now describe the different scheduling algorithms, which are the core part of the load balancer behavior.

3.1 Round-Robin

In this algorithm, the requests are sent to each server, one at a time in a circular way. Due to its simplicity, this algorithm is also the lightest with regards to CPU usage. When a Packet_In event happens, due to the first packet of the client request, the chosen server is always the next on a list of all the available servers in the network. The order of the servers in the list is always the same. This ensures that all the servers handle same number of requests, regardless of the load on each one, as implemented by Kaur et al. [12].

3.2 Direct Routing Based on Server Load

The server chosen in this algorithm is the server with least load. To simplify the meaning of load, we compare only current CPU load on the server machines.

Servers send their current CPU load, and the load balancer chooses the server with the least value. The server machines determine their current CPU load using the command mpstat. A UDP packet containing this load information is sent to some known ghost IP by a program running on the server machine at regular interval. The load balancer retrieves the udp packet, and stores the CPU load values for each server. When it is asked to choose a server where request should be sent, it replies with the server that has the least CPU load value.

3.3 Direct Routing Based on Server Connection

This algorithm chooses the server based on minimum active connections. In case of a tie, the server with the lowest identifier is chosen. The server machine runs a program that verifies how many connections it has using the command netstat. This command outputs the machine's current TCP connections. The program then counts how many lines this command outputs, and sends a User Datagram Protocol (UDP) packet destined to a ghost IP containing this information. The term ghost emphasizes that no machine on the network has that address assigned. The switch receives this packet and, since it does not have a matching flow entry for it, raises a Packet_In event. Since it is addressed to a specific ghost IP, the load balancer knows what it is and what to do with it. It reads the packet data to retrieve the number of active connections and associates it with the respective server. Servers send a UDP packet at regular interval, which means the load balancer has this information updated at regular interval.

4 Experimental Evaluation

For the experimental evaluation we created a real topology (Fig. 3) which consists of 4 computers. We implement the OpenFlow virtual switch and POX controller on one computer. Apache web server is implemented on other 2 linux computers (172.16.0.10, 172.16.0.31). One linux computer (172.16.0.51) act as a client on which we installed the load testing tool "Siege". Our load balancing implementation uses OpenFlow switch along with POX controller and two server machines and client connected to the ports of the OpenFlow switch. POX controller is started and connectivity between the openflow network and controller is verified. OpenFlow switch and controller exchange series of messages for connection establishment and setup. Our Load Balancer contains one virtual IP address. Client will send the requests to this virtual IP. This virtual IP is the address given to load balancer as well as to the apache web server hosts. The key components involved for experimental setup are: POX OpenFlow Controller, OpenFlow virtual switch, Siege load testing tool.

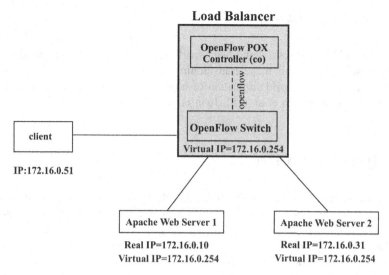

Fig. 3 Network topology

4.1 POX OpenFlow Controller

Kaur et al. [13] discusses about the POX that is the Python version of NOX. Its focus is on research and academia. NOXs core infrastructure was implemented in C++; however, it allowed applications to be written in either Python or C++. As development of NOX progressed, developers saw the need to build separate Python and C++ versions. So POX was forked from NOX, but the basic idea and framework remains the same [14].

4.2 OpenFlow Virtual Switch

It is a virtual or physical device used for packet forwarding in SDN architecture. These OpenFlow switches can be either pure or hybrid. In a traditional switch, the data path and the control path are tightly coupled into the same device. An OpenFlow switch decoupled the data plane from the control plane. The OpenFlow is the communication protocol between both planes. Separation of data path from control path allows easier deployment of new applications and simplified protocol management. SDN allows control path to be easily programmable.

4.3 Siege Load Testing Tool

Siege is an HTTP load testing and benchmarking tool. It has a similar interface to apache bench, which will make transitioning to the tool almost seamless. This allows for a more real-world simulation of how a user would use your system, as explained by Sukaridhoto et al. [15], Jung et al. [16].

4.4 Apache Web Server

Apache web is the most used web server. Apache web server is initially intended for Unix operating system. Apache is also available for other platforms such as Windows and Mac, as discussed by Hu et al. [17]. According to a Netcraft, 76 % of all Web sites on the Internet are using Apache web server, making Apache more widely used than all other Web servers.

We compared the Direct Routing Based on Server Load and Direct Routing Based on Server Connection with round robin strategy on the basis of total transactions per second, throughput and response time of the web server. In our case, we took the readings by sending load by using "Siege" load testing tool. Transactions per second,

Throughput and Response Time are better in our proposed algorithms. Figure 4 shows response time of servers. Here the number of concurrent users is represented on x-axis and response time is represented on y-axis. The Response time in Direct Routing Based on Server Load and Direct Routing Based on server Connections is better than round robin algorithm.

Figure 5 shows throughput of servers. Here the number of concurrent users is represented on x-axis and throughput in mbps is represented on y-axis. The Throughput in Direct Routing Based on Server Load and Direct Routing Based on server Connections is better than round robin algorithm.

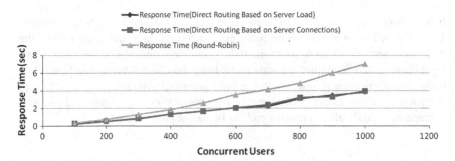

Fig. 4 Response time in direct routing based on server load, direct routing based on server connections and round-robin strategy

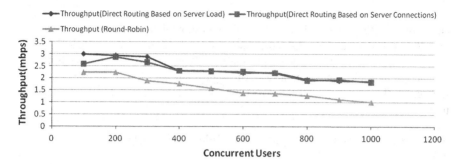

Fig. 5 Throughput in direct routing based on server load, direct routing based on server connections and round-robin strategy

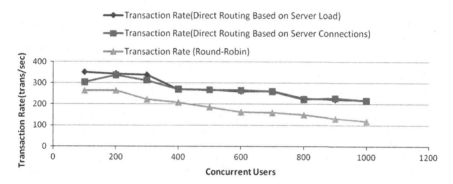

Fig. 6 Transaction rate in direct routing based on server load, direct routing based on server connection and round-robin strategy

Figure 6 shows Transaction Rate of servers. Here the number of concurrent users is represented on x-axis and Transaction Rate in trans/sec is represented on y-axis. The Transaction Rate in Direct Routing Based on Server Load and Direct Routing Based on server Connections is better than round robin algorithm.

5 Conclusion and Future Work

SDN load balancer solves many problems of traditional load balancers because traditional load balancer use dedicated hardware. That hardware is expensive and inflexible. But in SDN, OpenFlow device is converted into powerful load balancer by programming the SDN controller. Our main goal was to build a control application that performs server load balancing on SDN architecture, we were also able to attest that how simple it is to build and customize network control applications for such networks. We implemented load balancer using Direct Routing Based on

Server Load and Direct Routing Based on Server Connections algorithms and compared it with Round Robin strategy. Results show that our algorithms are better than Round Robin strategy. As is the case with most commercial load balancer, our load balancer can also be single point of failure. To eliminate this problem in future we can use multiple controllers instead of single controller. In case of one controller failing, another machine will take over the role of controller and continue routing traffic.

Acknowledgments We thanks Mr. Vipin Gupta of U-Net Solutions, Moga, India for his valuable help.

References

1. Cardellini, Valeria, Michele Colajanni, and S. Yu Philip. "Dynamic load balancing on web-server systems." *IEEE Internet computing* 3, no. 3 (1999): 28–39.
2. Ghaffarinejad, Ashkan, and Violet R. Syrotiuk. "Load Balancing in a Campus Network Using Software Defined Networking." In *Research and Educational Experiment Workshop (GREE), 2014 Third GENI*, pp. 75–76. IEEE, 2014.
3. Teo, Yong Meng, and Rassul Ayani. "Comparison of load balancing strategies on cluster-based web servers." *Simulation* 77, no. 5–6 (2001): 185–195.
4. Feamster, Nick, Jennifer Rexford, and Ellen Zegura. "The road to SDN: an intellectual history of programmable networks." *ACM SIGCOMM Computer Communication Review* 44, no. 2 (2014): 87–98.
5. Mendonca, Marc, Bruno Astuto A. Nunes, Xuan-Nam Nguyen, Katia Obraczka, and Thierry Turletti." A Survey of software-defined networking: past, present, and future of programmable networks." *hal-00825087* (2013).
6. Suzuki, Kazuya, Kentaro Sonoda, Nobuyuki Tomizawa, Yutaka Yakuwa, Terutaka Uchida, Yuta Higuchi, Toshio Tonouchi, and Hideyuki Shimonishi." A Survey on OpenFlow Technologies." *IEICE Transactions on Communications* 97, no. 2 (2014): 375–386.
7. Lara, Adrian, Anisha Kolasani, and Byrav Ramamurthy. "Network innovation using openflow: A survey." (2013): 1–20.
8. Wang, Richard, Dana Butnariu, and Jennifer Rexford. "OpenFlow-based server load balancing gone wild." (2011).
9. Koerner, Marc, and Odej Kao. "Multiple service load-balancing with OpenFlow." In *High Performance Switching and Routing (HPSR), 2012 IEEE 13th International Conference on*, pp. 210–214. IEEE, 2012.
10. Shang, Zhihao, Wenbo Chen, Qiang Ma, and Bin Wu. "Design and implementation of server cluster dynamic load balancing based on OpenFlow." In *Awareness Science and Technology and Ubi-Media Computing (iCAST-UMEDIA), 2013 International Joint Conference on*, pp. 691–697. IEEE, 2013.
11. Uppal, Hardeep, and Dane Brandon. "OpenFlow based load balancing." *University of Washington. CSE561: Networking. Project Report, Spring* (2010).
12. Kaur, S.; Singh, J.; Kumar, K.; Ghumman, N.S., "Round-robin based load balancing in Software Defined Networking," *Computing for Sustainable Global Development (INDIACom), 2015 2nd International Conference on*, vol., no., pp.2136,2139, 11–13 March 2015.
13. Kaur, Sukhveer, Japinder Singh, and Navtej Singh Ghumman. "Network Programmability Using POX Controller".
14. POX at https://openflow.stanford.edu/display/ONL/POX+Wiki.

15. Sukaridhoto, Sritrusta, Nobuo Funabiki, Toru Nakanishi, and Dadet Pramadihanto. "A comparative study of open source softwares for virtualization with streaming server applications." In *Consumer Electronics, 2009. ISCE'09. IEEE 13th International Symposium on*, pp. 577–581. IEEE, 2009.
16. Jung, Sung-Jae, Yu-Mi Bae, and Wooyoung Soh. "Web Performance Analysis of Open Source Server Virtualization Techniques." *International Journal of Multimedia and Ubiquitous Engineering* 6, no. 4 (2011): 45–51.
17. Hu, Yiming, Ashwini Nanda, and Qing Yang. "Measurement, analysis and performance improvement of the Apache web server." In *Performance, Computing and Communications Conference, 1999 IEEE International*, pp. 261–267. IEEE, 1999.

Building Stateful Firewall Over Software Defined Networking

Karamjeet Kaur and Japinder Singh

Abstract Current network architectures are ill suited to meet today's enterprise and academic requirements. Software Defined Networking (SDN) is a new way to Design, Build and Operate Networks. It replaces static, inflexible and complex networks with networks that are agile, scalable and innovative. The main idea is to decouple the control and data planes, allowing the network to be programmatically controlled. A key element of SDN architectures is the controller. This logically centralized entity acts as a network operating system, providing applications with a uniform and centralized programming interface to the underlying network. But it also introduces new security challenges. The challenge of building robust firewalls is the main challenge for protection of OpenFlow networks. The main problem with traditional firewall is that Network Administrator cannot modify/extend the capabilities of traditional vendor-specific firewall. Network Administrator can only configure the firewall according to the specifications given by the firewall vendor. To solve these problems we developed stateful firewall application that runs over SDN controller to show that most of the firewall functionalities can be built on software, without the aid of a dedicated hardware.

Keywords SDN · OpenFlow · Firewall · Packet filtering · Stateful firewall · Stateless firewall

1 Introduction

Firewall is used for preventing unauthorized access to and from the private network. Firewall examines all packets leaving or entering internal networks and block those who do not meet the defined security criteria. A firewall allows or rejects a specific

K. Kaur (✉) · J. Singh
Shaheed Bhagat Singh State Technical Campus, Ferozepur, India
e-mail: bhullar1991@gmail.com

J. Singh
e-mail: japitaneja@gmail.com

© Springer India 2016
S.C. Satapathy et al. (eds.), *Information Systems Design and Intelligent Applications*, Advances in Intelligent Systems and Computing 434,
DOI 10.1007/978-81-322-2752-6_15

159

type of information. Implementation of a stateful firewall is the mandatory part of an effective information security program.

Most businesses and institutions deploy firewall as the main security mechanism. A traditional firewall is placed at the boundary of public and private network. It prevents attacks and unauthorized access by examining all incoming and outgoing traffic. In Traditional Firewall deployments, all Insiders in the private network are considered trustworthy. So internal traffic is not examined and filtered by the firewall. The assumption that insiders are trustworthy is not valid these days since insiders can perform attacks on others by bypassing security mechanisms. The main problem with traditional firewall is that they use dedicated hardware. That hardware is expensive and inflexible. This can be an even greater burden if a network needs more than one, as is common. Network administrators can not add new features since traditional firewalls are vendor locked, difficult to program, as discussed by the Hu et al. [1, 2].

To solve the problems of traditional firewall, we created SDN application which turned simple OpenFlow device into powerful firewall. Our firewall application runs on an SDN controller. To achieve our goal we used the Ryu controller which is implemented in Python. In this paper, we focus on implementation of robust stateful firewall. Firewalls are the most widely used security mechanism for OpenFlow based networks.

Our main contribution is the development and evaluation of stateful firewall solution for software defined network.

- To implement Stateful firewall to keep track the state of network connections passing through it. The stateful firewall is programmed to differentiate between packets from different types of connections. Firewall will allow only those packets which match with known connection state, others will be rejected.
- To Test our stateful firewall application in real environment.
- To compare our stateful firewall with stateless firewall using HTTP (Hyper Text Transfer Protocol), ICMP (Internet Control Message Protocol) traffic and observe that our firewall is more secure because it is able to block the fake packets as compared to stateless firewall.

The outline of our paper is as follow. Section 2 contains background and related work. Section 3 describes implementation details. Section 4 covers Experimental Evaluation and Sect. 5 contains conclusion and future work.

2 Background and Related Work

Current networks are very complex and hard to manage. These networks consists of different types of devices such as routers, switches, firewalls, network address translators, load balancers, and intrusion detection systems. These devices run software that are typically closed and vendor specific. New network protocols goes through years of standardization efforts and interoperability tests. Network administrators have to configure each individual network devices as specified by the

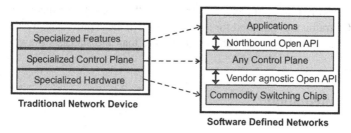

Fig. 1 Traditional and software defined network

vendor. Even more painful is that different products from the same vendor require different configuration. Traditional Networks has slowed down innovation, are very complex, and both the capital and operational expenses of running the network are high.

Mendonca et al. [3] discusses about the Software Defined Networking is emerging networking architecture in which the control plane is decoupled from data plane. Separation of control plane from data plane allows easier deployment of new applications and simplified protocol management as shown in Fig. 1. In SDN, the control plane is shifted into centralized place and shifted control plane is called SDN controller or OpenFlow controller. The network devices become simple packet forwarding devices that can be programmed via southbound interfaces such as ForCES (Forwarding and Control Element Separation), OpenFlow, as explained by Feamster et al. [4]. The research community and industry are giving significant attention to SDN [5]. The Open Network Foundation (ONF) has been created by service providers, network operators and vendors for promoting SDN and standardizing OpenFlow protocol. Although the concept of software defined networking is quite new, still it is growing at a very fast speed.

OpenFlow architecture basically consists of three components, OpenFlow switch, controller and Openflow protocol, as explained by Lara et al. [6]. OpenFlow protocol is used for communication between switch and controller through a secure channel. OpenFlow switch contains flow tables entries that consist of match fields, statistics, actions. When packet arrives at switch it is matched against rules. If match is found, corresponding action is performed. If no match is found, action is taken as specified in table miss entry. Controller is responsible for adding, updating or deleting flow entries, as discussed by Suzuki et al. [7].

Tariq et al. [8] successfully implemented layer 2 firewall by modifying layer 2 learning switch code in the POX controller. Limitation of this implementation is that this blocks only layer 2 traffic. Michelle et al. [9] Created simple user interface for the firewall. Their design looks at few header fields. Kaur et al. [10] implement stateless firewall that watch network traffic, and deny or drop packets based on source IP, destination IP, source MAC, destination MAC or other static parameters.

All previous work implemented stateless firewall that treats each packet in isolation. Stateless firewall is not able to determine whether packet is attempting to establish new connection or is a part of existing connection or just a fake packet.

Second problem of these implementations is that they were tested using only ICMP traffic. TCP traffic was not used for testing. To solve these problems, we design and implement a stateful firewall that is more secure than the stateless firewall.

3 Implementation

Firewall is used for preventing unauthorized access to and from the private network based on packet filtering rules. Firewall examines all packets leaving or entering internal networks and block those who do not meet the defined security criteria. When configuring filter rules for TCP (Transmission Control Protocol) and UDP (User Datagram Protocol) services such as File Transfer Protocol (FTP), Trivial File Transfer Protocol (TFTP), http, we have to allow traffic in both directions since these services are bi-directional. A TCP or UDP session involves two players, one is the client that initiates the session and the other is the server hosting particular service. As an example, Fig. 2 the rules are added in such a way so that traffic can cross the firewall between web client with address 192.168.0.2/24 and web server with address 172.16.0.31/16.

The server IP address, the client IP address, the server port also known as destination port and client port known as source port are the 4 attributes that defines a TCP or UDP session. Generally, the server port helps in identifying the type of service that is being offered. For example, port 80 is associated with web service and ports 20, 21 are associated with ftp service. Client port which is mostly greater than 1023 is dynamically chosen by client host's operating system. It also means that client's ports are mostly unpredictable and firewall has to allow all the source ports so that session can be successful. As a result, this type of approach introduces serious security problem. It enables malicious hosts to launch Denial of Service (DOS) attack by flooding the servers with unwanted traffic. The Fig. 2 shows network architecture used in experiment.

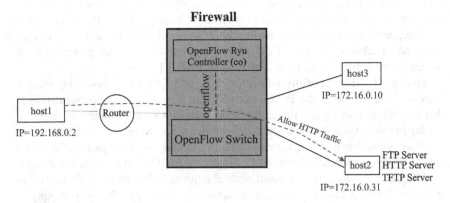

Fig. 2 Network topology

For the experimental evaluation we create a real topology which consists of 4 computers. We implement the OpenFlow switch and Ryu controller on one computer. On second computer we are implemented apache web server, FTP server and TFTP server. Other two systems act as clients on which we implemented Ostinato packet generation tool. Ryu controller is started and connectivity between the openflow network and controller is verified. OpenFlow switch and controller exchange series of messages for connection establishment and setup. We have got two hosts host1 with IP address 192.168.0.2/24 and a host2 with IP address 172.16.0.31/16 connected using of the OpenFlow switch.

3.1   Stateful TCP Packet Filtering

Stateful firewall handles the security issue by enabling connection tracking. The connection tracking keeps track of established TCP sessions. Firewall maintains entries of open TCP sessions in cache or in connection tracking table. An entry contains information regarding server ip address, client ip address, server port and client port. The firewall administrator does not have any information regarding client port at the time of configuration of rules. But at the time of connection setup, both server and client port information is available in TCP header. Stateful firewall allows traffic in both directions for packets belonging to existing TCP connection.

After connection establishment, the decision to allow or block subsequent packets will be based on contents of connection tracking table. When a subsequent packet reaches the firewall with flag ACK set and flag SYN unset, its information entry is checked in connection tracking table. If entry exists, the packet is allowed to pass through immediately. The packet gets rejected if no such entry is found.

On the completion of 3-way handshake, TCP connection state turns to ESTABLISHED state. On connection termination, the entry is removed from the connection tracking table. If TCP connection is inactive for a long time, a timeout value is kept to flush out the inactive entries from connection tracking tables and thus blocking the connection.

3.2   Stateful ICMP Packet Filtering

It is easy to track ICMP sessions. It involves 2 ways communication. For every response ICMP message, there should be corresponding ICMP request message. ICMP tracking is done on basis of Sequence number, Type Identifier, source address, destination address of reply and request messages. The sequence number and type identifier cannot alter in an ICMP session when returning a message to the sender. The sender matches each echo request with echo request by using these. These parameters should have the same values for echo request and echo reply. This is the only way for tracking ICMP sessions.

Upon receiving ICMP request packet, an entry is made in connection tracking table by stateful firewall. Stateful firewall will accept the echo reply if parameter value are the same as in request packet. Fake ICMP reply will be rejected since connection tracking table does not contain any entry the same parameter values.

4 Experimental Evaluation

The purpose of this evaluation is to ascertain the functionality of our Stateful firewall application. The key components involved for experimental setup are, Ryu OpenFlow Controller, OpenFlow virtual switch, Ostinato packet generation tool.

Ryu OpenFlow Controller
Monaco et al. [11] discusses about the Ryu controller that is a component-based OpenFlow controller in Software Defined Networking [11]. Ryu provides well defined Application programming interface that make it easy for developers to create new network control applications like Firewall, Load Balancer. Ryu supports different protocols for controlling network devices, such as OpenFlow, Netconf, OF-config, etc. Ryu supports OpenFlow versions 1.0, 1.2, 1.3. The code of Ryu is available under the Apache 2.0 license. Ryu is completely written in Python language.

OpenFlow Virtual Switch
An OpenFlow switch is a virtual or physical switch that forwards packets in a software defined networking (SDN) infrastructure. OpenFlow switches are either pure SDN switches or hybrid switches as explained by Bianco et al. [12].

Ostinato
Ostinato is GUI based network packet analyzer and generator tool. It is open source and cross platform tool. By using it, we can create and send several streams of packets at different rates having different protocols as explained by Botta et al. [13], Srivastava et al. [14].

Wireshark
Orebaugh et al. [15] discusses about the Wireshark that is a best available network protocol analyzer tool that is open source and multi-platform. It permits you to analyze information from stored captured file or using live network. Captured information can be scanned interactively, as explained by Sanders et al. [16].
The experiment includes two parts.

4.1 Stateful TCP Packet Filtering Testing

(1) We tested number of scenarios on firewall application. In first scenario we applied rule in which host1("192.168.0.2/24") was allowed to access the web

Table 1 Three-way handshake of web session

Source IP	Destination IP	Source port	Destination port	SYN	ACK
192.168.0.2	172.16.0.31	35591	80	1	0
172.16.0.31	192.168.0.2	80	35591	1	1
192.168.0.2	172.16.0.31	35591	80	0	1

Fig. 3 Fake TCP packet generated by ostinato packet builder

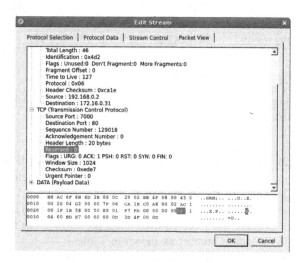

server at host2 ("172.16.0.31/16"). This means that web traffic between host2 web server and host1 web client is allowed.

(2) At host2, wireshark has been used for capturing 3 way TCP handshake packets during web session. Table 1 represents the three-way handshake web session.

(3) At Ostinato Packet Builder was used for sending a fake TCP packet from host1 to host2 as shown in Fig. 3. The fake packet pretended that connection to TCP port 80 is already established (SYN = 0, ACK = 1). The fake packet was having a source port that was different from the source port of current active session as shown in Table 2.

(4) Wireshark at host2 was not able to capture the fake packet sent by host1. This happened because fake TCP packet was blocked by the stateful firewall. Stateful firewall blocks TCP packets which do not belong to established TCP sessions. But in case of stateless firewall the fake packets are reached at web server. This means that stateful firewall is more secure than stateless firewall.

4.2 Stateful ICMP Packet Filtering Testing

(1) We tested number of scenario on firewall application. In this scenario we applied rule that allow host2 ("172.16.0.31/16") to ping host1 ("192.168.0.2/24").

Table 2 Fake TCP packet parameters

Source IP	Destination IP	Source port	Destination port	SYN	ACK
192.168.0.2	172.16.0.31	7000	80	0	1

Table 3 ICMP exchange packet parameters

Source IP	Destination IP	Type	Identifier	Sequence number
172.16.0.31	192.168.0.2	8	314	1
192.168.0.2	172.16.0.31	0	314	1

Table 4 Fake ICMP echo reply packet

Source IP	Destination IP	Type	Identifier	Sequence number
192.168.0.2	172.16.0.31	0	314	10

(2) Wireshark at host2 is used for capturing ICMP packets. Table 3 show the ICMP exchange packets between the two systems.

(3) Ostinato Packet Builder is used for sending a fake ICMP echo reply packet from host1 to host2. This fake packet pretended that ICMP echo request packet was received from host2 previously. The fake packet contains different sequence number as shown in Table 4.

(4) Wireshark at host2 is not able to capture the fake ICMP packet sent by host1. This happened because fake ICMP packet was blocked by the stateful firewall. Stateful firewall blocks fake ICMP echo reply packet. This means that stateful firewall is more secure than stateless firewall.

After stateful firewall implementation, Fig. 4 shows that latency got increased which means that firewall provide security with little more overhead than stateless firewall.

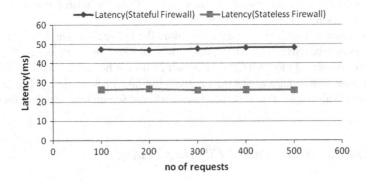

Fig. 4 Latency in Stateful and stateless firewall

5 Conclusion and Future Work

The popularity of SDN is increasing day by day. Although there exist many quality firewalls but they are very costly. Another limitation is that Network Administrator cannot modify/extend the capabilities of the traditional vendor-specific firewalls. They can only configure the firewall according to the specification given by the firewall vendor. We have implemented a stateful firewall and compared it with the stateless firewall and observed that our firewall is more secure (means able to block fake packets) than the stateless firewall. Our firewall is also able to block the SYN flooding attack by keeping a record of each connection.

Future direction can be to design and implementation Intrusion Detection system and combine it with our firewall for creating Intrusion Prevention system.

Acknowledgments We thanks Mr. Vipin Gupta of U-Net Solutions, Moga, India for his valuable help.

References

1. Hu, Hongxin, Wonkyu Han, Gail-Joon Ahn, and Ziming Zhao. "FLOWGUARD: building robust firewalls for software-defined networks." In *Proceedings of the third workshop on Hot topics in software defined networking*, pp. 97–102. ACM, 2014.
2. Hu, Hongxin, Gail-Joon Ahn, Wonkyu Han, and Ziming Zhao. "Towards a Reliable SDN Firewall." *Presented as part of the Open Networking Summit 2014 (ONS 2014)}* (2014).
3. Mendonca, Marc, Bruno Astuto A. Nunes, Xuan-Nam Nguyen, Katia Obraczka, and Thierry Turletti. "A Survey of software-defined networking: past, present, and future of programmable networks." *hal-00825087* (2013).
4. Feamster, Nick, Jennifer Rexford, and Ellen Zegura. "The road to SDN: an intellectual history of programmable networks." ACM SIGCOMM Computer Communication Review 44, no. 2 (2014): 87–98.
5. N. Feamster, "Software defined networking," Coursera, 2013. [Online]. Available: https://class.coursera.org/sdn-001.
6. Lara, Adrian, Anisha Kolasani, and Byrav Ramamurthy. "Network innovation using openflow: A survey." (2013): 1–20.
7. Suzuki, Kazuya, Kentaro Sonoda, Nobuyuki Tomizawa, Yutaka Yakuwa, Terutaka Uchida, Yuta Higuchi, Toshio Tonouchi, and Hideyuki Shimonishi. "A Survey on OpenFlow Technologies." IEICE Transactions on Communications 97, no. 2 (2014): 375–386.
8. Javid, Tariq, Tehseen Riaz, and Asad Rasheed. "A layer2 firewall for software defined network." In *Information Assurance and Cyber Security (CIACS), 2014 Conference on*, pp. 39–42. IEEE, 2014.
9. Suh, Michelle, Sae Hyong Park, Byungjoon Lee, and Sunhee Yang. "Building firewall over the software-defined network controller." In *Advanced Communication Technology (ICACT), 2014 16th International Conference on*, pp. 744–748. IEEE, 2014.
10. Kaur, K.; Kumar, K.; Singh, J.; Ghumman, N.S., "Programmable firewall using Software Defined Networking," *Computing for Sustainable Global Development (INDIACom), 2015 2nd International Conference on*, vol., no., pp. 2125, 2129, 11–13 March 2015.
11. Monaco, Matthew, Oliver Michel, and Eric Keller. "Applying operating system principles to SDN controller design." In *Proceedings of the Twelfth ACM Workshop on Hot Topics in Networks*, p. 2. ACM, 2013.

12. Bianco, Andrea, Robert Birke, Luca Giraudo, and Manuel Palacin. "Openflow switching: Data plane performance." In *Communications (ICC), 2010 IEEE International Conference on*, pp. 1–5. IEEE, 2010.
13. Botta, Alessio, Alberto Dainotti, and Antonio Pescapé. "A tool for the generation of realistic network workload for emerging networking scenarios." Computer Networks 56, no. 15 (2012): 3531–3547.
14. Srivastava, Shalvi, Sweta Anmulwar, A. M. Sapkal, Tarun Batra, Anil Kumar Gupta, and Vinodh Kumar. "Comparative study of various traffic generator tools." In *Engineering and Computational Sciences (RAECS), 2014 Recent Advances in*, pp. 1–6. IEEE, 2014.
15. Orebaugh, Angela, Gilbert Ramirez, and Jay Beale. *Wireshark & Ethereal network protocol analyzer toolkit.* Syngress, 2006.
16. Sanders, Chris. *Practical Packet Analysis: Using wireshark to solve real-world network problems.* No Starch Press, 2011.

A Proposed Framework to Adopt Mobile App in 'e-District' Projects to Move One Step Ahead for Successful Implementation

Manas Kumar Sanyal, Sudhangsu Das and Sajal Bhadra

Abstract Government of India (GI) have been driving the e-Governance projects rolling out program in India with serious note, keeping in mind to deliver Government services in real time at anywhere and at any time on the basis of citizen's needs. The primary focus of GI is to do proper transformation of Indian rural and urban society for adopting technology to facilitate Government services. GI is putting lot of effort and money for transferring India to Digital India considering the fact for shifting behaviors and expectation of common citizens on the matter of Government services. The reflections of changes among the Indian societies already have been accounted and it has been noticed that common citizens are showing tremendous interest to use e-Governance applications as front end interface for their daily Government interactions. In these initiatives, GI has deployed 'e-District' as one of the major project of Mission Mode Project (MMP) under National e-Governance plan of GI to offer citizens centric services which falls under district administration portfolio. The "e-District" project has been designed and build with based on three infrastructure pillars, the State Wide Area Network (SWAN) for communication, State Data Centre (SDC) for central data repository, and Common Service Centers (CSCs) for common citizens service counter. But, it has been observed that there is huge GAP in between Government expectations (services at door step) and its actual implementation of "e-District" projects. In the present implementation approach, essentially citizens have to attend KIOSK or Common Service Center (CSC) to avail Government services. Now-a-days, there is no way to deny that the acceptability of smart phone is growing and spreading and reaching to people very rapidly. Carrying this understanding, it could be anticipated that 'e-district' mobile apps could be solution to

M.K. Sanyal (✉) · S. Das · S. Bhadra
Department of Business Administration, Kalyani University,
Kalyani, India
e-mail: Manas_sanyal@rediffmail.com

S. Das
e-mail: iamsud@gmail.com

S. Bhadra
e-mail: Sajal.bhadra@gmail.com

© Springer India 2016
S.C. Satapathy et al. (eds.), *Information Systems Design and Intelligent
Applications*, Advances in Intelligent Systems and Computing 434,
DOI 10.1007/978-81-322-2752-6_16

169

minimize this GAP and Government services can reach out at citizens door steps very easily. In this study, author's main objective is to propose a framework to introduce mobile app for brining Government services at common citizen's door steps. This improvement and enhancement of existing 'e-District' project is supposed to ensure sustainability of e-Governance because it will change attitude and mentality of society towards acceptance of e-Governance in case government services, thus it will transform our society in one steps a-head for accepting and implementing digital Governance in India.

Keywords e-District · Network society · e-Governance · Mobile app

1   Introduction

On 18th May 2006, Government of India (GI) has taken ample initiatives for implementing e-Governance projects across the country through National e-Governance plan (NeGP) [1, 2]. The main focus of NeGP plan was to deliver all Government services to the common peoples at the possible nearest location with the help of common service centers (CSC) and kiosks. The primary expectation of GI from these initiatives is to bring more efficiency, transparency and reliability on Government service delivery where common people will bear very minimum cost to avail Government services. In NeGP plan, GI had decided to roll out 27 mission mode projects (MMP) and 8 components for facilitating Government services to the common people. Among all MPPs project, the 'e-District' project has been identified as major project for providing more citizens centric services. The 'e-District' project has been designed to cater all the Government services provided by district magistrate administration. The 'e-District' project has automated most of the approval workflows, did huge process improvement through digitization, and integrated all related departments for quick communication, ensured data security through Sate Data center (SDC) data repository. This application is very much comprehensive and internet based web site that acts as digital interface for Government's services portfolio. On a present note, all state across the country started to implement 'e-District' project. Few of them have been rolled out in pilot phase, few have got fully operational and rest have been planned to roll out during the year of 2015–16. The response and acceptability of 'e-District' project from all stake holders, especially from common citizens are un-doubtfully very impressive.

The main background and reason of this huge popularity and acceptability of 'e-District' was due to certain limitations in the legacy manual system. The few of them have been summarized in below [3, 4]:

- Applications submission process is very complex and require to maintain long queue
- No proper guideline and helpline to do application form fill up

- Application processing time takes long time
- Duplicate information required to provide in different departments
- No proper channel to track application status
- Don't provides answer of common citizens quarries as and when require.

1.1 "e-District" Project Architecture

The technical architecture of 'e-District' project have been developed by considering the fact that underline network will be facilitated by State Wide Area Network (WBSWAN), citizens will get necessary first level assistance form Common Service Center (CSC), and Data will be stored safely in State Data Center (SDC). According to National e-Governance Plan, GI has implemented these three projects to set up infrastructure backbone for implementing all others MMPs project across the country. So, all Government offices are interconnected (Fig. 1).

The main accelerator of e-district projects are [3]

- Common citizens are getting services at KIOKS and CSC where people have opportunity to get service outside of office time window
- Transparency have been improved with responsiveness and accountability

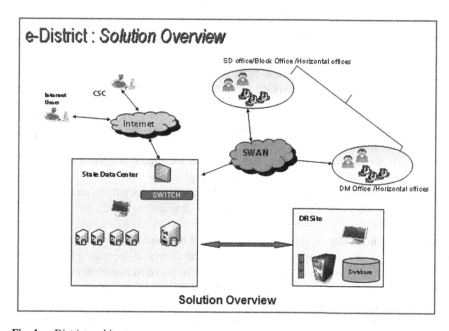

Fig. 1 e-District architecture

Sl.No.	Service Categories	Services
1	Certificates	Income
		Caste
		Domicile
2	RTI services & Grievance Redressal	RTI
		Grievance Redressal
3	Licenses	Blasting License
		Fatka / Fire Cracker License
		Fire Arm License
4	Industry Service	Issuance of EM-1
		Issuance of EM-2
		Credit Assistance under PMEGP
		EC Cum RC
		Subsidy
5	Pension	Old Age
6	Social Welfare	Integrated Child Development Scheme (ICDS)

Fig. 2 e-District services

- Common citizens can track the status without any limitation
- District administration have been automated like workflow have been implemented thus total lead time have been reduced
- All the related department have been integrated.

The major services which are delivering by 'e-District' project are shown in the Fig. 2 [3]

For issuing different certificates from 'e-District' web portal are following the below sequence of steps to facilitate services:

Step 1 Common citizens are required to attend at CSC or KIOSK with relevant required documents for submitting application towards availing Government services.
By the proposal of design, CSC and KISOK will be at nearest of citizen's residential location for easy accessibility.

Step 2 CSC or KIOSK operator will upload the relevant documents and will filled the necessary information in online application form on behalf of applicants.

Step 3 On successful submission of form, it will be routed to Government officers for review and approval. This application form may be routed to multiple officers where multi-level approvals are required in approval process.

Step 4 The application may be return back to CSC or KISOK operator if modification is required or get rejected.

Step 5 If everything get pass successfully then CSC or KIOSK operator will get download link for printing the certificates.

1.2 New Ideas in e-District Project Architecture

From the survey and literature, it has been pointed out very distinctly that 'e-District' project don't have any options to get/initiate the services from home which is adverse from objective of GI. GI want to reach out to all Indian citizens' door steps to provide Government services when it is rapidly advancing the technology space. With the growing population and increasing Smartphone penetration, India is going smart, mobile and digital.

Figure 3 is showing the mobile penetration in India [5]. The availability of low-price Smartphone and cut-throat price competition in internet tariffs are helping the huge population to become tech and net-savvy both form urban and rural India. Smartphone users find handy to use various mobile applications to execute various online transactions like Banking, online shopping, paying utility bills. A recent study says that the apps downloads in India is likely to increase from 1.56 billion in 2012 to 9 billion by 2015. The joint-study by Industry of India and Deloitte and the Associated Chambers of Commerce concluded that it would be a CAGR (Compound Annual Growth Rate) of around 75 %. It is also observed that the majority of apps are downloaded by the people between 16 and 30 years of age.

In this research paper, authors have proposed a solution framework for mobile apps for e-District application. Common citizens can use this mobile app to avail Government services.

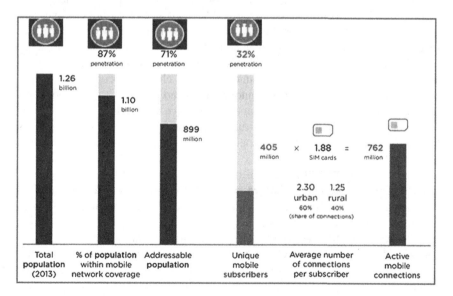

Fig. 3 Mobile penetration in India

2 Literature Review

GI has made IT strategy plan on 2006 to the interest of implementing ICT across the country for delivering Governance services under National eGovernance Plan (NeGP) The main moto of NeGP program is to "Make all Government Services accessible to the common people in his locality, through common service delivery outlets and ensure efficiency, transparency and reliability of such services at affordable costs to realize the basic needs of the common man". The 'e-District' project is one of the key citizens centric project among all others MMP's project and components of NeGP plan [2, 4, 6].

In a research book, Bagga and Gupta Piyush [7] have explained different best practices for implementing e-Governance project in India. In this Book, authors have done brainstorming for measuring different critical issues, findings proper capacity planning method, Users training and awareness, finding e-Governance projects challenges. According to authors, the most serious issues in Indian e-Governance implementation are (1) Building capacity for implementation, (2) Management of the complete project life cycle, (3) Enterprise architecture model, integration and interoperability, (4) Socio-political implementation of e-Governance.

As per GSM intelligence, 87 % of Indian Population is now within mobile coverage. Such a strong subscriber base along with one-to-one relationship of mobile device with subscriber allows Government and Government agencies to exploit this communication medium to provide innovative targeted services to its citizens. Mobile Marketing Association (MMA) has defined a set of guidelines and best practices that can be adhered to use the services of mobile operation for better Governance, especially in a developing country like India where other information channels have not grown that far [8]. Mobile communication and device can be extremely useful in bringing Government services closer to people and making communication more personalized. Innovation has no end and it is all about creativity and innovation to use this channel for effective use. Mobile services can be used to provide essential services to rural population with little investment. This channel can also be used to provide sophisticated services and application to urban subscribers [9].

Authors have explored different GI official sites and found that GI already has made framework for mobile Governance. To this endeavors, GI has launched Government official APP store intending to upload all the Government APPs in centralized repository. Already, lots of APPs have been uploaded for a common person downloads [10, 11]. The mobile framework is aimed to enable easy and round-the-clock access to the Government services by creating unique infrastructure including application development ecosystem for m-Governance in the country.

3 Research Methodology

This main backbone of this research tropic is to deliver new idea for enhancing the existing 'e-District' project in order to bring more usability and acceptability of the application. Authors have proposed to implement 'm-District' app for existing 'e-District' project through implementation and technical architectural design.

The main brainstorming of the research has been conducted by doing extensive past literature exploration and survey among IT Company's and CSC, KIOSK operators. CSC and KIOSK operators have helped to know the common citizen demands. Existing industrial project knowledge helped to get idea for designing the new mobile apps architecture in e-District project.

Android and Windows both kind of platform compatibility have been proposed for building two different APPS with the same functionality to support all category smart phone. HTML 5 using responsive UI will be as front end designing technology. JAVA, JQuery, JavaScript, Microsoft API and languages may be used to implement underline logic.

4 Mobile Application (m-District) for e-District

With the rapid expansion of 2G, 3G and now 4G mobile networks in India, mobile apps are getting popularity and are being implemented to deliver existing web based applications through user mobile devices for greater interest of user. Recently, the development of Mobile apps has got extra focus to transform our society to network society and this initiative is getting supported by evolution in network technology and mobile handsets both. Almost in every year, new generation network technology is get launched in market like 2G, 3G, 4G etc., on the other hand rapid changes is happening in mobile devices. Modern mobile devices have good display with high resolution, high capable in-build processor and high capacity memory including internal and external.

In this section, authors have provided an insight to develop mobile apps named 'm-District' to provide essential Government services as e-District is providing to both rural and urban population with little investment.

Figure 4 shows the solution overview of m-District. Considering the deep penetration of mobile and smartphones among Indian citizens, m-District will be a useful app to avail the e-District services on anytime and anywhere basis. This mobile application can be downloaded from Mobile Seva Appstore (https://apps. mgov.gov.in/index.jsp), hosted by ministry of Communication and Information Technology, GI. GI has hosted a number of useful mobile apps like mKrishan, mSwashtha, Sanskriti in Mobile Seva Appstore. Common citizens need to download these mobile application freely from Mobile Seva Appstore.

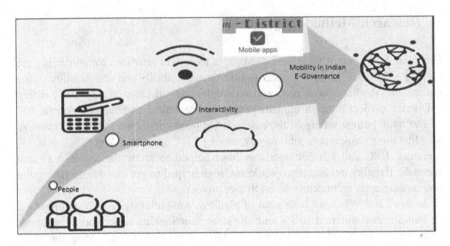

Fig. 4 Mobile apps (m-District) solution overview

Common citizens don't have to come to CSC or KIOSK or administrative offices to avail governance services. In the next section authors have detailed the Development and Implementation framework or step by step process for 'm-District'. It is a blend of well-known software development models like waterfall model, spiral model and prototyping models. In the Fig. 5, it has been outlines the step by step development and implantation framework. It is started with Apps Identification Phase which covers findings and planning of the idea to prepare functional and technical specification. The main objective of this phase is to brainstorm among a group of people and come up with new ideas related to 'm-District', a mobile apps, to provide services of e-District. We have conducted some brainstorming sessions among e-District experts, IT experts and Government representatives to gather the new ideas in details. The new ideas along with the each of the required functionalities minutely and handed over to the design team.

The next phase is Application Design Phase where several design related activities like feasibilities of various functionalities, selection of application platforms (like Android, windows 8), and detailed plan of work etc. are executed by the Application design team. Several discussions among e-District experts, Government representatives and application design team are required to be conducted so that all minute points are addressed in details. The outcome of this phase is Functional and Technical specification documents.

In Apps Development Phase, application coding is completed depending on the Technical specifications. Initially Development team tries to identify the parallel tasks which can be assigned and completed simultaneously. In this phase, initial apps prototype is created complying with the technical specification document. This initial prototype is revived as per the feedback of the e-District experts and Government representatives. It may require repeating the exercise until stakeholders are getting convinced by the developed prototype.

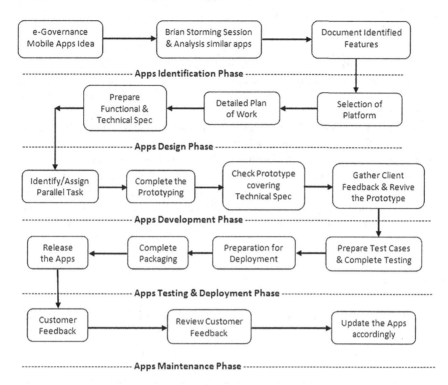

Fig. 5 Mobile apps (m-District) development and implementation framework

The next phase is Application Testing and Deployment Phase where the final prototype i.e. functionally robust application is tested and deployed for common users usage. All the functional test cases are prepared and tested minutely by the e-District experts before start deploying the system for the common users. For Deployment, there are some useful activities are required to complete like registering the application to Mobile Seva Appstore, checking rules and regulation for launching the application, refining the application by removing log files or comments, designing icons for the application etc.

The maintenance is a continuous process which continues with the use of the mobile application. User feedback should be collected time to time from common users and need to identify the required changes to be incorporated as change request (CR) or enhancement of existing functionalities. In this phase, continuous development and release deployment will be continued to introduce new functionalities, new user interfaces, performances tuning, security implementation etc.

5 Conclusion

In the era of network society, the acceptability of Mobile APPs and Smart Phone is growing rapidly among all rural and urban peoples due to enormous benefits. In this study, author have proposed a conceptual framework by utilizing Mobile APP and Smart Phone for delivering Government services at citizens' door step. By the implementation of proposed 'm-District' framework in India, GI will be able to minimize the GAP which found in between 'e-District' project service delivery model and real ground implementation. The 'm-District' will help to sustain 'e-District' project for long time because of better acceptability of the application to the common people. Also, it will help to grow network society in India and will ensure sustainability of e-Governance through transformation of our society in one steps a-head for accepting and implementing digital Governance in India.

References

1. Department of Electronics and Information Technology, M. o. (2006). National-e-governance-plan. Retrieved from http://deity.gov.in/content/national-e-governance-plan.
2. Department of Electronics and Information Technology, M. o. (2012). Home Page. Retrieved from NEGP: https://www.negp.gov.in/.
3. West Bengal Electronics Industry Development Corporation Limited. (n.d.). e_district. Retrieved from Webel India: http://www.webel-india.com.
4. Sanyal, M. K., Das, S., & Bhadra, S. K. (2014). E-District Portal for District Administration in West Bengal, India: A survey to identify important factors towards citizen's satisfaction. Procedia Economics and Finance, 510–521.
5. Alomari M. K., Sandhu, K., Woods, P. (2010), "Measuring Social Factors in E-government Adoption in the Hashemite Kingdom of Jordan", International Journal of Digital Society (IJDS), Volume 1, Issue 2, June.
6. FICCI. (2012). Project management roadmap for successful implementation of e-District projects. Kolkata: Federation of Indian Chambers of Commerce and Industry.
7. Bagga, R., & Gupta, P. (2009). TRANSFORMING GOVERNMENT: E-Governance initiatives in India. Hyderabad: The Icfai University Press.
8. Mitra R. K., Gupta M. P. (2007) "Analysis of issues of e-Governance in Indian police", Electronic Government, an International Journal 2007—Vol. 4, No. 1 pp. 97–125.
9. Kailasam R. (2010), "m-Governance …Leveraging Mobile Technology to extend the reach of e-Governance".
10. Department of Electronics and Information Technology, M. O. (2012). Framework for Mobile Governance from https://egovstandards.gov.in/sites/default/files/Framework_for_Mobile_Governance.pdf.
11. Department of Electronics and Information Technology, M. O. (2015). APPStore from https://apps.mgov.gov.in/popularapps.do?param=topapps.

A Semantic Framework to Standardize Cloud Adoption Process

Kumar Narander and Saxena Swati

Abstract The term Cloud computing defines one of the most popular internet based technology which benefits a business organization in terms of cost-affordability, transparency, efficiency, technical flexibility and working freedom. It enables a business organization to transfer its major share of computation, platform and software to a cloud so that it is managed by a cloud provider instead of an in-house team. Cloud transfer or adoption, however, has its own share of risks and challenges which are dealt in the present paper by using a semantic cloud adoption framework. This framework can act as an agreement between the provider and the adopting party so as to eliminate any future issue. A proper ans systematic follow-up of the presented framework ensures success and truct towards the cloud computing technology. Based on the presented framework, a sample comparison of various cloud providers is also given as an example for better understanding of one's preferences while using a cloud platform.

Keywords Cloud adoption · Framework · Challenges · Business organization · Risks

1 Introduction

Transferring business processes to cloud datacenters is the need of the hour. Streamlining work-load, cost-reduction, ease of management, global presence and a better quality control leads an organization to think about cloud adoption. There can be numerous reasons for cloud adoption by an organization, such as:

K. Narander (✉) · S. Swati
Babasaheb Bhimrao Ambedkar University (A Central University), Lucknow, India
e-mail: nk_iet@yahoo.co.in

S. Swati
e-mail: swatesaxena@gmail.com

© Springer India 2016   179
S.C. Satapathy et al. (eds.), *Information Systems Design and Intelligent Applications*, Advances in Intelligent Systems and Computing 434,
DOI 10.1007/978-81-322-2752-6_17

- Better insight and visibility: Today, Cloud computing is used extensively to analyze Big Data for better decision making, to share this geographically distributed data among different applications and to access and manage this ever-growing vast pool of Big data in cloud's storage for future need.
- Easy collaboration: To gain a competitive edge over other organizations in the market, cloud computing can be used to seamlessly integrate business development and operations. Also, cloud allows work to be accessed from anywhere, anytime so as to make collaboration easy and convenient.
- Support for various business needs: A wide variety of business needs such as storage, networking and data processing are supported by cloud datacenters.
- Rapid development of new products/services: Cloud provides a better and spontaneous understanding of the market, thus, enabling businesses to innovate new and better products and/or services.
- Visible and documented results: As a result of all the factors mentioned above, businesses adopting cloud are registering better market results such as cost reduction, increased efficiency and improved employee mobility.

Thus, in due time, cloud adoption, which was earlier considered to be a technical domain issue, soon became a business decision. However, cloud adoption comes with its share of challenges as well. Few of them are as follows:

- Losing Control: With the entire IT setup being taken by the third party cloud provider, many tasks are no longer handled by the organization, like recovering from a network issue, etc. Such scenarios lead to insecurity among businesses if there is no clear demarcation of responsibilities between an organization and a cloud provider.
- Security: Considered being a very burning issue, security leads to the fear of data theft especially in public clouds where there is a lot of vulnerability in data usage by different customers.
- Data Protection: Business data and specifically Big Data is being stored on geographically distributed cloud datacenters for better and timely access by different cloud customers. This, however, arises questions regarding policies that cloud providers employ to ensure the safety and compatibility of such valuable data.
- Service availability: Service downtime and poor performance are bigger challenges today when datacenters are expanding their storage and processing capacities. Communication and computation delays can have a substantial adverse effect on performance, thereby leading to drop in sales.
- Wrong choice of provider: Cloud adoption often begins with long-term contracts and compliance with a specific architectural platform. As a result, it becomes nearly impossible for businesses to switch cloud providers in the face of unsatisfactory performance or non-compliance to agreement.

The focus of this paper is to make cloud adoption a clean process while outlining clear responsibilities of both the parties, i.e., cloud provider and a cloud adopting business. We present a semantic framework which address all the risks and challenges mentioned above and provides its best possible solution. This framework establishes and maintains trust between the two involved parties and is crucial in the success of cloud computing technology.

The presented work is outlined in the following manner: Sect. 2 presents the related work, Sect. 3 introduces the cloud adoption framework and its modules, Sect. 4 gives a comparative study of various cloud providers based on the framework and Sect. 5 concludes the paper.

2 Related Work

Work presented in [1–5] have discussed various security issues in cloud computing. On one hand, Popović et al.'s [5] discussed cloud security controls and standards which focuses on the provider's interests and concentrates on cloud engineering, whereas Subashini and Kavitha [1] surveyed various security related risks specific to delivery models in the cloud arena. Also, a risk model for the cloud was developed by Kamongi et al. [4] which failed to cope with existing compliance standards. Statistical studies on the number of cloud providers who are adapting the cloud security standards is given in [6, 7] which raises questions on the cloud's capability of handling potential threats. Further it also outlines the feasible sources of concerns to cloud customers who have to make a choice among various cloud providers. The cloud computing reference architecture presented in [6, 8] categorizes the potential privacy and security policies with respect to the interest of a cloud provider. On the other hand, every aspect of the reference architecture follows the security compliance model. It is worth noting that these security controls are used to protect a cloud environment irrespective of the cloud delivery model and compliance standards are applied on these security controls. The IT compliance model [9] highlights network and IT infrastructure and electronic data processing. Compliance models implement rules and regulations across various components of IT to make them work harmoniously. Since an organization generally adopts a particular security control based on compliance models, it is imperative that transparency amongst the cloud service models, security controls and the compliance model is maintained so that consumers and end users receive a reliable data protection in the cloud.

3 Cloud Adoption Framework

This paper introduces a semantic framework to standardize the process of cloud adoption by large, medium or small scale businesses. It allows the cloud adopting party to fully assess the policies, practices and services offered by a cloud provider and helps in proper documentation which eliminates the chances of any doubt or regret once the agreement is done. Presented framework works well for all cloud

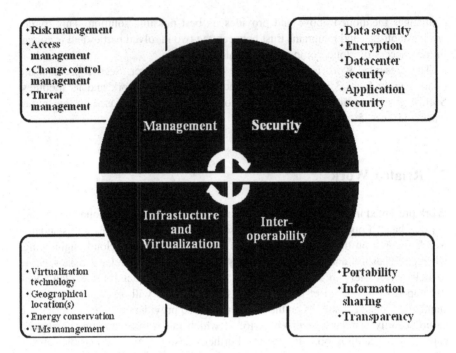

- Risk management
- Access management
- Change control management
- Threat management

Management

Security

- Data security
- Encryption
- Datacenter security
- Application security

Infrastucture and Virtualization

Inter-operability

- Virtualization technology
- Geographical location(s)
- Energy conservation
- VMs management

- Portability
- Information sharing
- Transparency

Fig. 1 Cloud adoption framework—an overview

delivery models: public, private and hybrid clouds. It also conforms to the standards of various service models in cloud computing, i.e., IaaS, PaaS and SaaS. Figure 1 given below shows the cloud adoption framework with its broad categories.

Figure 1 shows our presented framework which divides the whole adoption criteria into four broad domains, namely:

- **Management** This domain establishes an understanding of policies, procedures and measures taken by a cloud provider to manage business continuity, risks and possible threat situations. It also outlines how a provider is going to cope with change control management, operational maintenance and identity-access issues.
- **Security** This domain deals with various possible security factors such as data security measures ensured by the provider, datacenter security measure, application and interface security measures. It also deals with encryption and key lifecycle management issues within a datacenter as well as during communications among distributed datacenters.
- **Infrastructure and Virtualization** Infrastructure deals with human resource lifecycle management, maintenance of equipments and servers, cooling technologies employed, energy conservation methods used and providing complete knowledge to an organizaytion about the location of their data in case of multiple storage locations. Virtualization technology employed by a cloud

provider gives a clear idea to an organization about the level of performance expected. It also assess a cloud provider's capacity for future resources requirements.

- **Interoperability** This domain deals with portability and architectural issues that may arise in case of data transfer among different datacenters. Use of standard network protocols and maintaining audit logs is crucial for data integrity.

Above mentioned four domains and their sub-categories encompass all the major issues and risk factors which can come across while judging a cloud provider on its abilities and performance.

4 Comparison of Providers Based on the Framework

This section presents a comparative study of major cloud providers on different factors as outlined by the cloud adoption framework. Let us consider five hypothetical cloud providers as P1, P2, P3, P4 and P5, with their domain offerings as given below in Table 1.

Based on the example given in Table 1, we give a comparative analysis of these five cloud providers w.r.t. the adoption framework introduced in this paper.

Each Figs. 1, 2, 3, 4 and 5 gives a preferential indication towards a particular cloud provider on the basis of few selected sub-categories in the framework domains. For instance, Fig. 5 clearly shows cloud provider P3 as the preferred choice over other providers when network reliability is taken into consideration. Such analysis gives a clear insight to an organization about which cloud provider to attach its business to. A proper walk-through of the presented framework is useful

Table 1 Cloud providers and their services offerings (sample)

Framework domains		P1	P2	P3	P4	P5
Security	Downtime (h)	39.77	7.52	4.46	2.6	2.41
	Private online backup	No	No	Yes	No	Yes
Infrastructure and virtualization	Total instance options	24	35	15	10	14
	Max cores	40	32	49.12	16	52
Management	Resource over-provisioning	Yes	Yes	No	Yes	No
	Implementation complexity	Yes	No	No	Yes	Yes
Inter-operability	Network reliability (%)	28	54	67	9	33
	Cross-departmental analysis (%)	22	78	54	45	7

Fig. 2 Downtime (in hours)

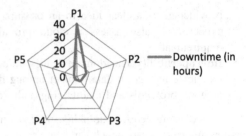

Fig. 3 Max cores

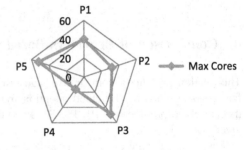

Fig. 4 Total instance options

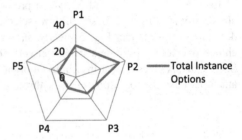

Fig. 5 Network reliability

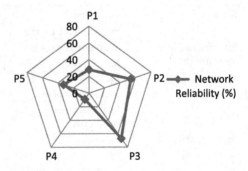

Fig. 6 Cross-departmental analysis

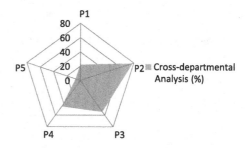

in better understanding between a cloud vendor and a business organization. Moreover, it eliminates any fear or apprehension in a business organization regarding cloud adoption challenges (Fig. 6).

5 Conclusion and Future Scope

Cloud computing is fascinating considering the numerous benefits it brings along with it. However, there is a lot of apprehension in the market regarding cloud adoption because of its inherent challenges and inflexibilities. To ease and streamline the process of cloud adoption, this paper presents a semantic framework which outlines various factors that must be considered between a cloud provider and a business organization. This framework touches every aspect of cloud adoption along with future scenarios and is suitable for all delivery and service models. A proper follow-up of the semantic framework brings trust among the parties involved and ensures success of cloud computing technology.

Future scope involves a systematic comparison of top cloud vendors using this framework to establish a detailed understanding of cloud adoption process in the real world.

References

1. S. Subashini, V. Kavitha, A survey on security issues in service delivery models of cloud computing, Journal of Network and Computer Applications, Volume 34, Issue 1, January 2011, Pages 1–11.
2. S. Ramgovind, M.M. Eloff, E. Smith, "The management of security in Cloud computing," Information Security for South Africa (ISSA), 2010, pp. 1,7, 2–4 Aug 2010.
3. T. Mather, S. Kumarswamy, S. Latif, Cloud Security and Privacy: An Enterprise Perspective on Risks and Compliance, O'Reilly Media, 2009.
4. P. Kamongi et al., Nemesis: Automated Architecture for Threat Modeling and Risk Assessment for Cloud Computing, ASE 2014.
5. K. Popović, Z. Hocenski, "Cloud computing security issues and challenges," MIPRO, 2010 Proceedings of the 33rd International Convention, pp. 344, 349, 24–28 May 2010.
6. NIST, NIST Cloud Computing Reference Architecture, 2011.

7. Cloud Security Alliance, 2013, The Notorious Nine: Cloud Computing Top Threats in 2013, pp. 8–21.
8. P. Mell, T. Grance, The NIST Definition of Cloud Computing, (Special Publication 800–145), 2011.
9. Privacy and data protection, Vol 7 Issue 4, IT compliance and IT security-Part 1, Dr. Jörg Hladjk, pp. 3–4.

Analysis of Efficiency of Classification and Prediction Algorithms (kNN) for Breast Cancer Dataset

G.D. Rashmi, A. Lekha and Neelam Bawane

Abstract Breast cancer is one of the regularly found cancer in India. In this paper the data mining techniques are used to provide the analysis for the classification and prediction algorithms. The algorithms used here are kNN classification algorithm and kNN prediction algorithm. The algorithms are used to find whether the tumour is either benign or malignant. Data sets are taken from the Wisconsin University database to find the success rate and the error rate. This is used to compare with the original success rate and the error rate.

Keywords Benign · Breast cancer · Classification · Malignant · Prediction

1 Introduction

Data mining is explained as the practice of deriving the information from huge sets of data. The data must be organized to extract useful information. It is an analytical process used to explore large amounts of data in search of some consistent patterns or common relationships [1]. Data mining techniques is used as a research tool for researchers to find, find patterns and relationships among large number of variables, which helps them to predict the result of a disease using the required datasets [2]. The paper provides analysis of Classification and Prediction data mining techniques.

Breast cancer is usually caused in women. It can occur both in male and female although breast cancer is rare in men [3]. Detection of the breast cancer in early

G.D. Rashmi (✉) · A. Lekha · N. Bawane
PES Institute of Technology, Bangalore, India
e-mail: deshpande.rashmi9@gmail.com

A. Lekha
e-mail: raoalekha@gmail.com

N. Bawane
e-mail: neelambawane@yahoo.com

© Springer India 2016
S.C. Satapathy et al. (eds.), *Information Systems Design and Intelligent
Applications*, Advances in Intelligent Systems and Computing 434,
DOI 10.1007/978-81-322-2752-6_18

187

stage is very helpful in identifying the type of tumour. The abnormal growth of extra cells are called tumour. There are two types of tumour malignant and benign. Benign tumour are not cancerous and they do not spread whereas malignant tumour are cancerous and invades to different parts of the body.

The data chosen for this experiment is the Breast Cancer dataset. The data used here is in the ARFF format also called as the attribute-relation file format. This dataset is taken from the Wisconsin University database. The dataset contains the attributes and the type of tumour. The experiment was carried out in MATLAB (matrix laboratory) which is a high-level language and interactional environment used by millions of engineers and scientists [4]. Matlab helps us to solve many numerical problems in small amount of time than the time it would take to write a program in a lower-level language. MATLAB helps in finer understanding and applying concepts in applications like engineering, living science and economics [5]. Data mining approaches are used for analysis. The techniques used here are the kNN Classification and kNN Prediction techniques.

1.1 Classification

Classification is one of the most common technique used in data mining. Classification is the process of identifying a set of functions which depicts and differentiate the data classes or concepts. The derived function is supported on the analysis of a set of training data where the class value is known [6]. The aim of the classification is to precisely classify the class value for each case in the data. Some examples are, a classification model could be used to identify loan applicants as low, medium, or high credit risks [7].

1.2 Prediction

Predicting the identity of one thing based purely on the description of another, related thing and it is used to derive the information from the data to predict trends and behaviour patterns. It is not necessarily the future events but just the unkowns [8].

The Sect. 2 is the literature survey which includes the details related to Breast Cancer, Classification and Prediction. Section 3 is the experimental setup which has the description of the dataset, the attributes that are present in the dataset and the algorithms used in the paper are specified. The Sect. 4 is about the methodology which contains the process and analysis of the algorithms. The final section is the conclusion which describes the end result.

2 Literature Survey

Breast cancer can be termed as a biologically heterogeneous disease due to the variations among the different types of lesions in terms of their size, growth and cellular composition. Clear difference between benign and malignant tissues becomes an important factor to cut down unnecessary biopsies [9]. To prevent biopsy which helps the physicians in differentiating between the benign and malignant tumuors, early diagnosis is required which must be accurate and a reliable procedure [2].

Mammogram was introduced in the year 1969 and it was the first digital step in detecting cancerous cell in breast tissues, it also has been a useful tool in early detection of breast cancer [10]. American College of Radiology (ACR), recommends annual mammography examinations should be carried out for all women who are above the age of 40 [11].

The most common aberrations found in mammograms are microcalcifications and tumours. The tumour can be either benign or malignant. They differ in shape, size, and contrast and are found in dense parenchymal tissues. The small difference between the malignant tissues and the background is the outcome of the poor radiologist contrast in mammograms with dense breast tissues. The patients who had undergone surgeries, radiations or breast implants makes the perception and automated detection of cancerous mammograms very difficult. This will result in missing 10–25 % of tumours even by the expert radiologists [12, 13].

Classification is one of the most important approach of data mining. It classifies the test data or the input data into its predefined classes. It is called as a supervised learning as the class values of the data are known. The objective of the classification is to build a classifier using some cases with one or many attributes to describe the objects or group of the objects. This classifier is then used to predict the group attributes of new inputs from the dataset based on the values of other attributes [14].

Prediction is the most important technique of data mining. It uses a set of classified examples to build a model which predicts the data and finds the relationship between the independent and dependent data [15]. The prediction techniques also discovers the relationship between independent data and relationship between dependent and independent data. There are different ways by which we can evaluate the performance of the algorithms [16].

3 Experimental Setup

3.1 Dataset

The Breast Cancer dataset is in ARFF (Attribute-Relation-File Format) and is taken from the Wisconsin University database. There are ten attributes for this Breast Cancer dataset including the class value (benign or malignant). The values of the attributes are in between 1 and 10 which tells the layers of penetration of the cancer into these cells [17]. There are a total of 683 instances out of which there are 444 benign instances and 239 malignant instances.

3.2 Attributes Description

- **Clump thickness** Non-cancerous cells are usually grouped in monolayers, whereas the cancerous cells are often grouped in multilayers [17, 18].
- **Uniformity of cell size/shape** Cancer cells differ in size and shape which is very important to determine whether the cells are cancerous or not [19].
- **Marginal adhesion** Normal cells stay together but cancer cells might lose this ability. Therefore the loss of adhesion is a sign of malignancy [17, 18].
- **Single epithelial cell size** It is similar to the uniformity of the cell size or cell shape. These cells are significantly enlarged and it can be a malignant cell [17, 18].
- **Bare nuclei** This term is used for nuclei which is not surrounded by cytoplasm that is the rest of the cell. They are mostly seen in benign tumours [17, 18].
- **Bland Chromatin** It describes the uniform texture of the nucleus seen in benign cells. The chromatin are usually larger in cancer cells [17, 18].
- **Normal nucleoli** The small arrangements seen in the nucleus is called as nucleoli. Nucleolus is very small even if it is visible in normal cells. The nucleoli will be more noticeable and more in number in cancer cells [17, 18].
- **Mitoses** The nucleus divides, this process is called cell division and it consists of four stages the prophase, metaphase, anaphase and telophase, normally resulting in two nuclei. However the abnormal cell division will cause tumour which may be cancerous [17].

3.3 k-Nearest Neighbour Classification (kNN) Algorithm

The most significant and simple classification method is the k-nearest-neighbour method. This can be used when there is little or no knowledge is known about the distribution of the data. kNN classification uses the distance method where distance between training data and testing data is calculated. The following steps are used for classification.

Step 1 The parameter k must be determined. Here k is the number of the nearest neighbours, k is a positive integer greater than zero.

Step 2 The next step is to calculate the distance between the test set and the training set using the Euclidean distance formula. $d = \sqrt{(x_2 - x_1)^2 + (y_2 - y_1)^2}$. The training set is specified with its respective class values as either benign or malignant.

Step 3   Once all the distance are calculated these distances are sorted and the nearest neighbours are determined using the kth minimum distance. All the selected neighbours which are less than or equal to k value are collected together that is if k = 3 then all the sorted distances greater than three are ignored.

Step 4   Once all the values are collected it is checked which class values is more in number. If the determined neighbours have more number of benign class then the given test set is classified to benign, if the neighbours have more number of malignant class then the given test set is classified as malignant [20].

3.4 k-Nearest Neighbour (kNN) Prediction Algorithm

In the k-nearest-neighbour prediction method, the training data set is used to predict the value of the class value for each member of a test dataset. The structure of the data consists of a various classes which are required for analysis, and also a number of additional attributes. For each row in the test data set that is the test set that has to be predicted, the k closest members that is the k nearest neighbours of the training data set is searched. The following steps are used for prediction.

Step 1   The parameter k must be determined. Here k is the number of the nearest neighbours, k is a positive integer greater than zero.

Step 2   The next step is to calculate the distance between the test set and the training set using the Euclidean distance formula. $d = \sqrt{(x_2 - x_1)^2 + (y_2 - y_1)^2}$. In kNN prediction technique the training set is will not have its class values specified.

Step 3   Once all the distance are calculated these distances are sorted and the nearest neighbours are determined using the kth minimum distance. All the selected neighbours which are less than or equal to k value are collected together that is if k = 3 then all the sorted distances greater than three are ignored.

Step 4   Once all the values are collected the total number of benign neighbours and the total number of malignant neighbours are calculated. Then the average of the benign values and the average of the malignant values are calcu-lated. The test set will be predicted to the class which has a larger average value [21].

3.5 Proposed System

The required arff files are selected for training set and testing set as input. These files are converted using weka interface, the application is executed and the result is displayed (Fig. 1).

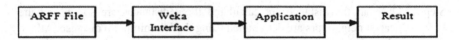

Fig. 1 Representation for proposed system

4 Methodology

The tools used in this application are Weka Interface 3.6 and MATLAB R2013a. The input given for the application is an .ARFF file. Weka interface is used for file conversion. A specified breast cancer dataset file should be used. The algorithms will not work for different kinds of arff files. The language used is MATLAB.

4.1 Input

The data used for analysis is taken from the Wisconsin University database. The algorithms accepts only the files which are in the arff format. Since matlab does not support the arff format weka interface is used for conversion. Once the file is converted it can be used in matlab for the execution. The user will select the required files for the input.

4.2 Process

The kNN classification algorithm will classify the test set to either malignant or benign. The prediction kNN algorithm will predict whether the test set is either malignant or benign. The output of the prediction algorithm will be same as that of the classification algorithm which will also help in verifying the result. The input for the testing set is chosen in random values. The success rate and error rate of the original test set is calculated. The success rate and error rate of random test values is computed and is checked with the original values.

4.3 Output

The output for both the classification and prediction techniques are Benign or Malignant. The comparison between the original data and the random data is shown in the form of graphs.

4.4   Case Study

There are seven input and output sets shown in the Tables 1, 2, 3, 4, 5, 6 and 7. For each set the success rate and the error rate is calculated. Respective graphs are shown for each input and output (Figures 2, 3, 4, 5, 6, 7 and 8).

Table 1 Observed input and output

Input 1
All 683 instances as input
Output 1
Malignant-35 %
Benign-65 %
Error rate-5 %
Success rate-95 %

Table 2 Observed input and output

Input 2
Random 25 data set as input for the testing set.
Output 2
Actual output Malignant-52 %, Benign-48 %
Expected output Malignant-35 %, Benign-65 %
Error rate-17 %, Success rate-83 %

Table 3 Observed input and output

Input 3
Random 25 data set as input for the testing set.
Output 3
Actual output Malignant-64 %, Benign-36 %
Expected output Malignant-35 %, Benign-65 %
Error rate-29 %, Success rate-71 %

Table 4 Observed input and output

Input 4
Random 50 data set as input for the testing set.
Output 4
Actual output Malignant-48 %, Benign-52 %
Expected output Malignant-35 % Benign-65 %
Error rate-13 %, Success rate-87 %

Table 5 Observed Input and Output

Input 5
Random 50 data set as input for the testing set.
Output 5
Actual Output Malignant-49 %, Benign-51 %
Expected Output Malignant-35 % Benign-65 %
Error rate-14 %, Success rate-86 %

Table 6 Observed Input and Output

Input 6
Random 300 data set as input for the testing set.
Output 6
Actual Output: Malignant-45 %, Benign-55 %
Expected Output: Malignant-35 %, Benign-65 %
Error rate-10 %, Success rate-90 %

Table 7 Observed Input and Output

Input 7
Random 300 data set as input for the testing set.
Output 7
Actual Output Malignant-48 %, Benign-52 %
Expected Output Malignant-35 %, Benign-65 %
Error rate-13 %, Success rate-87 %

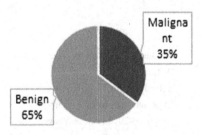

Fig. 2 Class distribution of the original data

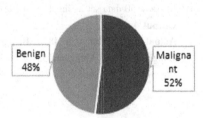

Fig. 3 Class distribution of the random data

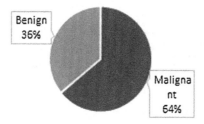

Fig. 4 Class distribution of the random data

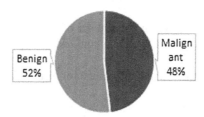

Fig. 5 Class distribution of the random data

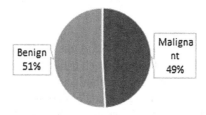

Fig. 6 Class distribution of the random data

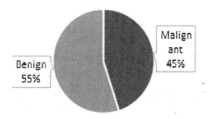

Fig. 7 Class distribution of the random data

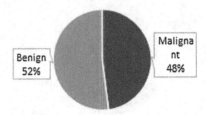

Fig. 8 Class distribution of the random data

4.5 Analysis

- The accuracy rate is poor when the number of instances taken as input is less.
- The accuracy improves and becomes consistent as the number of instances increases.
- When the inputs are random and the number of instances is very less the results are not very accurate.
- When the inputs are random and the number of instances is large the results does not match the expected results for all instances.

4.6 Result

The kNN algorithm has to be improved to work for small number of instances. The presence of outliers cannot be identified specifically through this algorithm. When compared to Naïve Bayes Classifier algorithm whose success rate increases as the number of instances increases, this algorithm does not have a consistent increase or decrease in efficiency. Both Naïve Bayes and kNN algorithm have a low success rate for small number of instances.

5 Conclusion

The paper provides analysis of Classification and Prediction data mining techniques. These results are based on the Wisconsin University Breast Cancer dataset. The kNN classification algorithm shows that the success rate is around 80–90 % and the error rate is 10–15 % when the data is randomized. While using the kNN prediction algorithm the success rate is around 80–90 % with the error rate being 10–15 %. Both the algorithms show the presence of outliers but does not identify the same. There is scope for improvement in the algorithms for various other combinations of inputs and to increase the success rate.

References

1. Data Mining Techniques, http://documents.software.dell.com/Statistics/Textbook/Data-Mining-Techniques.
2. "Using Data Mining Techniques for Diagnosis and Prognosis of Cancer Disease" International Journal of Computer Science, Engineering and Information Technology (IJCSEIT), vol. 2, no. 2, pp 55–66, April 2012, http://arxiv.org/.
3. National Breast Cancer Foundation, http://www.nationalbreastcancer.org/.
4. MATLAB-The language of Technical Computing, http://in.mathworks.com/products/matlab/.
5. Learning MATLAB7, MATLAB & SIMULINK by MathWorks ISBN: 0-9755787-6-6.
6. An Efficient Classification Approach for Data Mining by Hem Jyotsana Parashar, Singh Vijendra, and Nisha Vasudeva, International Journal of Machine Learning and Computing, vol. 2, no. 4, August 2012.
7. Classification, http://docs.oracle.com/cd/B28359_01/datamine.111/b28129/classify.htm#i100 5746.
8. Data mining prediction by Kevin Swingler (http://www.cs.stir.ac.uk/courses/CSC9T6/lectures/1%20Data%20Mining/4%20-%20Prediction.pdf).
9. "A Robust Polynomial Filtering Framework for Mammographic Image Enhancement from Biomedical Sensors," IEEE Sensors Journal, vol. 13, no. 11, pp. 4147–4156, November 2013.
10. "An Approach on Sequential Data Mining Algorithm for Breast Cancer Diseases Management" International Journal of Advanced Research in Computer Science and Software Engineering, vol. 3, issue 9, pp. 351–355, September 2013.
11. "Technical Advancements to Mobile Mammography using Non-Linear Polynomial Filters and IEEE 21451-1 Information Model," IEEE Sensors Journal, vol. 15, no. 5, pp. 2559–2566, May 2015.
12. "A Robust Approach for Denoising and Enhancement of Mammographic Breast Masses," International Journal on Convergence Computing, vol. 1, No. 1, pp. 38–49, 2013.
13. "A robust approach for denoising and enhancement of mammographic images contaminated with high density impulse noise" Int. J. Convergence Computing, vol. 1, no. 1, 2013.
14. "Data Mining Classification Techniques Applied for Breast Cancer Diagnosis and Prognosis", Indian Journal of Computer Science and Engineering, vol. 2, no. 2, pp. 188–195, April–May 2011, http://ijcsit.com/.
15. "Prediction of Higher Education Admissibility using Classification Algorithms", International Journal of Advanced Research in Computer Science and Software Engineering, vol. 2, no 11, pp. 330–336 November 2012.
16. "Data Mining Techniques for Weather Prediction: A Review", International Journal on Recent and Innovation Trends in Computing and Communication, vol. 2, no. 8, pp. 2184–2189, August 2014.
17. Prediction of Breast Cancer by Deepa Rao and Sujuan Zhao, Willammette University (Data Description), http://gsm672.wikispaces.com/Prediction+of+Breast+cancer.
18. Attribute Description, http://www.mcrh.org/Adhesions/anyone_know_meaning_following_terms_breast_cancer_analysis_2010124.htm.
19. Attribute Description—ijcit.com.
20. KNN Numerical Example, http://people.revoledu.com/kardi/tutorial/KNN/KNN_Numerical-example.html.
21. KNN for Smoothing and Prediction, http://people.revoledu.com/kardi/tutorial/KNN/KNN_TimeSeries.htm.

A Novel Genetic Algorithm and Particle Swarm Optimization for Data Clustering

Malini Devi Gandamalla, Seetha Maddala and K.V.N. Sunitha

Abstract Clustering techniques suffer from fact that once they are merged or split, it cannot be undone or refined. Considering the stability of the Genetic Algorithm and the local searching capability of Swarm Optimization in clustering, these two algorithms are combined. Genetic Algorithms, being global search technique, have been widely applied for discovery of clusters. A novel data clustering based on a new optimization scheme which has benefits of high convergence rate and easy implementation method is been proposed were in local minima is disregarded in an intelligent manner. This paper, we intend to apply GA and swarm optimization (i.e., PSO) technique to optimize the clustering. We exemplify our proposed method on real data sets from UCI repository. From experimental results it can be ascertained that combined approach i.e., PSO_GA gives better clustering accuracy compare to PSO-based method.

Keywords Genetic algorithms · PSO · Accuracy · CPU time

1 Introduction

Data mining is a detection process that allows users to comprehend the substance and relationships amid the data. Data architect/designer punctiliously defines entities and relationships from operational or data warehouse system. The conclusion of data

M.D. Gandamalla (✉) · S. Maddala
Department of Computer Science and Engineering, GNITS,
500008 Hyderabad, India
e-mail: gmalini12@gmail.com

S. Maddala
e-mail: seetha.maddala@gmail.com

K.V.N. Sunitha
College of Engineering for Women, BVRIT Hyderabad,
500090 Hyderabad, India
e-mail: k.v.n.sunitha@gmail.com

© Springer India 2016   199
S.C. Satapathy et al. (eds.), *Information Systems Design and Intelligent Applications*, Advances in Intelligent Systems and Computing 434,
DOI 10.1007/978-81-322-2752-6_19

mining can be used to intensify the efficacy of performance from the users. Data mining uses various techniques such as inductive logic programming, pattern recognition, image analysis, bioinformatics, spatial data analysis, decision support systems etc. for this kind of analysis. Among these methods, clustering is the most significant and extensively used method.

Clustering is utmost prevalent technique that tries to isolate data into dissimilar groups such that same-group data points are alike in its characteristics with respect to a referral point, where as data points of different-groups varies in its characteristics. Such dissimilar groups are labeled as clusters. These clusters consist of several analogous data or objects relating to a referral point [1].

Darwin's theory inspired Genetic algorithms development. GA is evolved to resolve solution to this problem. Algorithm begins with a set of solutions (chromosomes) known as population. Outcomes from a population are picked and used to structure a new population. This is aggravated by optimism, and the new population will enhance the old. Chosen solutions are used to form new solutions (offspring) which are chosen according to their fitness values—the more opposite will lead to more chances of reproduction [2].

Particle Swarm Optimization (PSO) is form of population based algorithm. This algorithm can be implemented in achieving a self evolution system in fish schooling behavior or bird flocking. Even though the search process is not random it searches for optimum solution. According to the diverse problems, it decides the search method by the fitness function. PSO needs smaller parameters to decide the solution, compared to other evolutionary techniques [3]. PSO has a steady convergence spirit with great computational efficiency and is easily implemented. Genetic algorithm, were in each 'individual' of the generation stands for a feasible solution to the problem, coding distinct algorithms/parameters should be evaluated by a fitness function. GA operators cause mutation (change of random position of the chromosome) and crossover (change of slices of chromosome between parents). An extremely capable evolutionary based clustering method by PSO is provided to find the near optimal solution in search space to trounce the previous problems [4].

In original data, clustering aims at identify and mining significant groups. A novel particle swarm optimization with genetic algorithm is an optimization task which runs after the genetic algorithm, later terminates in order to improve the value of fitness function. From genetic algorithm, final point is taken as initial point in this function.

2 Genetic Algorithm

Conceptual representations of realistic solutions are represented by GA endeavor with a population of individuals. Solution of the performance is represented and measured by individuals assigned fitness function. Improved result yields superior fitness value. This population evolves to enhanced solutions. The progress starts with population of absolutely unsystematic personnel and iterates into new cohort.

Fitness function of every individual is measured in each generation. Individuals elected stochastically to form a new fitness function from current population (based on their fitness), and customized by way of operators transformation and crossover [5].

For specific problems, clarification of GA represents as an individual chromosome. After which it characterizes component of realistic result of the crisis. Consequently, search space cleans solution space were in each viable solution is defined by a distinct chromosome. To form the preliminary populations a set of chromosomes are arbitrarily selected from the search space. Subsequently, sorting of individuals is done in a spirited method, depending on their fitness which is gauged by an overt objective function value. To achieve a latest cohort of chromosomes the genetic operators like mutation, crossover and selection are then applied in sequence. Projected quality of entire chromosomes is enhanced compared to earlier cohort. Process needs to be recurred till terminating condition is satisfied, and hence concluding solution is reached with the finest chromosomes of the last generation [6].

2.1 Genetic Algorithm Operators

The fundamental genetic algorithm operators are crossover and transformation. These two operators work together to discover and take advantage of the search space by creating latest variants in the chromosomes. It is confirmed that mutation operator play the similar important role as that of the crossover.

2.2 Crossover

Fundamental role of the genetic algorithms is that crossover (recombination) plays best design and implementation of vigorous evolutionary systems. Through GA, individual are epitomized by fixed-length strings and crossover works on pairs of individuals (parents) to construct new strings (offspring) by exchanging segments from the parent strings. Traditionally, crossover point's count (which decides no. of fragments exchanged) has been stable at a very low constant value of 1 or 2 [7].

2.3 Mutation

Mutation is a common operator which is used to evaluate and preserve variety in the population by finding latest points in the search space. A random change is made to the values of some locations in the chromosome, when a chromosome is chosen for mutation [8].

Tournament selections and roulette-wheel selection are most commonly used selection mechanism. Deterministic tournament selection is used in this paper, which chooses arbitrarily one with top fitness is selected from two individuals from population. This benevolent method is very effective and simple to execute. Applying the genetic algorithm to precise optimization problem, need to stipulate a design of solutions in individuals, fitness function and also classical operators like crossover and transformation [9]. These subsequent stages summarize the genetic algorithm functioning:

1. Genetic algorithm instigates with constructing an unsystematic initial population
2. Then it constructs a series of innovative populations. On every step, it uses individuals in the current generation to construct the next population.

   a. By computing the fitness value, it tallies each member of the current population.
   b. To recondition them to more usable range of values, it gauges the raw fitness scores.
   c. Some individuals in the current population which have inferior fitness are selected as best. Hence finest individuals are conceded to next population.
   d. Then, it creates children from parents. By making random changes to a single parent—mutation or by uniting the vector entries of a pair of parents-crossover, children are produced.
   e. It replaces present population with children to form the next cohort.

   Generations When number reaches value of generations it stops working.
   Time limit After running for some seconds which equals to time limit, the algorithm stops.
   Fitness limit When value of fitness function at supreme point in the present population is minimal or equals fitness limit, algorithm stops functioning.

3 PSO Technique

Population is called swarm, which is a population-based optimization technique. Stochastic optimization method based on swarm intelligence is followed by PSO. Basic idea is every particle signifies a dormant answer which updates according to its own experience and also that of its neighbors. Individuals in as swarms, advance to best experiences of its neighbors, present velocity and the prior experience. By adjusting trajectories of moving points in m-dimensional space, the PSO searches for problem domain [10].

Optimal solution can be governed by the interactions of individual particles which are in motion. The connectivity between the individual particles is established by the position, velocity, best performance of individuals and their neighbors.

3.1 Notations for PSO

1. The position of the ith particle in D-dimensional object space with a swarm size n, is projected, xi = (xi1, xi2, ..., xiD).
2. The best previous position (i.e., the position giving the best function value pBest) of the ith particle is recorded and represented by pi = (pi1, pi2, ..., piD).
3. The change in position (velocity) of the ith particle is Vi = (Vi1, Vi2, ..., ViD).
4. The position of the best particle of the swarm (i.e., the particle with the smallest function value) is denoted by index gBest.
5. The particles are then manipulated according to the following equations.

$$V_{id}(t+1) = V_{id}(t) + C_1 * \text{rand} * (\text{pBest}(t) - x_{id}(t)) + C_2 * \text{rand} * (\text{gBest}(t) - x_{id}(t));$$

$x_{id}(t+1) = x_{id}(t) + V_{id}(t+1)$ where d = 1, 2, ..., D and i = 1, 2, ..., n. V: Velocity, C1, C2: positive constant parameters = 1.49 and rand: random numbers between (0, 1).

4 Genetic Algorithm with PSO

Evolutionary techniques have the following strategy:

1. Initial population is generated randomly
2. Considering the fitness function value for each object, which depends on the distance to the optimum solution?
3. Based on fitness function values reproduction of the new population.
4. If optimum solution is met, then stop. Otherwise go back to step 2.

This procedure means that PSO distributes several common points with GA. These two algorithms begin with a set of an arbitrarily formed population, which has fitness value to evaluate the population. Both techniques revise the given population and look for the most favorable solution with unsystematic techniques. Both systems do not guarantee success. The flip side, PSO will not have genetic operators like crossover and mutation. Update the particles among themselves with their internal velocity. Memory utilization is considered vital to this algorithm [11].

Information sharing is notably dissimilar when compared with genetic algorithm in PSO. In GAs, sharing of information among chromosomes takes place which triggers a complete population movement resembling single group movement to best possible solution space. With PSO, information can be passed to other objects by gbest (or lbest) solution. The progress expects the finest result. In local version of most cases, all the particles tend to converge to the best solution rapidly when compared with GA. Degrade in regulation of its speed and direction indicates PSO algorithm suffering from the partial optimism [12].

The novel evolutionary technique unites reward of GA and PSO to offer an enhanced PSO_GA which is a hybrid method. This enhanced hybrid PSO_GA technique is evaluated with PSO and GA methods with a set of standard mathematical functions [13]. This enhanced hybrid PSO_GA technique is superior to individual evolutionary techniques. To make advantage of both PSO and GA, we combined the both algorithms. Thereby, efficient results are obtained.

Procedure for PSO with GA

1. Apply GA on the datasets for clustering.
2. The results are passed to the PSO for execution.
3. PSO is applied on the output obtained from GA.
4. Repeating step 2 and 3 until the convergence is achieved

5 Results and Discussion

In this paper, the experimental results on four real datasets were discussed. Basically two algorithms were implemented in this paper, i.e., PSO and Hybrid PSO_GA. Results were behavior on four datasets from the UCI machine learning repository: Iris, Wine, Lenses and Hayes-Roth. These results show the relationship between PSO and PSO_GA. The properties of the dataset are described as:

5.1 Data Sets

Iris plant data set: Which is comprised of 150 data points, include 50 instances of all the three types of iris plant [14].

Wine dataset: UCI Machine Repository is source from which Wine dataset is derived, consists of 178 data points and holds information for types of wines matching 14 integer and real attributes. Every given wine data point is confined into 3 classes.

Lenses dataset: UCI Machine Repository is source from which Lenses dataset is Hayes-Roth dataset is derived, consists of 160 instances along with 5 categorical attributes. Every Hayes-Roth data point categorized to 3 classes.

5.2 PSO

Once the data sets are given as input to the algorithms, a cluster center matrix comes as output along with the CPU time and object function value. In order to show the results there has been a comparison done between the OFV and CPU time with the

Table 1 Analysis of PSO algorithm based on OFV for four datasets

Datasets	OFV
Iris	−107.0286
Wine	−154.3372
Lenses	−186.7309
Hayes-Roth	−186.3407

Table 2 Analysis of PSO algorithm based on time for four datasets

Datasets	Time
Iris	33.1000 s
Wine	33.342 s
Lenses	33.199 s
Hayes-Roth	33.345 s

clusters. By taking various cluster values the comparison is done between cluster and time and also between cluster and OFV for PSO algorithm for all the datasets.

In Tables 1 and 2 and Figs. 1 and 2 depicts that the PSO technique implemented on various datasets with OFV and time. When compared to all datasets, lenses datasets give low fitness and time with PSO. The PSO based clustering techniques suffer from various limitations when dealing with multidimensional data, which may lead to inaccurate solution. Clustering techniques are applied in the form of smart hybrid systems, where more than two techniques are judiciously combined to exploit the strong point of all combined algorithms.

5.3 PSO_GA

In Tables 3 and 4 and Figs. 3 and 4 it can be stated that iris, wine datasets exhibits the proposed, GA_PSO is better algorithm in terms of OFV and it takes slightly more time than PSO but the difference in OFV compensates the time factor. As the

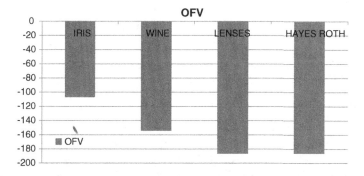

Fig. 1 Comparison between the OFV on four datasets with PSO

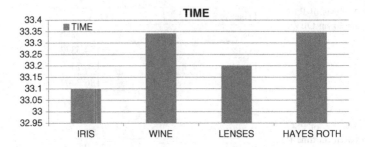

Fig. 2 Analysis of PSO_GA algorithm on four datasets based on Time

Table 3 Analysis of PSO_GA algorithm on four datasets based on OFV

Datasets	PSO	GA_PSO
Iris	−107.8089	−147.2694
Wine	−154.3372	−181.9505
Lenses	−186.7309	−51.3864
Hayes-Roth	−186.3407	−32.4162

Table 4 Analysis of PSO_GA algorithm on four datasets based on time

Datasets	PSO	GA_PSO
Iris	32.928	34.084
Wine	33.342	35.6567
Lenses	33.199	31.33067
Hayes-Roth	33.345	31.3485

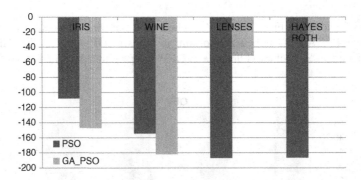

Fig. 3 Analysis of PSO_GA algorithm on four datasets based on OFV

Lenses dataset having less number of instances and Hayes-Roth dataset is having repeated values in their instances, GA_PSO is not suitable. Faster convergence rate because when stagnation of PSO occurs, even though the solution is worse GA

Fig. 4 Analysis of PSO_GA algorithm on four datasets based on time

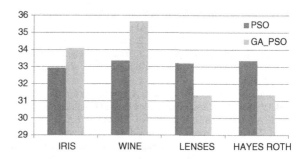

diversifies the particle position which makes PSO_GA extra flexible and robust. Distinct to standard PSO, PSO_GA is extra consistent in giving better fitness solution with reasonable computational time. Experimental results are examined with benchmarks functions and outcomes specify proposed PSO_GA model betters the standard PSO.

6 Conclusion

Once a merge or a split is committed, it cannot be undone or refined that is the main limitation of clustering techniques. To prevail over the given merged data to a high-level view, general grouping of dimensions into like behaviors. Global optimization algorithm mainly indulges Particle swarm optimization (PSO). The PSO based clustering techniques suffer from various limitations when dealing with multidimensional data, which may lead to inaccurate solution. Solving optimization problems on the natural selection is the best method to defeat this GA. GA endeavours best point in population which later reaches to optimal solution with rapid convergence rate. The end results are the fusion of PSO_GA whose performance is notably enhanced when compared to former algorithms for these data sets.

References

1. H. Stahl, "Cluster Analysis of Large Data Sets", In W. Gaul and M. Schader, editors Classification as a Tool of Research, pp. 423–430, Elsevier, Amsterdam (1986).
2. Santhosh Peddi, Alok Singh, Grouping Genetic Algorithm for Data Clustering, Springer Berlin Heidelberg, Swarm, Evolutionary, and Memetic Computing, Lecture Notes in Computer Science, Volume 7076, pp. 225–232 (2011).
3. Jayshree Ghorpade-Aher, Vishakha Arun Metre, PSO based Multidimensional Data Clustering: A Survey, International Journal of Computer Applications (0975–8887), Volume 87, No. 16, (2014).
4. K. E. Parsopoulos and M. N. Vrahatis, "Recent approaches to global optimization problems through particle swarm optimization," Natural Computing. An International Journal, vol. 1, no. 2–3, pp. 235–306(2002).

5. A. Sibil, N. Godin, M. R'Mili, E. Maillet, G. Fantozzi Optimization of Acoustic Emission Data Clustering by a Genetic Algorithm Method, Springer-Verlag, Journal of Nondestructive Evaluation, Volume 31, Issue 2, pp 169–180(2012).
6. Lleti, R., Ortiz, M.C., Sarabia, L.A., et al.: Selecting variables for k-means cluster analysis by using a genetic algorithm that optimises the silhouettes. Anal. Chim. Acta 515, pp. 87–100 (2004).
7. AnutoshPratap Singh, Jitendra Agrawal, Varsha Sharma An Efficient Approach to Enhance Classifier and Cluster Ensembles Using Genetic algorithms for Mining Drifting Data Streams, IJCA (0975–8887), Vol. 44, No. 21(2012).
8. M. Imran, H. Jabeen, M. Ahmad, Q. Abbas, and W. Bangyal, "Opposition based PSO and mutation operators," in Proceedings of the 2nd International Conference on Education Technology and Computer (ICETC '10), pp. V4506–V4508 (2010).
9. Tansel Özyer, Reda Alhajj, Parallel clustering of high dimensional data by integrating multiobjective genetic algorithm with divide and conquer, Springer US, Vol. 31, pp. 318–331, 2009.
10. M.Imran, R. Hashim, and N. E. A. Khalid, "An overview of particle swarm Optimization variants," Procedia Engineering, vol. 53, pp. 491–496 (2013).
11. Painho, M., Fernando, B.: Using genetic algorithms in clustering problems. In: Proceedings of the 5th International Conference on GeoComputation (2000).
12. Garai, G., Chaudhury, B.B.: A novel genetic algorithm for automatic clustering. Pattern Recognition Letters 25, 173–187 (2004).
13. Jones D, Beltramo, Solving partitioning problems with genetic algorithms. In Proceedings of the Fourth International Conference on Genetic Algorithms, pp. 442–449 (1991).
14. Nirmalya Chowdhury, Premananda Jana Finding the Natural Groupings in a Data Set Using Genetic Algorithms, Springer Berlin Heidelberg, Applied Computing, Lecture Notes in Computer Science Volume 3285, pp. 26–33(2004).

Design Issues of Big Data Parallelisms

Koushik Mondal

Abstract Data Intensive Computing for Scientific Research needs effective tools for data capture, curate them for designing appropriate algorithms and multidimensional analysis for effective decision making for the society. Different computational environments used for different data intensive problems such as Sentiment Analysis and Opinion Mining of Social media, Massive Open Online Courses (MOOCs), Large Hadron Collider of CERN, Square Kilometer Array (SKA) of radio telescopes project, are usually capable of generating exabytes (EB) of data per day, but present situations limits them to more manageable data collection rates. Different disciplines and data generation rates of different lab experiments, online as well as offline, make the issue of creating effective tools a formidable problem. In this paper we will discuss about different data intensive computing tools, trends of different emerging technologies, how big data processing heavily relying on those effective tools and how it helps in creating different models and decision making.

Keywords Data intensive computing · High dimensional data · Scalable framework · Data science · Machine learning · Semi-stochastic · X-informatics · Big data

1 Introduction

Different domains of computing seek to understand the digital data of the all-relevant fields in the light of information science to design a model, which fits the available data, with the help of necessary scientific tools. The paradigms of Science move through different phases viz. empirical, theoretical and computational. Now we are in the era of data exploration and are trying to create a unified

K. Mondal (✉)
Indian Institute of Technology, Indore, India
e-mail: gemkousk@gmail.com

© Springer India 2016
S.C. Satapathy et al. (eds.), *Information Systems Design and Intelligent Applications*, Advances in Intelligent Systems and Computing 434,
DOI 10.1007/978-81-322-2752-6_20

209

framework for theory, experiment or observation, modelling or simulation of theory and data driven experiments. In the present scenario, emerging technologies focusing on productivity growth with the help of computational and data exploration paradigms of scientific research, as described in [1], depicted in Fig. 1. The main aim of data science or big data modelling is to gain insights into data through computation, statistics and visualization.

A data scientist is a combination of statistician, programmer, storyteller, coach and artist. Volume, Velocity, Variety, Veracity and Value—these five "V" has often been used to describe the issues at hand in big data world. We are now look into the data fairly late in the pipeline after it has processed through different computational methods. Different tools, techniques and technologies are used in processing scientific experiments and thus we can easily termed the computational science as data intensive computing. Big data referred to the data driven analysis approach on these "big" amount of data to find out how to achieve technical and scientific values that will help us to take crucial decision of providing better services and generate economic value. As we are using these datasets in a machine-learning framework thus we have see that it must be fit in some suitable optimization algorithm, in-memory or out-of-the-core capacity of the machine to handle large chunk of data to process and thus, huge computational requirements are there, to analyze those large scientific datasets.

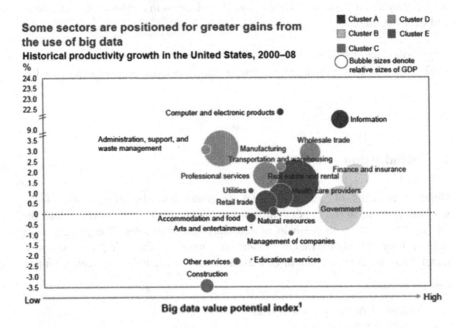

Fig. 1 Productivity growth due to use of big data [1]

2 Related Work

In recent years, many people engaged in developing frameworks for data intensive computing applications for scientific datasets. We are all now witnessed three-tier development viz.

a. Distributed platform level that includes accelerating devices for data processing, storage devices optimization and security and privacy management;
b. Business Intelligent level that includes Analysis, Data management and Visualization; and
c. Application design optimization that includes optimized software stack for providing business solutions.

These three tiers effectively build the stack of frameworks for data storage e.g. distributed file system such as Hadoop distributed File System [2], data processing such as MapReduce [3] and data manipulation using query language such as Pig [4] and Hive [5]. Some vertex-centric models are also used in various graph based distributed frameworks such as Hama [6], GraphLab's GraphChi [7], ApacheGiraph [8]. These frameworks are also help to build domain specific programming libraries such as Mahout [9] and Pegasus [10]. The Mahout library gives a collection of machine-learning algorithms and the Pegasus framework offers collections of graph mining algorithms under Hadoop for large graphs. Some detailed theory for the MapReduce framework is given in [11]. The operators in MapReduce model i.e. "map" and "reduce" operators are based on classical list-manipulation primitives introduced in LISP. The MapReduce environment offered that ability to correctly express computationally iterative algorithms which is generally involved in large number of data movement. Bu et al. [12] proposed Haloop to solve this problem. Haloop is able to handle iterative environment by engaging superior task scheduler, which is able to cache frequent data used in the environment. Even after this MapReduce framework unable to handle large collections of data intensive, iterative algorithms. Bulk Synchronous Parallel framework (BSP) [13], developed by Valiant, aimed to give a theoretical base for parallel designed algorithms which is able to calculate communications and cost of synchronization with computation for total model cost. BSP framework usually follows three iterative steps viz. computational value, communication value and synchronization.

The underlying mathematical models those are available in the machine learning domain for efficient parallel, iterative computation are described in [12, 14, 15]. The new challenges that driven the recent research is to model such kind of hypothesis functions which will able to predict unseen data accurately. The new trend demands criteria like—large-scale, stochastic, non-linearity, parallel, good generalized environment that have opened a wide area for research. The broad two mathematical categories for designing large-scale parallel and iterative mathematical machine learning models are batch and stochastic. In general, batch method selects a huge number of training datasets to build an objective function to minimize sum of the error terms using a standard gradient-based method.

3 Parallel Design Approach

Most of the design approach described in Sects. 3.1 and 3.2 are already in [16, 17]. However, we recall the design approaches for the completeness and easy understanding of the results. We will discuss in three phases: First, how to design mathematical model for parallel environment; second, programming models available as of now and lastly different computational mechanisms.

3.1 Mathematical Model

The mathematical model for parallel environment is quiet different from sequential in nature. It is difficult to parallelize stochastic gradient model compare to batch methods [18]. Thus we have to look for algorithms that will work in both the environments efficiently. The new generation algorithms developed an environment to encounter both, batch and stochastic, in some interesting manner. It is sometimes called 'semi-stochastic'. A typical implementation chooses a single training set (x_i, z_i) in arbitrary manner and revised the parametric values of the model through iterative way. The size of the datasets, $|S_k|$, gradually increased through iterative manner, which is given by

$$w_{k+1} = w_k - \alpha_k \frac{1}{|S_k|} \sum_{i \in S_k} \nabla l(h(w_k; x_j), z_j) \tag{1}$$

Here α_K is a step-length parameter, l is the loss function that calculates the derivatives between the data and the prediction where w_K is a random variable. This method emulates the stochastic gradient method at first but evolve toward batch gradient model. This approach enjoys work complexity as well as that of stochastic gradient model provided that $|S_k|$ increases geometrically with k.

3.2 Programming Models

We want to bring parallelism in a parallel programming environment either in terms of data parallelism or task parallelism with help of communication channels, concurrency, synchronization, locality/affinity management. These parallelisms offered us low level, control-centric, easy to implement environment. These parallelisms suffer from too many distinct notations for expressing parallelism, lots of user-managed details and locality. Over the time, different programming languages are developed to handle the different memory models like shared memory, distributed memory, in-memory, out of the core memory etc. efficiently. Most of the new generation programming languages is based on C, C++, Fortran and Java.

The following table will help us to understand hardware parallelism in a better way:

3.3 Computational Mechanisms

Speed up in the performance of parallel computation is largely depends upon by how much of the job must be performed in sequentially. Amdahl's Law characterizes the speedup S that can be achieved by n processors collaborating on an application where p is the fraction of the job that can be executed in parallel and unit time for a single processor to complete the job. The main challenge to speed up in processing large scientific datasets is mainly depends upon the data movement between disk and main memory and parallel and non-parallel blocks in a program. To reduce the data movements between main memory and disk, creating and managing standard datasets for machine learning environment are crucial.

It is important to understand out of different programming models described in Table 1, MPI, used in distributed memory model, is the key player for parallelism process. The main focus is on who owns which data and what communication is needed to process that data. Different frameworks such as Hadoop, Programming with Big Data in R (pbdR) [19], RHadoop [20] and KNIME [21] are efficiently present working environment in this direction.

Let us now discuss MapReduce model that is sufficiently reducing the effort required for large-scale data processing with the help of efficiently designed software for big data. The MapReduce model consists of two phases: map and reduce. In the map phase, a list of key-value pairs are generated/gathered for a specific task and in the reduce phase, computation is performed on all of the key-value pairs from the map phase and the result is saved. This allows for task-parallel execution, where compute node can pick up map and reduce tasks to perform.

In the other hand pbdR is gaining much attention to the scientific community for its different libraries, viz. pbdBASE, pbdDEMO, pbdMPI, pbdDMAT, pbdMPI etc. for distributed big data programming and manipulation. Traditional RDBMS is efficient to handle organized and structure data sets, and its query language is de facto for DDL, DML and DCL queries for both developer and end-users but there is hardly any design aspect available for unstructured data, which are dynamic in nature. To overcome these sharing model based approach and its limitations [22], big data framework offers a bouquet of NoSQL (Not only SQL) environments that

Table 1 Hardware parallelism

Type of H/W parallelism	Programming model	Unit
Instruction level vectors/threads	Pragmas	Iteration
Intra-node/multicore	OpenMP/pthreads	Iteration/task
Inter-node	MPI	Executable
GPU/accelerator	CUDA/OpenCL/OpenAcc	SIMD function/task

deals with a very specific approach to solve rather specific unstructured massive dynamic datasets. KNIME is such an environment that handles structure and unstructured data efficiently. Another important aspect that we have to keep in mind during model design/selection is bulk synchronous and bulk asynchronous processing. In bulk synchronous machine learning paradigm, all the computations are completed in phases and then the messages are sent to the next phase. It is simpler to build, like MapReduce model. We need not have to worry about race conditions as barrier guarantees data consistency and make it simpler fault-tolerant. If we compare it with the matrix domain problem, it is similar to Jacobi iteration. Slower convergence for many machine-learning problems is the main issue with synchronous bulk processing. In the other hand, asynchronous bulk processing is mainly used to address graph-parallel execution. It is harder to build as race conditions can happen all the time. We have to take care of fault tolerance issues and need to implement scheduler over vertices as vertices see latest information from neighbours. It is similar to Gauss-Seidel iteration. The convergence rate is quite high in asynchronous bulk processing.

4   Results and Discussions

We have performed our experiments with different datasets in pbdR and KNIME workflow under different virtual machines and accessing HDFS via Hive for big data environment. Our main aim is to extract features from large datasets even if there exist small overlapping subsets in the dataset. The workflow design for data analysis is in Fig. 2.

Sample outputs snippets for moderate size Iris dataset, as it comes in interactive table and scatter plot, are depicted in Figs. 3 and 4 respectively.

Most of the MapReduce like framework works optimally when the algorithm is embarrassingly parallel/data parallel and able to decomposed into large numbers of

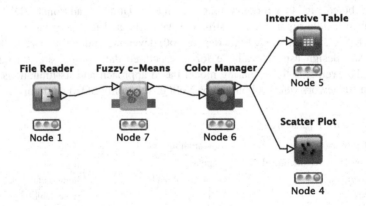

Fig. 2 Workflow design for data analysis

Row127	l	3	4.9	1.8	Iris-virginica				cluster_1
Row128	ı	2.8	5.6	2.1	Iris-virginica				cluster_0
Row129	?	3	5.8	1.6	Iris-virginica				NoiseCluster
Row130	ı	2.8	6.1	1.9	Iris-virginica				NoiseCluster
Row131	)	3.8	6.4	2	Iris-virginica				NoiseCluster
Row132	ı	2.8	5.6	2.2	Iris-virginica				cluster_0
Row133	3	2.8	5.1	1.5	Iris-virginica				cluster_1
Row134	l	2.6	5.6	1.4	Iris-virginica				NoiseCluster
Row135	?	3	6.1	2.3	Iris-virginica				NoiseCluster
Row136	3	3.4	5.6	2.4	Iris-virginica				cluster_0
Row137	ı	3.1	5.5	1.8	Iris-virginica				cluster_0
Row138		3	4.8	1.8	Iris-virginica				cluster_1
Row139	)	3.1	5.4	2.1	Iris-virginica				cluster_0
Row140	?	3.1	5.6	2.4	Iris-virginica				cluster_0
Row141	)	3.1	5.1	2.3	Iris-virginica				cluster_0
Row142	3	2.7	5.1	1.9	Iris-virginica				cluster_1
Row143	3	3.2	5.9	2.3	Iris-virginica				cluster_0
Row144	?	3.3	5.7	2.5	Iris-virginica				cluster_0
Row145	?	3	5.2	2.3	Iris-virginica				cluster_0
Row146	3	2.5	5	1.9	Iris-virginica				cluster_1
Row147	;	3	5.2	2	Iris-virginica				cluster_0
Row148	?	3.4	5.4	2.3	Iris-virginica				cluster_0
Row149	)	3	5.1	1.8	Iris-virginica				cluster_1

Fig. 3 Interactive table output

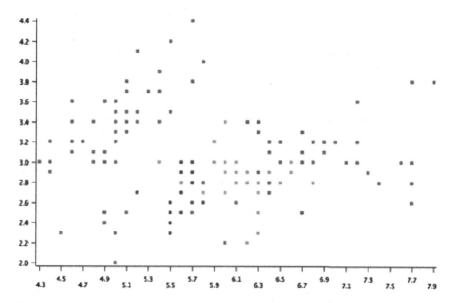

Fig. 4 Scatterplot of different clusters

independent computations. The MapReduce abstraction fails when there exists
computational dependencies in the data. For example, MapReduce can be used to
extract features from a massive collection of images but unable to represent computation that depends on small overlapping subsets of images. MapReduce doesn't
provide a mechanism to directly encode iterative computation. Thus it is obvious
that a single computational framework is unable to express the all the large-scale

features like unstructured, dynamic, asynchronous, iterative computation etc. that is common to many machine learning algorithms. We are capable of processing 200 millions row (approx.) to optimize the performance with these tools and techniques.

5 Conclusion

Big data analysis with different quantitative techniques and tools efficiently helped end users to take effective decisions making. In this paper, we have discussed different computational techniques that are useful in designing different models of big data parallelisms. It is obvious that a single computational framework is unable to express all the large-scale features like dynamic, asynchronous, iterative computation etc., that is common to many machine learning algorithms.

References

1. Big data: The next frontier for innovation, competition, and productivity, McKinsey Global Institute, May 2011.
2. Apache Hadoop. http://hadoop.apache.org.
3. J. Dean, S. Ghemawat, MapReduce: simplified data processing on large clusters, Communications of ACM 51, Vol. 1, 2008, pp. 107–113.
4. C. Olston, B. Reed, U. Srivastava, R. Kumar, and A. Tomkins, "Pig latin: A Not-So-Foreign Language for Data Processing," Proc. of the SIGMOD International Conference on Management of Data. ACM, 2008, pp. 1099–1110.
5. Apache Hive. https://hive.apache.org.
6. S. Seo, E.J. Yoon, J. Kim, S. Jin, J.-S. Kim, S. Maeng, Hama: an efficient matrix computation with the mapreduce framework, in 2nd International Conference on Cloud Computing Technology and Science, IEEE, 2010, pp. 721–726.
7. Y. Low, J. Gonzalez, A. Kyrola, D. Bickson, C. Guestrin, J.M. Hellerstein, Graphlab: a new parallel framework for machine learning, in: Conference on Uncertainty in Artificial Intelligence (UAI), Catalina Island, California, 2010.
8. Apache Giraph, http://giraph.apache.org.
9. Apache Mahout, http://mahout.apache.org/.
10. U. Kang, C.E. Tsourakakis, C. Faloutsos, Pegasus: a peta-scale graph mining system implementation and observations, in the Ninth IEEE International Conference on Data Mining, ICDM'09, IEEE, 2009, pp. 229–238.
11. M.F. Pace, BSP vs MapReduce, Procedeeings of Computer Science, Vol. 9, 2012, pp. 246–255.
12. Y. Bu, B. Howe, M. Balazinska, M.D. Ernst, HaLoop: efficient iterative data processing on large clusters, Proceedings of VLDB Endowments, Vol. 3 (1–2), 2010), pp. 285–296.
13. L.G. Valiant, A bridging model for parallel computation, Communnications of ACM, Vol. 33 (8), 1990, pp. 103–111.
14. L. Bottou and O. Bousquet, The tradeoffs off of large scale learning, Advance Neural Information Process Systems, Vol. 20, 2008, pp. 161–168.
15. R.H. Byrd, G.M. Chin, J. Nocedal, Y.Wu, Sample size selection in optimization methods for machine learning, Mathematical Programming 134:1, 2012, pp. 127–155.

16. Koushik Mondal, Big Data Parallelism: Issues in different X-Information Paradigms, Elsevier Procedia Computer Science, Special Issue on Big Data, Cloud and Computing Challenges, ISSN 1877–0509, Vol. 50, pp. 395-400, 2015.
17. Koushik Mondal, Paramartha Dutta, Big Data Parallelism: Challenges in different computational Paradigms, IEEE Third International conference on Computer, Communication, Control and Information Technology, ISBN: 978-1-4799-4446-0, 2015.
18. B.Recht, C.Re, S. Wright,F.Niu, Hogwild: A lock-free approach to parallelizing stochastic grapient descent, Advance Neural Information Processing System,s Vol. 24, 2011, pp. 693–701.
19. Drew Schmidt and George Ostrouchov, Programming with Big Data in R, user 2014 summit, http://r-pbd.org/.
20. RHadoop: Open Source Project, http://projects.revolutionanalytics.com/rhadoop/.
21. KNIME: Open for Innovation Project, https://www.knime.org/.
22. Bonnet L., Laurent A., Sala M., Laurent B., Sicard N., Reduce, You Say: What NoSQL Can Do for Data Aggregation and BI in Large Repositories, 22nd International Workshop on Database and Expert Systems Applications, pp. 483–488, 2011.

Revised ECLAT Algorithm for Frequent Itemset Mining

Bharati Suvalka, Sarika Khandelwal and Chintal Patel

Abstract Data mining is now a day becoming very important due to availability of large amount of data. Extracting important information from warehouse has become very tedious in some cases. One of the most important application of data mining is customer segmentation in marketing, demand analyzes, campaign management, Web usage mining, text mining, customer relationship and so on. Association rule mining is one of the important techniques of data mining used for discovering meaningful patterns from huge collection of data. Frequent item set mining play an important role in mining association rules in finding interesting patterns among complex data. Frequent Pattern Itemset Mining from "Big Data" is used to mine important patterns of item occurrence from large unstructured database. When compared with traditional data warehousing techniques, MapReduce methodology provides distributed data mining process. Dataset can be found in two pattern one is horizontal data set and another one is vertical data set. Tree based frequent pattern MapReduce algorithm is considered more efficient among other horizontal frequent itemset mining methods in terms of memory as well as time complexity. Another algorithm is ECLAT that is implemented on vertical data set and is compared with my proposed revised ECLAT Algorithm. As a result the performance of ECLAT Algorithm is improved in proposed algorithm revised ECLAT. In this paper will discuss improved results and reasons for improved results.

Keywords Big data analytics · MapReduce · ECLAT · BFS · FP tree · FP growth

B. Suvalka (✉) · S. Khandelwal · C. Patel
Geetanjali Institute of Technical Studies, Udaipur, India
e-mail: bharatiisuvalka@gmail.com

S. Khandelwal
e-mail: sarikakhandelwal@gmail.com

C. Patel
e-mail: smilingchintal@gmail.com

© Springer India 2016
S.C. Satapathy et al. (eds.), *Information Systems Design and Intelligent Applications*, Advances in Intelligent Systems and Computing 434,
DOI 10.1007/978-81-322-2752-6_21

1 Introduction

1.1 Big Data

Nowadays data is not only growing in its size but also increasing in variety and variability. The big data is defined as database that can store, manipulates, analyze and manage huge amount of data whose size is beyond the ability of traditional database tools. The big data and its attributes can be explained via 3V's that is volume, variety, varsity and velocity [1]. The volume refers to the amount of data that can be processed at the same time. Big data have ability to process and store terabytes of data. The frequency by which data is generated by some real time application refers to velocity of the big data. The variety includes different formats of data that is structured, semi structured and unstructured data.

Big Data Analytics is the method to analyze large amount of data to uncover unknown relation, hidden patterns and other useful information about the data [2]. Big data analytics is proved to be beneficial in many areas like business analysis, marketing strategies, customer reviews to make useful decisions for the enterprises. The main objective of big data analytics is to help different enterprises to make better business decisions and other users to analyze huge volumes of transaction data as well as other data which is left by Business Intelligence programs. Predictive analytics for further launching the further business strategies [3].

One of such big data analytics is frequent item set mining. In frequent item set mining the set of items purchased together are examined. This analysis helps the enterprise and stores to launch the schemes that give together packs or money saver offers to tempt the customers. The main purpose of frequent itemset mining is analyse transaction data of the supermarket. This is used to examine the behavior of customer in terms of products purchased. There are various algorithms that are used for frequent item set mining [4]. These algorithms can be implemented on horizontal data format as well as on vertical data format [5].

1.2 Frequent Item Set Mining

Horizontal item set mining takes data in horizontal format and that have two basic algorithms. Most fundamental algorithm for frequent item set mining is Apriori algorithm it converts the database in the form of graph before processing. This algorithm makes multiple passes over the data and it uses multiple iterative approach and BFS to explore itemsets [6].

Apriori algorithm goes through the iterative approach and it scans database for many passes and take database as graph that increases its time complexity. Considering the drawbacks of apriori algorithm FP growth algorithm was introduced that uses divide and conquer approach. It compresses the database to tree structure called FP Tree that is frequent pattern tree instead of graph [7].

Apriori algorithm and FP growth algorithm were used for horizontal data base and it is found that comparing and scanning capability of vertical data set is less than horizontal so vertical frequent item set mining techniques were introduced.

Then ECLAT algorithm was implemented for frequent data mining. In this paper will discuss the drawbacks of ECLAT and how the proposed algorithm revised ECLAT is used to reduce the time complexity and its graph and improved results are also being explained.

1.3 ECLAT Algorithm

ECLAT Equivalence class transformation was developed to mine frequent item sets efficiently using vertical data format [8]. The length of transaction set in the item set is simply the support count of an item. Starting with k = 1, all the k-item set are used to form the candidate (k + 1) item sets. The intersection of the TID set is used to compute item sets. This process repeats for all transactions.

```
ECLAT TIDset (T, β)
Input: T is vertical data set and β is support value.
Output: List of frequent item sets
  1.  Repeat step  for i=1 to Li != NULL
      (a)  Initialize Pᵢ={Li ͵T(Li) fall itemsets of T}
               Li refers to the item name
               T(Li) refers to the transactions to that particular transactions.
      (b)  i=i+1
  2.  Repeat  for j=1 to Pj != NULL
      (a)  Xj = Lj and Yj = T(Lj)
      (b)  J=j+1
  3.  Set value of  i=1 and  k=1
  4.  Repeat for ( Xj != NULL) and  (Yj != NULL)
      (a)  J=i+1
      (b)  Nij = Xi ∪ Xj  and T(Nij) Yi ∩ Yj
      (c)  If count of T(Nij) >β then
               Bk= Nij and  k=k+1
      (d)  i=i+1
  5.  Print (Bk)
  6.  End
```

This ECLAT algorithm have complexity $O(k(k + 1))$ where k is the no of items. I had gone through the ECLAT and found some scope to reduce complexity in this paper I had proposed the revised ECLAT in which the time complexity is reduced. In revised ECLAT after scanning the data a lexicography algorithm is being applied and then the data having number of transaction count less than that of support count than those items are removed from the list so after that the scanning time is being reduced. Time of process is less for revised ECLAT than that of ECLAT. Improved results and graph are being explained in this paper.

1.4 Revised ECLAT Algorithm

In Revised ECLAT algorithm, first step is loading the data from database to the data structure. Then the data is being sorted and arranged in descending order. The transactions having transaction count less than the support count are directly be omitted from the data structure and then the algorithm is implemented.

Revised ECLAT TIDset (T, β)
Input: T is vertical data set and β is support value.
Output: List of frequent item sets
1. Repeat step for i=1 to Li != NULL
   (a) Initialize P_i ={Li T(Li) fall itemsets of T}
          Li refers to the item name
          T(Li) refers to the transactions to that particular transactions.
   (b) i=i+1
2. Arrange Pi in lexicography order according to the transaction count of each transaction set corresponding to that item
3. Remove all the item set where count of T(Li) < β
4. Repeat for j=1 to Pj != NULL
   (a) Xj = Lj and Yj = T(Lj)
   (b) J=j+1
5. Set value of i=1 and k=1
6. Repeat for (Xj != NULL) and (Yj != NULL)
   (a) J=i+1
   (b) Nij = Xi ∪ Xj and T(Nij) Yi ∩ Yj
   (c) If count of T(Nij) > β then
          Bk= Nij and k=k+1
   (d) i=i+1
7. Print (Bk)
8. End

1.5 Revised ECLAT Algorithm Without Sorting

Further, the time can be reduced further by eliminating the sorting algorithm from the algorithm and making one to one comparison and directly omitting the item set which is purchased in transaction whose count is less than the support count. The support count is also a taken directly without computation depending upon the scheme to be launched. The support count is decided prior and then it is being taken with inputs. By eliminating the sorting step the processing time is reduced and a good improvement is being observed. The algorithm after improvement is as follow.

Revised ECLAT1 TIDset (T, β)
Input: T is vertical data set and β is support value.
Output: List of frequent item sets
1. Repeat step for i=1 to Li != NULL
    (a) Initialize P$_i$={Li .T(Li) fall itemsets of T}
        Li refers to the item name
        T(Li) refers to the transactions to that particular transactions.
    (b) i=i+1
2. Arrange Pi in lexicography order according to the transaction count of each transaction set corresponding to that item
3. Remove all the item set where count of T(Li) < β
4. Repeat for j=1 to Pj != NULL
    (a) Xj = Lj and Yj = T(Lj)
    (b) J=j+1
5. Set value of i=1 and k=1
6. Repeat for (Xj != NULL) and (Yj != NULL)
    (a) J=i+1
    (b) Nij = Xi ∪ Xj and T(Nij) Yi ∩ Yj
    (c) If count of T(Nij) > β then
        Bk= Nij and k=k+1
    (d) i=i+1
7. Print (Bk)
8. End

The intersection of transactions and union of itemsets are taken for each transaction with another. When no further intersection is possible or the intersection results in null set then again the support value is compared with the transaction count of each item set and then the transaction set having maximum count will be a printed as resultant itemset and that itemset is called as frequent itemset.

2 Results

2.1 Observation Table

Table 1 shows the time difference analyzed by different algorithm in different format having different data volume.

2.2 Observation from Graph

From the graphs it is being observed that as the data volume increases the processing time also increases along with it. Different algorithm uses different technique and the processing time changes. For Apriori algorithm the processing time observed is highest then the ECLAT algorithm was implemented and that algorithm have less processing time as compared to Apriori. The proposed algorithm enhanced ECLAT have processing time less than both Apriori and ECLAT Algorithm. In enhanced ECLAT as the data volume was increasing the processing

Table 1 Experimental result shows time required to process the data

S. no.	Data volume	Horizontal data format	Vertical data format		
		Apriori algo. (s)	ECLAT (s)	Enhanced ECLAT (s)	Enhanced ECLAT without sorting (s)
1	200 MB	1.48	1.28	1.06	1.02
2	500 MB	2.17	1.47	1.15	1.08
3	750 MB	2.37	2.08	1.52	1.55
4	1 GB	2.54	2.27	2.17	2.03
5	1.25 GB	3.08	3.02	2.53	2.37
6	1.5 GB	4.52	4.28	4.02	3.04
7	1.75 GB	5.08	4.38	4.23	3.52
8	2 GB	5.27	4.42	5.2	4.28

time was also increasing and when the data volume was 2000 MB the processing time of ECLAT is 4.42 s and for enhanced ECLAT the time increases to 5.2 s. (Fig. 1).

The clear comparison can be observed from the bar graph given below. Processing time for different technique for different data volume is being compared and clearly observed that the Enhanced ECLAT1 Algorithm in which sorting is being eliminated gives the best result among all other algorithms (Fig. 2).

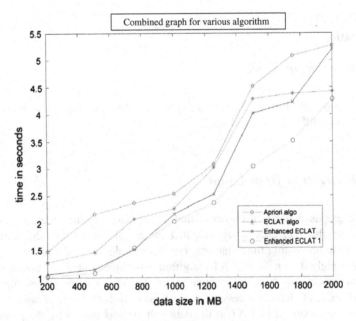

Fig. 1 Different plot for different algorithm for different data format

Fig. 2 Bar graph comparison of different algorithm for different data volume

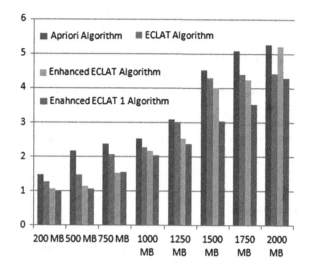

2.3 Inference

From the observation table, graphs and bar graph it is being observed that as size of data increases the processing time of algorithm also increases. The process time of Apriori Algorithm for horizontal data format is more than that of ECLAT algorithm. ECLAT algorithm is used for vertical data set. The frequent item set mining of data in horizontal format take time more than that of data in vertical data format. After going through ECLAT algorithm new algorithm Enhanced ECLAT is being proposed and observed that processing time is less than that of ECLAT algorithm. In enhanced ECLAT as the data volume was increasing the processing time was also increasing and when the data volume was 2000 MB the processing time of ECLAT is 4.42 s and for enhanced ECLAT the time increases to 5.2 s. The sorting algorithm was eliminated and data is one to one compared and the data with transaction count less than that of support count is directly eliminated from the data structure and then the algorithm is implemented. It is being observed that the Enhanced ECLAT without sorting algorithm give us the best results.

3 Limitation

Limitation of revised ECLAT algorithm is that with increasing the size of data the processing time will not give significant changes. When data volume is small then revised ECLAT is most efficient algorithm but with increase in size of data volume it gives the change but much refined.

4 Conclusion

Frequent data mining is one of the most important big data analytics. In frequent data mining the analyst search the products purchased together and then it is used to find strategies and consider customer reviews. This helps enterprises to launch their business intelligence strategies. This paper explained frequent data mining on vertical data format using ECLAT algorithm and then proposed a new algorithm named revised ECLAT that took less processing time as shown in result table and graphs.

References

1. Bharati Suvalka, Sarika Khandelwal, Siddharth Singh Sisodiya "Big Analytics using meta machine learning" in international journal of innovative research in science engineering and technology in August 2014.
2. Tilmann Rabl, Mohammad Sadoghi, Hans-Arno Jacobsen "solving big data challenges for enterprise application Performance management".
3. Noll, Michael G. "Running hadoop on ubuntu linux (single-node cluster)." Mar- 2013 [Online]. Available: http://www.michael-noll.com/tutorials/running-hadoopon-ubuntu-linux-single-nodecluster/.
4. Lam, Chuck. Hadoop in Action. Manning Publications Co., 2010.
5. Han, Jiawei, Micheline Kamber, and Jian Pei. Data mining: concepts and techniques. Morgan kaufmann.
6. Zikopoulos, Paul, and Chris Eaton. Understanding big data: Analytics for enterprise class hadoop and streaming data. McGraw-Hill Osborne Media.
7. Zikopoulos, Paul, Krishnan Parasuraman, Thomas Deutsch, James Giles, and David Corrigan. Harness the Power of Big Data The IBM Big Data Platform. McGraw Hill Professional.
8. Shvachko, Konstantin, Hairong Kuang, Sanjay Radia, and Robert Chansler. "The hadoop distributed file system." In Mass Storage Systems and Technologies (MSST), 2010 IEEE 26th Symposium.

User-Interface Design Framework for E-Learning Through Mobile Devices

Ankita Podder, Tanushree Bhadra and Rajeev Chatterjee

Abstract User-interface plays an important role in e-learning. Good and user-friendly interfaces help in better understanding of the system and reduces the cognitive load of a learner during the process of learning. The advancement in mobile technology has influenced a number of users to learn using mobile and handheld device that provides not only flexibility of learning but also allow them for learning at any-time and any-where or on the go. This has already created a requirement of a user-interface design framework for mobile devices that can help the designers to design e-learning interface for the same. This paper proposes a user-interface design framework after understanding of the requirements at the level of a user for the mobile and handheld devices.

Keywords E-learning · User interface · Mobile devices

1 Introduction

Software product development starts with requirement specification. This requirement specification hardly provides any clue on the requirement of user interface. This leads to a situation where products and services were developed without

A. Podder (✉) · T. Bhadra · R. Chatterjee
National Institute of Technical Teachers' Training & Research,
Block-FC, Sector-III, Salt Lake City, Kolkata 700149, India
e-mail: decentdithi.podder@hotmail.com

T. Bhadra
e-mail: bhadra.tanu92@gmail.com

R. Chatterjee
e-mail: chatterjee.rajeev@gmail.com

© Springer India 2016
S.C. Satapathy et al. (eds.), *Information Systems Design and Intelligent
Applications*, Advances in Intelligent Systems and Computing 434,
DOI 10.1007/978-81-322-2752-6_22

providing attention to user interface design. The popularity of mobile platform has created a requirement of porting the applications over the mobile devices. As we are aware that mobile devices have small and a personalized display that have altogether a different user interface. This creates a requirement for development of user-interface design framework for the small and handheld devices that are mobile and personalized. The design framework should have basic functional requirements of e-learning system like reusability, interoperability, accessibility, durability etc. User interface design framework is proposed to be based on the studies on different pedagogical approach, basic principles of computer science and software development, instructional system design, and communication protocols.

The paper is organized as follows. In Sect. 2, we discuss Review of the Existing Works. In Sect. 3, we briefly explain Our Proposed Design Approach. In Sect. 4, Results and comparisons are given. In Sect. 5, Conclusion and Future Scope are given.

2 Review on Existing Work

Adnan et al. [1] focus on good interface design by using appropriate visual elements (*image, shape, color, text* etc.) that facilitates user experience in interaction and navigation.

Faghih et al. [2], proposed three golden rules to design interface by considering user in control activity, reducing the user's work load and making the interface consistent.

Barzegar et al. [3] suggest User Interface design by dividing those education environments into (a) Personal Page; (b) Virtual class (either in Synchronous Mode or Asynchronous Mode); (c) Communicative Part (Communication Protocol between teacher and learner); (d) Library (Course-repository).

Das and Chatterjee [4] define a framework for designing the User Interface Framework based on Synchonous and Asynchronous mode of e-learning.

Bahadur et al. [5] have proposed visualization techniques that describe arrangement and organization of various e-learning element.

Guralnick [6] considers user experience in interaction as a crucial part of courseware acceptance and effectiveness.

Georgiev and Georgieva [7] define a scope for designing user interface for several applications for the mobile devices such as cell phones, smart phones, PDAs and super phones and also consider design issues in mobile devices: (a) various screen resolution in mobile devices, (b) navigation design in mobile devices.

3 Our Proposed Design Approach

3.1 A General Overview of Our Proposed Framework

In our User-interface design approach for mobile devices, we decompose the overall screen layout in four different modules: (a) Course details module, (b) Learner resources, (c) Content area and (d) Navigation that is shown in Fig. 1.

(a) **Course Details Module**: This module consists of:

    (i) *Menu*: This option facilitates user by providing control over courseware through the sub-fields like home, settings, course repository, notice board, virtual classroom.

    (ii) *Logo*: Contains institutional logo.

    (iii) *Curriculum*: It describes all the courses and course-modules as well as works as courseware progress indicator that gives opportunity to view successfully completed chapters, due courses and also the chapters or courses where improvement is required.

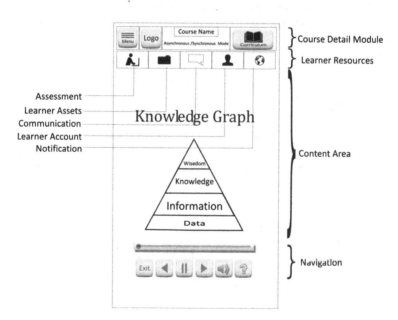

Fig. 1 Our proposed user-interface framework

(b) **Learner Resources**: This module involves:

    (i) *Learner Account*: This provides access to the learners to manage their own details like Personal details, Profile picture, Language, and Notifications.

    (ii) *Communication*: Learner can communicate to the teacher, instructor or other students through various ways such as having discussion, live chat, any query, report problem, raise hand options. However, the raise hand option is only available in synchronous mode learning.

    (iii) *Learner Assets*: It provides learners the options to find their chapter notes (in various formats such as .pdf or as recorded video), solved problems, and obtain certificates and transcripts.

    (iv) *Assessment*: This option is designed to serve the purpose of learner assessment. The assessment procedure can be in form of practice test, assignments and quizzes. A grade is associated regarding each test.

(c) **Content Area**: The contents are delivered in either Asynchronous mode or Synchronous mode of learning. The content area covers the presentation, white-board, the title, teacher/instructor details, live lecture of the teacher/instructor, animation, video, etc.

(d) **Navigation**: The navigation contains navigation buttons like pause/play, backward, forward, volume control of audio, exit, and help at the bottom of the screen (Table 1).

Table 1 A proposed user-interface design framework

User interface modules	Components		Icon	Functionality	Pedagogy
1. Course details module	Menu	Home	Yes	Learner dependent (registered student)	System assisted
		Settings			
		Course repository			
		Notice board			
		Virtual classroom			
	Logo		Yes	Institution dependent	System assisted
	Curriculum	Courses and course modules	No	Instructional designer dependent	Cognitive effect
		Completed modules	No	Learner dependent	Pacing effect
		Incomplete modules			
		Try to improve modules			

(continued)

Table 1 (continued)

User interface modules	Components		Icon	Functionality	Pedagogy
2. Learner resources	Learner account	Profile picture	Yes	Learner dependent (registered student)	System generated
		Language			
		Personal details			
		Timeline			
		Notification			
	Communication	Discussion	Yes	Learner dependent	Pacing effect
		Live chat			
		Any query			
		Report problem			
		Raise hand			
		Write message			
	Learners' assets		Yes	Learner dependent	System generated
	Assessment	Practice test	Yes	Time dependent	Knowledge Testing
		Assignments			
		Quizzes			
3. Content area	Course content		No	Interactive, instructor dependent	Modality and contiguity effect
4. Navigation	Pause/play		Yes	Time dependent	Standard identifiable icons
	Backward				
	Forward				
	Sound				
	Exit				
	Help				

3.2 An Algorithmic Approach Towards the Asynchronous and Synchronous Modes of Operations

In our mobile based e-learning approach we consider the overall framework with respect to the basic two mode of learning: (a) Asynchronous Mode Learning and (b) Synchronous Mode Learning.

3.2.1 Algorithm

```
Begin //Begin with one particular action
{switch ( action)
{ case 1: switch ("Communication" option) {
      case "Discussion" option: //user decide to discuss
//anything;
      case "Live Chat" option: // user decide for live
//chat to anyone; A chat box will appear
      case "Any Query" option: //user have some query;
      case "Report Problem" option: //user want to report
//any problem
      case "Write Message" option: //user want to send any
// message to teacher or instructor

case 2:  switch("Curriculum" option)
{ case "Successfully Completed" option: //user can view
//the successfully completed modules
   case "Due Courses" option: //user can view the chapter
//names that are not completed yet
   case "Try to Improve" option: //user can view the
//chapters where further improvement is needed if any

case 3:  Select the "Notification" icon //it requires to
//notify some information to user by expanding the
//notifications

case 4:  switch("Learners' Assets" option)
{  case "Chapter Notes" option: //user can collect
//chapter notes by downloading
   case "Solved Problem" option: //user can collect
//solved problem by corresponding to download option
   case "Certificates" option: //user can collect
//certificates of each course by download

case 5: switch("Assessments " option)
{  case "practice test" option: // user has to take
//practice tests;
   case "Assignment" option: // user has to perform or
//submit assignments regarding the course
   case "Quizzes" option: //user is informed to response
//answers to quizzes;
} End.
```

3.2.2 Asynchronous Mode of Learning

In Asynchronous Mode the learning is performed through web based content delivery mechanism given in Fig. 2. The managements of various process like registration, content deployment, content delivery, certificates and transcripts, etc. are done using a specialized application commonly known as Learning Management System (LMS). Each learner can learn at a pace and time convenient to himself or herself. Multimedia contents are delivered in asynchronous time. The learner can communicate with the Instructor through online chat, email, and messaging system.

3.2.3 Synchronous Mode of Learning

In Synchronous Mode the learning is performed through live lectures or it can also be through recorded videos given in Fig. 3. Here we consider an extra feature in the

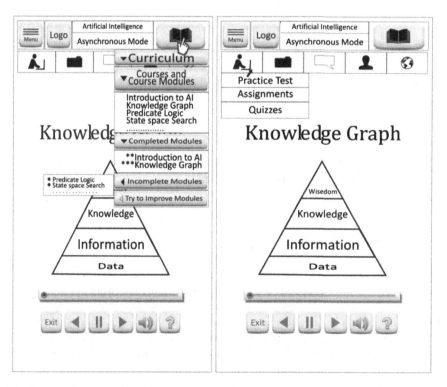

Fig. 2 Asynchronous mode of learning

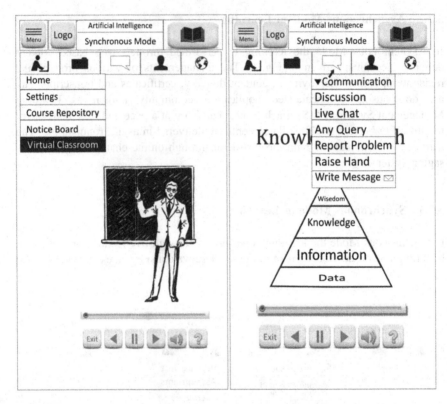

Fig. 3 Synchronous mode of learning

area of communication that user can interact with the teacher or lecturer by raising
hand and teacher provides response to the hand.

4 Results and Comparisons

The proposed framework is compared with the existing research work and is given
in Table 2. We have proposed a user interface design framework for e-learning
system used in mobile devices. The proposed framework may become a guideline
for user interaction in the e-learning system and may be used in both synchronous
and asynchronous e-learning mode. This may also reduce the overall cognitive load
of a learner and may help to increase learner expectation from the e-learning system
which effect in the end result of the learner performance.

Table 2 Comparison of proposed framework with existing research works

S. no.	Parameter	Existing research works	Our proposed framework	Remarks
1	Visual impact	Consider visual elements as important criterion [1]	Focus on designing interface using standard icon	Better for mobile devices
2	Design principle	Segregate design into different sub-section based on access mode [3]	Segregate design into different modules on user control and navigation	Flexible user interface enhance performance
3	Design flexibility	Design the User Interface Framework is based on desktop format [4]	Describes functionality and pedagogy aspect in various screen size in mobile devices	Flexible design reduces cognitive load
4	Accessibility	Standard design framework is supported in PC/Desktop environment	Framework design is flexible enough to support various screen formats and navigation in mobile environment.	Support ubiquitous learning that enhances accessibility

5 Conclusion and Future Scope

Since user interface design plays a great role in software project development so it may be concluded that if we want to design more responsive and effective e-learning system, we may uses a user interface design framework as a standard guideline for user interface design. In future, we may propose a framework that may enhance the performance of a user with respect to cognitive performance and interaction which ensures better user experience.

References

1. Adnan, A.S., Ahmad, M.A.R., Zaidi, A., Sofi, M.: The Utilisation of Visual Elements on Interface Design of E-learning. Proceeding of IC-ITS 2015 e-ISBN:978-967-0850-07-8, pp. 273–279. International Conference on Information Technology & Society, Kuala Lumpur, Malaysia (8–9 June 2015).
2. Faghih, B., Azadehfar, M.R., Katebi, S.D.: User Interface Design for E-Learning Software. The International Journal of Soft Computing and Software Engineering [JSCSE]. vol. 3, pp. 786–794. No. 3, Special Issue: The Proceeding of International Conference on Soft Computing and Software Engineering 2013 [SCSE'13], San Francisco State University, CA, U.S.A. Doi: 10.7321/jscse.v3.n3.119 (March 2013).
3. Barzegar, S., Shojafar, M., Keyvanpour, M.R.: Improvement User Interface in Virtual Learning Environment. International Journal of Digital Society (IJDS), vol. 1, pp. 221–229. (3, September 2010).
4. Das, S., Chatterjee, R.: A Proposed Systematic User-Interface Design Framework for Synchronous and Asynchronous E-learning Systems. © Springer India 2015 J.K. Mandal et al.

(eds.), Information Systems Design and Intelligent Applications, Advances in Intelligent Systems and Computing 340, pp. 337–347. DOI 10.1007/978-81-322-2247-7_35 (2015).
5. Bahadur, S., Sagar, B.K., Kondreddy, M.K.: User Interface Design With Visualization Techniques. International Journal of Research in Engineering & Applied Sciences http://www. euroasiapub.org. vol. 2, pp. 51–62. Issue 6 (June 2012), ISSN: 2249-3905.
6. Guralnick, D.A.: User Interface Design for Effective, Engaging E-Learning. pp. 1–10.
7. Georgiev, T., Georgieva, E.: User Interface Design for Mobile Learning Applications. e-Learning'09. pp. 145–150 (2009).

Application of Three Different Artificial Neural Network Architectures for Voice Conversion

Bageshree Sathe-Pathak, Shalaka Patil and Ashish Panat

Abstract This paper designs a Multi-scale Spectral transformation technique for Voice Conversion. The proposed algorithm uses Spectral transformation technique designed using multi-resolution wavelet feature set and a Neural Network to generate a mapping function between source and target speech. Dynamic Frequency Warping technique is used for aligning source and target speech and Overlap-Add method is used for minimizing the distortions that occur in the reconstruction process. With the use of Neural Network, mapping of spectral parameters between source and target speech has been achieved more efficiently. In this paper, the mapping function is generated in three different ways, using three types of Neural Networks namely, Feed Forward Neural Network, Generalized Regression Neural Network and Radial Basis Neural Network. Results of all three Neural Networks are compared using execution time requirements and Subjective analysis. The main advantage of this approach is that it is speech as well as speaker independent algorithm.

Keywords Artificial neural network · Discrete wavelet transform · Packet decomposition · Spectral transformation · Speech transformation

1 Introduction

Speech is the most significant ancient way to communicate and express oneself. Recent development has made it possible to use speech in applications such as security systems, mimicry, female to male speaker transformation, generating

B. Sathe-Pathak (✉) · A. Panat
Priyadarshani College of Engineering, Nagpur, India
e-mail: bvpathak100@yahoo.com

A. Panat
e-mail: ashishpanat@gmail.com

S. Patil
Cummins College of Engineering, Pune, India
e-mail: shalakapatils@gmail.com

© Springer India 2016
S.C. Satapathy et al. (eds.), *Information Systems Design and Intelligent Applications*, Advances in Intelligent Systems and Computing 434,
DOI 10.1007/978-81-322-2752-6_23

237

cartoon voice for different game applications etc. For most of these applications, speech transformation technique plays a most vital role in achieving the desired goal. Basically, it is a difficult task to determine an optimal method for speech transformation that will maintain intelligibility of speech with minimum distortion. So, different speech conversion systems that employ different methods exist.

The earlier studies with reference to vocoding applications have used formant frequencies for modification and representation of the spectrum of vocal tract to achieve speech transformation [1]. Single-scale methods like interpolation of speech parameters and modeling using Formant Frequencies (FF) for the speech [2] are proposed. Linear Prediction Coding (LPC) using Cepstrum Coefficients (CC) as features is proposed in [3] and Segmental Codebooks (SC) [4], besides others, suffer due to absence of information during the extraction of coefficients of the formants and the excitation signal. Hence, all these methods are more appropriate for Speech Recognition or Classification purpose.

It is widely accepted that some successful Speech Transformation methods, like time-scaling, still needs improvement in quality of pitch modified speech signal. Furthermore, most of the current Voice Conversion systems work mainly on the sentence level and their extension to lengthy speech should also consider the information of the speaking style. This indicates that even if Voice Conversion is successful in transforming speaker identity for a sentence, it fails for longer speech. On the other hand, the algorithms developed for modeling and transformation of prosody parameters provide the required framework for obtaining high quality transformed output. Time and Frequency Domain methods of Pitch Synchronous Overlap Add (PSOLA) which are commonly used methods for pitch and duration parameters scaling are the most widely used techniques [1, 5]. The study in [6] designs and accumulates a phonetically balanced set of small sized parallel speech database to construct conversion function. The Gaussian Mixture Bigram Model (GMBM) is normally adopted as a conversion function to characterize the spectral and temporal behavior of the speech signal. This study uses Speech Transformation, Representation using Adaptive Interpolation of weiGHTed spectrum (STRAIGHT) algorithm, to estimate the spectrum and pitch contours. This investigation [7], depends on a well known sinusoidal model of the signal. The analysis algorithm behaviors like a windowed signal, and produces a sum of sinusoids of the time-varying representations. The sinusoidal modeling is a set of triples of frequency, magnitude and phase parameters describing each frame, with a residual noise component. In [8], the transformation is achieved by using Arabic utterances from source and target speakers. Gaussian Mixture Model (GMM) is used to develop a conversion function for transforming the spectral envelope which is described by Mel Frequency Cepstral Coefficients (MFCC). Mousa [9] deals with pitch shifting method for voice conversion using pitch detection of the signal by Simplified Inverse Filter Tracking (SIFT) and changes are made according to the target pitch period by the time stretching technique using Pitch Synchronous Over Lap Add algorithm (PSOLA).

In the proposed algorithm, in order to avoid artifacts and discontinuities while reconstruction of a speech, multi-scale spectral transformation technique along with the efficient use of Overlap and Add method for reconstruction of transformed signal is used. This particular technique uses Discrete Wavelet Transformation using wavelet packet decomposition, in an utmost try to preserve the naturalness along with the intelligibility of a speech signal [10]. The packet decomposition is accompanied with Neural Network which particularly assists in developing a mapping function between source and target speech signal. For implementation of our project, we have used the proposed Voice Conversion algorithm to accomplish conversion of source emotion to target emotion in order to achieve Emotion Transformation of a speech signal.

The organization of the paper is as follows: Sect. 2 explains the overall block diagram of proposed algorithm. Section 3 gives the emphasis on Training phase in proposed algorithm, Sect. 4 provides insight in Testing phase of algorithm, Sect. 5 elaborates three different Neural Networks mentioned earlier, that are used for transformation purpose, Sect. 6 shows the experimental results obtained in testing phase, and finally, Sect. 7 gives conclusions that are drawn based on subjective evaluations.

2 Proposed Algorithm

The standard single scale transformation techniques produce artifacts during reconstruction of transformed signals. In contrast to this, proposed approach describes an algorithm based on multi-scale spectral transformation technique which achieves reconstruction of a speech signal in its most natural form while keeping its intelligibility untouched. While implementing this algorithm, firstly, both source and target speech signals are pre-processed and then the Discrete Wavelet Transform is computed over processed speech signals. As the duration or length of source and target files differ, based on different characteristics of speech signal, it is required to align the features before generating a mapping function between source and target signals. This alignment is achieved using Dynamic Frequency Warping technique in frequency domain. These aligned features are given as input to Neural Network. Then, this trained Neural Network is used for the transformation phase. Finally, the transformed speech is reconstructed from modified output coefficients of Neural Network. For the implementation of this algorithm, the database has been self created by ten female trained speakers who are drama actresses. This database is consists of total 400 sentences. Figure 1 displays the flowchart of proposed algorithm.

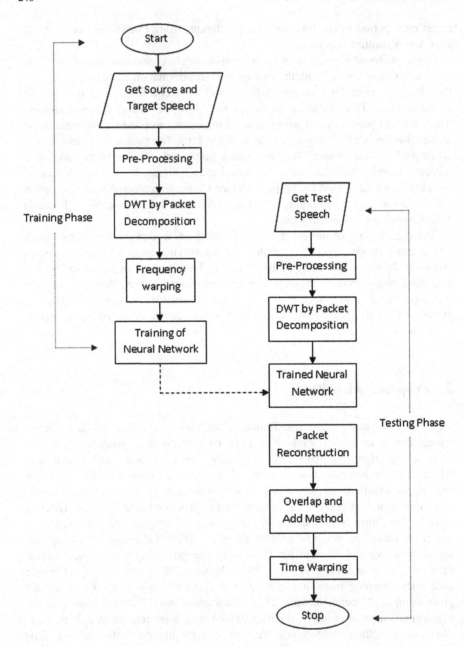

Fig. 1 Flowchart of proposed speech transformation algorithm

3 Training Phase

In the training phase of proposed approach, 80 % of all source and target speech sentences from the database are used for training purpose. Figure 1 indicates the block diagram of training and testing phase. As seen in the block diagram, each speech file is pre-processed using the techniques, such as, Voice Activity Detection, Framing and Windowing. For Voice Activity Detection, Energy content of speech signal is considered as a reference. Due to Framing, the otherwise non-stationary speech signal is considered as stationary for the entire duration of one particular frame [11, 12]. After Framing, Windowing using Hamming Window is accomplished. These pre-processing techniques have been carried out for both source as well as target speech files.

The next step after pre-processing is to achieve five level wavelet packet decomposition of source and target speech. Packet decomposition is a generalized form of wavelet decomposition which offers a richer signal analysis [10]. After decomposition, terminal node (32) coefficients of wavelet packet tree are normalized and coefficients from first 25 terminal bands are considered for further processing, based on the maximum energy content.

The process is repeated for all speech files that are under consideration for training purpose. Once again, the database of these normalized coefficients is created for all source and target speech files that have been considered for training purpose. Finally, the above mentioned 3 different Neural Networks are trained, with this newly created database.

4 Testing Phase

In testing phase, the Pre-processing and Wavelet Packet Decomposition is carried out as shown in block diagram of Fig. 1. The terminal nodes are further normalized and the normalized coefficients of first 25 bands are given to trained Neural Network. The obtained transformed coefficients from Neural Network are de-normalized and are used for packet reconstruction. The reconstructed wavelet packets are then overlap added in order to reconstruct the transformed speech signal. Using Overlap Add method, the discontinuity, which might be occurring otherwise, is removed and a distortion free speech is obtained [13]. As a final step, reconstructed speech signal is further time warped with database target signal as a means of incorporating required characteristics in transformed signal more efficiently.

5 Applications of Different Neural Network Architectures

For this proposed approach, three different types of Neural Networks have been used to generate three different transformation functions which will be performing a mapping between source and target speech signals. In this section, Feed Forward Neural Network, Radial Basis Network and Generalized Regression Neural Network are explained.

5.1 Feed Forward Neural Network (FFNN)

It is a type of Back propagation Neural Network. In this network, the first layer has weights that come from the inputs. Each of the next layers has a weight which comes from its previous layer. The last layer gives the network output. This network works on the basis of 'Back propagation' algorithm. Figure 2 shows the performance of network in proposed algorithm, in terms of Mean Square Error. The three lines in the plot indicate training, validation and testing vector distribution. 60 % of the input signals are used to train the network. 20 % of the inputs are used to validate how well the network is generalized and remaining 20 % for testing by the Neural Network. For our input data, the Mean Square Error is 0.16532 at 0th epoch. Figure 3 shows the training plot of FFNN. Standard back-propagation network like FFNN uses a gradient descent algorithm, were the network weights move along the negative gradient of the performance function as seen in Fig. 3. The step size 'mu' reduces after each successful step and locates the minimum of the function, based on the Levin Berg-Marquardt algorithm, used for training in FFNN. The networks error is reduced as training of the vectors continues as long as the training reduces the network's error. The training is stopped only after the network

Fig. 2 Performance of feed forward neural network

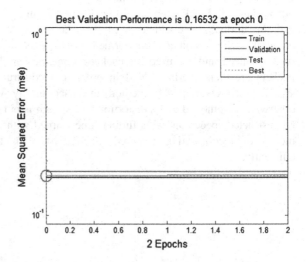

Fig. 3 Training plot of feed
forward neural network

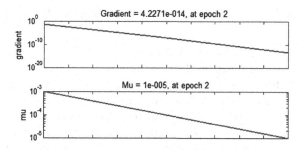

memorizes the training set. This technique automatically avoids the problem of over-fitting, which plagues many optimization and learning algorithms.

5.2 Radial Basis Neural Network (RBN)

A special case of feed forward network is the Radial Basis Neural Network. This maps nonlinearly the input space to the hidden space. This is followed by linear mapping of hidden space to output space [14, 15]. The function used to design RBN takes the input vectors and target vectors (in matrix form), from newly created feature database. The function then returns a network with biases and weights in such a way that the outputs are exactly same as target speech signals. The training of RBN is shown in Fig. 4. It shows how the performance (MSE) reduces and reaches to a value of 0.07969 in 875 Epochs (iterations).

Fig. 4 Training plot of radial
basis neural network

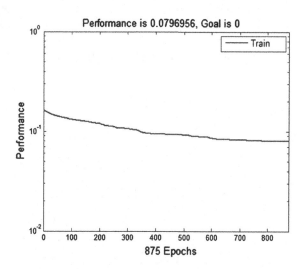

5.3 Generalized Regression Neural Network (GRNN)

Generalized Regression Neural Network is a special case of Radial Basis Network. It is often used for function approximation purpose. The structure of this Neural Network is relatively simple and static with 2 layers, namely Pattern and Summation layers. Pattern layer can also be termed as Radial Basis layer while Summation layer is a special linear layer. Once the input passes through each unit in the pattern layer, the relationship between the input and the output response is memorized and stored. Working of this network is similar to working of Radial Basis Network.

6 Result and Evaluation

It is observed that Feed Forward Neural Network takes the least execution time than RBN and GRNN. For conducting Subjective evaluation, an informal listening test was carried out with 20 subjects who were ME students having proper under-standing of the subject Speech Processing. According to the subjects, the results for all the three NNs were satisfactory. The Mean Opinion Score (MOS) was evaluated on a scale of 0–5, where 0 stands for 'Unsatisfactory' and 5 for 'Excellent'. This MOS is based on 2 judging factors that are: Clarity and Intelligibility. Table 1 shows the respective Mean Opinion Scores obtained for considered Neural Network Architectures.

All the three Neural Networks, considered so far, are further compared on the basis of their individual execution time. Computation time is measured on a computer with specifications: Random Access Memory (RAM) of 6.00 GB, Processor: Intel(R) Core (TM) i5-2450M CPU @ 2.50 GHz. For execution of proposed algorithm, MATLAB R2009b software was used.

Table 2 specifies individual computational time required by three Neural Networks while generating the Transformed speech signal from Test speech signal.

Table 1 Mean opinion scores for the three neural networks

	RBN	FFNN	GRNN
MOS	3.7	4.1	3.8

Table 2 Computational time (s)

Emotions	RBN	FFNN	GRNN
Angry	828.129	127.85	134.90
Happy	272.31	106.86	115.29
Sad	292.06	136.49	148.54

7 Conclusion

As seen from the tables, it can be concluded that the Feed Forward Neural Network gives best transformed output with respect to intelligibility and distortion as compared to RBN and GRNN. Table 2 depicts that FFNN requires least execution time when compared with the other two Neural Networks. Because of Neural Network, efficient spectral mapping function is generated which helps in achievement of proper transformation of source speech to target speech. With the use of Wavelet Packet Decomposition, the naturalness of speech is preserved while reconstruction of signal. The main advantage of our algorithm is that it is Speech Independent as well as Speaker Independent. Once the Neural Network is trained with source and target inputs, it can perform efficient transformation for any new input signal.

References

1. Oytun Turk, "New Methods for Voice Conversion," M.S. dissertation, Bogazici University, Dept. of Electrical and Electronics Engineering, 2003.
2. Pei-Chih Su and Pao-Chi Chang, "Pitch-Scale Modification Based on Formant Extraction from Resampled Speech," Dept. of Communication Engineering, National Central University, Proceedings of the 2005 workshop on Consumer Electronics and Signal Processing (WCEsp 2005).
3. Nisha. V.S and M. Jayasheela, "Speaker Identification Using Combined MFCC and Phase Information," International Journal of Advanced Research in Computer and Communication Engineering Vol. 2, Issue 2, February 2013, ISSN: 2319-5940.
4. Levent M. Arslan, " Speaker Transformation Algorithm using Segmental Codebooks (STASC)," 1999 Elsevier Science, Speech Communication 28 (1999) 211–226.
5. H. Valbret, E. Moulines, J. P. Tubach, "Voice Transformation using PSOLA Technique," Acoustics, Speech and Signal Processing, 1992. ICASSP-92., IEEE International Conference 1992.
6. Chi-Chun Hsia, Chung-Hsien Wu, Jian-Qi Wu, "Conversion Function Clustering and Selection Using Linguistic and Spectral Information for Emotional Voice Conversion", IEEE transactions on computers, vol. 56, no. 9, pp. 1245–1254, September 2007.
7. Carlo Drioli, Graziano Tisato, Piero Cosi, and Fabio Tesser, "Emotions and Voice Quality: Experiments with Sinusoidal Modeling", ISCA speech organisation archive, VOQUAL'03, Geneva, August 27–29, 2003.
8. Rania Elmanfaloty, N. Korany, El-Sayed A. Youssef, "Arabic Speech Transformation Using MFCC in GMM", International Conference on Computer and Communication Engineering (ICCCE 2012), 3–5 July 2012, Kuala Lumpur, Malaysia, pp. 734–738, 978-1-4673-0479-5/12/$31.00 ©2012 IEEE.
9. Allam Mousa, "voice conversion using pitch shifting algorithm by time stretchingwith psola and re–sampling", pp. 57–61, journal of electrical engineering, vol. 61, no. 1, 2010.
10. Rodrigo Capobianco Guido, Lucimar Sasso Vieira, Sylvio Barbon Junior, Fabricio Lopes Sanchez, Carlos Dias Maciel, Everthon Silva Fonseca, Jose Carlos Pereira, "A neural-wavelet architecture for voice conversion," Neurocomputing: An International Journal, Vol 71(1–3), Dec, 2008. pp. 174–180.
11. Dr. Shaila D. Apte, "Speech and Audio Processing", Wiley India, 1st edition.
12. Ian Mcloughlin, "Applied Speech and Audio Processing: With Matlab Examples," Cambridge University Press, 2009.

13. Vivek Vijay Nar, Alice N. Cheeran, Souvik Banerjee, "Verification of TD-PSOLA for Implementing Voice Modification," International Journal of Engineering Research and Applications (IJERA), Vol. 3, Issue 3, May–Jun 2013, ISSN: 2248-9622.
14. Jagannath Nirmala, Suprava Patnaik, Mukesh Zaveri, Pramod Kachare, " Multi-scale Speaker Transformation using Radial Basis Function," International Conference on Computational Intelligence: Modeling Techniques and Applications (CIMTA) 2013.
15. Jagannath Nirmal, Suprava Patnaik, Mukesh Zaveri, and Pramod Kachare, "Complex Cepstrum Based Voice Conversion Using Radial Basis Function," Research Article, ISRN Signal Processing, Volume 2014 (2014), Article ID 357048.

Foreground Background Segmentation for Video Processing Through Dynamic Hand Gestures

C. Swapna and Shabnam S. Shaikh

Abstract The main idea of the paper is to apply the dynamic hand gestures for the video in-painting technique of video processing, as video in-painting is an advanced enhancement technique in multimedia environment. The mouse operations of a computer are controlled by the implemented gesture events. The extraction and replacement of object i.e., foreground, background segmentation in the video frames are achieved through dynamic hand gesture using few hardware devices. The algorithm used for the object replacement in video is pixel based rather than the patch based which allows faster in-painting, improving the overall outline image quality.

Keywords Human computer interaction · Video in-painting · Segmentation · Gesture recognition

1 Introduction

The incompliance of basic input devices such as the mouse and keyboard are limiting the creativity and capabilities of the user. The aim of this paper is to develop a real-time system which is capable of understanding commands given by hand gestures and expand the ways that people are able to interact with the computers. Such kind of system have applications in virtual reality [1], human robot interaction, healthcare, computer games, and industrial control and so on.

Interaction of hand with object or without object is known as hand gesture. The main focus of this paper was to use real time hand gestures captured by the camera as frame of video which is very user friendly. The hand gestures can be static

C. Swapna (✉) · S.S. Shaikh
Computer Engineering Department, AISSMS College of Engineering, Pune, India
e-mail: chillalswapna@gmail.com

S.S. Shaikh
e-mail: shabnamfsayyad@gmail.com

© Springer India 2016
S.C. Satapathy et al. (eds.), *Information Systems Design and Intelligent Applications*, Advances in Intelligent Systems and Computing 434,
DOI 10.1007/978-81-322-2752-6_24

247

gestures or dynamic gestures. The constant gestures i.e. Static gestures, independent of the movement of hand or dynamic gestures require motion of the object and are based on the trajectory which is formed during the motion. These gestures can be recognized either by the Sensor based or Vision based technology.

In sensor based approach different types of sensors [2, 3] such as data gloves and "LEDs", color markers are used on hand. The use of external hardware damages the natural motion of the hand and complex gestures are hard to determine. In Vision based approach the bare hands were used rather than the color bands. In bare hand recognition the natural motion of the hand is detected through various types of skin detection algorithms [4, 5] and by background subtraction algorithms. The proposed method uses a YCbCr color model for the detection of hand and Finite state machine [1] for gesture recognition.

The events handled by the recognized hand gestures are applied for the video processing. The selection of the object and segmentation from a stream of video are important in video in-painting technique. After object selection, the next step is to extract and replace the object in a frame of video. Many of the approaches [6, 7] are proposed for the removing of the foreground object and to fill the region with background segmentation when the object is constant in video frames. The filling of space in foreground can be done by the Region based [6], Patch based [8], Pixel based [7] method.

In this system in-painting approach in-paints the damaged areas of an image using data from the current frame and neighboring frame using the concept of spatiotemporal extension of video. A pixel based approach of PIX-MIX [8] algorithm is used for filling the gap.

The organization of the paper is as follows in Sect. 2 the related work of hand gestures and video in-painting is studied and the architectural diagram with description of modules is presented in Sect. 3. In Sect. 4 the results are shown and Sect. 5 gives the conclusion.

2 Related Work

Most of the recent work related to hand gesture include glove based technique and vision based techniques. Keskin et al. [9] used color gloves for the recognition of the gestures of the hand. For hand gesture recognition used the Hidden Markov Model (HMM). The start and end points of the gestures were hard to determine. So the author proposed an algorithm Baum-Welch algorithm with HMM. Lee-Cosioa et al. [10] have used Dynamic Bayesian Network of Artificial neural network application for the recognition of hand gestures. The authors used the Wii remote which will rotate in X, Y, Z directions for gesture recognition. After recognizing the gestures of data, this data was normalized by k-means and filtered by Fast Fourier transform algorithm to increase the performance and computational cost. Ren et al. [2] used a Kinect Sensor to capture even a small color and depth data. Hand is detected by the camera then the black color gloves are used to find the finger tips. Introduced Fingers Earth Movers algorithm which is used to track only the finger points discarding the whole hand. Peng et al. [11] proposed a model such a way that hand gestures are used for information retrieval. For skin detection the authors used the YCbCr color model, and for detecting and tracking the hand CAMSHIFT algorithm is used. For gesture recognition Principal Component Analysis is used for gesture recognition. Dominio et al. [12] used Kinect sensor for the depth information. In the first step of hand gesture recognition, from the background the hand samples are segmented in palm, wrist and arm regions by applying the Principal Component analysis (PCA). Multi class Static Vector Machine (SVM) is used to recognize the hand gestures. In this paper for skin detection the YCbCr color model with local adaptive technique is used for the range of skin color and Finite state machine for the gesture recognition.

Most of the video in-painting algorithms depend upon the region based technique. Lee et al. [8] used the region segmentation for their proposed robust exemplar based image in-painting algorithm. Robust in-painting is done by the segmentation map which increases the performance. Neelima et al. [6] introduced a method for extracting a big object from the digital image frames. Block based sampling process with the combination of the Best first algorithm is used to fill the gap of the removed object background. Camplani et al. [13] used Kinect sensors to retrieve the depth data. Foreground and Background segmentation methods are proposed without considering the color data. The combination of classifiers were introduced to fill the gap based on average weight of the color and depth data. Camplani et al. [14] used a region segmentation and introduced a background subtraction method. This background subtraction method is used with a combination of region based foreground prediction method. Depending on the scene at pixel at region level at the foreground the object is detected by the background method. Tang et al. [15] have a proposed a model by considering a Spatio-temporal continuity for the restoration of old images. In the temporal information the gaps are filled for the extra reference from the sequence of neighboring video frames. The main disadvantage of patch based is the edges of the gap cannot be in-painted properly so in this proposed method a pixel based approach is used.

3 Architecture of Proposed System

The proposed architecture shown in Fig. 1 contains hand gesture recognition to perform the operations on objects in the video frames. The proposed system consist of mainly three modules. 1. Hand gesture recognition. 2. Input video 3. Video in-painting method.

Fig. 1 Architecture of proposed system

Fig. 2 Gesture recognition
system

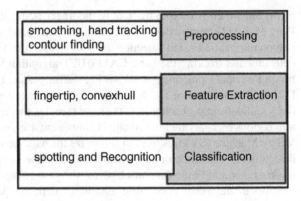

3.1 Hand Gesture Recognition

The system recognizes dynamic hand gestures captured by the camera. The hand
gesture recognition module mainly consist of three stages as shown in the Fig. 2.

1. Preprocessing.
2. Feature Extraction.
3. Classification.

3.1.1 Preprocessing

The foremost and important step in preprocessing is hand detection. The proposed
method implements skin detection with YCbCr color model. The difficulty with the
background subtraction was the segmentation of the foreground and background
was less accurate and time taken for the training was more compared to skin
detection in real time. The luminance and chrominance components are prominent
choice for skin detection in real time because of light conditions. YCbCr color
model with adaptive local technique is used to acquire the range for skin color.
Thresholding is applied that converts an image into black and white which is
followed by morphological operation of XOR. This operation enables to obtain a
single component for further processing to classify the gestures. After thresholding
skin pixels are set to 1 and other pixels are set to 0 and the image is converted to
binary image. To remove noisy data from binary image, median filter is used.
Median filter is a non-linear digital filtering technique that stores the edges while
removing noise.

The next step is to find the contour of an image. A contour is a list of points that
represent a curve in an image. The characteristic points of the contour region decide
the structural features of the hand discarding the other skin colored objects. One
approach is to find the high curvature points. Suzuki's algorithm [16] is used for
determining the contour of the hand region.

3.1.2 Feature Extraction

For comparison between the events or patterns the good features from the contour should be extracted. Finger tips, and angle between the fingers with respect to the middle finger and thumb are used to extract features. Fingertips are with the X-axis and Y-axis which has the largest Y-point is the middle finger. With respect to the middle finger and the angle the other fingers are extracted with the fingertips. The use of concept of convex hull [16] and the concave points obtained from the hull extract the features of the hand.

3.1.3 Classification

For recognition and determining the hand gestures the Finite state machine is used. Bounding box and the total line length of the hand region is measured. The 4 mouse events are handled by recognizing the 5 gestures. These mouse events are determined by assigning the numbers. If any one finger is opened then it work as a "left click", for the "right click" two fingers are opened, three fingers are opened for the "double click" and four fingers are opened for the "mouse move". For the reset all the five fingers are opened. In the "mouse move" the sequence of states are identified with "mouse move left", "mouse move right", "mouse move up", " mouse move down" and "mouse move reset". The following snapshots in the Fig. 3a indicate the detection of the hand, Fig. 3b indicate the "left click", Fig. 3c the "mouse move".

The below snapshots shows the result for the hand detection left click of the mouse and Mouse move. In Fig. 3a Hand detection the five patches takes the range of skin color. In Fig. 3b, c the binary image is shown and the events are handled for the "Left click" and "Mouse move".

The blue line or arc which is around the hand is contour image and the green dots shows the finger tips formed by the convex hull. The red line shows the bounded box area of hand region.

(a)   **(b)**   **(c)**

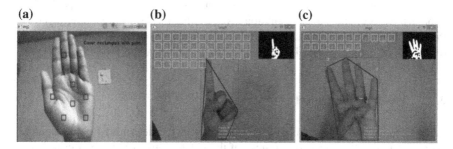

Fig. 3 a Hand detection. **b** Left click. **c** Mouse move

3.2   Input Video

The mouse events which are handled by the finite state machine are used for the selection of the video from the disk and for video in-painting algorithm. Then the frames are extracted by using the Squalar library from the video.

3.3   Video In-Painting

In multimedia applications the video in painting technique is most important and attractive method. In the proposed method simple video stream manipulation would be a static image in the current frame followed by the interpolation of neighboring video frames. This system is used to handle the static elements using synthesized key frame as a model of reference. It mainly consist of three steps. 1. Selection of object 2. Tracking of object 3. Video pipelining.

3.3.1   Selection of Object

The frame for the video in-painting is selected by the hand gesture. For the object selection many approaches are studied that includes the combinations of mean-shift, graph, painting based techniques gives segmentation results slower for video streams. In this system contour algorithm of [17] is used for tracking and selection of object. The application of fingerprints [18] is used that are distributed on rough user defined contour. The n fingerprints are compared inside the rough user defined contour and tested upon similarity. All these pixels of an object are considered to build a binary mask. After selecting the object for the in-painting the concept of clustering is used for the exact boundaries of the object. As there are many methods such as k-means, hierarchical, but they need to define number of clusters in advance. So a clustering method [18] with Nearest Neighbor Field (NNF) approach with highest probability of matching pixel is considered.

This segmentation is combined with multi-resolution approach [18]. This approach starts at the coarsest resolution layer and forward to the next finer layer. At this finer layer according to the fingerprint dissimilarity only the border pixels of an object are investigated. A dilation filter is used to remove the tiny gaps.

3.3.2   Tracking the Object

After selecting the object which is to be removed, the determined object contour should be tracked in successive frames of the video. A two phase contour tracking approach [18] is used. Homography based contour tracking is used in first phase while new contour is refined and adjusted with undesired object area in the second phase. The strongest contour points are tracked between two frames by pyramid base motion detection for homography determination [17]. The final synthesis mask is determined after the second phase, while the homography is calculated in the first phase. The mask is based on the object contour and where as homography is based on strong correspondences between contour point.

3.3.3 Video Pipelining

Through the object selection method a precise contour is found and synthesized mask is invoked. The frame I is divided into two sets target and base region with T and B.

$$I = TUB. \tag{1}$$

All pixels from target region are replaced by the pixels defined in base region. F determine the mapping between the target region and the base region and is given by

$$F = T \rightarrow B. \tag{2}$$

F has been determined the target pixels are replaced by the base pixels in creating the final image. The synthesis image is stored as the key frame F_k. Define the calculated homography as a reference model F_r for each new frame F_n with binary synthesis Mask M_n. For all desired pixels ($M_n(p) = 1$) the current frame pixels are copied for undesired pixels ($M_n(p) = 0$) and are replaced by the key frame F_k interpolation information. This interpolation is defined by homographies. The concept of mapping forwarding by homography H_k [18] for the reference model is given by

$$F_r(P) = \begin{cases} F_n(P), M_n(P) = 1 \\ F_k(H_k \cdot P), M_n(P) = 0 \end{cases} \tag{3}$$

H_k is used to reduce the number of iterations to find a transformation for the new frame. Then the new object is replaced after filling the gap with the base region. The object selection and replacement of object is done through the events which are handled by the gestures. Figure 4a shows the frame of video in this the half of the mirror is replaced with the other object of Oral B it is shown in Fig. 4b i.e., after replacement of object.

4 Results

The results of hand detection compared with the other papers are shown in Table 1.

Fig. 4 a The frame before in-painting through hand gestures. **b** The frame after in-painting through hand gestures

Table 1 Comparison of hand detection accuracy

S. no	Authors	Algorithm used	Static/dynamic	Accuracy (%)
1	Peng et al. [11]	PCA	Dynamic	93.1
2	Ren et al. [2]	Finger earth movers classification	Dynamic/black belt	93.2
3	Gharaussie et al. [9]	HMM	Dynamic	93.84
4	Dominio et al. [12]	SVM	Dynamic	92.5
5	Proposed method	FSM	Dynamic	94–96

For the hand gesture detection, each event is tested by the five persons for the 10 times and the accuracy for the each event is shown in Table 2 and the Overall percentage for the events handled ranges from 94 to 96 %.

The results for video In-painting with Pix-Mix algorithm through this events are shown in the Table 3.

Table 2 Results for the hand gesture recognition

Events detected	No. of persons	Input given output		% of accuracy
Left click	5	50	47	94
Right click	5	50	46	92
Mouse move	5	50	47	94
Reset	5	50	48	96

Table 3 Results for the video in-painting through the hand gesture

Input	Videos	No. of frames	% output video played	Time required to replace the object in the video (min)
At seashore	3	300	92	50
At the road side	3	150	93	25
Inside the building	3	90	95	20
Outside the building	3	60	95	10

The input videos are taken from different locations for testing the accuracy ranges from 92 to 95 % the time required for the larger number of frames of video is greater compared to less frames of video.

5 Conclusion

A model for the foreground background segmentation of an object in a sequence of frame of video through dynamic hand gesture is proposed. Finite state machine is used for hand gesture recognition with YCbCr color model for the skin detection is used. A Pix-Mix algorithm for the video in painting is used and is a pixel based approach. The results of preprocessing of the hand gestures and the event handling with snapshots is shown. The main idea behind this proposed method is to apply the dynamic hand gestures which are captured by the camera for the object selection and replacement of object in a frame of video in video processing and compare the results with object replacement video to the original video. The proposed method works well with minimal hardware requirements but under some background condition when detecting the hand through the camera. The Future scope include applying the hand gesture for video in-painting at the object selection with high accuracy with minimal background issue.

References

1. Alper Aksa, Orkun ztrk, and Tansel zyer, "Real-time Multi-Objective Hand Posture/Gesture Recognition by Using Distance Classifiers and Finite State Machine for Virtual Mouse Operations", IEEE Electrical and Electronics Engineering (ELECO), 2011 7th International conference on 1–4 Dec 2011.
2. Zhou Ren, Junsong Yuan, Jingjing Meng, Zhengyou Zhang. "Robust Part-Based Hand Gesture Recognition Using Kinect Sensor". IEEE Transactions on Multimedia, Vol. 15, No. 5, August 2013.
3. Popa, M. "Hand Gesture recognition based on Accelerometer Sensors". IEEE Networked Computing and Advanced Information Management (NCM), 2011.

4. Guoliang Yang, Huan Li, Li Zhang, Yue Cao. " Research on a Skin Colour Detection Algorithm Based on Self-Adaptive Skin Colour Model". IEEE Communications and Intelligence Information Security (ICCIIS), 2010.
5. Lei Yang, Hui Li, Xiaoyu Wu, Dewei Zhao, Jun Zhai. "An algorithm of skin detection based on texture". IEEE Image and Signal Processing (CSIP), 2011.
6. N. Neelima, M. Arulvan, B. S Abdur, "Object Removal by Region Based Filling Inpainting". IEEE transaction 978-1-4673-5301/2OI3 IEEE.
7. Jan Herling, and Wolfgang Broll, "High-Quality Real-Time Video Inpainting with PixMix", IEEE Transcations On Visualization and computer graphics, Vol. 20, No. 6, June 2014.
8. Jino Lee, Dong-Kyu Lee, and Rae-Hong Park, "A Robust Exemplar-Based Inpainting Algorithm Using Region Segmentation". IEEE Transactions on Consumer Electronics, Vol. 58, No. 2, May 2011.
9. M.M Gharaussie, H. Seyedarabi. "Real time hand gesture recognition for using HMM". IEEE Transaction 2013.
10. Blanca Miriam Lee-Cosioa, Carlos Delgado-Mataa, Jesus Ibanezb. "ANN for Gesture Recognition using Accelerometer Data". Elsevier Publications, Procedia Technology 3 (2012).
11. Sheng-Yu Peng, Kanoksak Wattanachote, Hwei-Jen Lin, Kuan-Ching Li, "A Real-Time Hand Gesture Recognition System for Daily Information Retrieval from Internet", IEEE Fourth International Conference on Ubi-Media Computing, 978-0-7695-4493-9/11 © 2011.
12. Fabio Dominio, Mauro Donadeo, Pietro Zanuttigh, "Combining multiple depth-based descriptors for hand gesture recognition", Elsevier, Pattern recognition Letters 2013.
13. Massimo Camplani, Luis Salgado, "Background foreground segmentation with RGB-D Kinect data: An efficient combination of classifiers". published at Elsevier J. Vis. Commun. Image R. 25 (2014) 122–136.
14. Massimo Camplani, Carlos R. del Blanco, Luis Salgado, Fernando Jaureguizar, Narciso García, "Multi-sensor background subtraction by fusing multiple region-based probabilistic classifiers", published at Elsevier Pattern Recognition Letters 2013.
15. Nick C. Tang, Chiou-Ting Hsu, Chih-Wen Su, Timothy K. Shih, and Hong-Yuan Mark Liao, "Video Inpainting on Digitized Vintage Films via Maintaining Spatiotemporal Continuity", published at IEEE Transactions on Multimedia, Vol. 13, No. 4, August 2011.
16. Ogata. K Futatsugi. K, "Analysis of the Suzuki-Kasami algorithm with SAL model checkers", Software Engineering Conference 12th Asia Pacific. Published in 2005.
17. J. Herling and W. Broll, "Advanced Self-Contained Object Removal for Realizing Real-Time Diminished Reality in Unconstrained Environments," Proc. IEEE Ninth Int'l Symp. Mixed and Augmented Reality (ISMAR'10), pp. 207–212, Oct. 2010.
18. Jan Herling, and Wolfgang Broll, "High-Quality Real-Time Video Inpainting with PixMix", IEEE Transaction On Visualisation of Computer Graphics, Vol. 20, No. 6, June 2014.

Load Encroachment Detection Algorithm for Digital Signal Processor Based Numerical Relay

Shanker Warathe and R.N. Patel

Abstract In this paper we have described the cause of Indian grid failure during the stressed condition and developed the algorithm for numerical relay to prevent the unforeseen blackout. Indian power grid is consists of five regional grids and one of the largest power supply utility in the world. There was a major grid disturbance occurred in the Northern Region of India causes major shut down in many state leading to a entire blackout in eight states. The regional wise power import, export and generation give the idea of stressed condition and skewed power generation among the grid. The voltage profile of regional buses also indicates the insufficient reactive power compensation. In this work the impact of increase of load and loss of generation on angular separation is observed and the performance of developed algorithm for numerical relay is tested for different stressed condition and gives the expect result.

Keywords National grid · Numerical relay · Load encroachment · Algorithm · Digital signal processor · Synchro-phasors measurement unit · Global time reference

1 Introduction

The Indian black outs occurred on 30th and 31st July 2012, due to multiple outages between western and northern grid and the cascading tripping of the tie lines between western and eastern grids [1]. This situation gives the first alarm to take major steps to prevent the unforeseen collapses of the Indian electric grids. The numerical relay embedded with digital signal processor (DSP) is possible solution

S. Warathe (✉)
Department of Electrical Engineering, Government Polytechnic Durg, Durg, India
e-mail: sb.govt@gmail.com

R.N. Patel
Department of Electrical and Electronic Engineering, SSCET, Bhilai, India

© Springer India 2016
S.C. Satapathy et al. (eds.), *Information Systems Design and Intelligent Applications*, Advances in Intelligent Systems and Computing 434,
DOI 10.1007/978-81-322-2752-6_25

257

for prevention of unforeseen blackout [2]. The digital signal processor (DSP) based relays provides high speed precision calculation for the mathematical algorithm involved in the processor. These relays are capable of performing complex processing faster and with higher accuracy [3–7]. The disturbance recorders provide additional information of phasor angle by phasor measurement unit (PMU) measurement and reported in paper [8]. The IEEE C37.118 standard on the synchrophasors outlines certain stringent requirements in terms of how accurately measure the load angle with reference to the global time–the coordinated universal time and how to report the phasor information [9]. The standard also specifies the Total Vector Error allowed in evaluating the phasor for different compliance level to allow interoperability between different vendor PMUs. To avoid the blackout situation many techniques are available in literature. In paper [10], the author proposed the relaying algorithms to detect virtual faults prefer not to use communication links to design and to base its decision on values of current and voltage at the relay location [11] while on the other hand some other techniques [12–14] use new technologies based on remotely measured data like synchronized phasor measurements (SPM) and fiber optic communications.

The malfunction in zone three protections of impedance relays with offset mho-characteristics causes cascading failures of network as seen in several previous blackouts in the world [11]. This malfunction could be due to the increase of the electrical load level to the limit that the relay interprets the system voltage and current into impedance that its value appears to the relay as if it is a fault while it is not, e.g. load encroachment.

1.1 Indian Power Grid

The synchronously connected Indian grid is shown in Fig. 1, it consists of NEW Grid comprising of the Northern, Eastern, Western and North-Eastern Grids are meeting a demand of about 75,000.00–80,000.00 MW and the Southern Grid which is connected to NEW Grid asynchronously, is meeting a demand of about 30,000.00 MW. The backbone of Indian national grid is formed by the 400 kV transmission line and the newly 765 kV transmission lines [1]. In the month of July 2012, due to failure of monsoon and high electrical loading in the Northern region of India was the main cause of over drawing bulk amount of electrical power from the neighbouring regional grids like Western and Eastern as shown in Fig. 2 by blue colour import bar, the Fig. 2 show the plot of field data on the incident day. The electrical power demand was less in the Western and Eastern region due to sufficient rain and it was under drawing condition as clearly shown in Fig. 2 by blue colour export bar. The major grid disturbance in month of July 2012 was occurred due to heavily loaded tie lines between East to North and West to North grid of India to feed the over draw power and skewed power generation balance among the regional grids as shown in Fig. 2, by generation and demand bar. This disturbance causes blackout in the entire Northern region of India covering neighboring eight

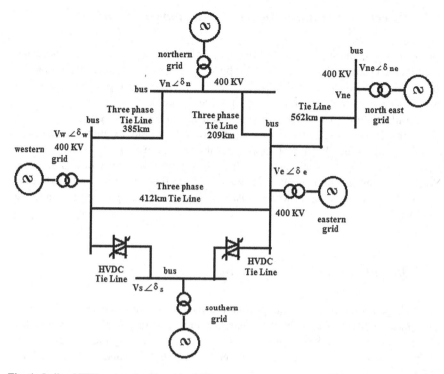

Fig. 1 Indian NEW regional grids with tie lines

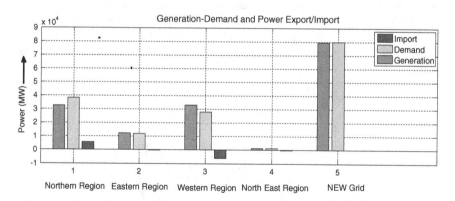

Fig. 2 Generation demand and exchange of power

states. The NEW grid frequency prior to the incident was 49.68 Hz. The National demand prior to the incident was nearly 99700.00 MW and the demand being met in the Northern Region of India was 38000.00 MW [1].

1.2 Power Generation—Demand and Import/Export

The power generation, import and export on the day of incident just before the disturbance is shown by the Fig. 2. On the x axis regional grids are shown and on the y axis the power is plotted in MW. The region wise power generation, demand and import/export are shown by green, pink and blue bar.

The illustration clearly shows the demand is more than the generation in the Northern regional grid and power is imported from the neighboring Western regional grid, in the Western region generation is more than the demand in the region so power is exported as shown by blue bar from the Western region to the Northern region and tie lines are over loaded. In the Eastern region and North East region the generation of power is just meeting the demand of power in there region [1].

1.3 Regional Bus Voltage Profile

The voltage profile of the Indian grid before the grid disturbance on the 30th July and 31st July 2012 is shown by the illustration Fig. 3, voltage in kV is shown on the y axis and region wise bus is shown on x axis. On both the day the voltage pattern for Northern region is similar except one node voltage is high and the range is 380–415 kV but bus voltages in Eastern region above the rated voltage and it shows the Eastern region is lightly loaded. In the Western region the voltage profile on both days are just opposite to the each other and voltage range is wide this shows at some buses the load is lightly connected and on other buses the load is heavily connected. It shows imbalance power generation among the region and inter region. The voltage profile also shows the inadequate reactive power compensation in the grid [1].

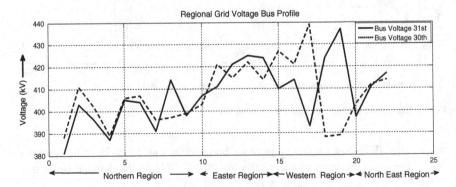

Fig. 3 Voltage profile in four regions on 30th and 31st July 2012

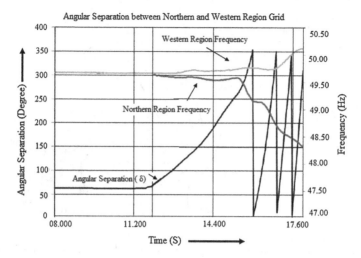

Fig. 4 Angular separation between Northern and Western regions during disturbance

1.4 Angular Separation Between Northern and Western Region

The frequency profile and angular separation of the Northern region and the Western region is shown in Fig. 4, on the day of grid disturbance the frequency of both the region is nearly 49.68 Hz and constant before the incident occurred at 13.00 h [1]. After the incident the frequency of Northern region is decrease as shown by the blue curve, this shows there was heavy loading and reduction in power generation but simultaneously the frequency in Western region shown by red curve increases it shows more generation in the region and lightly loaded regional grid. Same time it clearly shows the angular separation between Western and Northern region is increases linearly shown by the black curve. Nearly at time 15.40 h there was number of generators tripped in Northern region causes blackout.

2 Load Encroachment

During the peak demand the line current is increases and voltage is decreases as a result the apparent impedance decreases. If the load demand is continuously increasing the overloading on the transmission line also increases and sometimes it reaches beyond the loading capacity of transmission line and the apparent impedance decreases below the setting of mho relay zone three protection. The value of impedance causes overloading approaches in the protective zone of mho relay is called load encroachment. The impedance relay used for distance protection of transmission line also measuring this apparent impedance of load called as

Fig. 5 Load encroachment
characteristics on R-X plot

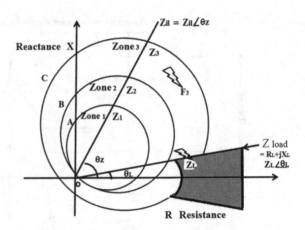

load-impedance. The magnitude of this apparent impedance of load varies with the variation in load. The relay measures the impedance value which is inside the relay's characteristics shape. As a result, the distance relay, see the value of impedance below the pre set value of impedance and consequently issue the trip command to circuit breaker. This is a highly undesirable operation of relay because a heavily loaded line has been taken out of service causes heavy power swing and other disturbances, when there is no real fault exists and this condition lead to black out situation. The impedance of large load is low as compared to impedance of faulty line. So the protective relay must be made so selective to discriminate between impedance of higher load and the impedance of fault condition. Figure 5 shows the load encroachment as the load increases the equivalent of load resistance decreases and fall within the zone 3 of mho relay characteristics.

3   Simulation Model and Result

The Indian grid simulation model is shown in Fig. 6; the western region grid is represented by Thevenin's equivalence source of 220.0 kV, 50.0 Hz. The 33024 MW power feed to the national grid. Similarly the northern region grid, eastern grid and north east grids are represented by Thevenin's equivalence source of 220.0 kV, 50.0 Hz and feeding the electrical power 32636.0, 12452.0 and 1367.0 MW to the national grid. The inter regional first tie lines is three phase transmission line of 385.0 km from northern region to western region, second tie lines between east region to northern region of 209.0 km, third tie line length between north east to east region is 562.0 km and the fourth inter tie lines between western region to east region is 412.0 km.

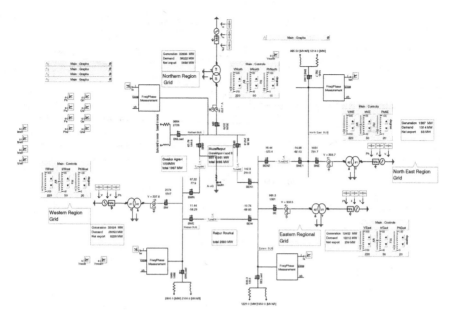

Fig. 6 Simulation model in PSCAD

Table 1 Electrical load and an angle separation between regional grid buses

Northern region load (MW)	Northern bus voltage angle (θ°)	Western bus voltage angle (θ°)	Eastern bus voltage angle (θ°)
793.2	31.48	31.78	31.66
1254.0	32.27	31.86	30.68
1704.0	33.05	31.95	30.71
2153.0	33.83	32.04	30.73
2596.0	34.6	32.12	30.76
3033.0	35.38	32.21	30.78
3465.0	36.14	32.29	30.81

The simulation is run for 5.0 s to observe the impact of increasing the load on the angular separation between regional grids. The observations are tabulated in Table 1, for three regional grid buses voltage angles by varying the electrical load in northern region from 793.20 to 3465.00 MW and illustrated by Fig. 7, it clearly shows the increasing the electrical load in the northern region the angular separation also increases linearly between northern and western region. Figure 8 shows the angular separation for loss of power generation in western region and Fig. 9 shows the loss of generation in northern region.

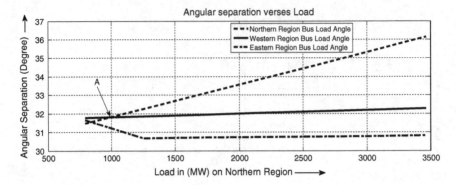

Fig. 7 Angular separation verses electrical load

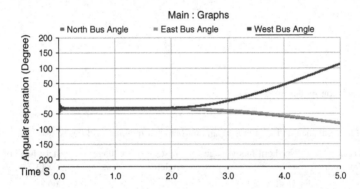

Fig. 8 Angular separation between regional buses during loss of power in western region

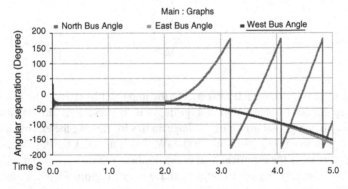

Fig. 9 Angular separation between regional buses during loss of power in northern region

Fig. 10 Algorithm for
numerical relay

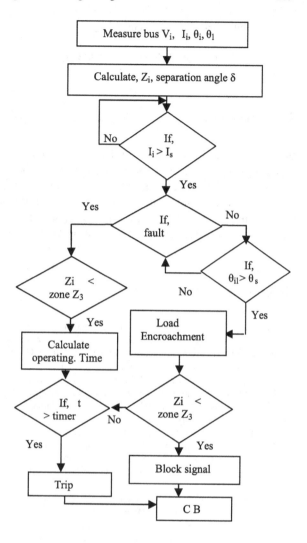

3.1 Numerical Relay Algorithm

On the basis of above observation and illustration the algorithm for DSP based
numerical relay is developed, shown in Fig. 10, the algorithm is tested for stressed
load condition and loss of generation, when the load angle between buses is
exceeded beyond the pre set value of load angle between regional grid buses the
relay will issue the block command and prevent the mal functioning of relay due to
load encroachment.

Suppose there is n number of buses in the grid and load angle for Ith bus and lth
bus is calculated as Eq. 1, measured voltage, current and line impedance are given
as V_i, I_i, Z_i.

Load angle between bus i and l is given by

$$\text{Load angle between bus } i \text{ and } l \text{ is given by } \delta = (\theta_i - \theta_l) \tag{1}$$

4 Conclusion

In this work the Indian power grid is divided into five regional grids and represented by Thevenin's equivalent source and the model was simulated for three stressed load condition, when the load is continuous increases the angular separation also increases and numerical algorithm detect the such condition and issuing the blocking command when the angular separation value exceed its pre set value and prevent the malfunction of relay during load encroachment.

References

1. Report of the Grid Disturbances on 30th July and 31st July 2012 by POSOCO (2012).
2. Goh Y. L., Ramasamy A. K., Nagi F. H., Abidin. A. Z., Evaluation of DSP based Numerical Relay for Over current Protection, International Journal Of Systems Applications, Engineering & Development, Issue 3, Volume 5, pp. 396–403, (2011).
3. Jingxuan H., McLaren P. G., Dirks E., Wang P., and Bisewski B., A graphical block platform for real-time DSP-based digital relay development, Power System Technology, Proceedings. Power Con 2002. International Conference on Volume 1, pp. 265– 269, (2002).
4. Chul-Hwan Kim, Myung-Hee Lee, Aggarwal R. K., Johns A. T., Educational Use of EMTP MODELS for the Study of a Distance Relay Algorithm for Protecting Transmission Lines, IEEE Transactions on Power Systems, Vol. 15, No. 1, (2000).
5. Jonsson, M.; Daalder, J.E., An adaptive scheme to prevent undesirable distance protection operation during voltage instability, Power Delivery, IEEE Transactions on, vol. 18, no. 4, pp. 1174–1180, (2003).
6. Biswas B., Bhattacharya K. D., Purkait P., and Das J. K., Design of a Numerical Adaptive Relay Based on Memory Mapped Technique, Proceedings of the International Multi Conference of Engineers and Computer Scientists Vol II, pp. 1484–1488, Hong Kong, (2009).
7. Darwish H. A., and Fikri M., Practical considerations for recursive DFT implementation in numerical relays, IEEE Trans. on Power Delivery, vol. 22, pp. 42–49, (2007).
8. Phadke A. G., Thorp J. S, Synchronized Phasor Measurements and Their Applications, Springer, pp. 213, (2008).
9. IEEE Standard for Synchrophasors for Power Systems, IEEE C37.118, (2005).
10. Ahmad Farid Abidin, Azah Mohamed, Afida Ayob, "A New Method to Prevent Undesirable Distance Relay Tripping During Voltage Collapse", European Journal of Scientific Research, Vol.31 No.1, pp. 59–71, (2009).
11. Seong-Il Lim; Chen-Ching Liu; Seung-Jae Lee; Myeon-Song Choi; Seong-Jeong Rim, Blocking of Zone 3 Relays to Prevent Cascaded Events, IEEE Transactions on Power Systems, vol. 23, no. 2, pp. 747–754, (2008).
12. Machowski Jan, Janusz W. Bialek, James R. Bumby, Power System Dynamics: Stability and Control, 2nd Edition, John Wiley & Sons Ltd., pp. 121, (2008).

13. Nan Zhang; Kezunovic, M., A study of synchronized sampling based fault location algorithm performance under power swing and out-of-step conditions, Power Tech, 2005 IEEE Russia, vol., no., pp. 1–7, 27–30, (2005).
14. Apostolov A., Tholomier D., Richards S., Zone 3 Distance Protection—Yes or No?, 60th Annual Georgia Tech Protective Relaying Conference (2006).

K-Nearest Neighbor and Boundary Cutting Algorithm for Intrusion Detection System

Punam Mulak, D.P. Gaikwad and N.R. Talhar

Abstract Intrusion detection system is used for securing computer networks. Different data mining techniques are used for intrusion detection system with low accuracy and high false positive rate. Hicuts, HyperCuts, and EffiCuts are decision tree based packet classification algorithm which performs excellent search in classifier but requires high amount of memory. So in order to overcome these disadvantages, new approach is provided. In this, we present a hybrid approach for intrusion detection system. Boundary Cutting Algorithm and K-Nearest Neighbor using Manhattan and Jaccard coefficient similarity distance is used for high detection rate, low false alarm and less memory requirement. KDD Cup 99 dataset is used for evaluation of these algorithms. Result is evaluated using KDD CUP 99 dataset in term of accuracy, false alarm rate. Majority voting is done. This approach provides high accuracy and low memory requirements as compare to other algorithm.

Keywords Intrusion detection system · Supervised learning · Unsupervised leaning · Data mining

1 Introduction

Network security has become more and more important because of the rapid development of network services and sensitive information on the network. Although different security technologies are available for securing computer network, there are many undetected attacks. IDS are used to find any malicious attack on network. Therefore, intrusion detection system has become a research area.

P. Mulak (✉) · D.P. Gaikwad · N.R. Talhar
Computer Department, AISSMS College of Engineering, Kennedy Road, Pune, India
e-mail: punammulak@gmail.com

D.P. Gaikwad
e-mail: dp.g@rediffmail.com

N.R. Talhar
e-mail: nrtalhar@gmail.com

© Springer India 2016   269
S.C. Satapathy et al. (eds.), *Information Systems Design and Intelligent Applications*, Advances in Intelligent Systems and Computing 434,
DOI 10.1007/978-81-322-2752-6_26

Intrusion detection system is mainly divided into three types [1]: Host based; Network based and Hybrid Intrusion detection system. Host based IDS is deployed at host level. All the parts of dynamic behavior and state of computer system are analyzed by a host based IDS. Network based IDS observe all the packets into the network and find the threats which affects network. Hybrid IDS is combination of both network and host based IDS. Supervised and unsupervised learning approach is used in intrusion detection system. Supervised learning predicts unknown tuples from training dataset. Hidden structure from unlabeled data can be found using unsupervised learning.

Different data mining and packet classification techniques are provided for intrusion detection system. Most of data mining techniques give high misclassification cost and false positive rate. So due to this reason quality of product may degrade. Hicuts, HyperCuts, and EffiCuts are decision tree based packet classification algorithm which performs excellent search in classifier but requires high amount of memory.

In this paper, a new method is introduced by combining data mining and packet classification approach to reduce memory requirement for rule matching and increase intrusion detection rate. Two proximity measures: Jaccard coefficient and Manhattan Distance and boundary cutting packet classification algorithm is used in proposed framework for high detection rate, low false alarm and less memory requirement. Jaccard coefficient requires less computation and provides high accuracy. Jaccard Coefficient and Manhattan distance helps to improve quality of classification algorithm. KDD Cup 99 dataset is used for evaluation of these algorithms. Classification is done based on voting criteria. Result is evaluated using KDD CUP 99 dataset in term of accuracy, false alarm rate. This combined approach provides high accuracy and low memory requirements than other algorithm.

Rest of the paper is organized as follows: Sect. 2 presents related work for intrusion detection system. Section 3 describes theoretical background of used algorithm. Section 4 provides proposed system. Section 5 gives experimental result of the system. And Sect. 5 present conclusion of paper.

2 Related Work

Different mechanism is used for intrusion detection system. In paper [2], author used random forest to reduce tedious rule formation for network traffic. Active routers are identified by using anomaly based detection. Packets are checked by using active router and random forest is used to detect attacks. Misuse and anomaly detection are performed by using random forest. In paper [3] author used decision tree data mining algorithm for intrusion detection and then compared performance with support vector machine. Both algorithms are evaluated using KDD cup 99 dataset. Decision tree algorithm gives better performance than SVM. Neural network is used for intrusion detection system and classification of attacks [4]. Author used multilayer Perceptron for intrusion detection system in offline approach.

In order to improve the detection and classification rate accuracy and achieve high efficiency, author proposed a multi-layer hybrid approach [5]. Three layers are implemented. In first layer, feature selection is done using principal component analysis algorithm. Anomaly detector is generated by using genetic algorithm which differentiates normal and abnormal behaviors. Classification is done using Different classifiers such as naive Bayes, multilayer perceptron neural. Bhattacharyya introduced clustering based classification method for Network intrusion detection system [6]. Similarity function is used to divide a set of labeled training data into clusters. Methods are provided for clustering, training and prediction. TUIDS Intrusion data set is used to evaluate proposed system.

Yanya and Yongzhong present multi-label k-Nearest Neighbor algorithm with semi supervised and multi-label learning approach [7]. K-nearest neighbor algorithm is used to indentify unlabeled data. Then Maximum a posterior principle is used to determine the label set for unlabeled data. Overall performance of intrusion detection system is increased. Hierarchical clustering is used for IDS and achieved 0.5 % false positive rate which is evaluated on KDDCup99 [8]. Te-Shun Chou introduced hybrid classifier systems [9] using fuzzy belief KNN for intrusion detection system. Overall detection rate is 93.65 %.

3 Theoretical Background

3.1 K-Nearest Neighbour Algorithm

K-nearest neighbor is instance based learning. It is the lazy learner. Lazy learners do more work at the time of classification than training [10]. Firstly all training tuples are stored. K-NN algorithm calculates the distance between test tuple $z = (x', y')$ and all the training tuples $(x, y) \in D$ which helps to decide its nearest neighbor list, D_z. Test tuple is classified based using majority voting class.

Algorithm

---

1. Determine k as number of nearest neighbors. D is the set of training tuples.
2. **for** each test example $z = (x', y')$ do
3. Compute $d(x', x)$, calculate the distance between z and every training tuple, $(x,y) \in D$
4. Select D_z, the set of k closest training example to z.
5. Classify test example based on majority voting
$$y' = argmax \sum_{(xi,yi) \in D_z} I(v = yi)$$
6. **end for**

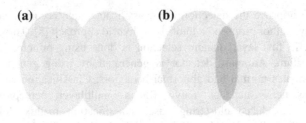

Fig. 1 The region of union and intersection between two sets P and Q. **a** PUQ. **b** PnQ

Different proximity measures are used to calculate the distance between training tuple and test tuple. It is describe as follows:

A. **Jaccard Coefficient Distance**

The similarity between sample sets is calculated by using Jaccard coefficient [11]. As shown in figure, the region of Intersection $(P \cap Q)$ and Union $(P \cup Q)$ between these two sets can be measured according to set theory (Fig. 1). The Jaccard coefficient can be given as below:

$$JC(P,Q) = \left| \frac{P \cap Q}{P \cup Q} \right| \tag{1}$$

where P and Q are two different sets. Jaccard distance finds dissimilarity between two different sets. It is complementary of Jaccard coefficient (JC). The Jaccard distance can be given as below:

$$JC(P,Q) = 1 - JC(P,Q) = \frac{|P \cup Q| - |P \cap Q|}{|P \cap Q|} \tag{2}$$

B. **Manhattan Distance**

It is the distance between two points measured along axes. The distance between each data point and training tuples is calculated using the Manhattan distance metric as follows [12]:

$$Dist_{xy} = \left| X_{ik} - X_{jk} \right| \tag{3}$$

3.2 Boundary Cutting

Packet classification is one of essential function in network security. Hicuts, HyperCuts, EffiCuts are packet classification algorithm which perform excellent search but requires huge storage requirements. Boundary cutting (BC) algorithm is

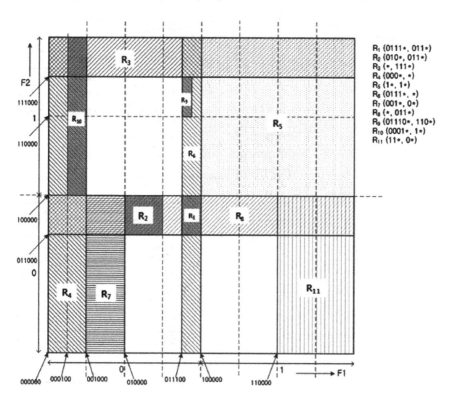

R₁ (0111*, 011*)
R₂ (010*, 011*)
R₃ (*, 111*)
R₄ (000*, *)
R₅ (1*, 1*)
R₆ (0111*, *)
R₇ (001*, 0*)
R₈ (*, 011*)
R₉ (01110*, 110*)
R₁₀ (0001*, 1*)
R₁₁ (11*, 0*)

Fig. 2 Prefix plane for rules

used to overcome these disadvantages. BC finds the space that each rule covers. And then perform cutting according to the space boundary [13].

Consider two dimensional (2-D) planes which consist of first two prefix fields. An area is used to present the rule. Area that each rules cover in prefix plane are given in following Fig. 2.

In case of Hicuts algorithm, it cuts the spaces into subspaces using one dimension per step. Space and binth are used. A space factor (spfac) is space measure function used to give the number of cuts for selected field. The binth is a function which predetermined number of rules. Process starts from root node, and then bits of a header field are checked that is set at each interval node until a leaf node is reached to condition. But in this algorithm, cutting is based on a fixed interval and partitioning is ineffective. In order to balanced required memory size and search. In boundary cutting algorithm, each rule of starting and ending boundaries can be used for cutting. Decision tree for boundary cutting algorithm is given in Fig. 3. Binth is set to 3. Fixed intervals are not used for cuts at each internal node of Boundary cutting decision tree. This decision tree algorithm finds for subspace in which packet belongs and compares the header of input packets for

Fig. 3 Decision tree using
boundary cutting algorithm

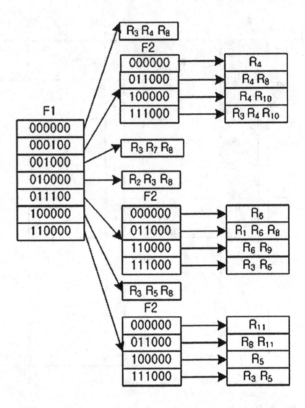

entire fields to rules belonging the subspace. Unnecessary cuttings are avoided.
Rules replication is also reduced.

3.3 Normalization Techniques

The dataset available may contain some row data. Attributes of dataset have different data types. Range of these attributes may vary widely. The attributes are
scaled to fit into a specific range. All the values are divided by the maximum value
in the dataset. This leads to normalization. Different methods are provided for
normalization [14].

1. Min-Max Normalization: This method performs linear transformation on original data. It maps a value of v_i in the range of new_min_A, new_max_A.

$$v_i' = \frac{v_i - min_A}{max_A - min_A}\left(new\_\max_A - new\_\min_A\right) + new_{min_A} \qquad (4)$$

2. Z-score normalization: The value for attribute A is normalized by using mean
   and standard deviation of A. Equation for Z-score normalization is given below:

$$v_i' = \frac{v_i - A}{\sigma_A} \tag{5}$$

3. Decimal Scaling: Normalization is performed by moving the decimal point of value of attribute A. Result of this method is in between -1 and 1.

$$v_i' = \frac{v_j}{10^j} \tag{6}$$

4 Proposed System

The proposed system is a intrusion detection system which is developed using two proximity measures and packet classification algorithm i.e. Jaccard Coefficient based classification, Manhattan distance and Boundary cutting algorithm respectively as shown in Fig. 4. Majority voting has been done to increase the overall accuracy.

System is presented with authentication. User need to first login to the system.

Here, KDDCup99 dataset is used in system. This dataset is given as input to training module. KDD dataset is divided into different files depending on attack type like training3attack, training4attack, training5attack, training6attack and so on. Attributes of KDD dataset have different data types. Range of these attributes may vary widely. So there is need to perform normalization. Different methods are provided for normalization. Among them min-max normalization method is more effective. So here min-max normalization is performed to fit the attributes of dataset into specific range. KDD dataset contain some string values like services, flags, etc. which is not understand by classifiers. So these string values are converted into numeric value. After normalization, system is trained using three classifier i.e. K-NN using two proximity measures and boundary cutting algorithm. All the training tuples are stored in nearest neighbor classifier and value of k is defined. When new test tuple is given then distance between all training tuples and test tuple is calculated using Jaccard and Manhattan distance separately. All the tuples are sorted in ascending order. And majority voting is done to assign class label to test tuple. This majority

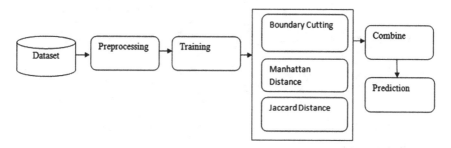

Fig. 4 The architecture of the proposed intrusion detection system

voting is depends on closeness of test tuple and training tuples. In case of boundary cutting algorithm, decision tree is build using boundary cutting algorithm. K-dimension header fields are considered for BC. Here, k = 10 means first 10 attributes of dataset are used for cutting. Algorithm cuts the space into smaller sub-regions, one dimension per step. Bucket size is taken as 16. Space factor is 2.0. Decision tree is constructed according cuts to header fields. Cutting is stopped when no space found in header fields. Depth of this tree is 5. Classification of dataset is done using decision tree. Cross Validation is used for testing. Main goal of cross validation is to check how accurately a predictive model will predict the unknown data. In cross validation, dataset is partition into subsets. One subset is given to training and other for testing. This partition of dataset is called folds. Maximum 10 folds cross validation can be done. Here, 3-fold cross validation is used because it provides more accuracy. KDD dataset is divided into three subsets. This division of dataset is done randomly. First two parts is provided for training and other part is applied for testing against trained algorithms. Combined approach is also provided. In combine approach majority voting is done. It is the process by which multiple classifiers are combined to solve a particular problem intelligently to improve the performance of a model. In system, predictions of these three algorithms are compared and majority of predictions are return. This method improves the accuracy of system.

5 Experimental Result

The proposed intrusion detection system using majority voting of BC algorithm and K-NN is evaluated and tested using KDDCup99 dataset. The experiments are performed on Laptop with 2 GB RAM and Pentium Dual Core processor. The performance of the proposed system is evaluated in term of classification accuracy. It is found that the classification accuracy of the proposed combined approach is increased up to 99 %. Following graph shows attack wise classification accuracy of the system (Fig. 5).

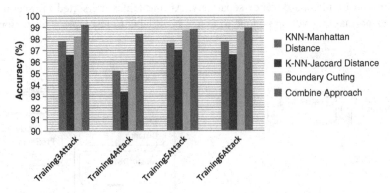

Fig. 5 Accuracy graph

6 Conclusion

In this paper, the architecture of general intrusion detection system have been studied and presented. The Hicuts algorithm is mostly used for packet classification in router. This algorithm can give better search performance, but it takes more memory for large rules. To overcome this problem, the BC algorithm is used with K-NN approach. The intrusion detection system has implemented using the combination method of BC algorithm and K-NN. The experiments have performed on separate algorithms and combined approach. It is found that the combined method provides better classification accuracy as compared to individual approaches. KNN using Jaccard and Manhattan distance gives more accuracy, sensitivity and specificity than other distance measures. The combined approach exhibits low false positive rate as compared to individual approaches. It is found that overall performance of the proposed classifier is increased due to the combination of these two approaches. The main disadvantage of the system is that it takes more time to generate rule from training dataset.

References

1. Punam Mulak, Nitin Talhar.: Novel Intrusion Detection System Using Hybrid Aprroach, International Journal of Advanced Research in Computer Science and Software Engineering, Volume 4, Issue 11, November (2014).
2. Prashanth, V. Prashanth, P. Jayashree, N. Srinivasan.: Using Random Forests for Network-based Anomaly detection at Active routers, IEEE International Conference on Signal processing, Communication and networking, Madras Institute of Technology, Anna University Chennai India, and Jan 4–6, (2008).
3. Sandhya Peddabachigari, Ajith Abraham, Johnson Thomas.: Intrusion Detection Systems Using Decision Trees and Support Vector Machines, IEEE.
4. Mehdi MORADI and Mohammad ZULKERNINE.: A Neural Network Based System for Intrusion Detection and Classification of Attacks.
5. Amira Sayed A. Aziz Aboul Ella Hassanien Sanaa El-Ola Hanafy M.F. Tolba.: Multi-layer hybrid machine learning techniques for anomalies detection and classification approach, IEEE, (2013).
6. Prasanta Gogoi, B. Borah and D. K. Bhattacharyya.: Network Anomaly Identification using Supervised Classifier, Informatica 37 93–7 (2013).
7. Yanyan Qian, Yongzhong Li.: An Intrusion Detection Algorithm Based on Multi-label Learning, Workshop on Electronics, Computer and Applications, IEEE, (2014).
8. R. W.-w Hu Liang and R. Fei.: An adaptive anomaly detection based on hierarchical clustering, Information Science and Engineering (ICISE), 2009 1st International Conference on, Changchun, China, Dec. 2009, pp. 1626–1629.
9. Zhao Ruan, Xianfeng Li, Wenjun Li.: An Energy-efficient TCAM-based Packet Classification with Decision-tree Mapping, IEEE, (2013).
10. Pang-Nang Tan, Michael Steinbach, Vipin Kumar.: Data Mining.
11. Rajendra Prasad Palnatya, Rajendra Prasad Palnaty.: JCADS: Semi-Supervised Clustering Algorithm for Network Anomaly Intrusion Detection Systems, IEEE, (2013).
12. Archana Singh, Avantika Yadav, Ajay Rana.: K-means with Three different Distance Metrics International Journal of Computer Applications, Volume 67–No.10, April (2013).

13. Hyesook Lim, Nara Lee, Geumdan Jin, Jungwon Lee, Youngju Choi, Changhoon Yim.: Boundary Cutting for Packet Classification, IEEE/ACM TRANSACTIONS ON NETWORKING, VOL. 22, NO. 2, APRIL, (2014).
14. Jiawei Han, Micheline Kamber, Jian Pei.: Data Mining concepts Technologies, Third Edition Elsevier.

Analysis of Electromagnetic Radiation from Cellular Networks Considering Geographical Distance and Height of the Antenna

S. Venkatesulu, S. Varadarajan, M. Sadasiva, A. Chandrababu and L. Hara Hara Brahma

Abstract In this article, an analysis of the electromagnetic radiation emitted by Cellular Networks and other wireless communication devices located in urban and rural areas, is carried out by taking the different exposure situations into account. The emitted RF radiation power values are estimated through the derived theoretical analysis. As per the FCC, TRAI and DOT standards, it is less hazardous if the observed radiation is below −30 dBm. From these standards, the compliance distance from the cellular base stations antenna can be determined. The EM radiation emitted from the different RF sources is analyzed by computing the radiation power theoretically at given distance from the cellular base station antenna. This analysis is further enhanced by considering the some more parameters like distance of separation between the situated antennas, frequency of operation besides the parameter of distance from the RF source. This improved analysis adds accuracy in determining the compliance distance beyond which the emitted radiation is not significant.

Keywords Electromagnetic radiation · Field intensities · Power density · Received power · Cellular networks

S. Venkatesulu (✉) · M. Sadasiva · A. Chandrababu
Department of ECE, YITS, Tirupati, India
e-mail: v.sugali7@gmail.com

M. Sadasiva
e-mail: m.sadasiva@gmail.com

A. Chandrababu
e-mail: chandrababunit@gmail.com

S. Varadarajan
ECE, SV University, Tirupati, India
e-mail: varadasouri@gmail.com

L. Hara Hara Brahma
Department of ECE, KEC, Kuppam, India
e-mail: brahmahari426@gmail.com

© Springer India 2016
S.C. Satapathy et al. (eds.), *Information Systems Design and Intelligent Applications*, Advances in Intelligent Systems and Computing 434,
DOI 10.1007/978-81-322-2752-6_27

279

1 Introduction

As per the pressing need, there is a tremendous increase in the usage of cellular mobile communications and broadcasting systems. Cellular mobile communication was started to be widely utilized in the 21 century all over the world [1–3]. All the cell towers together, transmit several tens to hundreds of watts of power, which causes severe health problems. Because of the augmenting number of mobile phone users, the numbers of base stations, which enable mobile phones to connect to other mobile phones, are to be increased to provide a better communication. Therefore, base stations are to be mounted closer to each other.

In India, there are about 5.5 lakh cell phone towers at present to meet the communication services; the number will rise to 15 lakh towers by 2020 [4–8]. A lot of these towers were located near the residential and office buildings to provide good quality service of mobile communication to the users. The cellular base station is designed in such a way that it covers a distance at which the received signal strength is optimum for proper communication. Continuously receiving of these harmful signals causes serious health hazards to the people [9, 10].

The electromagnetic radiation is categorized as ionizing and non-ionizing radiations depending on the ionizing nature of the signal. The ionizing radiation has sufficient energy to pull the bound electrons from the orbit of an atom such that it becomes an ionized atom which may cause severe health problems. On the other hand, the non-ionizing radiation does not possess the enough energy to ionize (change) the atoms. But continuous exposure to non-ionization radiation may break bonds between the atoms which in turn causes health problems.

The daily usage of radio frequency (RF) devices like mobile phones, is increasing continuously. In general, high levels of electromagnetic radiation sources are found in medical applications and at certain workplaces. Medical devices used for magnetic resonance imaging, diathermy, hyperthermia, various kinds of RF ablation, surgery, and diagnoses may cause high levels of electromagnetic fields at the patient's position or locally inside the patient's body [11–13].

2 Analysis Power Density and Radiation Level of the Individual Base Stations

The field intensities, radiated power density and radiated power of an RF sources [14, 15] are computed by using the following equation

$$E = \frac{5.47\sqrt{ERP}}{d} \tag{1}$$

$$H = \frac{E}{\eta} \tag{2}$$

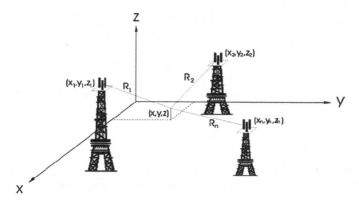

Fig. 1 A system of cellular base stations represented in 3D coordinate system

$$P_d = E \times H \tag{3}$$

$$P_r = 7161.97 \frac{P_d G_r}{f^2 \, (MHz)} \tag{4}$$

where d is the distance from the RF source in meters, f is the operating frequency of the RF source in MHz, ERP (effective radiated power) is the effective power of the transmitter with maximum input power applied to the antenna in Watts, P_d is the power density observed at distance d from the transmitter, P_r is the received power at distance d, and η is the characteristic impedance of the medium in Ohms.

Analysis Power Density and Radiation Level of set of Cellular Base Stations
The analysis of EM radiation of the cellular base station is further focused by considering the radiation of the base station antennas which are separated by less distance. Assume that there are n antennas as illustrated in Fig. 1. The ith one may be characterized by its power density (P_{d_i}), and coordinates relative to a certain origin (x_i, y_i, z_i), where $i = 1, 2,..., n$.

The electric field intensity due to ith antenna whose Effective Radiative Power (ERP$_i$) is given by

$$E_i = \frac{5.47\sqrt{ERP_i}}{d_i} \tag{5}$$

where d_i is the distance between the cellular base station (x_i, y_i, z_i) and observing point is (x, y, z) is given by

$$d_i = (x - x_i)^2 + (y - y_i)^2 + (z - z_i)^2 \tag{6}$$

The magnetic field intensity of the ith antenna is

$$H_i = \frac{E_i}{\eta} \tag{7}$$

The EM radiation power density of the ith antenna is estimated using the expression

$$P_{d_i} = E_i \times H_i \tag{8}$$

The observed received power using the receiving antenna with gain G_r for the ith base station which operating at the frequency f_i. is computed using

$$P_{r_i} = 7161.97 \frac{P_{d_i} G_r}{f_i^2 \, (\text{MHz})} \tag{9}$$

The received total power is the sum of the powers contributed by all the individual base station antennas, and is represented mathematically as,

$$P_{total} = \sum_{i=1}^{n} P_i \tag{10}$$

Generally a function of coordinate variables, it may be safely treated as a constant to simplify calculations. The electromagnetic power density (P_{d_i}) due to the ith radiating antenna at point (x, y, z) located in the far field free space is given by the general equation. Where d_i is the distance from the RF source in meters, f_i is the operating frequency of the RF source in MHz, ERP$_i$ (effective radiated power) is the effective power of the transmitter with maximum input power applied to the antenna in Watts, P_{d_i} is the power density observed at distance d_i. From the transmitter, P_{r_i} is the received power at distance d_i, and η is the characteristic impedance of the medium in Ohms.

3 Assessment of Electromagnetic Radiations from Different Base Stations

From the theoretical derivations presented in the Eqs. (1–10), the electric, magnetic field intensities, power density, the total radiated power are calculated at the various distances from a cellular base station operating with particular Effective Radiated Power (ERP) and at the particular operating frequency. The following tables shows the evaluated results for electric and magnetic field intensities observed from the different base stations operating at different frequencies, determined at different distances. From the results, it is obvious that the values of electric and magnetic

Table 1 2/2/2 GSM single carrier f = 1800 MHz

ERP	d (m)	E (V/m)	H (A/m)	P_d (W/m^2)	P_r (W)	P_r (dBm)
22 W	10	2.565	6.80×10^{-3}	0.01744	0.000122	−9.136
	20	1.282	3.400×10^{-3}	0.004359	3.04×10^{-5}	⁻15.17
	30	0.855	2.267×10^{-3}	0.001939	1.35×10^{-5}	−18.70
	40	0.641	1.700×10^{-3}	0.001089	7.61×10^{-6}	−21.19
	50	0.513	1.52×10^{-3}	0.000779	5.44×10^{-6}	−22.64
	60	0.427	1.132×10^{-3}	0.000483	3.37×10^{-6}	−24.72
	70	0.366	9.708×10^{-4}	0.000355	2.48×10^{-6}	−26.06
	80	0.320	8.488×10^{-4}	0.000272	1.90×10^{-6}	−27.21
	90	0.285	7.559×10^{-4}	0.000215	1.50×10^{-6}	−28.23
	100	0.2565	6.80×10^{-4}	0.000174	1.22×10^{-6}	−29.13
	150	0.171	4.535×10^{-4}	7.75×10^{-5}	5.41×10^{-7}	−32.67
	200	0.128	3.39×10^{-4}	5.80×10^{-5}	4.05×10^{-7}	−33.92
	250	0.085	2.70×10^{-4}	2.75×10^{-5}	1.92×10^{-7}	−37.17
	300	0.073	2.25×10^{-4}	1.91×10^{-5}	1.33×10^{-7}	−38.76
	400	0.057	1.697×10^{-4}	1.08×10^{-5}	7.54×10^{-8}	−41.23
	500	0.046	1.352×10^{-4}	6.80×10^{-6}	4.75×10^{-8}	−43.23
	600	0.040	1.120×10^{-4}	5.40×10^{-6}	3.01×10^{-8}	−45.09
	800	0.032	8.50×10^{-5}	2.72×10^{-6}	1.89×10^{-8}	−47.21
	900	0.028	7.55×10^{-5}	2.15×10^{-6}	1.50×10^{-8}	−48.22
	1000	0.025	6.80×10^{-5}	1.70×10^{-6}	1.18×10^{-8}	−49.25
	1200	0.021	5.66×10^{-5}	1.19×10^{-6}	8.31×10^{-9}	−50.80
	1300	0.019	5.23×10^{-5}	9.94×10^{-7}	6.94×10^{-9}	−51.58
	1400	0.018	4.85×10^{-5}	8.74×10^{-7}	6.1×10^{-9}	−52.14
	1500	0.017	4.53×10^{-5}	7.71×10^{-7}	5.38×10^{-9}	−52.69
	1600	0.016	4.25×10^{-5}	6.8×10^{-7}	4.74×10^{-9}	−53.23
	1700	0.015	4×10^{-5}	6×10^{-7}	4.19×10^{-9}	−53.77
	1800	0.014	3.77×10^{-5}	5.29×10^{-7}	3.69×10^{-9}	−54.32
	1900	0.013	3.58×10^{-5}	4.65×10^{-7}	3.24×10^{-9}	−54.88
	2000	0.012	3.4×10^{-5}	4.08×10^{-7}	2.84×10^{-9}	−55.45

field intensities decreases with increase in the distance and hence the power density and received power.

Table 1 shows the calculated radiated power at the distances ranging from 10 to 2000 m for 2/2/2 GSM Single carrier with 22 W of ERP operating at f = 1800 MHz frequency. From the table, it can be depicted that the radiation of the base station is very high up to a distance (d) of 100 m at which the radiation power is above −30 dBm (danger level of radiation values are indicated by red color), indicating it as a dangerous zone and beyond that distance, the radiated power decreases gradually. The safety level of radiation is the one which is below −30 dBm and are indicated in green color in the following table and graphical plots.

It will be more clear, if the data is interpreted graphically, than the tabular representation of the data consisting the computed radiation power values for the distance values up to 1 km, are represented graphically (using MATLAB 7.6) in the following Fig. 2a–c.

Figure 2a shows the graph plotted between the calculated radiation power values and the distances upto 1 km, for the different cellular base stations with ERPs of 63, 40, 36 and 28 W. From this figure, it is clear that, greater the ERP of the cellular base station, the more is radiated power from it. Also, the radii of the zones affected by hazardous RF radiation (danger level of radiation are indicated by red color), of the different cellular base stations are 300, 180, 140 and 120 m respectively. The total radiated power for these four base stations (assumed that they are colocated in a place) is hazardous with in the distance of 440 m from the base stations.

Figure 2b shows the the the graph plotted between the calculated radiation power values and the distances upto 1 km, for the different cellular base stations with the maximum possible ERP values of with 63, 58, 40 and 36 W. From this Fig. 2d, it is clear that, greater the ERP of the cellular base station, the more is radiated power from it. Also, the radii of the zones affected by hazardous RF radiation, of the different cellular base stations are 340, 180, 160 and 80 m respectively. The total radiated power for these four base stations (assumed that they are colocated in a place) is hazardous with in the distance of 420 m from the base stations.

The Fig. 2c shows the graph plotted between the calculated radiation power values and the distances upto 10 km, for the different cellular base stations with the minimum possible ERP values of with 36, 28, 22 and 20 W. It is obvious that, greater the ERP of the cellular base station, the more is radiated power from it. Also, the radii of the zones affected by hazardous RF radiation, of the different cellular base stations are 800, 600, 400 and 300 m respectively. The total radiated power for these four base stations (assumed that they are colocated in a place) is hazardous with in the distance of 1.8 km m from the base stations.

From the Fig. 2a–c, the following observations are made:

1. The greater the ERP of the cellular base station, the more is the radius of the zone affected with hazardous level of radiation.
2. The total radiation of the colocated cellular base stations is more than that of the individual base station.
3. The total radiation of the base stations excited with high ERPs affects the wide area.

So far, the theoretical analysis of radiated power from a cellular base station is limited to distance of 2 km only. But, the communicating device like mobile phone can be operated even −90 dBm received signal power level, at which the range may be beyond 2 km. Hence, there is a need to study the radiation powers of the cellular base station with different configuration for the distances more than 2 km. Figure 3 represent the graphs plotted (using MATLAB 7.6) between the radiation power of cellular base station and the distance up to 10 km.

Fig. 2 **a** Computed radiation power (dBm) of different cellular base stations with 63, 40, 36 and 28 W of ERPs operating at different frequencies. **b** Computed radiation power (dBm) of different cellular base stations with 63, 58, 40 and 36 W of ERPs (maximum possible ERP values) operating at different frequencies. **c** Computed radiation power (dBm) of different cellular base stations with 36, 28, 22 and 20 W of ERPs (minimum possible ERP values) operating at different frequencies

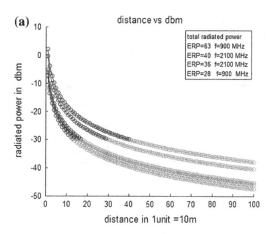

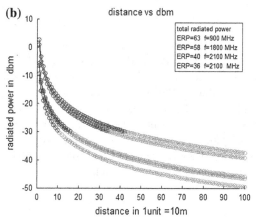

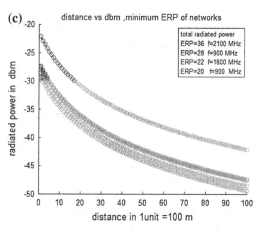

Fig. 3 Computed radiation power (dBm) of different cellular base stations with 63, 50, 40 and 36 W of ERPs (higher ERP values) operating at different frequencies

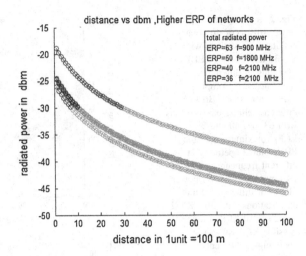

Figure 3 shows the computed radiation power values of different cellular base stations with excited ERPs 63, 50, 40 and 36 W for the distance upto 10 km. It is obvious from the Fig. 3 that the radius of the dangerous zone wherein the radiation is more than −30 dBm, are more than 1 km, for the cellular base stations operating with high ERPs. The total radiated power for these four base stations (assumed that they are colocated in a place) is hazardous with in the distance of 3 km from the base stations. However, the received powers from the different cellular base stations at 10 km distance are −56, −63, −65 and −66 dBm and these powers are sufficient to be received by a mobile unit to perform communication satisfactorily.

4 Assessment of Electromagnetic Radiations from Different Cellular Base Stations Considering Geographical Distance and Antenna Height

The compliance distance beyond which the safety level of RF radiation exists, is calculated precisely, by taking the Geographical distance between the base station antennas and the height of the antenna into account. The simulated results shown below in the Fig. 4a, b.

Figure 4a depicts the radiation pattern of the four different cellular base stations with ERPs 24, 16, 24 and 36 W, with 1800, 900, 2100 and 1900 MHz frequencies of operation. Here, the distance from each cellular base station is computed in the 3D coordinate system. The received power at a particular point is computed as algebraic sum of the observed powers of the individual base stations. From the figure, it is obvious that the radiation is heavy in the vicinity of cellular base station and gradually decreases with distance. Though the second base station operated with an ERP of 16 W, the radiation caused by it is more because of its operating

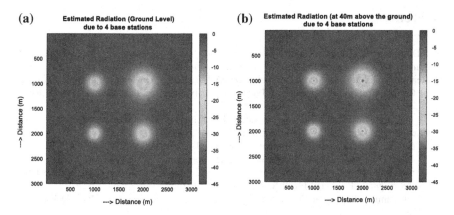

Fig. 4 **a** Radiation pattern (ground level) of the four different cellular base stations with 1 km distance of separation. **b** Radiation pattern (at the height of 40 m above the ground) of the four different cellular base stations with 1 km distance of separation

frequency 900 MHz is lower value than that of the others. The radiation is not distributed equally by the cellular base stations because of difference in the ERPs and operating frequencies. Hence, the radiation at the point which is at equidistance from all the four base stations is not equally contributed by the individual antenna. Similarly, Fig. 4b shows the radiation pattern of the same cellular base stations at the height of the 40 m above the ground. As the height of the observing point is increased, it is nearer to the cellular base station antenna, and hence it experiences more radiation than that of the point in ground level, and the same thing is obvious from the graph.

5 Conclusion

The research work in this article is concentrated on the assessment of the RF radiation level in the premises of a cellular base station. From the theoretical derivation of the received power from an RF source, the received power is computed at given distances from the cellular base station with different excited Effective Radiated Powers (ERP) operating at a particular frequency. The computed radiation power values at the given distance help in determining the distance where the radiation is hazardous. This analysis is further intensified by computing radiated power considering both geographical distance and the height of the antenna.

In this work, the radiated power values are computed by considering the antenna is isotropic, but in the practical it is not possible. So, one can compute the radiated power by considering the directive antenna instead of isotropic antenna. This calculations would provide more accurate radiation distribution pattern of the given set of antennas.

References

1. P.D. Inskip, R. Tarone, 'Cellular-Telephone Use And Brain Tumors', *The New England Journal Of Medicine*, Vol. 344,, No. 2, 2001, pp. 79–86.
2. Report of the Inter-Ministerial Committee on EMF Radiation, Government of India, Ministry of Communications & Information Technology Department of Telecommunications, 2010. 1–50.
3. Mobile communication radio waves & safety, Department of the telecommunications ministry of communication & IT, Government of India, 2010. 1–35.
4. Installation and configuration: Horizon II macro cell report.2007, 1–215.
5. Mobile Telecommunications and health research programme (MTHR) Report 2007. 1–345.
6. Bio-initiative Report, A Rationale for a Biologically-based Public Exposure Standard for Electromagnetic Fields (ELF and RF), 2007, 2012. 1–1457.
7. Girish Kumar.; Report on Cell Tower Radiation Submitted to Secretary, DOT, Delhi.2010, 1–50.
8. Kumar, N.; and Kumar, G.; "Biological effects of cell tower radiation on human body", ISMOT, Delhi, India, pp. 678–679, Dec. 2009.
9. Mallalieu Kim.; "Report on BVI Radiation Measurement Project", June 2010 Communication Systems Group, Department of Electrical and Computer Engineering, The University of the West Indies, 2010, 1–355.
10. Liu, W.C.; and Hsu, C.F.; IEE Electronics Letters, 41(3), 2005, 390–391.
11. Ray, K.P.; Hindawi Publishing Corporation International Journal of Antennas and propagation, 11(6), 2008, 1–8.
12. Ashok Kumar Kajla.; Vaishali.; and Vivek Kumar.; (IJMCTR), 1(8), 2013, 21–30.
13. Sabah HawarSaeid.; Proceedings of the 2013 International Conference on Electronics and Communication Systems, 2013, 87–90.
14. Venkatesulu. S, et al, "Printed Elliptical Planar Monopole Patch Antenna for wireless communications" International Conference on Green Computing, Communication and Conservation of Energy (ICGCE) 978-1-4673-6126-2/13/$31.00_c 2013. IEEE Xplore Digital Library, 2013, 314–319.
15. S. Venkatesulu et al, 'Theoretical Estimation of Electromagnetic Radiation from Different Cellular Networks' i-manager's journal on Embedded Systems, ISSN: 2277-5102, Vol. 3, No. 3, Oct – 2014, pp. 1–7.

Revitalizing VANET Communication Using Bluetooth Devices

G. Jaya Suma and R.V.S. Lalitha

Abstract Nonetheless, the number of accidents has continued to expand at an exponential rate. Due to dynamic nature of VANET positions, there will be considerable delay in transmission of messages to destination points. With attention to emergency message transmission incidentally, the data dissemination techniques are to be refined for fast and accurate transmission of messages. In order to track the information of vehicle positions proximity sensors plays crucial role in the real time world. Since the sensing capability of proximity sensors is limited, sometimes there may be information loss due to its poor quality. To enhance the high precision accuracy, we require an efficient methodology to collect information instantaneously and storing it in a local database for further use. In this paper, An App is developed using App Inventor tool to collect data from near by nodes using Bluetooth. The dictum of customizing Bluetooth communication through this App, is to store the communication held between the two communication parties in a local database for further analysis in case fake message transmission occurred. The vehicle information is collected by Bluetooth device available in the Android mobiles ensembles the accuracy of information collected. In collecting neighborhood information, Bluetooth provides adequate solutions in fast collection of data. Evidently this mechanism reduces loss in collection of information from near by nodes and provides means for establishing effective way of communication among VANET nodes. In this paper, a comparative analysis is made with the sensor communication to the Bluetooth communication in connection with the communication capabilities.

Keywords Proximity sensors · Bluetooth technology · Location information · Rank correlation coefficient · V2V communication

G. Jaya Suma (✉)
JNTUK-Vizianagaram, Vizianagaram, India
e-mail: gjscse@gmail.com

R.V.S. Lalitha
SSAIST, Surampalem, India
e-mail: rvslalitha@gmail.com

© Springer India 2016
S.C. Satapathy et al. (eds.), *Information Systems Design and Intelligent Applications*, Advances in Intelligent Systems and Computing 434,
DOI 10.1007/978-81-322-2752-6_28

1   Introduction

Mobile Ad hoc Networks are formed by spontaneous connection of wireless nodes using wi-fi [1–3]. These are used in short range networks. To increase the transmission range they will be connected with GSM technology or through Internet for communicating with nodes that are in other networks. In this paper, to establish communication across VANET nodes, Bluetooth features of the Android mobiles are used for data transmission initially. The state of art of this paper is organized in the following manner. Section 2 describes about the previous work. In Sect. 3, automatic message transmission using sensors is discussed. In Sect. 4, shows the simulation results of sensor communication. And in Sect. 5, system architecture in establishing Bluetooth communication is discussed. In Sect. 6, simulation results Bluetooth communication using Android mobiles are discussed. In Sect. 7, analyzing Bluetooth communication with Sensor communication is discussed. And finally in conclusion, the significance of incorporating Bluetooth communication in VANETs is discussed.

1.1   Vehicle-to-Vehicle (V2V) Communication

V2V establishes communication between vehicles on the road for sharing data across them [4–6]. This is mainly essential during occurrences of accidents and where occurrences of traffic jam are frequent. This facilitates in providing information to drivers for taking necessary remedial measures. Normally, this is done with the help of GSM communication to transmit emergency information over the VANET. The information as seen by the vCar1 sends to vCar2 via GSM communication as shown in Fig. 1. During the case of accidents, Breakdown car(vCar) causes obstruction on the road. Hence vCar1 and vCar2 communicates via GSM about the Breakdown car which creates in crossing over it. Hence, we need to find a solution that provides effective communication to inform others either to take diversion or for moving that car from the middle of the road in less time, so that it will not effect much traffic jam.

Fig. 1 Conventional messaging through GSM to other vehicles in the lane because of obstruction ahead

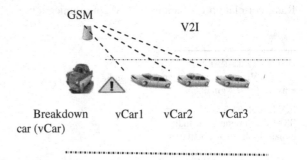

The chief characteristic of VANET users is Location information which is used to track the position of the vehicle on the road. As already Android mobiles are featured with Geographic Positioning System (GPS), tracking of vehicles is convenient with these mobiles. Also, these mobiles are equipped with sensors, which lead to combine the Location Information with the information sensed by the sensors, so as to give a beneficial solution in delivering emergency messages [7].

2 Previous Work

VANETs communication became popular for being its safety alerts in case of accidents, entertainment, data sharing etc. Tamer Nadeem discussed about providing traffic information using GPS to drivers and also about broadcasting of messages [8] over VANET. In 2007, Andreas Festag et al. implemented of Vehicle-to-Vehicle communication using C as software and a special type of embedded system as hardware and then tested on VANET [5]. Josiane et al. gave detailed analysis and experimentation about Routing and Forwarding in VANETs in 2009. Keun Woo Lim defined Vehicle to Vehicle communication using sensors placed on the road for sensing vehicular information through Replica nodes [7].

3 Automatic Message Transmission by Sensors

The use of sensors [7] in Vehicular Ad hoc networks plays a predominant role in maintaining information i.e. sent to and from the vehicles by storing information in the local database. This paper is mainly focused on three aspects namely (i) Transmission of Emergency messages through Web Server using Proximity sensors and Google Cloud Messaging (GCM) service, which allows transmission to nearby Android mobiles (range based) automatically as shown in Fig. 2. (ii) All the VANET node positions viewed by Web Server on the Google Map as in Fig. 3. The following procedure is adopted to send information to VANET users about the accident object: Track latitude and longitude positions of the VANET users using GPS services. Update the location information of the VANET users after every 20 s using AsyncTask that runs in background as VANET is dynamic in nature. Send updated coordinates along with the device id which is obtained using GCM service to server for storing in a local database and the location information of the VANET users is shown in Fig. 4 using Google Map APIs.

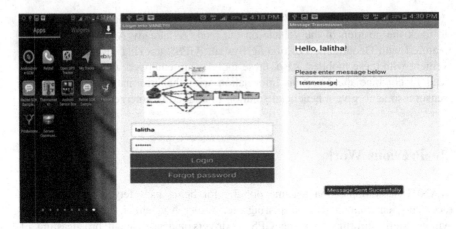

Fig. 2 Message communication by the mobile sensors to transmit urgent message using GCM by accessing Web server

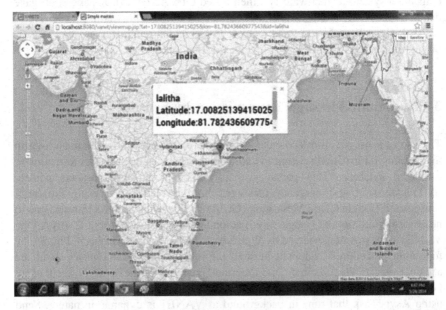

Fig. 3 Vehicular location information traced by Web server on Google map

4 Simulation Results

To perform communication across VANET nodes, register with the Web Server and then send the message to other nodes in the network using GCM services. Automatic transmission of messages by sensors in Android mobiles through Web server using GCM Services using Sensor Communication app is as shown in Fig. 2.

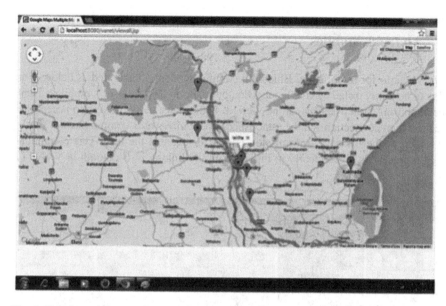

Fig. 4 Vehicle position traced by Web server on Google map

5 System Architecture

The system architecture of the Bluetooth communication across VANETs is shown in Fig. 5. In this connection, each vehicle is equipped with an Android mobile and is installed with VANET_vie_Bluetooth app. The messages that are sent and received from other nodes will be sent to Web Server for future analysis.

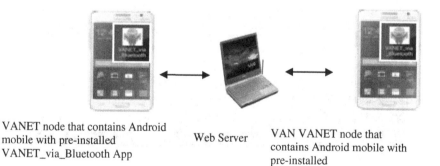

VANET node that contains Android mobile with pre-installed VANET_via_Bluetooth App   Web Server   VAN VANET node that contains Android mobile with pre-installed VANET_via_Bluetooth App

Fig. 5 System architecture

294   G. Jaya Suma and R.V.S. Lalitha

6 Simulation Results of Bluetooth Communication Using Android Mobiles

The scenario for establishing Bluetooth communication is illustrated as below. For this to be done, install VANET_via_Bluetooth app in the two user's mobiles. Then neighbouring nodes nearby can be viewed. If the other party, willing to accept the request then connection will be established. There afterwards communication starts. As soon as the message transmission is over disconnect the link. This communication process is given clearly in Fig. 6. With the linking of Bluetooth in VANET,

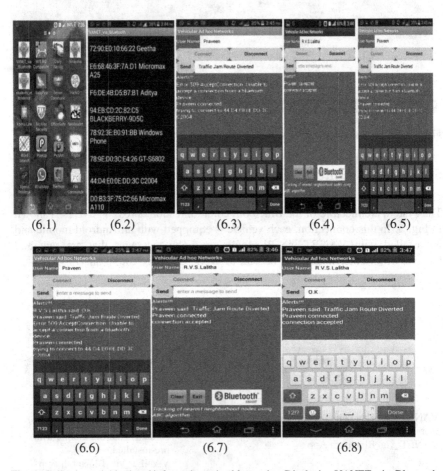

(6.1)   (6.2)   (6.3)   (6.4)   (6.5)

(6.6)   (6.7)   (6.8)

Fig. 6 Collection neighborhood information using bluetooth. **a** Displaying VANET_via_Bluetooth app in android mobile. **b** List of VANET nodes near by. **c** and **d** Pairing of two VANET nodes. **e** and **f** Accepting to communicate by two VANET nodes. **g** Transmission of message by VANET node. **h** Reception of message by VANET node

reliable message communication is established to nodes that are very nearer by. These nodes need to be linked to the node that are at farther distances for fast spread of messages.

7 Analyzing Bluetooth Communication with Sensor Communication

The key characteristic of both Bluetooth and Sensor communication is how well the messages are transmitted with in a particular period of time. For this to analyze compute the Rank Correlation Coefficient for message transmission using Bluetooth communication and Sensor communication. The Rank Correlation Coefficient varies between +1 and −1 if no duplicates exist in the experimental values. Hence, recording the no. of messages transmitted at time t1, t2, t3, t4 and t5 by Bluetooth as well as Sensors at distance of 10 m, the following observations are made as shown in Table 1.

```
If dᵢ=xᵢ-yᵢ, then dᵢ=-1,-1,1,1 and 1.
    ∑dᵢ²=1+1+1+1+1=5
Rank Correlation coefficient,ρ=1-[6∑dᵢ²/(n(n²-1)]  where  n
is no. of trials.

=1-[6*25/5*24]

=-.25
```

The correlation coefficient varies w.r.t to the distance the messages to be transmitted. Since the distance to be communicated is short, the effect of transmission makes no difference. If the distance is more, then Rank Correlation Coefficient varies abruptly, as the sensing range is different.

Comparisons made between Bluetooth communication and Sensor communication across VANETs. Figure 7 is drawn showing the neighborhood coverage of both Bluetooth and Sensors. The Green line is computed based on the objective functions computed with respect to the sensing capability of the sensor. The points above Green line can be sensed only by sensors and the points below will be sensed either by sensor or by Bluetooth or even by both (Table 2).

Table 1 The no. of messages transmitted successfully at time T1, T2, T3, T4 and T5 out of 10 messages

Type of communication technology used	T1	T2	T3	T4	T5
Bluetooth	9	8	5	6	7
Sensors	8	7	6	7	8

Fig. 7 Graph showing the coverable nodes based on fitness function computed

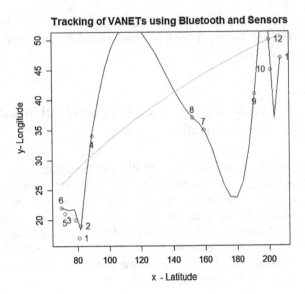

Tracking of VANETs using Bluetooth and Sensors

Table 2 Comparisons made between Sensor communication and Bluetooth communication

Type	Sensor communication across VANETs	Bluetooth communication across VANETs
Communication methodology implemented	Android sensor communication	Using app inventor tool
Transmission capacity	Up to 100 m	Up to 1 m
Connectivity	Reachable to all the nodes within the range	Reachable to nodes within the range and willing to accept connection
Signal interference	Moderate	Appropriate as it covers all the nearest neighborhood nodes if the distance is very short
Communication between two nodes	Once message transmitted, it reaches to all VANET nodes that are with in the range	Based on the paired and connection accepted nodes (max. 2)
Accuracy	Overall performance high	Works efficiently in small range

8 Conclusions

In this paper, Bluetooth communication is adapted across VANET nodes. This functions well to send messages to near by nodes. In addition to that, this can be linked with sensor communication for transmitting information to farther nodes automatically. Since, VANET communication requires abrupt information about the incident happened, this kind of linking, functions well in spreading data all over the network.

References

1. Fan Ye, Haiyun Luo, Jerry Cheng, Songwu Lu, Lixia Zhang.: A TwoTier Data Dissemination Model for Largescale Wireless Sensor Networks. UCLA Computer Science Department, Los Angeles, CA 9000951596, fyefan, hluo, chengje, slu, lixiag@cs.ucla.edu.
2. B. Tavli.: BroadcastCapacityofWirelessNetworks. IEEECommunicationsLetters, vol. 10, no. 2, pp. 68–9 (2006).
3. Nataragan M et al: ENERGY EFFICIENT DATA DISSEMINATION IN WIRELESS SENSOR NETWORKS USING MOBILE SINKS. Asst Prof., Department of Computer Science, Jackson State University, Jackson, MS 39217, USA, Ph.D. Student, Department of Computer Science, Iowa State University, Ames, IA 50011, USA, Assoc Prof., Department of Computer Engineering, Jackson State University, Jackson, MS 39217, USA.
4. A. Benslimane.: Localization in Vehicular ad-hoc networks. Proceeding of the 2005 System Communication, 2005.
5. Andreas Festag, Alban Hessler, Roberto Baldessari, Long Le, Wenhui Zhang, Dirk Westhoff.: VEHICLE-TO-VEHICLE AND ROAD-SIDE SENSOR COMMUNICATION FOR ENHANCED ROAD SAFETY.NEC Laboratories Europe, Network Research Division Kurfursten-Anlage 36, D-69115 Heidelberg.
6. Zhang, ZingZhao, Guohong Cao.: Data Dissemination in Vehicular Ad hoc Networks (VANET).
7. Keun Woo Lim, Woo Sung Jung, Young-BaeKo.: Multi-hop Data Dissemination with Replicas in Vehicular Sensor Networks. Graduate School of Information and Communication Ajou University, Suwon, Republic of Korea.
8. Tamer Nadeema.: TrafficView.: Traffic Data Dissemination using Car-to-Car Communication. Department of Computer Science, University of Maryland, College Park, MD, USA.

Authors Biography

Dr. G. Jaya Suma Is Associate Professor and H.O.D, in the Department of Information Technology, JNTUK-University College of Engineering, Vizianagaram. She received her Ph.D from Andhra University. Her current interesting research includes Data Mining, Soft computing, Pattern Recognition and Mobile Computing. She has more than 15 years of experience in teaching. She is Nominee for JNTUK for Smart Campus. She is Life Member in IEEE, CSI and ISTE. She published many papers in International Journals and Conferences.

R.V.S. Lalitha Is currently working towards her Ph.D in JNTUK-University, College of Engineering, Kakinada. Her research includes Mobile Computing and Soft Computing.

Virtual Manipulation: An On-Screen Interactive Visual Design for Science Learning

Sanju Saha and Santoshi Halder

Abstract Interactivity in e-learning environment is an innovative approach in teaching-learning. Predominantly theoretical justification of interactive learning environment has been discussed on the basis of the process of visual and auditory information in the memory system emphasizing the result oriented perspective in the sense that they have given importance on computer response to learner action rather than learner activity and engagement in computer programming. However, by definition interactivity is described as 'to act'. In this view point the present research attempts to explore the effectiveness of enactment by manipulating virtual features in interactive visualization when compared with visual animation. To investigate the effectiveness of different visual condition researchers have developed two different types of instructional module (interactive virtual manipulation and animated visual). Total 360 students have been selected to implement the study with different matching criteria. MANOVA is conducted to find out the group difference in different condition. Result showed a momentous mean difference in different condition i.e., in virtual manipulation (execution of action) condition where student perform virtually in the on-screen object better than animation (observed action) in respect to various learning outcome. Result is discussed critically from several theoretical focal points.

Keywords E-learning · Virtual manipulation · Enactment · Interactive design · Multimedia learning · Motor encoding

S. Saha (✉)
Department of Education, University of Calcutta, Alipore Campus,
1 Reformatory Street, Kolkata 700027, India
e-mail: sanju_saha@yahoo.co.in

S. Halder
University of Calcutta, Alipore Campus, 1 Reformatory Street,
Kolkata 700027, India
e-mail: santoshi_halder@yahoo.com

© Springer India 2016
S.C. Satapathy et al. (eds.), *Information Systems Design and Intelligent Applications*, Advances in Intelligent Systems and Computing 434,
DOI 10.1007/978-81-322-2752-6_29

1 Introduction

Learning with computer based on-screen learning environment is a growing phenomenon since the use of computer as a mode of teaching learning. Nevertheless, researchers have been focusing on interactivity in visual instruction which is not less important than linear visual instruction such as video tape and static animation because learning is not simply a process of information transmission, rather students should become actively engaged for deep learning [1, 2]. However, predominantly theoretical justification of visual instruction specially on multimedia and interactive on-screen instructional environment mainly focus on cognitive process of visual and audial sequence in the memory system, so to say a dual coding approach [3], Cognitive Theory of Multimedia Learning (CTML) [4] and Integrated Theory of Text and Picture Comprehension [5]. All mentioned theories have emphasized mainly on encoding of auditory and visual information with two separate channel help through specific process (selection, organization and integration) for achieving meaningful learning. There arises major question regarding previous theoretical justification that whether only visual or auditory information processing mainly effect the cognitive system for effective multimedia or interactive onscreen instructional visual design or there is other factors as well. As more recently a number of contemporary theories in the area of pure cognitive psychology have discussed the relationship between physical activities and learning. One of the most important theories 'Multimodal Theory of Memory' have emphasized simultaneously dual channel information processing (visual and audial encoding) on 'enactment' (Motor encoding) in the memory system. In Multimodal Theory of Memory, Engelkamp [6] also described that motor encoding facilitates the subsequent retrieval of information and that there four modality-specific components such as a nonverbal input (visual) and output (enactment) system and a verbal input (hearing and reading) and output (speaking and writing) system connecting conceptual system. Notably, many previous empirical study has given major emphasize on outcome oriented perspective in the sense that they have given importance on computer response to learner action rather than learner activity and engagement in computer programme. This research illuminates the multimodal theory and enactment conception radiating a distinctively different domain such as incorporating virtual enactment in instructional visualization modeling as virtual manipulation. That is relatively unique contribution in this research especially in India where this research is conducted (Discussed in Sects. 2 and 3).

2 Enactment in Cognitive Psychology and Interactive Multimedia Environment: Research Review

Researches on pure cognitive psychology have pondered over that 'enactment' have a significant impact on cognition. Cook [7] through his research established the fact that body movement can help the student learn and retain conceptual information. Moreover, another interesting finding by Glenberg [8] showed improving students' comprehension level by manipulating relevant toys in reading passage. However, there have been number of previous research directly related with the present research. Engelkamp and Zimmer [9] has shown a positive effect when participant performed themselves (self-performed) as compared to observer's other performance (simply observing). Notably, a previous findings related to the research establishes that enactment is effective not only when learner performed with real object but also found the effectiveness of the use of imaginary object [10]. Further, another research reinforces the present study that actual performance enhances the retention power than imaging or planning [11]. However, a contradictory result has also been found showing both actual performance and imaginary performance being effective while the imaginary performance groups have read the action phase text twice without performance [7]. One can explain this finding both in multimodal point of view that motor action itself needs planning and activating the conceptual level as compared with mimic action which originated from sensory level [12].

Besides, many empirical studies incorporated various interactive features in on-screen learning environment but they have given major emphasis on outcome oriented perspective in the sense that they have given impotency of computer response to learner action rather than learner activity and engagement in computer programming [13]. For example recently, Glenberg [14] compared three condition in instructional visualization (read text, manipulates screen object, read and re-reading) have found that effect of on-screen manipulation condition is superior to reading condition. Notably, this study emphasized on design of instructional media (e.g., object manipulates or not) ignored the learner activity and engagement. Only limited number of studies has described interactive instruction from the motor activity perspective. Study by [12] examined the effect of four different types of interactive (iconic, symbolic, look and listen) condition and have found that iconic (dragging) interactivity is superior than other three conditions in free recall and recognition tasks and describes this result from the enactment focal point. However, for meaningful learning there is need to emphasize different knowledge domain which is ignored in these researches. Nevertheless, previous research on enactment or participant performance has been conducted mainly on real situation and major emphases have been given on free recall and recognition task. However, present researcher have gone step further to explore this theoretical assumption on virtual manipulation performance in computer based instructional environment and enrich previous research by measuring effectiveness of this performance in factual, conceptual and rules and principal knowledge domain.

3   Research Design

3.1   Objective of the Study

To investigate the effect of virtual manipulation (interactive visualization as compared with Animated instruction) on student achievement of learning objectives (factual, conceptual, and rules and principle knowledge).

3.2   Hypothesis

H0: There will be no significant difference in the student achievement of different learning objective (Factual, Conceptual and Rules and principle knowledge) with respect to various instructional visualizations (interactive with virtual manipulation and animation).

3.3   Participant

Present study was conducted on Central Board of Secondary School (CBSC) in Kolkata. Most of the students belonged to lower-middle-class families. From 500 students, 360 students were strictly matched on the following criteria:

- Sufficient knowledge of English for reading text comprehension.
- Scored 10 or greater in computer proficiency test developed by researcher.
- Age ranged from 13 to 16 (mean age 15.02 years and SD = 2.36).

3.4   Measurement Instrument

General Information Schedule: General Information Schedule comprised of student demographic information and Socio economic status (Parental education, income and occupation).

   Computer Proficiency test: To match experimental and control group, a computer proficiency test was developed by the researchers. The main objective of this test was to measure how efficient they were to use different functions of the computer specially mouse, keyboard and computer screen. Reliability of this test was measured as 0.84.

   Prior Knowledge Test (pre-test as covariate): The Prior knowledge test originally developed by the researchers Dwyer [15], consisted of 36 multiple-choice questions on human physiology. For this study purpose the test was re-standardized and validated by Kuder-Richerdson estimation and by content validation. The objective of this test was to measure student's previous knowledge regarding human physiology. Reliability of the prior knowledge test was 0.89 indicating very high reliability.

Criterion Measures Test (post-tests): The three criterion tests (Identification test, Terminology test and Comprehension test as discussed below) used in this study was the adopted version of the original tool developed by [15] Each test consisted of twenty multiple-choice questions worth 1 point.

Identification test: The main objective of identification test was to measure student's factual knowledge about content material used for the present study. This test measured student ability to identify the names and positions of the parts. Students have to identify the parts of the heart indicated by the numbered arrows on a heart outline drawing.

Terminology test: The main objective of terminology test was to measure conceptual knowledge of student about content material used for the study. The terminology test measured student knowledge of specific facts, terminologies, and definitions. Students answered the multiple-choice questions selecting the answer that best described different parts of the heart.

Comprehension test: The main objective of comprehension test was to measure student's rules and principle knowledge about content material used for the study on the topic (human heart). Rules and principle knowledge learning of students on the given module (human heart) refers to those cause-and-effect or co-relational relationships that are used to interpret events or circumstances.

3.5 Development of Instructional Module and Learning Material

Instructional content material of this study is adapted from a color-coded, paper-based booklet developed by the researchers Dwyer [15] on the topic 'human heart 'containing five units: (1) the heart's structure; (2) the veins and arteries; (3) the valves of the heart; (4) the blood flow through the heart; and (5) the phases of the heart cycle. This content was chosen as it allows the evaluation of different levels of learning objectives. This topic is selected after consultation with experts in the subject.

Illustration of Instructional Module
For the purpose of the study following two separate instructional modules was developed by the researchers:

Virtual manipulative interactive visualization condition: Under this condition the above mentioned instructional content was framed in 20 different slides. Each frame introduced the structure and function of human heart. Extreme left sides of each frame had text and the right side had a corresponding virtual manipulative graphical and programmed instruction elaborating the text. Over every

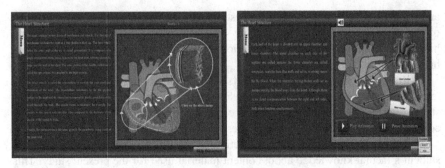

Fig. 1 *Left side* image is the screen shot of virtual manipulation with action phase screen design and *right side* image is the screen short of animated visual with action phase screen design

manipulative graphic there were some action phases (instruction given). Student needs to read the text and work as per the given action phase. In each frame the user can hear an audio corresponding to the text and action phases.

Animated Condition: Akin to virtual manipulation condition there were instructional content framing 20 different slides. Each frame introduced the learner structure and function of human heart presented in animated video (function of human heart) along with some particular button (play, pause, and stop) (Fig. 1).

4 Reliability and Validity of All Three Criterion Tests

The KR 20 reliability results were all above 0.80, which is satisfactory level of reliability. The validity scores for the test was 0.86 for Identification test, 0.81 for Terminology test and 0.85 for Comprehension test indicating high reliability for the three criterion tests [16].

5 Results Analyses and Interpretation

The overall objective was to find out the effectiveness of various instructional visualization (virtual manipulation in interactivity and animation) conducted in two phases.

First Phase, Covariate Data Analysis: Prior Knowledge Test on the Physiology
A variance of analysis was conducted on the physiology test scores to determine if there was a significant difference among the treatment groups on their prior knowledge regarding subject matter (Human heart).

The result of the ANOVA analysis indicated that there is no significant differences among the treatment groups on the test (Table 1) score $F = 1.28$, $\rho = 0.27$. Result indicated that the participants were approximately equal in their prior

Table 1 ANOVA result for tests of between-subjects effects (prior knowledge test and three criterion tests)

Criterion test	Sum of squares	df	Mean square	F	Sig.
Identification + terminology + comprehension	68.60	8	8.58	1.28	0.27

Note Each criterion test consist of 20 items

knowledge on the content material used in the study and therefore any results of treatment effects would not be attributed to the difference in participants' prior knowledge.

Second Phase: Results of MANOVA

As more than one dependent variable was used in conjunction with the independent variable, a multivariate analysis of variance (MANOVA) was conducted to analyze the effect of treatment material in instructional visualization in the student achievement of learning of educational objectives (overall effect i.e., factual, conceptual, and knowledge of rules and principles) through computer based instruction visualization method (Table 2).

From the above table we found that there was a significant main effect of instructional visualization (Wilks' Lambda = 0.75, F (3/38.39) and $\rho = 0.00 > 0.05$) in the three criterion test (identification, terminology and comprehension) and the multivariate effect size was estimated at 0.24, which is large [17] implying that 24.0 % of the variance in the canonically derived dependent variable was accounted by instructional visualization. This significance MANOVA result and percentage of partial eta square was sufficient to do univariate follow-up ANOVAs that helped to further isolate exactly where the significant and interesting mean differences were found (Table 3) [18].

Subsequent univariate tests or exploratory follow-up analysis using ANOVA (Table 3) result indicated significant differences in achievement among students who received different conditions of instructional visualization on the three criterion test (Identification test F (1/358) = 113.56 and $\rho = 0.00 < 0.05$, $\eta p^2 = 0.24$,

Table 2 Represents analysis with all criterion test (identification, terminology and comprehension test) indicating MANOVA results using Pallai's trace and Wilks' lambda

Effect	Tests	Value	F	Sig.	ηp^2
Instructional visualization	Wilks' lambda	0.75	38.39	0.00[a]	0.24

Note [a]Mean difference significance and each of the criterion tests contains 20 items

Table 3 Test between subject effect instructional visualization on three criterion test

Experimental group	Test by treatment	df	F	Sig.	Partial Eta square
Instructional visualization	Identification	1	113.56	0.00[a]	0.24
	Terminology	1	11.84	0.00[a]	0.03
	Comprehension	1	5.63	0.01[a]	0.16

Note [a]Mean difference significance at 0.05 level in the criterion tests (contains 20 items)

Table 4 Presents the adjusted means and standard errors for different types of instructional visualization condition on three criterion test

Dependent variable	Instructional visualization	Mean	Std. error	95 % confidence	
				Lower	Upper
Identification	Animation	9.672	0.204	9.271	10.073
	Virtual manipulation	12.744	0.204	12.344	13.145
Terminology	Animation	11.911	0.256	11.408	12.414
	Virtual manipulation	13.156	0.256	12.653	13.658
Comprehension	Animation	12.867	0.238	12.398	13.335
	Virtual manipulation	13.667	0.238	13.198	14.135

Terminology test $F (1/358) = 11.84$, $\rho = 0.00 < 0.05$, $\eta p^2 = 0.03$, Comprehension test $F (1/358) = 5.63$, $\rho = 0.01 < 0.05$, $\eta p^2 = 0.16$). This significant ANOVA result on the three criterion test indicate the need to explore which of the specific groups of instructional visualization differed viz, virtual manipulation in interactive visualization, and animated visual. To further identify the differences (Table 4) adjusted means and standard errors for type of instructional visualization on three criterion test were done.

From (Table 4) we found that student who used virtual manipulation in interactive instructional visualization outperformed the students who used animated visualization in identification (12.74), terminology (13.15) and comprehension (13.66) tests which measured factual, conceptual and rules and principal knowledge.

6 Significance of the Study

Implication for Future Researcher of Educational Technology
Theoretically, present research result is highly significant for contributing a new conceptual understanding paving the way for on-screen learning environment that not only verbal or visual component but also planning, observing and execution of motor action is a significant component of the instructional visualized learning environment.

Present study extends and applies previous multimodal theory in instructional visual instruction. The result of this research introduces a theoretical approach to thinking more systematically regarding the different types of visual instruction and their impacts on learning outcomes from the enactment or motor activity focal point. This major contribution can be helpful for forthcoming researcher of educational technology and instructional designer to design an educational multimedia learning material.

Implication in the Field of Teaching and Learning

One of the major practical advantages is that in the classroom environment one is not able to produce various abstract concepts. By adding virtual manipulation features in instructional visualization one can present all types of abstract and real world object in computer based laboratory environment by visual simulation. Last but not least, it's really tough for a developing country like India to create a real lab environment in each classroom. Virtually performed environment can be helpful for its cost efficiency and flexibility to produce abstract and complex concept with minimum expense in those schools of rural area where there is no laboratory.

Acknowledgments The authors acknowledge the funding supports from Rabindranath Tagore center for Human Development Studies (RTCHDS), IDSK, University of Calcutta partly for funding the research project.

References

1. Halder, S., Saha, S., & Das, S.: Development of instructional visualization based module and exploring its effect on teaching-learning: An Interactive learner control approach in teaching learning. World Journal on Educational Technology, 6 (3), 308–316. (2014).
2. Halder, S., Saha, S., and Das, S.: Computer based Self- pacing Instructional Design Approach in Learning with Respect to Gender as a Variable, in the book Intelligent Syst., Computing, Vol 340, Mandal et al (Eds): Information Systems Design and Intelligent Applications, Springer (2015).
3. Paivio, A.: Dual coding theory: Retrospect and current status. Canadian Journal of Psychology, 45, 255–287 (1991).
4. Mayer, R. E., & Chandler, P.: When learning is just a click away: Does simple user interaction foster deeper understanding of multimedia messages? Journal of Educational Psychology, 93 (2), 390–397. doi:10.1037//0022-0663.93.2.390, (2001).
5. Schwan, S., & Riempp, R.: The cognitive benefits of interactive videos: Learning to tie nautical knots. Learning and Instruction, 14(3), 293–305. doi:10.1016/j.learninstruc.2004.06. 005. (2004).
6. Engelkamp, J.: Memory for actions. Hove, East Sussex, UK: Psychology Press. (1998).
7. Cook, S., Mitchell, Z., & Goldin-Meadow, S.: Gesturing makes learning last. Cognition, 106, 1047–1058. http://dx.doi.org/10.1016/j.cognition.2007.04.010, (2008).
8. Glenberg, A. M., Goldberg, A. B., & Zhu, X.: Improving early reading comprehension using embodied CAI. Instructional Science, 39(1), 27–39. doi:10.1007/s11251-009-9096-7, (2011).
9. Engelkamp, J., & Zimmer, H. D.: Sensory factors in memory for subject-performed tasks. ActaPsychologica, 96 (1–2), 43-60. doi:10.1016/S0001-6918(97)00005-X, (1997).
10. Engelkamp, J., & Krumnacker, H.: The effect of cleft sentence structures on attention. Psychological Research, 40(1), 27–36. doi:10.1007/BF00308461, (1978).
11. Nilsson, L., Nyberg, L., Klingberg, T., Åberg, C., Persson, J., & Roland, P. E.: Activity in motor areas while remembering action events. NeuroReport, 11(10), 2199–2201. doi:10.1097/00001756-200007140-00027, (2000).
12. Schwartz, R. N., Plass, J. L.: Click versus drag: User-performed tasks and the enactment effect in an interactive multimedia environment. Computers in Human Behavior 33, 242–255. doi: http://dx.doi.org/10.1016/j.chb.2014.01.012, (2014).
13. Trninic, D., & Abrahamson, D.:Embodied artifacts and conceptual performances. In J. v. Aalst, K. Thompson, M. J. Jacobson, & P. Reimann (Eds.), (2012).

14. Glenberg, A. M., Gutierrez, T., Levin, J. R., Japuntich, S., &Kaschak, M. P.: Activity and Imagined Activity Can Enhance Young Children's Reading Comprehension. Journal of Educational Psychology, 96(3), 424–436. doi:10.1037/0022-0663.96.3.424, (2004).
15. Dwyer, F. M.:Strategies for improving visual learning. State College, PA: Learning Services. (1978).
16. Anastasi, A., & Urbina, S.:Psychological testing. Upper Saddle River, NJ: Prentice Hall.
17. Cohen, J. (1988). Statistical power analysis for the behavioral sciences. Hillsdale, NJ: L. Erlbaum Associates. (1997).
18. Thompson, B.: Foundations of behavioral statistics: An insight-based approach. New York: Guilford Press. (2006).

Bundle Block Detection Using Genetic Neural Network

Padmavathi Kora and K. Sri Rama Krishna

Abstract Abnormal Cardiac beat identification is a key process in the detection of heart ailments. This work proposes a technique for the detection of Bundle Branch Block (BBB) using Genetic Algorithm (GA) technique in combination with Levenberg Marquardt Neural Network (LMNN) classifier. BBB is developed when there is a block along the electrical impulses travel to make heart to beat. The Genetic algorithm can be effectively used to find changes in the ECG by identifying best features (optimized features). For the detection of normal and Bundle block beats, these Genetic features values are given as the input for the LMNN classifier. ECG, Bundle Branch Block, Genetic Algorithm, LMNN classifier.

Keywords ECG · Bundle branch block · Genetic algorithm · LMNN · Classifier

1 Introduction

Electro-Cardiogram is used to access the electrical activity of a human heart. The diagnosis of the heart ailments by the doctors is done by following a standard changes. In this project our aim is to automate the above procedure so that it leads to correct diagnosis. Early diagnosis and treatment is of great importance because immediate treatment can save the life of the patient. BBB is a type of heart block in which disruption to the flow of impulses through the right or left bundle of His, delays activations of the appropriate ventricle that widens QRS complex and makes changes in QRS morphology. The changes in the morphology can be observed through the changes in the ECG. Good performance depends on the accurate detection of ECG features. The waveform changes in the different types of ECG beats was shown in the Fig. 1. Detection of BBB using ECG involves three main steps: preprocessing, feature extraction and classification. The first step in preprocessing mainly concentrates in removing the noise from the signal using filters. The

P. Kora (✉) · K. Sri Rama Krishna
GRIET, Hyderabad, India

© Springer India 2016
S.C. Satapathy et al. (eds.), *Information Systems Design and Intelligent Applications*, Advances in Intelligent Systems and Computing 434,
DOI 10.1007/978-81-322-2752-6_30

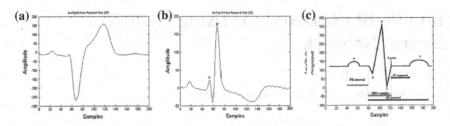

Fig. 1 **a** LBBB, **b** RBBB, **c** Normal signals

next step is the detection of 'R' peak and then based on the 'R' peak ECG file is segmented into beats. A total of 200 samples before and after the 'R' peak are considered as one ECG beat. The set of these beat samples around the 'R' peak is considered as features. The last step is the classification of the ECG file into different heart diseases.

The samples (features) that are extracted from each beat contains non uniform samples. The non uniform samples of each beat are converted into uniform samples of size 200 by using a technique called re-sampling. The resampled single ECG beat is shown in Fig. 1c. In the previous studies [1] morphological features are extracted for clinical observation of heart diseases. The feature extraction using traditional or morphological techniques generally yield a large number of features, and many of these might not be significant. Therefore, the common practice is to extract key features useful in classification. Instead of using traditional [2] feature extraction techniques, this paper presents GA as the feature extraction technique. Feature extraction in the present paper is based on the extracting key features using nature inspired algorithm called GA. In recent years many models are developed based on the evolutionary behaviors of living beings and have been applied for solving the sensible real world issues. Among them, Genetic Algorithm (GA) may be a population based optimization technique. GA is used extensively as a model to solve many engineering applications [3, 4]. In Recent years, GA has been applied with success to some engineering concepts like, harmonic estimation [5], optimum management [6], reduction machine learning [7] and transmission loss [8] and so on.

Here in this paper GA technique used as the feature optimization technique. The proposed GA scheme has been compared Particle Swarm Optimization (PSO) [9, 10] which is a popular algorithm for optimization of ECG features with respect to the following performance measures like convergence speed, and the accuracy in the final output. GA feature classification using neural network learning was implemented in [11]. The ECG classification flow diagram is shown in the Fig. 2.

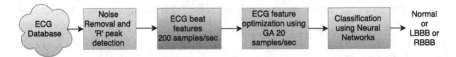

Fig. 2 ECG classification flow diagram

2 Pre Processing

2.1 Data Collection

The data for the classification was taken from the MIT BIH arrhythmia database [12]. Which consists of 5 normal, 3 LBBB and 3 RBBB patients data at 360 Hz Sampling rate of 1 h duration.

2.2 Noise Removal, R Peak Detection and Beat Segmentation

Sgolay filtering was used to remove the baseline wander present in the signal as shown Fig. 3a The R peaks of the ECG signal are detected as shown in Fig. 3b. Distance between two R peaks is called RR interval. 2/3 of the RR interval samples to the right of R peak and 1/3 of the RR interval samples to the left of R peak were considered as one beat as in Fig. 3c. Each beat after segmentation was re sampled to 200 samples.

3 Genetic Algorithm (GA)

GA [13] models the process of the evolution in nature. The method of GA can be applied to solve many real-world problems. GA follows three steps: Reproduction, Crossover, Mutation to generate a next breeding from the current generation. They are as shown in Fig. 4. The main purpose of using GA in this paper is to minimize the features (dimensionality) [14] before performing the classification operation [15]. There are many optimization algorithms available literature, among them GA resolves the optimization problems using the process of evolution and proved to be an auspicious one. The first step in GA is to evaluate the fitness of each candidate in

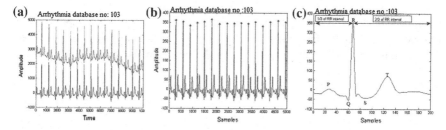

Fig. 3 Filtering, peak detection, segmentation

Fig. 4 GA flowchart

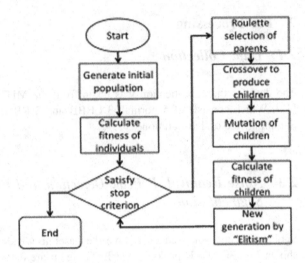

the population. Then the individuals with highest fitness values will enter into the next iteration as population. GA can used for choosing the optimal features, in which the selected data offers very high classification accuracy.

4 ECG Feature Extraction Using GA

GA reduces the size of population by mimicking the evolution process of living beings. Below matrix shows that there are N population each having six features. Using GA the size of population can be reduced by taking the population with good fitness values only.

$$Data = \begin{bmatrix} population\,1 & f1 & f2 & f3 & f4 & f5 & f6 \\ - & f1 & f2 & f3 & f4 & f5 & f6 \\ - & f1 & f2 & f3 & f4 & f5 & f6 \\ - & f1 & f2 & f3 & f4 & f5 & f6 \\ population\,N & f1 & f2 & f3 & f4 & f5 & f6 \end{bmatrix}$$

where f1, f2, f3, f4, f5 are the features of the each population. In this paper the GA is used for reducing the features (not for reducing the population). So we have taken the transpose of above data so that rows of the below matrix represents features.

$$Data^T = \begin{bmatrix} population\ 1 & - & - & - & population\ N \\ f1 & f1 & f1 & f1 & f1 \\ f2 & f2 & f2 & f2 & f2 \\ f3 & f3 & f3 & f3 & f3 \\ f4 & f4 & f4 & f4 & f4 \\ f5 & f5 & f5 & f5 & f5 \\ f6 & f6 & f6 & f6 & f6 \end{bmatrix}$$

Here our aim is to reduce features (rows) and keeping the size of population size same as the above. After applying the GA the features are reduced as shown in the below matrix

$$Data^T = \begin{bmatrix} population\ 1 & - & - & - & population\ N \\ f2 & f2 & f2 & f2 & f2 \\ f1 & f1 & f1 & f1 & f1 \\ f3 & f3 & f3 & f3 & f3 \end{bmatrix}$$

The features with the lowest fitness values die and remaining features are placed in the descending order of their fitness. Again find the transpose of the above matrix as

$$Data = \begin{bmatrix} population\ 1 & f2 & f1 & f3 \\ - & f2 & f1 & f3 \\ - & f2 & f1 & f3 \\ - & f2 & f1 & f3 \\ - & f2 & f1 & f3 \\ population\ N & f2 & f1 & f3 \end{bmatrix}$$

These reduced feature are given as input for the Neural Network so that its convergence speed and final accuracy can be increased. The ECG beats after segmentation are re-sampled to 200 samples/beat. Instead of using morphological feature extraction techniques, in this paper GA is used as the feature extraction technique. By using GA the ECG beat features are optimized to 20 features. The GA gives optimized features (best features) for the classification.

ECG beat samples before optimization = [1 2 3 ....200];

The reduced features using GA = [41 14 198 17 189 139 22 81 177 1 171 82 134 40 49 38 80 86 129 138].

5 Classification of ECG

5.1 Levenberg-Marquardt Neural Network (LM NN)

In this work for the detection of Bundle Branch Block LM back propagation Neural Network was used. This Neural Network provides rapid execution of the network to be trained, which is the main advantage in the neural signal processing applications.

6 Results

- Count of Normal beats used for classification-9,193.
- Count of RBBB beats user for classification-3,778.
- Count of LBBB beats user for classification-6,068.
- Total number of beats used for classification-19,039.
- Count of correctly classified beats-18,800.
- Total misclassified beats-239.

GA and PSO Features are classified using LM NN classifier and the results are shown in the Table 1. In the training we applied multilayer NN, and checked the network performance and decide if any changes to be made to the training process, or the data sets, the network architecture. The training record is checked with 'trainlm' Matlab function. The property training indicates the iteration is up to the point, where the performance of the validation reached a minimum. The training continued for 16 iterations before the stop. The next step is validating the network, a plot of epochs Versus Mean Squared Error (MSE) is as shown in the Fig. 5, which shows the relationship between the number of epochs of the network to the MSE. If the training is perfect the network outputs and the targets are be exactly equal, but that is rare in practice. Receiver Operating Characteristic (ROC) is a plot of Sensitivity verses Specificity as in Fig. 6. The points on the ROC plot corresponds a Specificity/sensitivity pair representing a particular parameter. The normal and abnormal classes can be clearly distinguished using the measure of area under the curve.

Table 1 Classification results

Classifier	Sensitivity (%)	Specificity (%)	Accuracy (%)
PSO+SCG NN	83.9	71.2	77.9
PSO+LM NN	83.1	80.8	84.7
GA+SCG NN	76.4	72.12	74.07
GA+LM NN	96.97	98	98.9

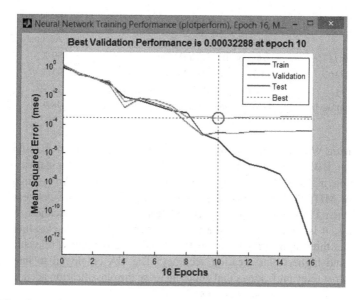

Fig. 5 Neural network training performance plot

Fig. 6 Performance comparison of different classifiers with GA features

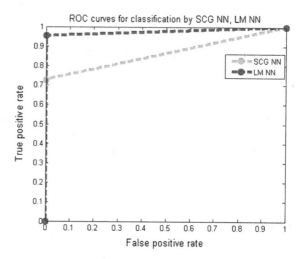

7 Discussion

The proposed GA is compared against other three BBB detection algorithms such as Wavelet Transform (WT), Continuous wavelet Transform (CWT), Wavelet transform and Probabilistic Neural Network (PNN) in terms of related features selected from the original database and classification accuracy obtained from different classifiers using Matlab software. The work in [16], explores an experimental

Table 2 Comparative study for detection of BBB

Studies	Approach	Accuracy (%)
Ceylan et al. [17]	Wavelet transform (WT)	99.2
Kutlu et al. [16]	Continuous wavelet transform (CWT)	97.3
Yu et al. [18]	WT and probabilistic neural network	98.39
Proposed approach	GA and neural network	98.9

study of using WT for extracting relevant features and KNN based classifier for the detection of BBB. The work presented in [17], uses morphological features for classification using SVM. The work proposed in [18], uses Arrhythmia dataset taken from MIT/BIH repository and 20 morphological and wavelet features are extracted then PNN is used for supervised learning and classification. From the experiments, it is concluded that the proposed GA with LMNN classifier outperformed other three algorithms with selection of minimal number of relevant features from GA and highest classification accuracy. The GA employed to intelligently select the most relevant features that could increase the classification accuracy while ignoring noisy and redundant features (Table 2).

8 Conclusion

The results showed that the proposed GA method extracts more relevant features for ECG analysis. These features that are extracted from GA are fed into a LM NN to classify the beats as the normal or LBBB or LBBB. The LM NN clearly distinguishes the Left and Right bundle blocks by taking features from the GA. The GA is compared to PSO algorithm. If this procedure helps us to automate a certain section or part of the diagnosis then it will help the doctors and the medical community to focus on other crucial sections. This has also increased the accuracy of diagnosis.

References

1. Padmavathi, K., and K. Sri Ramakrishna.: Classification of ECG Signal during Atrial Fibrillation Using Autoregressive Modeling: Procedia Computer Science 46 (2015): 53–59.
2. Padmavathi, K., and K. Krishna.: Myocardial infarction detection using magnitude squared coherence and Support Vector Machine. International Conference on Medical Imaging, m-Health and Emerging Communication Systems (MedCom): IEEE, (2014): pp. 382–385.
3. Karakatič, Sašo, and Vili Podgorelec.: A survey of genetic algorithms for solving multi depot vehicle routing problem.: Applied Soft Computing 27 (2015): 519–532.
4. Srinivas, Nidamarthi, and Kalyanmoy Deb.: Muiltiobjective optimization using non dominated sorting in genetic algorithms.: Evolutionary computation 2.3 (1994): 221–248.

5. Coury, Denis Vinicius.: Programmable logic design of a compact Genetic Algorithm for phasor estimation in real-time.: Electric Power Systems Research 107 (2014): 109–118.
6. Lim, Ting Yee, Mohammed Azmi Al-Betar, and Ahamad Tajudin Khader. "Adaptive pair bonds in genetic algorithm: An application to real-parameter optimization." Applied Mathematics and Computation 252 (2015): 503–519.
7. Gopalakrishnan, Hariharan, and Dragoljub Kosanovic.: Operational planning of combined heat and power plants through genetic algorithms for mixed 0âÄ§1 nonlinear programming.: Computers and Operations Research 56 (2015): 51–67.
8. El-Fergany, Attia A., Ahmed M. Othman, and Mahdi M. El-Arini.: Synergy of a genetic algorithm and simulated annealing to maximize real power loss reductions in transmission networks.: International Journal of Electrical Power and Energy Systems 56 (2014): 307–315.
9. Sadeghzadeh, H., M. A. Ehyaei, and M. A. Rosen.: Techno-economic optimization of a shell and tube heat exchanger by genetic and particle swarm algorithms.: Energy Conversion and Management 93 (2015): 84–91.
10. Soleimani, Hamed, and Govindan Kannan.: A hybrid particle swarm optimization and genetic algorithm for closed-loop supply chain network design in large-scale networks.: Applied Mathematical Modelling 39.14 (2015): 3990–4012.
11. Das, Manab Kr, Dipak Kumar Ghosh, and Samit Ari.: Electrocardiogram (ECG) signal classification using s-transform, genetic algorithm and neural network.: Condition Assessment Techniques in Electrical Systems (CATCON), 2013 IEEE 1st International Conference on IEEE, 2013.
12. Moody, George B., and Roger G. Mark.: The MIT-BIH arrhythmia database on CD-ROM and software for use with it.: Computers in Cardiology 1990, Proceedings. IEEE, 1990.
13. Goldberg, David E., and John H. Holland.: Genetic algorithms and machine learning.: Machine learning 3.2 (1988): 95–99.
14. Raymer, Michael L.: Dimensionality reduction using genetic algorithms.: Evolutionary Computation, IEEE Transactions on 4.2 (2000): 164–171.
15. Martis, Roshan Joy.: The Application of Genetic Algorithm for Unsupervised Classification of ECG.: Machine Learning in Healthcare Informatics: Springer Berlin Heidelberg, 2014. 65–80.
16. Kutlu, Yakup, Damla Kuntalp, and Mehmet Kuntalp.: Arrhythmia classification using higher order statistics.: Signal Processing, Communication and Applications Conference, 2008. SIU 2008. IEEE 16th. IEEE, 2008.
17. Ceylan, Rahime, and YÃijksel ozbay.: Wavelet neural network for classification of bundle branch blocks.: Proceedings of the World Congress on Engineering. vol. 2. No. 4. 2011.
18. Yu, Wei.: Application of multi-objective genetic algorithm to optimize energy efficiency and thermal comfort in building design.: Energy and Buildings 88 (2015): 135–143.

Tools for Data Visualization in Business Intelligence: Case Study Using the Tool Qlikview

Ayushi Shukla and Saru Dhir

Abstract This research paper, discusses the different data visualization tools that are implemented for Business Intelligence by different organizations for the purpose of business analysis. Even though all tools are basically used for the same purpose, that is, data visualization, exploration and analysis, but each tool are very different from the other. Here we draw comparisons between all the tools and see that which tool proves to be most efficient and most effective in fulfilling the user demands. A case study on a project university analysis application is also discussed using the business intelligence and data visualization tool Qlikview. Hypothetical data of the last 5 years are used to analyze the placement record of both the universities. At the end of the case study a comparative analysis is shown in between two universities and analyze the performance of both the universities.

Keywords Business intelligence (BI) · Tools · Data visualization · Qlikview

1 Introduction

Business intelligence (BI) is the set of techniques and tools for the transformation of raw data into meaningful and useful information for the purpose of Business Analysis. Identifying new opportunities and implementing an effective strategy based on insight can help in making effective business decisions that work in the favor of the company and help in its growth. The main goal of BI is to allow quick understanding and interpretation of huge amounts of data.

According to the future prediction of 2020, 50 times of the amount of data will generate worldwide in 2011 and there will be 75 times the amount of information sources [1]. Within these data lies great chances for business profit. But to benefit

A. Shukla · S. Dhir (✉)
Amity University Uttar Pradesh, Uttar Pradesh, India
e-mail: sdhir@amity.edu

© Springer India 2016
S.C. Satapathy et al. (eds.), *Information Systems Design and Intelligent Applications*, Advances in Intelligent Systems and Computing 434,
DOI 10.1007/978-81-322-2752-6_31

319

from these opportunities, business analysts should have all the data in their minds at all times which is possible through BI technologies.

BI technologies to deal with large amounts of raw data to help recognize and design some good business strategies using Data Visualization. Data visualization is the process of presenting the data in such form that it is easy to understand by the user. It can be represented in graphical forms like bar graphs, gauze charts, radar charts, etc. The main goal is to display the data or the information in a lucid way to the users such that analysis can be carried out with the help of visualization. Data analysis is not just important for any business, but it is the lifeline of all businesses. If the data are not analyzed, then the life of the organization will definitely be small and it will not be able to make much progress even while it is in the market.

Hence, business intelligence is implemented for making profitable business decisions using data analysis with the help of data visualization tools.

Major Benefits of Data Visualization tools are better decision making, data and knowledge sharing across organizations, provides the end user with the ability to work on his own, increases rate of interest, saves time reduces the pressure on IT [2].

2 Literature Review

Business Intelligence is a way of applying intelligence to support the decision-making methodology and is an outcome of the gradual growth of existing systems that support business decision making [3]. Although there is not much academic research on BI, but the main focus is on the literature of vendor and organization factors.

A report was generated on the detailed description of data visualization and explained everything about it, right from what it is to how it works for even its significance [4]. Another paper tells how the best visualization is possible and also gives a good account of the different charts that are used for visualization [5].

According to the report [6] Business Intelligence (BI) system adds value to the business and also helps in making more efficient managerial decision. A paper on the Qlikview business discovery platform and the ways in which Qlikview is different from other BI and data visualization tools [7]. This paper is basically for information employees and decision makers in understanding the uniqueness of qlikview. A collection of tools and techniques was presented which are useful for making the effective visualization and analyzing patterns [8].

Cognos [9] tells about IBM Cognos being an enterprise-class platform which is open and is built on a service-focused architecture and how it gains its capital from the infrastructure that is already in existence, adjusts to the changes quickly and gives extensibility for quite a long period of time.

2.1   Tools Used for Data Visualization in Business Intelligence

Visualizations help us see the things that were not obvious before. Even with large volumes of data, patterns can be spotted easily and in a matter of seconds. Visualizations convey information in a standard manner and make it easy to share ideas with others. All the data visualization tools provide a way of displaying data, no matter how big that data is, in such a way that it can be quickly analyzed and understood. There are many such tools in the market these days. Some of them are R, Spotfire, Tableau, Qlikview, Cognos, Profitbase, Sisense etc. For example, R is a complex visualization tool having a statistical package used to parse large data sets and one that takes a while to understand, but has a strong community and package library, with more and more being produced. Other tools are explained below:

2.1.1   Qlikview

QlikView is one of the most easy to understand and flexible Business Intelligence tools for turning data into knowledge. As the name suggests, QlikView is a combination of quick and click which means that it is fast as well as easy to use. Using Qlikview one can visualize data, search for some specific required data, design reports and view different trends in the data just at a single glance for unprecedented business insight. There are approximately 24,000 organizations in the world that have employed Qlikview. For example, this is the Business Intelligence tool being used by the University of St Andrews.

2.1.2   Spotfire

Spotfire was a business intelligence company and bought by TIBCO in 2007 [10]. TIBCO Spotfire is analytic software designed majorly for data exploration. It enables you to analyze the data using predictive and complex statistics in the analysis as well as discover and depict critical insights in the data. Spotfire makes use of in-memory function and an effective and user friendly platform to develop greatly interactive representation of the charts. It has a data analysis environment that is visually effective and lets the users inquire, envision and look up for information in actual time. It has customers and users from a variety of industries ranging from Life sciences, Energy, Financial Services, Government, Healthcare and many others.

2.1.3 Tableau

Tableau is a business intelligence and data visualization tool that helps in visualizing and creating interactive reports that can help the organization in connecting and sharing their data. It's easy enough for any person who knows Excel to quickly grasp it and at the same time efficient enough to represent even the most complicated analytical problems. Designing and displaying one's data findings only takes seconds.

2.1.4 Cognos

IBM Cognos is an integrated as well as web based Business Intelligence tool by IBM. It has many features like reporting of results, monitoring of the patterns and analysis of information [2]. This BI tool used for reporting offers different ways to do so. Firstly we can do the reporting with "guided analysis and prompted reports", which helps in quickly analyzing the organization and also in selecting or filtering those areas that need to be seen from the top down view. Second method of reporting is with the help of dashboards shown in the reports to get a quick idea in just a single glance (Table 1).

3 Case Study

A project implemented is discussed as the case study using the Business Intelligence and data visualization tool called Qlikview. The project was based on University Analysis Application. This application is helpful for drawing inter and intra university comparisons. It helps in judging the performance of the university over the years based on how the students have performed and how have their placements been. The project was implemented using the data of two universities. The data were assumed and it contained the fields: University, course, branch, student, email id, pass out year, cgpa and placement. The data from the aforementioned database was loaded into Qlikview and different charts were designed to draw various comparisons within a university and also between the two different universities for analysis of the data. Using these charts we try to get an idea of the performance of the universities and also decides the university with best performance.

Figure 1 shows the Dashboard of the application. Essential charts that include key findings, impact analysis, history and trends and give an overview of the application are built into the dashboard. This dashboard tells about the placement trends for the two universities over the years and also their cumulative grades on an average and which university has performed better than the other.

Figure 2 represents the (KPI) Key Performance Indicators. It uses the measures as the keys to judge the performance of the given university. Here the KPIs are the

Table 1 Comparison table between various data visualization tools

Features	Qlikview	Spotfire	Tableau	Cognos
Vendor	Qliktech	TIBCO	Tableau	Cognos
Year of launch	1993	1997	2003	2005
Implementation time	Low (fastest to implement)	Average	Average	High (10 times more than Qlikview)
Implementation cost	Average	High	High	High (6 times more than Qlikview)
Dashboard support	Excellent (best for dashboards)	Very good (best for dashboards)	Good	Good (special ability to create widgets in which user can drag and drop his reports, charts)
Interactive visualization	Excellent	Very good	Very good	Good
Visual drill down	Excellent (drill down is possible till transaction level)	Very good	Good	Good (drill down is possible till aggregated data level)
Application Customization	Easy (the 'look and feel' type of customizations can be done with a few clicks, even by end users)	Average	Average	Difficult (it requires creation of different cubes to get different views of the data)
Scripting language used	VB script and set analysis	Iron python	VizQL (Visual Query Language)	Javascript
Special feature	If you want a unified view of your data and pick Qlik sense	If you want easy-to-use visualizations to choose Spotfire	If you just want pretty visualizations to choose Tableau	If you want to create highly complex custom applications choose Cognos

placement status of students of various courses as well as branches into different companies for both the universities.

Figure 3 shows the Branch Wise Analysis which tells about the number of students placed in different companies on the basis of different branches. The university and the year for which this data needs to be analyzed and can be selected from the dimension list boxes provided within the tool.

Figure 4 shows comparative analysis in between two universities, which provides various combinations between different branches and courses for a selected year.

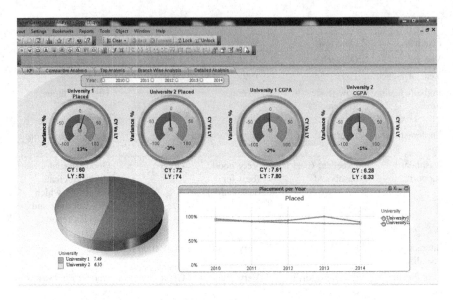

Fig. 1 Dashboard for university analysis application

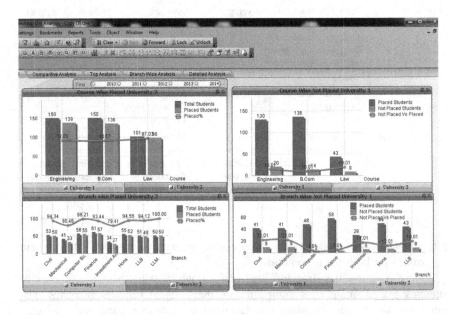

Fig. 2 Key performance indicator

Using Qlikview as the data visualization tool, all the charts could be designed very easily and efficiently. It provides us with all types of analysis just with the click of a button. For example, if someone wants to view a particular data, Qlikview

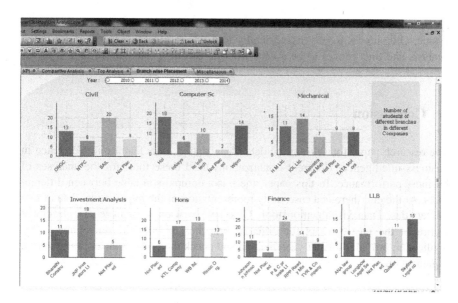

Fig. 3 Branch wise analysis

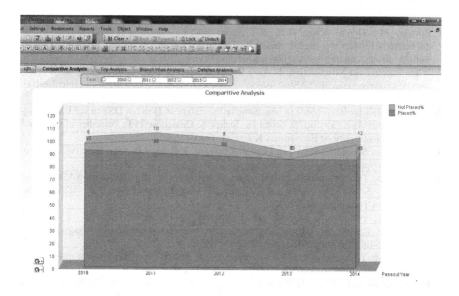

Fig. 4 Comparative analysis

simply filters that data from all the charts based on the selection that the user made. In the above application, if the branch Mechanical is selected from the Branch list box and the year 2013 is selected from the Year list box, then the information related to mechanical branch in the year 2013 is filtered out from the sheets of the application and all the charts display the results of this particular selection only. Hence, Qlikview is a very user friendly tool which greatly helps to make quick analysis and thus proves to be a very efficient Business Intelligence tool.

4 Conclusion

The research paper focused on the data visualization tools that are implemented in business intelligence. Different organizations used these tools for the analyses of business performance. In this paper, there is a comparison table between different tools. At the end there is a case study representation of the hypothetical data of two universities. Data visualization tool- Qlikview, is used to analyze the placement record of the past 5 years for different branches of both the universities. This application provides greater visualization and check the performance of the both the organizations.

References

1. Boman, M., Digital Cities, EIT ICT Labs, pp. 12, www.eitdigital.eu/fileadmin/files/publications/130426_wp_digital_cities.pdf; University of St. Andrews, Scotland, Official website, www.st-andrews.ac.uk/qlikview/whatisqlikview (2012).
2. White Paper: Data Visualization: Making Big Data Approachable and Valuable, SAS, pp. 2, www.sas.com/content/dam/SAS/en_us/doc/whitepaper2/sas-data-visualization-marketpulse-106176.pdf (2012).
3. Gray, P., Business Intelligence: a new name or the Future of DSS? in DSS in the Uncertainty of the Internet Age. Eds Bui T., Sroka H., Stanek S. Goluchowski J., Publisher of the Karol Adamiecki University of Economics, Katowice. (2003).
4. 7 Things You Should Know About Data Visualization, Educause learning initiative, http://go.wisc.edu/787fpp (2009).
5. White paper: Data Visualization Techniques-From Basics to Big Data with SAS® Visual Analytics, SAS, http://www.sas.com/offices/NA/canada/downloads/IT-World2013/Data-Visualization-Techniques.pdf (2012).
6. Gibson, M., Arnott, D., Jagielska, I: Evaluating the Intangible Benefits of Business Intelligence: Review & Research Agenda", http://citeseerx.ist.psu.edu/viewdoc/download?doi=10.1.1.94.8550&rep=rep1&type=pdf (2004).
7. White Paper: What makes QlikView unique-A Qlik White Paper, http://www.qlik.com/us/explore/resources/whitepapers/what-makes-qlikview-unique (2014).
8. Bertini, E., Univ. of Konstanz, Konstanz, Germany; Tatu, A., Keim, D.,: Quality Metrics in High-Dimensional Data Visualization: An Overview and Systematization, Visualization and Computer Graphics, IEEE Transactions on Volume:17, Issue:12, http://ieeexplore.ieee.org/xpl/login.jsp?tp=&arnumber=6064985&url=http%3A%2F%2Fieeexplore.ieee.org%2Fxpls%2Fabs_all.jsp%3Farnumber%3D6064985 (2011).
9. White Paper: The IBM Cognos platform: Meeting the needs of both business users and IT, IBM Software Group, http://www.5xtechnology.com/Portals/92116/docs/ibm_cognos_platform.pdf (2009).
10. BusinessIntelligence.com, businessintelligence.com/bi-vendors/tibco-spotfire.

Application of Support Vector Machines for Fast and Accurate Contingency Ranking in Large Power System

Soni Bhanu Pratap, Saxena Akash and Gupta Vikas

Abstract This paper presents an effective supervised learning approach for static security assessment of power system. The approach proposed in this paper employs Least Square Support Vector Machine (LS-SVM) to rank the contingencies and predict the severity level for a standard IEEE-39 Bus power system. SVM works in two stage, in stage 1st estimation of a standard index line Voltage Reactive Performance Index (PI_{VQ}) is carried out under different operating scenarios and in stage II (based on the values of PI) contingency ranking is carried out. The test results are compared with some recent approaches reported in literature. The overall comparison of test results is based on the, regression performance and accuracy levels. Results obtained from the simulation studies advocate the suitability of the approach for online applications. The approach can be a beneficial tool to fast and accurate security assessment and contingency analysis at energy management centre.

Keywords Artificial neural network · Contingency analysis · Performance index (PI) · Static security assessment · Support vector machines (SVMs)

Please note that the LNCS Editorial assumes that all authors have used the western naming convention, with given names preceding surnames. This determines the structure of the names in the running heads and the author index.

S.B. Pratap (✉) · G. Vikas
Department of Electrical Engineering, Malaviya National
Institute of Technology, Jaipur 302017, India
e-mail: er.bpsoni2011@gmail.com

G. Vikas
e-mail: vgupta.ee@mnit.ac.in

S. Akash
Department of Electrical Engineering, Swami Keshvanand
Institute of Technology, Jaipur 302017, India
e-mail: aakash.saxena@hotmail.com

327

1   Introduction

Contemporary power system is a complex interconnected mesh having many utilities of different nature at generation, transmission and distribution ends. Complexity of power system is increasing with every passing day due to exponential increase in population and increased load demands. Under these conditions a question mark appears on the reliable operation of the power system [1]. The utilities at different ends are operating at their operating and security limits. To make system extremely reliable the offline studies under different operating scenarios for occurrence of probable contingencies is the field of interest. Contingency analysis is carried out by screening and sensitivity based ranking methods [2–12]. In ranking methods the study is based on the calculation of standard indices namely MVA, Voltage reactive etc. However, in screening methods different load flow methods are employed namely DC load flow, Ac load flow and local solution methods. Sensitivity based methods are efficient but inaccurate on the other hand screening methods are less efficient. Artificial Neural Networks (ANNs) [2–4], hybrid decision tree model [5], Cascaded Neural Network [6, 8], hybrid neural network model [7] Radial Basis Function Neural Network (RBFNN) [9–11] have been applied to estimate critical contingencies and rank them in order of occurrence and severity. Most of the approaches employed supervised learning approach for detection and understanding the complex behavior of power system under different operating condition. Performance Index (PI) has been considered as an output and responsible denominator for explaining the power system state. Two critical issues related with neural nets are identified and these are 1. Large offline data set preparation and time associated for collection 2. The efficiency of the net to train with the given data set. In these approaches [1–12] neural nets employed as regression agent and classifier. The efficiency of the net depends upon the selection of appropriate features (inputs), neurons and hidden layers. In these experiments optimal feature selection was extracted by optimization methods and correlation coefficients [2].

Recent years Support Vector Machines (SVMs) are employed in many approaches to classify the data and as a regression tool. Least Square SVM is used for prediction of the optimum features and classification of power quality events [12]. Fault identification and classification carried by SVM [13]. This paper presents application of LS-SVM to rank and classify the contingencies in large power network.

In view of above literature review following are the objectives of this research paper.

(i) To develop a supervised learning based model which can predict the performance index based on voltage and reactive power for a large interconnected standard IEEE 39 bus test system under a dynamic operating scenario.

(ii) To develop a classifier which can screen the contingencies of the power system into three states namely not critical, critical and most critical.

(iii) To present the comparative analysis of the reported approaches with the proposed approach based on accuracy in prediction of the PI. The following section contains the details of the performance index based on line voltage reactive flow.

2 Contingency Analysis

Contingency evaluation is essential to know the harmful/less harmful situations in power network. Without knowing the severity and the impact of a particular contingency, preventive action cannot be initiated by the system operator at energy management center [9]. Contingency analysis is an important tool for security assessment. On the other hand, the prediction of the critical contingencies at earlier stage, which can present a potential threat to the system stability (voltage or rotor angle) helps system operator to operate the power system in a secure state or initiate the corrective measures. In this paper line outages at every bus in New England system are considered as a potential threat to the system stability. Performance Index (PI) methods are widely used for determination in contingency ranking [2, 7, 8, 11].

2.1 *Line Voltage Reactive Performance Index (PI$_{VQ}$)*

On the basis of literature review, it is observed that system stress is measured in terms of bus voltage limit violations and transmission line over loads. System loading conditions in a modern emerging power system are dynamic in nature and impose a great impact on the performance of the power system. An index based on Line VQ flow is determined to estimate the extent of overload.

$$PI_{VQ} = \sum_{1}^{NB} \left(\frac{W_{Vi}}{M}\right) \left[\frac{V_i - V_i^{sp}}{\Delta V_i^{Lim}}\right]^M + \sum_{1}^{N_G} \left(\frac{W_{Gi}}{M}\right) \left[\frac{Q_i}{Q_i^{max}}\right]^M \tag{1}$$

where $\Delta V_i^{Lim} = V_i - V_i^{max}$ for $V_i > V_i^{max}$, $V_i^{min} - V_i$ for $V_i < V_i^{min}$, $V_i S_i^{post}$ is the post contingency Voltage at the ith bus, V_i^{max} the maximum limit of voltage at the ith bus, V_i^{min} the maximum limit of voltage at the ith bus, N_B the number of buses in the system, W_{Vi} the real non-negative weighting factor (=1), M(=2n) is the order of the exponent for penalty function. The first summation is a function of only the limit violated buses chosen to quantify system deficiency due to out-of limit bus voltages. The second summation, penalizes any violations of the reactive power constraints of all the generating units, where Q_i is the reactive power produced at bus i, Q_i^{max} the maximum limit for reactive power production of a generating unit,

N_G the number of generating units, W_{Gi} is the real non-negative weighting factor (=1). The determination of the proper value of 'n' is system specific. The optimum integer value 'n' for this paper is taken as 4.

3 Support Vector Machine

Recently the mappings and classification problems are handled well by the Artificial Neural Networks (ANNs). Two basic properties of neural nets make themselves different from other conventional approaches these properties are 1. Learning from the training samples 2. A unique property to adapt according to new environment. In LS-SVMs the input data is mapped with high dimensional feature space with the help of kernel functions [12, 13]. By using kernel functions the problem can be mapped in linear form. The least square loss function is used in LS-SVM to construct the optimization problem based on equality constraints.

The least squares loss function requires only the solution of linear equation set instead of long and computationally hard quadratic programming as in case of traditional SVMs. LS-SVM equation for function estimation can be written as shown in Eq. (2).

$$y(x) = \sum_{k=1}^{N} \alpha_k K(x, x_k) + b \tag{2}$$

where α_k is the weighting factor, x are the training samples and x_k are the support vectors, b represents the bias and N is the training samples. The RBF kernel function for the proposed SVM tool can be written by Eq. (3) (Fig. 1).

$$K(x, x_k) = \exp\left(-\frac{\|x - x_k\|^2}{2\sigma^2}\right) \tag{3}$$

In the present simulation work LS-SVM is interfaced with MATLAB and dataset of 14,000 different operating conditions along with line outages of each line is considered. The values of PI obtained from standard Newton Raphson (NR) methods are used for training purpose. The least square estimator uses the optimize values of σ (kernel width) larger the value more will be the width of kernel. This value indicates that system is more global and near to a linear system. Unlike neural network SVM trains in less time and possess no hidden layers.

Fig. 1 Data generation for contingency analysis

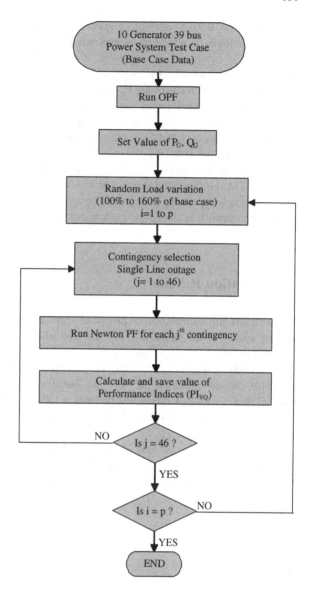

4 Proposed Methodology

Data generation is an important task in supervised learning approach. In this study a rich data of 14,000 samples is employed to train, test and validate the networks. Following are the steps involved in the process.

i. A large no of load patterns are generated by randomly perturbing the real and reactive loads on all the buses and real and reactive generation at the generator buses.

ii. The features are selected as per [2]. Total 11 features are chosen for training purpose. These features are P_{g10}, Q_{g1}, Q_{g2}, Q_{g3}, Q_{g4}, Qg_5, Q_{g7}, Q_{g8}, Q_{g9}, Q_{g10} and Q_{d14}. A contingency set for all credible contingencies are employed. N − 1 contingencies are the most common event in power system. Single line outages are considered for each load pattern and the value of index is stored in iteration of the simulation.

iii. The obtained values of the index are normalized between 0.1 and 0.9 to train the SVM. Further the binary classification is done to train the classifier.

iv. The system operating state contingency type and the regression performance of the network is stored for each operating scenarios.

5 Simulation Results

The Simulink implementation of proposed approach has been tested over IEEE-39 bus test system (New England) [14]. The modeling of the system and simulation studies are performed over Intel® core™, i7, 2.9 GHz 4.00 GB RAM processor unit. Bus no. 39 has taken as slack bus. For line contingency 14,000 patterns are generated, which includes the 46 line outages and different loading patterns (300) (Table 1). Out of these 200 patterns are those where NR method failed to converge. The comparative results for the performance of the PI are shown in Table 2.

From Table 2 it can be judged that the LSSVM possess lower values of Mean Square Error (MSE) and high values of regression coefficient (R). It is empirical to judge that often the performance of the ranking methods is questioned due to wrong

Table 1 Performance index for classification

Class	I (Non-critical)	II (Critical)	III (Most critical)
PI range	<0.2	0.2–0.8	>0.8

Table 2 Comparative performance of different neural networks

Type of regression agent	MSE	R
Elman backprop	0.7062e−3	0.96253
Cascaded FNN	0.8091e−3	0.9837
FFNN	0.8755e−3	0.9822
FFDTDNN	0.6830e−3	0.98457
Layer recurrent	0.7752e−3	0.9844
NARX	0.646e−3	0.98846
LSSVM	0.575e−3	0.9915

Table 3 Sample result of PI_{VQ} calculation and contingency analysis

Outage no.		1345	7811	9014	587	2984
Line no.		6–7	8–9	9–39	25–26	20–34
PI_{VQ}	NR	0.8421	0.5846	0.3876	0.4232	0.1201
	Elman backdrop	0.8143	0.651	0.4231	0.3647	0.1345
	Cascaded FBNN	0.8001	0.4322	0.387	0.4515	0.1141
	FFNN	0.8436	0.5561	0.4015	0.3484	0.122
	FFDTD	0.8015	0.5334	0.4312	0.3486	0.1546
	Layer recurrent	0.8451	0.5457	0.3342	0.2247	0.134
	NARX	0.8245	0.3475	0.3015	0.3846	0.1426
	LS-SVM	0.8425	0.5901	0.387	0.4231	0.121
Class	LS-SVM	III	II	II	II	I
	NR	III	II	II	II	I

detection or miss-ranking of a critical contingency. From Table 2 it is concluded that the PI prediction by LSSVM is more accurate as compared with the reported approaches. Classifier performance of different approaches is shown in Table 3. The high values of R show that how efficient machine is to learn the relationship between output and input. It is observed that with a slight difference in the values misranking of many contingencies can take place. Such misrankings are the pitfall associated with the learning methods. However, the results obtained from LS-SVMs are quite promising. Further on the basis of values associated to the indexes the classification of the contingencies is also exhibited in Table 3. The analysis associated with the classification is explained with the help of confusion diagram. Confusion matrix is an error matrix, the rows represent the instances in predicted class and column represents instances in actual class. Table 3 shows the cumulative performance of LS-SVM as a regression agent and a classifier. Table 1 shows the classes of contingencies as per calculated index values.

Following points are emerged from Table 3:

(i) It is observed that for line outage 6–7 the values of PI from NR method is 0.8421. For this operating scenario, this contingency is a potential threat to the system stability and critical in nature. The value predicted by backdrop method is 0.8413, Cascaded FBNN is 0.8001 and 0.8015 FFDTDNN. LS-SVM predicted the value which is very near to the NR method and that is 0.8425.

(ii) It is also observed that for line 20–34 the contingency is neither severe nor critical hence the values predicted by all network topologies fall in the same range. However, the value predicted by LSSVM is quite close to the original values.

(iii) The classes identified by the NR method are same with the class identified by LS-SVM. It is worth to mention here that in this simulation study total 14,000 cases were simulated. Out of these cases 1232 cases identified as critical contingencies and 10,243 were identified as most severe critical contingencies.

Fig. 2 Confusion matrix for
PI_{VQ} of LS-SVM network

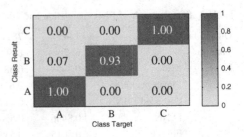

The classification is handled by LS-SVM is verified through NR and presented in a lucid manner for some contingencies.

(iv) It is concluded that both classification and prediction parts are performed by SVM with selected features. The efficacy of the proposed method is compared with contemporary types of neural networks. It is pragmatic to say that LS-SVM proved as a better supervised approach for real power system problem. Confusion matrix for the index detection is shown in Fig. 2. Confusion matrix is a classical way to determine the accuracy of the classifier. From Fig. 2 it can be observed that for Class C (III) 100 % cases were identified. For class B (II), 93 % cases were identified i.e. from 1232 cases 1146 cases are identified by the classifier. 7 % remaining cases were identified as a contingency A.

The neural nets are trained several times and the final results are taken from the best performed networks. Table 3 shows the calculated PIs for different neural networks (Feed forward back drop, NARX Layer Recurrent) and proposed approach. It is observed that SVM is able to classify the contingency efficiently. It is also observed that the estimation of the PIs under different outages is also carried out in a very effective manner by SVM as compared with other approaches.

6 Conclusions

This paper proposes a supervised learning model based on least square loss function with RBF Kernel function to estimate the contingency ranking in a standard IEEE-39 bus system. Following conclusion can be drawn with this work.

(a) Comparative analysis of existing learning base approaches are carried out for a large interconnected power system under dynamic operating conditions. It is observed that neural nets of different types exhibit their quality to act as a regression agent. However, the best regression results based on MSE and R values are observed by LS-SVM. The numerical results obtained for the index calculation advocated the efficacy of the propose approach.

(b) In second part the classification of the contingencies are carried out by LS-SVM. A binary classifier is obtained with three binary classes based on the values of voltage reactive index. The performance of the SVM as a classifier is exhibited through confusion diagram. It is concluded that SVM shows a better efficiency to classify the contingencies.

(c) The proposed approach is suitable for online application. The operator at energy management centre can easily get the details of the contingency and severity of the same with the help of these offline tested results. Study on larger system with multiple contingencies lays in the future scope.

References

1. J.C.S. Souza, M.B. Do Coutto Filho, M.Th. Schilling, Fast contingency selection through a pattern analysis approach, Electric Power Systems Research, Volume 62, Issue 1, Pages 13–19 (2002), ISSN 0378–7796, http://dx.doi.org/10.1016/S0378-7796(02)00016-0.
2. Verma, Kusum, and K. R. Niazi. "Supervised learning approach to online contingency screening and ranking in power systems." International Journal of Electrical Power & Energy Systems Vol. 38, no. 1, pp. 97–104 (2012).
3. Swarup KS, Sudhakar G. Neural network approach to contingency screening and ranking in power systems. Neurocomputing Vol. 70, pp. 105–118 (2006).
4. Niazi KR, Arora CM, Surana SL. Power system security evaluation using ANN: feature selection using divergence. Electr Power Syst Res Vol. 69, pp. 161–167 (2004).
5. Patidar N, Sharma J. A hybrid decision tree model for fast voltage screening and ranking. Int J Emerg Electr Power Syst Vol. 8, No. 4, Art. 7 (2007).
6. Singh SN, Srivastava L, Sharma J. Fast voltage contingency screening and ranking using cascade neural network. Electr Power Syst Res Vol. 53, pp. 197–205 (2000).
7. Srivastava L, Singh SN, Sharma J. A hybrid neural network model for fast voltage contingency screening and ranking. Electr Power Energy Syst Vol. 22, pp. 35–42 (2000).
8. Singh R, Srivastava L. Line flow contingency selection and ranking using cascade neural network. Neurocomputing, Vol. 70, pp. 2645–50 (2007).
9. Devraj D, Yegnanarayana B, Ramar K. Radial basis function networks for fast contingency ranking. Electr Power Energy Syst Vol. 24, pp. 387–95 (2002).
10. Refaee JA, Mohandes M, Maghrabi H. Radial basis function networks for contingency analysis of bulk power systems. IEEE Trans Power Syst, Vol 18, No. 4, pp. 772–8 (1999).
11. Jain T, Srivastava L, Singh SN. Fast voltage contingency screening using radial basis function neural network. IEEE Trans Power Syst Vol. 18, No. 4, pp. 1359–66 (2003).
12. Hüseyin Erişti, Özal Yıldırım, Belkıs Erişti, Yakup Demir, Optimal feature selection for classification of the power quality events using wavelet transform and least squares support vector machines, International Journal of Electrical Power & Energy Systems, Volume 49, Pages 95–103, (2013) ISSN 0142–0615, http://dx.doi.org/10.1016/j.ijepes.2012.12.018.
13. Sami Ekici, Support Vector Machines for classification and locating faults on transmission lines, Applied Soft Computing, Volume 12, Issue 6, Pages 1650–1658 (2012), ISSN 1568–4946, http://dx.doi.org/10.1016/j.asoc.2012.02.011.
14. Power system test cases at http://www.ee.washington.edu/pstca/.

A Novel Circular Monopole Fractal Antenna for Bluetooth and UWB Applications with Subsequent Increase in Gain Using Frequency Selective Surfaces

Divyanshu Upadhyay, Indranil Acharya and Ravi Prakash Dwivedi

Abstract In this paper a novel monopole circular fractal disk is analyzed in details. The antenna is mounted on a FR4 substrate having a dimension of 30×35 mm^2. A 50 Ω impedance matched microstrip line is used as feeding mechanism for the antenna. The antenna exhibits stable radiation patterns and has a bandwidth of 9.64 GHz. A U-shaped slot having appropriate dimensions is etched from the patch in order to make the antenna compatible for Bluetooth applications. In the later section, different types of Frequency Selective Structures (FSS) are introduced for enhancing the gain of the antenna. A gain of 8.94 dB is observed with the introduction of slot type FSS structures beneath the ground plane. All the analysis of the antenna is done in HFSS 2013.

Keywords Fractal antenna · Bandwidth · UWB technology · Notch · U-shaped slot · Frequency selective surface · Gain · Return loss

1 Introduction

In 2002 Federal Communication Commission (FCC) authorized the unlicensed use of 7.5 GHz bandwidth (from 3.1 to 10.6 GHz) as the official band for Ultra-wideband applications [1]. Since its inception, there has been a tremendous

D. Upadhyay (✉) · I. Acharya · R.P. Dwivedi
Vellore Institute of Technology, School of
Electronics Engineering (SENSE), Chennai, Tamil Nadu, India
e-mail: divyanshuu78@gmail.com

I. Acharya
e-mail: indranil907@gmail.com

R.P. Dwivedi
e-mail: raviprakash.dwivedi@vit.ac.in

© Springer India 2016
S.C. Satapathy et al. (eds.), *Information Systems Design and Intelligent Applications*, Advances in Intelligent Systems and Computing 434,
DOI 10.1007/978-81-322-2752-6_33

337

demand for antennas operating in this band because of its inherent advantages of being compact, having low power consumption, high data rate within short distance, phase linearity and stable omni-directional radiation pattern [2–5]. UWB technology is widely used in medical imaging, breast cancer detection, indoor/outdoor communication, radars and military surveillance. Many antennas have been designed for UWB applications. But the main challenge lies in its miniaturization. A perpetual tradeoff exists between the achievable bandwidth and the antenna physical dimension. Antennas with high dielectric constant values reduce its resonant frequency but at the cost of a reduced bandwidth [6]. Fractal geometry, a concept of mathematics can be extended for designing UWB antennas because of its several unique features. They have a unique property of space-filling meaning which implies that large apparent electrical lengths can be obtained within a limited given volume. Since the radiation pattern of an antenna is immensely affected by its electrical length, this efficient packing by fractal concept serve to be quite indispensable. Although inherently narrowband, some fractal antennas do have wideband characteristics like the snowflake fractal antenna for C-band frequency applications [7]. A trademark feature of snow flake antennas is that they have an impedance bandwidth of 49 %. Another interesting property of fractal antennas is self-similarity that leads to antennas operating at multiple frequencies thus serving as a potential candidate for UWB applications as specified by FCC [8]. Notable in this regard is the Desecrates Circle Theorem (DCT) which led to the discovery of novel UWB antennas which are compact in size [9]. The self-similarity approach discussed above is utilized here in case of a circular monopole disk fractal antenna.

In the first part, a novel monopole fractal antenna is studied. The designed antenna is compact in size and exhibits wideband characteristics. Moreover with the incorporation of certain incisions in the patch, the antenna is made to operate in the ISM band (2.36 GHz). In order to mitigate interference between co-existing systems in the UWB range (viz. WiMAX, C band, WLAN etc.), a notch is created mainly to discard WLAN frequency (5.15–5.35/5.725–5.825 GHz). The later section of the paper is mainly concerned with enhancing the gain of the antenna without interfering the wide bandwidth. The concept of Frequency Selective Surface (FSS) is adopted in this regard. FSS structures have unique characteristics of suppressing surface waves and acting as a high impedance surface that reflects in phase plane waves. But in order to obtain maximum reflection, the position of the FSS with respect to the antenna becomes indispensable. In this paper, three different FSS surfaces viz. Jerusalem cross structure, Tri-pole FSS and slot-type structures are employed and a comparative analysis is carried out in HFSS 2013.

1.1 Basic Antenna Structure

The geometric representation of the antenna is shown in Fig. 1. The optimized dimensions are shown in Table 1. The antenna under consideration is placed on

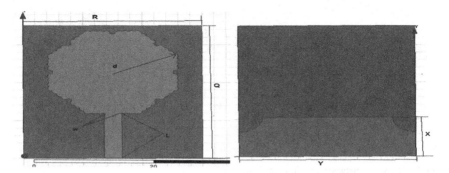

Fig. 1 Structure of the proposed antenna

Table 1 Optimized dimensions of the basic proposed antenna

Dimensions of the antenna	Optimized value (in mm)
R	30
Q	35
L	11.86
w	3
d	11.347
X	10.5
Y	30

FR4 epoxy substrate of thickness 1.6 mm. A 50 Ω microstrip line having length (L mm) and width (w mm) is used as feed and matching section.

Two circles each having radius of 4 mm is removed from the sides of the modified ground plane. This provides a better impedance matching and results in a wider bandwidth formation.. The basic structure consists of a circular patch having radius 7 mm. In the first iteration, eight circles each having radius 2.7475 mm is placed on the radiating patch thereby leading to an increase of 1.58 in the overall circumference value. In the second iteration, five circles each having radius 1.3737 mm is placed upon each smaller circle. In the third iteration, three circles each of radii 0.68685 mm is paced upon the smaller circles. This process can be repeated till infinity, but for the antenna under consideration the above procedure is constrained to three iterations only. The transition between the different iterations is demonstrated in Fig. 2.

1.2 Results Obtained

Figures 3 and 4 show the return loss and the gain of the antenna. It can be observed from the S_{11} plot that the antenna resonates at frequencies 3.43, 5.79, 8.76 and 11.37 GHz having return loss of −40.57, −34.74, −27.29 and −39.75 dB

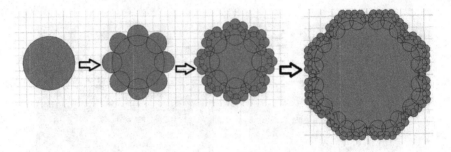

Fig. 2 Transition between the different iterations

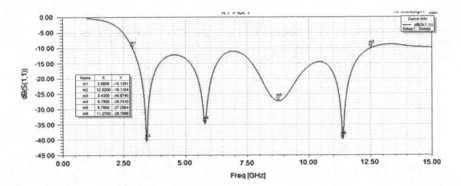

Fig. 3 Return loss of the planar monopole antenna

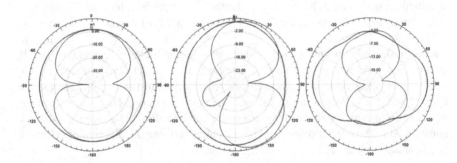

Fig. 4 Radiation pattern of the antenna at different frequencies

respectively. A wide bandwidth of 9.64 GHz is obtained between 2.88 GHz. At 3.5 GHz a gain of 1.83 dB is obtained.

The radiation pattern of the antenna is displayed in Fig. 4 having resonant frequencies at 3.5, 6 and 9 GHz respectively.

Radiation patterns are obtained by varying theta (θ) and phi (ϕ) angles. Here, only theta values are varied but phi remains constant to zero value.

1.3 Creating Notch on the Basic Antenna for Discarding WLAN Frequencies

In this case an inverted U shaped incision is etched out from the patch of the normal antenna. Figure 5 shows the geometric representation of the antenna with an inverted U-shaped slot.

The dimensions of the notch slot are represented as: t = 8.5 mm, g = 0.6 mm, p = 2 mm and z = 3 mm. By varying the length of the slot (t) different variation in the stop band can be obtained. As can be inferred from Fig. 6, with increase in the length of the slot, an increase in the bandwidth of the stop band is obtained.

2 Modified Antenna for ISM Band (Bluetooth) and UWB Applications

In order to make the antenna compatible for ISM band as well as UWB applications, a U-shaped slot is inserted from the patch. Figure 7 shows the geometric representation of the modified antenna. The dimensions of the slot are as follows: n = 0.5 mm, j = 4.2 mm, u = 7 mm, s = 17.4 mm.

Fig. 5 Basic antenna with inverted U-shaped slot

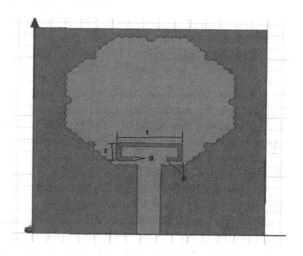

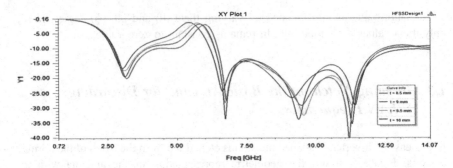

Fig. 6 S_{11} plot showing variation of the stop band with change in the dimension of slot length (t)

Fig. 7 Geometric representation of the modified antenna

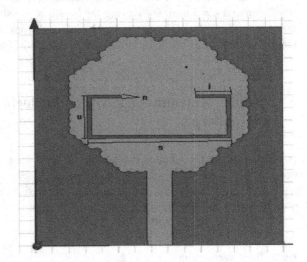

2.1 Results Obtained

Figure 8 shows transmission coefficient of the antenna. From the S_{11} plot, it can be obtained that the antenna resonates at frequencies 2.4, 3.91, 5.78, 7.93 and 11.44 GHz having return loss of −14.82, −22.28, −26.38, −34.08 and −27.76 dB respectively. A wide bandwidth of 9 GHz is obtained between 3.4 and 12.4 GHz.

The radiation pattern of the modified antenna is shown in Fig. 9 at frequencies 3.5, 6 and 9 GHz respectively.

The gain obtained is 1.60 dB at a frequency of 3.5 GHz. Here the gain is obtained by varying θ and ϕ values respectively.

The surface current distribution of the modified antenna is shown in Fig. 10. It can be observed that the incorporation of the slot in the patch, current density around its upper corners increases. This leads to more propagating modes and at the

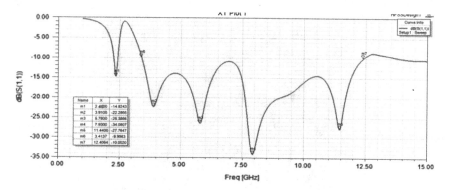

Fig. 8 Return loss of the modified antenna

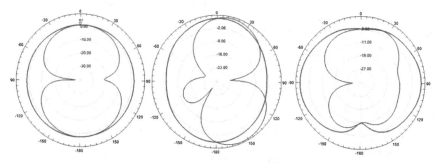

Fig. 9 Radiation pattern of the modified antenna at different frequencies

same time creates a null (at approximately 2.55 GHz) as the currents in the exterior and the interior of the slot are oppositely directed.

3 Frequency Selective Surface Design

The proposed UWB circular fractal antenna is basically a low gain antenna and the gain is considered as matter of great concern for wireless application. The two important techniques, Electromagnetic Band Gap (EBG) and the Frequency Selective Surface (FSS), are used for enhancing the gain of the antenna. The Frequency Selective Surface (FSS) is an important alternative to EBG due to its compactness and low profile configuration [10]. A Jerusalem cross-pair FSS having dimensions of unit cell 8.4 mm × 8.4 mm and the gap between each unit cell is taken as 1 mm. The structure of this type of FSS is shown in Fig. 11.

A Tripole-pair FSS is the combination of various unit cells of double periodical arrangements of tripoles and the gap between each unit cell is taken as 1 mm. The structure of this type of FSS is shown in Fig. 12.

Fig. 10 Current distribution on the modified antenna

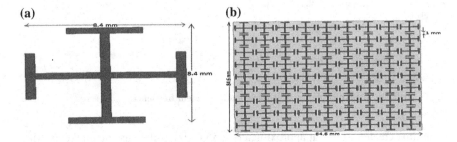

Fig. 11 Jerusalem cross-pair FSS **a** Unit cell **b** 9 × 9 FSS

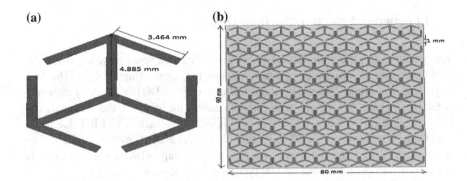

Fig. 12 Tripole-pair FSS **a** Unit cell **b** 9 × 9 FSS

(a)   **(b)**

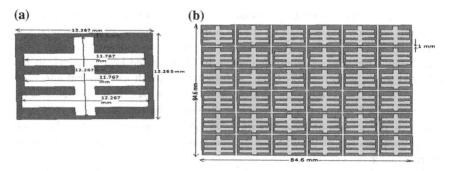

Fig. 13 Slot-pair type FSS **a** Unit cell **b** 6 × 6 FSS

A new slot-pair type FSS is introduced with its unit cell dimensions of 13.267 mm × 13.265 mm and gap between the unit cells is taken to be 1 mm. By etching various slots in the unit cell, a capacitive effect is created and this effect gives rise to lower resonance frequency as compared to the above FSS structures (Fig. 13).

A common substrate, Rogers RT/duroid 5880™ is chosen for these FSS structures (dielectric constant value: 2.2 and thickness 0.8 mm).

3.1 Results Obtained

Figure 14 shows the comparative return loss plots of UWB circular fractal antenna with different FSS structures. The FSS structures were placed at height of h = 21.635 mm.

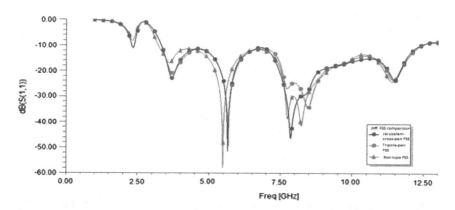

Fig. 14 Comparative return loss plots of UWB circular fractal antenna with different FSS structures

From the above comparative plot of return loss, it is shown that the coupling between the patch and the FSS layer affects the characteristics of the antenna. The coupling effect occurs due to the reflection of the waves, which are radiated from the antenna structure, in certain frequency bands from the FSS layer. Then, the reflected waves from the FSS layer fall back on the patch antenna and affect the current distribution on the patch. The electromagnetic waves, which are radiated by the antenna, act as an excitation source to illuminate the FSS layer. The FSS layer forces the distribution of these EM waves in the space and controls the phase and hence the unit cells of the FSS get excited and the whole structure works as an aperture antenna [11].

The radiation pattern plots of UWB circular fractal antenna with different FSS structures are shown in Figs. 15, 16 and 17 at resonant frequencies of 3.5, 6 and 9 GHz.

Table 2 shows the gain comparison between different FSS structures used in UWB circular fractal antenna at a chosen frequency of 3.5 GHz. It is shown that slot-pair type FSS has resulted in gain enhancement of the antenna by a factor of 2 (approx.) as compared to other FSS structures. This type of FSS has shown compactness and more capacitive loading effect than other FSS.

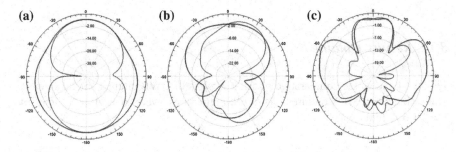

Fig. 15 Radiation patterns of UWB circular fractal antenna with Jerusalem cross-pair FSS at **a** 3.5 GHz, **b** 6 GHz, **c** 9 GHz

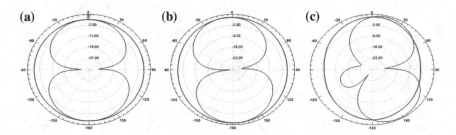

Fig. 16 Radiation patterns of UWB circular fractal antenna with Tripole-pair FSS at **a** 3.5 GHz, **b** 6 GHz, **c** 9 GHz

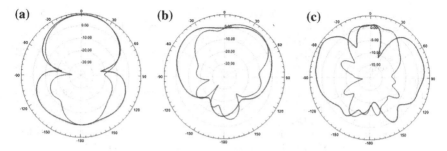

Fig. 17 Radiation patterns of UWB circular fractal antenna with Slot-pair type FSS at **a** 3.5 GHz, **b** 6 GHz, **c** 9 GHz

Table 2 Gain Comparison between different FSS structures

FSS Structures	Gain (in dB)
Jerusalem-cross pair type	4.77
Tripole-pair type	4.93
Slot-pair type	8.94

4 Conclusion

In this paper a novel fractal monopole antenna is analyzed in details. The antenna is compact and has a bandwidth of 9 GHz and exhibits stable radiation patterns thus serving as a potential candidate for UWB applications. Moreover with the insertion of a U-shaped slot, the same antenna is made to work for Bluetooth applications and at the same time retaining its ultra-wideband features. In order to mitigate interference between co-existing systems, an inverted U-shaped slot is inserted in the patch near the feed line which efficiently rejects WLAN frequencies. The later section of the paper is mainly concerned with increasing the gain of the antenna by placing different types of FSS structures at an optimal distance beneath the antenna. A comparative analysis is carried out in which it is found that the slot-type FSS gives the most acceptable gain enhancement of 8.94 dB from 1.6 dB. The designed antenna can be effectively used for short range communication, medical spectroscopy and various other bio-medical applications like detecting breast cancer.

References

1. FCC, First Report and Order 02–48. February 2002.
2. J. Liang, C. C. Chiau, X. Chen, and C. G. Parini, "Study of a printed circular disc monopole antenna for UWB systems", IEEE Trans. Antennas Propag., vol. 53, no. 11, pp. 3500–3504, Nov. 2005.
3. J. Liang, L. Guo, C.C. Chiau, X. Chen and C.G. Parini, "Study of CPW fed circular disc monopole antenna for ultra wideband applications", IEE Proc.-Microw. Antennas Propag., Vol. 152, No. 6, December 2005.

4. Mohamed Nabil Srifi, Otman El Mrabet, Fransisco Falcone, Mario Sorolla Ayza, and Mohamed Essaaidi, "A novel compact printed circular antenna for very ultra-wideband applications", Microwave And Optical Technology Letters, Vol. 51, No. 4, April 2009.

5. L. Wang, W. Wu, X.W. Shi, F. Wei, and Q.L. Huang, "Design of a novel monopole UWB antenna with a notched ground", Progress In Electromagnetics Research C, Vol. 5, 13–20, 2008.

6. J. Kula, D. Psychoudakis, W. Liao, C. Chen, J. Volakis, and J. Halloran,"Patch-antenna miniaturization using recently available ceramic substrates", IEEE Antennas Propag. Mag., vol. 48, no. 6, pp. 13–20, Dec. 2006.

7. B. Mirzapour and H.R. Hassani,"Size reduction and bandwidth enhancement of snowflake fractal antenna", IET Microw. Antennas Propag., Vol. 2, No. 2, March 2008.

8. Min Ding, Ronghong Jin, Junping Geng, And Qi Wu, "Design of a CPW-fed ultra-wideband fractal antenna", Microwave And Optical Technology Letters, Vol. 49, No. 1, January 2007.

9. Salman Naeem Khan, Jun Hu, Jiang Xiong, and Sailing He, "Circular fractal monopole antenna based on Descartes Circle Theorem for UWB application", Microwave And Optical Technology Letters, Vol. 50, No. 6, June 2008.

10. Nagendra Kushwaha, Raj Kumar, "Design of Slotted ground Hexagonal microstrip patch antenna and gain improvement with FSS screen", Progress in Electromagnetic Research B, vol. 51, pp 177–199, 2013.

11. Subash Vegesna, Yanhan Zhu, Ayrton Bernussi, Mohammad Saed, "Terahertz Two-layer Frequency selective surfaces with improved Transmission characteristics" IEEE Transactions on Terahertz science and Technology, vol. 2, no. 4, pp. 441–448, july 2012.

Application of Clustering for Improving Search Result of a Website

Shashi Mehrotra and Shruti Kohli

Abstract The paper identifies the scope of improvement in the search result of a website using the clustering technique. Search option is extensively used at almost every website. The study uses hybrid clustering approach for grouping the search results into the relevant folders for efficient analysis. Every clustering algorithm has some advantages and disadvantages. Some of the most commonly used clustering algorithm are experimented on same data set. The paper analyzed some research where clustering is being used for improving web elements in various way. Cross-validation method is adopted for the experiments, and performance parameters namely, relevance, speed and user satisfaction are considered for the evaluation.

Keywords Clustering algorithm · Cross-validation · Distance matrix proximity

1 Introduction

Due to the massive increase in the use of internet and huge electronic data access, some efficient technique is needed for efficient analysis of web data. Data analysis is required in various ways. It is required to group these data or objects for effective analysis, where clustering is very useful. People encounter a large amount of stored information and analyze it for various uses.

All the website uses the search option, and the search results are displayed usually in a list form, which contains title and snippet. Only a few results are relevant to the user for many reasons such as one word can be used for different objective. Results grouping in the meaningful folder will facilitate user to search

S. Mehrotra (✉) · S. Kohli
Birla Institute of Technology, Mesra, India
e-mail: sethshashi11@gmail.com

S. Kohli
e-mail: kohli.shruti@gmail.com

© Springer India 2016
S.C. Satapathy et al. (eds.), *Information Systems Design and Intelligent Applications*, Advances in Intelligent Systems and Computing 434,
DOI 10.1007/978-81-322-2752-6_34

349

relevant result in a quick manner, where clustering can be useful. Clustering is dividing of data into groups of objects, where each group, contains the data that are similar in nature and dissimilar to the objects of other groups [1]. The purpose of clustering is to organize a collection of data items into groups. The organization is in such a way so that items belonging to a cluster are more similar in nature to each other than they are to items in the other clusters [2]. In cluster analysis, objects in a group are divided into some homogeneous subgroups on the basis of a measure of similarity [3].

Clustering technique is one of the most challenging tasks of handling diverse information available on the web.

1.1 Motivation

Search option used excessively at almost every website. To improve search result is needed to retrieve relevant information and efficient access. Grouping the search result into a set of clusters will be useful, which can be achieved by clustering [4]. Clustering is an important procedure in a variety of fields, yet cluster analysis is a challenging problem, as many factors play an important role. Same algorithm with different parameters, using different presentation or using different similarity measure may generate different output. Some words have more than one meaning. Thus, hybrid clustering approach can be used to overcome some of the problem such as the number of clusters need to be defined initially and semantically relating the word.

1.2 Objective

The main objective of the study is to improve search result display of a website by proposing a novel model. The model uses hybrid clustering approach, adding new objective functions for the genetic algorithm and using the concept of the ontology. The model can handle large volume of data as well as high dimensional features.

1.3 Web Based Application of Clustering

Some web-based applications of clustering such as search result, to reduce information overload problem [5], to cluster web pages [6], social network analysis [7]. It is also used for website clustering [8].

The paper is organized as follows: Sect. 1 is an introduction to clustering and what are the web based applications of clustering. Section 2 talk about basic classification of clustering. Section 3 covers the challenges faced by clustering algorithms. Section 4 discuss some applications of clustering to solve some of the challenges. Section 5 describe the research methodology used. Section 6 discuss the experimental results, and Sect. 7 is the conclusion.

2 Clustering Algorithms

Broadly clustering techniques can be classified as hierarchical clustering and partition clustering. Hierarchical clustering approach group the data objects with a sequence of partitions while partition clustering directly divides data objects into some pre-specified number of clusters without the hierarchical structure [3]. Hierarchical clustering is commonly used in medical imagining, word wide web in social networking analysis, marketing research for grouping shopping items [9].

3 Challenges

There are some of the challenges, that affect the clustering, such as identification of distance measure, selection of the number of clusters, the result may be affected by the order of tuples arranged, the database may contain nominal, ordinal, binary type of data. These attributes need to be converted to categorical type [10], visualizing the large result visualizing is the critical challenge [11], and there is no clarity in discrimination of some words.

Our study will focus on the issue, visualizing large result, considering no clarity in discrimination of some words.

4 Applications of Clustering to Solve Some of the Above Challenges

The paper describe some research work where clustering is being focused on grouping web elements in a various way, and make it more efficient for analysis.

Search Result Clustering (SRC)

This approach helps users to access the search result efficiently. The search engine returns clustering of the short text into a list of folders, summarize the context of the searched keyword with in the result pages. It has some issue to implement it efficiently and effectively. The short version of each snippet must be used for grouping otherwise it will take too long time to download the result pages. The size of the cluster should be reasonable, neither too small nor too big and the number of clusters should be limited to facilitate a fast glance of the topic. Labeling should be meaning full so that via folder labels, it allows the user an efficient and effective browsing of the search results [12]. Traditional clustering algorithms cannot address the above specific requirements.

Topical Clustering of search Result [12]

Topical clustering of search results proposed to move to graph-of-topic paradigm from big-of-words. They wanted it to implement so that search engines return snippets with high accuracy. A note is added to every snippet represented by Wikipedia pages. Then the graph is build that consists of two types of node: one is annotated topics, and the other is the snippets that search engines return.

WebRat: Supporting Agile Knowledge Retrieval Through Dynamic, Incremental Clustering and Automatic Labelling of Web Search Result Sets [13]

WebRat is for visualizing and refining search result sets. Documents matching a query are clustered and visualized as a counter map of islands.

Website Clustering from Query Graph Using Social Network Analysis [8]

The system has extracted site relation from the query logs of the search engine, to detect some sensitive website, usually which are illegal such as pornographic or politically sensitive material. This function is performed by clustering website from query log using social network analysis.

Particle Swarm Optimization Based Hierarchical Agglomerative Clustering [14]

A new method of clustering, named Hierarchical Particle Swarm Optimization (HPSO) by using partition and hierarchical clustering approach Is used to tackle the problem of hierarchical agglomerative clustering through hierarchical particle swarm intelligence. The proposed algorithm used the swarm intelligence, keeping in view the efficiency of swarm-based clustering approach and problems in the existing clustering approach.

Clustering Technique on Search Engine Dataset Using Data Mining Tool [15]

Unlabeled document collections are becoming increasing common and mining such databases becomes a major challenge. The major issue is to retrieve good websites from the larger collections of websites.

Analytics tool *Keyword Similarity Measure Tool* (KSMT) [16]

Kohli et al. presented a tool, that aims to take care of the limitations of similar keywords in the report and improves the data accuracy; thus optimizing the report. It aims to provide a consolidated view and content analysis, by combining the matrices with the similar content analysis such as bounce rate and visits. KSMT algorithm was used to measure keyword similarity and combine keyword based on the factor of similarity.

Feature Selection From High Dimension Web data [17]

The paper presented a strategy to reduce the dimension of the attribute. They used FIFA World cup data set for experimental testing to show how the feature selection procedure find the most relevant variable.

Semantic Web to Improve Search Result Through Exploration and Discovery [18]

The author focuses on the semantic web using the concept of ontology, which focus the dictionary for the term used for search and find the relative relationship between the term searched and find in the dictionary.

5 Research Methodology

Our model will take search result as input. Pre-processing will be performed to clean the data and to remove stop words. Hybrid clustering approach will be applied using the concept of the ontology.

Performance evaluation will be performed to evaluate our approach over the data set. The parameters; relevancy, speed and user satisfaction will be considered for the performance evaluation (Fig. 1).

6 Experimental Analysis

In this section, comparative analysis of few of the most commonly used clustering algorithms, such as K-means, EM and canopy is performed. Experiments were conducted using file SearchOrganiCompetitors from the Similar web data set.

The study considered two parameters for comparison, i.e., accuracy and time taken for training and testing.

Data Sets

Data were collected from Similar Web Data repository. Following is the data sets used for experiments.

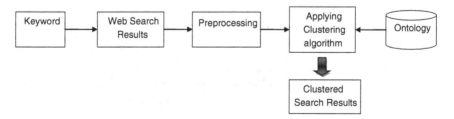

Fig. 1 Workflow of our hybrid approach

Data File: SearchOrganiCompetitors

Fiile description: No of Instances: 3338.

No of attributes: 5.

Attribute names: Domain, Category, Global Rank, Affinity, and AdSense.

The Fig. 2 represents the clusters generated by K-Means, EM, and Canopy algorithms respectively using SearchOrganiCompetitors dataset. The last figure which shows the cluster generated by using Canopy algorithm. It shows overlapping of data between clusters while the figures generated using K-Means and EM shows less overlapping of data.

Fig. 2 Generated clusters

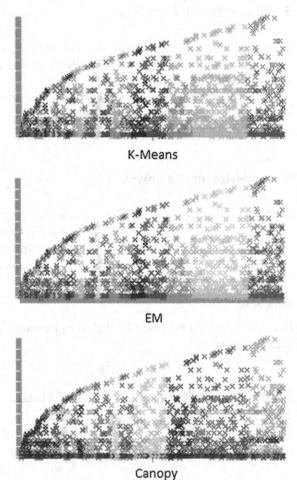

Clusters of Categories

K-Means

EM

Canopy

7 Conclusion

Cluster analysis is an interdisciplinary subject. Cluster analysis examines unlabeled data by either constructing a hierarchical structure or forming a set of groups, according to a pre-specified number. A massive increase in the usage of internet and web data, the grouping of web element is often needed for the efficient analysis. Clustering plays an important role for grouping. Clustering is used to group diversified web element. Thus, the role of clustering is challenging here. Many researchers have used clustering for web element optimization and to overcome some of the existing challenges, still improvement is needed.

Future work Our future work will focus on improving search result of a website, using the clustering approach. Looking at the experimental result, and efficiency of K-Means, proposed hybrid clustering algorithm will be an improvement of K-Means using the genetic algorithm. Semantic web using the concept of the ontology will be used to retrieve search result more relevant and meaning full. System will also analyze the user satisfaction and effect of user behavior.

References

1. Rai, Pradeep, and Shubha Singh. "A survey of clustering techniques." International Journal of Computer Applications 7, no. 12 (2010): 156–162.
2. Grira, Nizar, Michel Crucianu, and Nozha Boujemaa. "Unsupervised and semi-supervised clustering: a brief survey." A review of machine learning techniques for processing multimedia content, Report of the MUSCLE European Network of Excellence (FP6) (2004).
3. Xu, Rui, and Donald Wunsch. "Survey of clustering algorithms." Neural Networks, IEEE Transactions on 16, no. 3 (2005): 645–678.
4. Jiawei, Han, and Micheline Kamber. "Data mining: concepts and techniques." San Francisco, CA, itd: Morgan Kaufmann 5 (2001).
5. Yang, Nan, Yue Liu, and Gang Yang. "Clustering of Web Search Results Based on Combination of Links and In-Snippets." In Web Information Systems and Applications Conference (WISA), 2011 Eighth, pp. 108–113. IEEE, 2011.
6. Guo, Jiayun, Vlado KeÅ¡elj, and Qigang Gao. "Integrating web content clustering into web log association rule mining." In Advances in Artificial Intelligence, pp. 182–193. Springer Berlin Heidelberg, 2005.
7. Wu, Hui-Ju, I-Hsien Ting, and Kai-Yu Wang. "Combining social network analysis and web mining techniques to discover interest groups in the blogspace." In Innovative Computing, Information and Control (ICICIC), 2009 Fourth International Conference on, pp. 1180–1183. IEEE, 2009.
8. Wang, Weiduo, Bin Wu, and Zhonghui Zhang. "Website clustering from query graph using social network analysis." In Emergency Management and Management Sciences (ICEMMS), 2010 IEEE International Conference on, pp. 439–442. IEEE, 2010.
9. Namratha M, Prajwala T R, A Comprehensive Overview Of Clustering Algorithms in Pattern Recognition, IOR Journal of Computer Engineering, Volume 4 Issue 6 (Sep-Oct. 2012).
10. Qiao, Haiyan, and Brandon Edwards. "A data clustering tool with cluster validity indices." In Computing, Engineering and Information, 2009. ICC'09. International Conference on, pp. 303–309. IEEE, 2009.

11. Ilango, V., R. Subramanian, and V. Vasudevan. "Cluster Analysis Research Design model, problems, issues, challenges, trends and tools." International Journal on Computer Science and Engineering 3, no. 8 (2011): 2926–2934.
12. Scaiella, Ugo, Paolo Ferragina, Andrea Marino, and Massimiliano Ciaramita. "Topical clustering of search results." In Proceedings of the fifth ACM international conference on Web search and data mining, pp. 223–232. ACM, 2012.
13. Granitzer, Michael, Wolfgang Kienreich, Vedran Sabol, and G. Dosinger. "Webrat: Supporting agile knowledge retrieval through dynamic, incremental clustering and automatic labelling of web search result sets." In Enabling Technologies: Infrastructure for Collaborative Enterprises, 2003. WET ICE 2003. Proceedings. Twelfth IEEE International Workshops on, pp. 296–301. IEEE, 2003.
14. Alam, Shafiq, Gillian Dobbie, Patricia Riddle, and M. Asif Naeem. "Particle swarm optimization based hierarchical agglomerative clustering." In Web Intelligence and Intelligent Agent Technology (WI-IAT), 2010 IEEE/WIC/ACM International Conference on, vol. 2, pp. 64–68. IEEE, 2010.
15. Ahmed, MD Ezaz, and Preeti Bansal. "Clustering Technique on Search Engine Dataset using Data Mining Tool." In Advanced Computing and Communication Technologies (ACCT), 2013 Third International Conference on, pp. 86–89. IEEE, 2013.
16. Kohli, Shruti, Sandeep Kaur, and Gurrajan Singh. "A Website Content Analysis Approach Based on Keyword Similarity Analysis." In Proceedings of the 2012 IEEE/WIC/ACM International Joint Conferences on Web Intelligence and Intelligent Agent Technology-Volume 01, pp. 254–257. IEEE Computer Society, 2012.
17. Menndez, Hctor, Gema Bello-Orgaz, and David Camacho. "Features selection from high-dimensional web data using clustering analysis." In Proceedings of the 2nd International Conference on Web Intelligence, Mining and Semantics, p. 20. ACM, 2012.
18. Li, Laura Dan. "InfoPlanet: Visualizing a semantic web to improve search results through exploration and discovery." In Professional Communication Conference (IPCC), 2012 IEEE International, pp. 1–7. IEEE, 2012.

Performance Evaluation of Basic Selfish Node Detection Strategy on Delay Tolerant Networking Routing Protocols

Upendra B. Malekar and Shailesh P. Hulke

Abstract A Delay Tolerant Network (DTN) is a network of regional networks with characteristics of intermittent connectivity, opportunistic contacts, scheduled contacts, asymmetric data rates and high error rates. Main applications of DTN are Satellite communication, interplanetary communication and underwater communication networks. Selfish nodes may present in the network which degrades the performance of DTN. DTN nodes are resource constraints is a root cause of selfishness problem. DTN nodes have limited battery power, limited buffer space and limited computational resources. Hence, every DTN node tries to save their resources to achieve more lifetimes in the network. Selfishness of DTN nodes can collapses the well designed routing scheme in DTN and jeopardizes the whole network. Hence, selfish nodes in DTN should be detected and punished to increase the performance of DTN. Basic selfish node detection strategy is used to detect the selfish nodes in DTN, which when used with different DTN routing protocols behave differently in terms of number of selfish node detection. Thorough analysis is done on Basic selfish node detection strategy when used with different DTN routing protocols and found that Direct Delivery router detects least number of selfish nodes and Spray and Wait router detects highest number of selfish nodes in DTN.

Keywords DTN · Selfishness · Basic selfish node detection

U.B. Malekar (✉)
Department of Information Technology, MIT, Pune, India
e-mail: upendramalekar95@gmail.com

S.P. Hulke
Department of Computer Engineering, College of Engineering, Pune, India
e-mail: hulkesp13.comp@coep.ac.in

© Springer India 2016
S.C. Satapathy et al. (eds.), *Information Systems Design and Intelligent Applications*, Advances in Intelligent Systems and Computing 434,
DOI 10.1007/978-81-322-2752-6_35

357

1 Introduction

DTN is a network which is mainly operated where connectivity between the nodes is not continuous. DTN architecture and protocols are specially designed for long delays between the nodes. DTN assume that nodes in the network are self motivated and help each other in the bundle forwarding. The DakNet project of India is a typical example of DTN [1]. Jiang et al. [2] gives different types of Incentive mechanisms and focuses on the selfishness problem of DTN. Zhu et al. [3] proposed a probabilistic misbehavior detection strategy which is a modified version of Basic selfish node detection strategy. This approach detects the selfish nodes in DTN and incentive strategy stimulates the DTN nodes in bundle forwarding. Characteristics of DTN [4] which is the backbone of DTN given below:

Intermittent Connectivity Intermittent connectivity is the major issue in DTN which causes communication problems between the nodes. This characteristic distinguishes DTN from other types of networks.

Asymmetric Data Rates DTN nodes are heterogeneous in nature and hence they have different data rates according to their design parameters.

Long or Variable Delay As DTN is operated over the long distances like satellite communication there exists a long and variable delays between the nodes.

High Error Rates The communications between the nodes have long distances hence there is higher probability of adding the noise in the signals due to environmental conditions and many other factors.

2 Selfishness Problem of DTN

DTN assumes that the nodes help in forwarding the bundles for other nodes. In real scenario this assumption does not holds good and hence, the well-designed routing scheme does not perform well. There are two main reasons for selfishness problem of DTN nodes:

- Nodes are operated by rational entities
- Nodes are resource constraints with limited battery power, buffer space and computational power

As nodes are operated by rational entities like human and organizations, bundle forwarding decision of rational entities is the source selfishness problem. These entities may not forward the bundles to save their resources for longer life in the network. Co-operation probability (Pc) and decision probability plays a vital role in bundle forwarding of DTN. There is impact of velocity and bundle holding time on bundle dropping event in vehicular delay tolerant network [5]. There are four types of DTN nodes present in the network when coming to bundle forwarding as mentioned below:

Type I: DTN nodes that co-operate in bundle forwarding, Pc = 1
Type II: DTN nodes that do not co-operate in bundle forwarding, Pc = 0
Type III: DTN nodes that co-operate in bundle forwarding with co-operation probability $0 < Pc < 1$.
Type IV: DTN nodes that accept bundles which are originated for themselves but drop others.

A. *DTN nodes that co-operate in bundle forwarding (Type I)*

Nodes in DTN co-operate in bundle forwarding for other nodes and obey the DTN assumption. Type I nodes earn more credits and reputation values from the incentive schemes. These nodes are the best choice in bundle forwarding as they are healthier than other nodes in the network. The co-operation probability of Type I nodes is always 1 i.e. Pc = 1 and greater than any other nodes in the network.

B. *DTN nodes that do not co-operate in bundle forwarding (Type II)*

Nodes in DTN do not co-operate in bundle forwarding as nodes disobey the DTN assumption. Type II nodes earn less credits and reputation values from the incentive schemes. These nodes are the least choice in bundle forwarding as they are poor than other nodes in the network. The co-operation probability of Type II nodes is always 0 i.e. Pc = 0 and less than any other nodes in the network.

C. *DTN nodes that co-operate in bundle forwarding with co-operation probability $0 < Pc < 1$ (Type III)*

Nodes in DTN co-operate in bundle forwarding as nodes obey the DTN assumption. Type III nodes earn moderate credits and reputation values from the incentive schemes. These nodes are the fair choice in bundle forwarding as they are better than other Type II nodes in the network. The co-operation probability of Type III nodes is always $0 < Pc < 1$.

D. *DTN nodes that accept bundles which are originated for themselves but drop others (Type IV)*

Nodes in DTN do not co-operate in bundle forwarding for other nodes and disobey the DTN assumption. Type IV nodes earn less credits and reputation values from the incentive schemes. These nodes acts as a sink node in DTN as in-degree of bundle absorption is the number of bundles originated for a specific Type IV node in the network and the out-degree of bundle forwarding of such nodes is 0 as they do not take part in bundle forwarding.

3   Basic Selfish Node Detection Strategies

The Basic misbehavior detection algorithm [3] is used to detect the selfish nodes in DTN.

Let, SND = Selfish node detection
M = Message
MFR = Set of message forwarding requests
MF = Set of message forwarded
CN = Set of contacted Nodes
NHC = Set of next hop node chosen for message forwarding

Procedure SND

   for each M ∈ MFR **do**

   if M ∉ MF && CN! = 0 **then**

   return 1

   else if M ∈ MF && NHC ⊈ CN **then**

   return 1

   end if

   end for

   return 0

end Procedure

4 Performance Evaluations

In this paper performance analysis of selfish node detection is proposed. ONE simulator [6] is used to perform the simulation of DTN. In this section experimental setup and performance analysis of DTN is done and results are calculated which are given below (Fig. 1, Table 1).

Different scenarios of DTN routers with different simulation time are considered for calculating the results. Analysis for Epidemic router, MaxProp router, Prophet router, Direct delivery router and Spray and wait router are done by performing the ONE simulation from 1,000–10,000 ms and calculate the number of selfish DTN nodes present in the network. Selfish Ratio (SR) is a ratio of total number of selfish DTN nodes to the total number of DTN nodes is a parameter used for performance analysis (Tables 2 and 3, Fig. 2).

$$SR = \frac{Total\ number\ of\ selfish\ DTN\ nodes}{Total\ number\ of\ DTN\ nodes}$$

From the above results and graph Spray and Wait router has the highest average selfish ratio as compared to other routing protocols. Maximum numbers of selfish nodes are detected, when simulating with Spray and Wait router. MaxProp router and Prophet Router has average selfish ratio and Direct Delivery router has lowest selfish ratio.

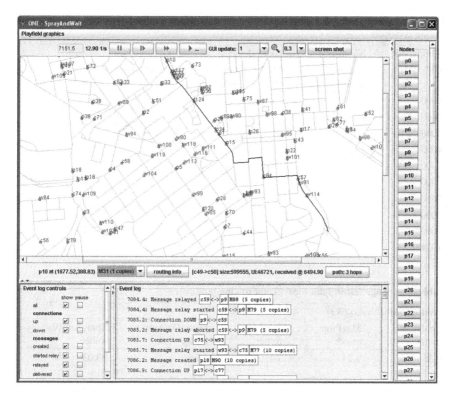

Fig. 1 Vehicular DTN considered for simulation

Table 1 ONE simulation setting

Parameters	Values
Simulation area	5,000 m * 5,000 m
Number of DTN nodes	60
Duration of simulation	1000–10,000 ms
Router	MaxProp/spray and wait/direct delivery/epidemic/prophet etc.
Velocity of DTN nodes	15 m/s
Transmission range	100 m
Buffer size	5 Mb
Holding time to wait next node	0–180 min
Bundle size	500 kB–1 Mb
Bundle TTL	300 min

Table 2 Number of selfish nodes detected by basic selfish node detection strategy for different routing protocols

Simulation time (ms)	Number of selfish nodes detected				
	Epidemic router	MaxProp router	Prophet router	DirectDelivery router	SprayAndWait router
1000	2	6	2	1	2
2000	4	8	4	0	6
3000	7	10	8	0	9
4000	2	4	9	0	9
5000	5	8	5	1	10
6000	6	3	8	4	7
7000	1	4	14	0	9
8000	4	9	3	0	11
9000	6	2	5	1	13
10,000	5	7	8	6	10

Table 3 Average selfish ratio for different routing protocols

Average selfish ratio (%)				
Epidemic router	MaxProp router	Prophet router	DirectDelivery router	SprayAndWait router
6.36	9.24	10	1.96	13.03

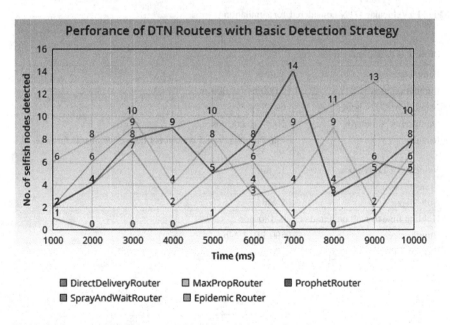

Fig. 2 Performance of DTN routers with basic selfish node detection strategy

5 Conclusion

There are four types of DTN nodes present in the network which behave differently based on co-operation probability. Performance evaluation is done for basic selfish node detection strategy on delay tolerant networking routing protocols, when all four types of DTN nodes are present in the network. In the above analysis it is found that different routing protocols behave differently for selfish node detection approach. Direct Delivery router with selfish node detection approach detects least number of selfish nodes while Spray and Wait router with selfish node detection approach detects highest number of selfish nodes in the network.

References

1. Pentland A., Fletcher R., Hasson A., "DakNet: rethinking connectivity in developing nations," Computer, vol. 37, no. 1, pp. 78, 83, Jan 2004.
2. Xin Jiang; Xiang-Yu Bai, "A survey on incentive mechanism of delay tolerant networks," Wavelet Active Media Technology and Information Processing (ICCWAMTIP), 2013 10th International Computer Conference on, pp. 191, 197, 17–19 Dec 2013.
3. Haojin Zhu, Suguo Du, Zhaoyu Gao, Mianxiong Dong and Zhenfu Cao, "A Probabilistic Misbehavior Detection Scheme towards Efficient Trust Establishment in Delay-tolerant Networks," IEEE Transactions On Parallel And Distributed Systems, 2014.
4. F. Warthman (March 2003). Delay Tolerant Networks (DTNs): A Tutorial http://www.dtnrg.org/docs/tutorials/warthmn1.1.pdf.
5. Upendra Malekar, Lalit Kulkarni, "Impact of Velocity and Bundle Holding Time on Bundle Dropping Event in Vehicular Delay Tolerant Network," 2015 International Conference on Industrial Instrumentation and Control (ICIC) College of Engineering Pune, India, May 28–30, 2015, ©2015 IEEE.
6. Ari Keränen, Jörg Ott and Teemu Kärkkäinen, 'The ONE Simulator for DTN Protocol Evaluation', SIMUTools 2009, Rome, Italy. Copyright 2009 ICST ISBN978-963-9799-45-5.

Dynamic Texture Recognition: A Review

Deepshika Tiwari and Vipin Tyagi

Abstract Dynamic texture recognition is a very important part of texture analysis. It mainly consists of recognition of moving texture that exhibits certain form of temporal stationarity. There are a good number of approaches developed by different research groups for dynamic texture recognition. This paper, analyze various dynamic texture recognition approaches and categorized into one of the four major groups: Discriminative based methods, Model based method, Flow based method and finally, Transform based methods that use wavelet based features to represent the dynamic texture. This survey critically evaluates various state-of-the-art dynamic texture recognition methods in order to provide a comprehensive modelling of dynamic texture.

Keywords Dynamic texture · Discriminative method · Model based method · Flow based method · Transform based method

1 Introduction

Dynamic texture (DT) or temporal texture is an image sequence that contains spatially repetitive, time-varying visual pattern with some unknown spatiotemporal extent. In dynamic texture, the concept of repetitive pattern of traditional image texture extended to spatiotemporal domain. Dynamic textures are natural processes that exhibit stochastic dynamics such as smoke, ripples at the surface of water, flowing river, windblown vegetation etc.

In machine vision, the concept of texture dated back to early fifties when Gibson [1] discussed visual phenomena such as texture gradient, color, motion, illumination, edges shapes, surfaces etc. In early sixties, Julesz [2] introduced the statistical characterization of textures. Following that, there has been extensive study done in the area

D. Tiwari · V. Tyagi (✉)
Jaypee University of Engineering and Technology, Raghogarh, Guna, MP, India
e-mail: dr.vipin.tyagi@gmail.com

© Springer India 2016   365
S.C. Satapathy et al. (eds.), *Information Systems Design and Intelligent Applications*, Advances in Intelligent Systems and Computing 434,
DOI 10.1007/978-81-322-2752-6_36

of 2D texture analysis, recognition and synthesis. Temporal texture analysis is comparatively a new area. The study of temporal texture started in early nineties when Nelson et al. [3] published the pioneer work on visual motion recognition. Authors observed that objects are easier to identify when they are moving than they are stationary and defined certain statistical spatial and temporal features. Based on the observation and analysis of moving objects, Nelson et al. [4] classified visual motion into three categories: temporal texture, motion events and activities. Temporal texture shows statistical regularity but have unknown spatial and temporal extent such as windblown tree or grass, ripple on the water, turbulent flow in cloud pattern etc.

This survey mainly focuses on the first visual motion category i.e. temporal texture. The main aim of spatiotemporal texture classification is to assign unknown temporal texture to one of the known texture class. Dynamic texture classification plays an important role in various computer vision applications like Remote monitoring and Surveillance applications such detection of fire or smoke, traffic monitoring, crowd analysis and management, Security applications, face recognition etc. Zhao et al. [5] addressed the following major issues in dynamic texture classification:

(a) Combine appearance feature with motion feature
(b) Perform local processing to catch the transition information in space and time
(c) Defining features which are insensitivity to illumination variations
(d) Defining features which are robust against image transformations such as rotation
(e) Multi-resolution analysis
(f) Computation simplicity

Chetverikov et al. [6] have also given a survey on dynamic texture recognition in which author categorized the dynamic texture recognition into five classes. Their survey poorly categorised and some classes contained only one or two studies whereas other classes contain discussion on large no of studies. Therefore, this survey paper attempts to reorganize the classes of dynamic texture classification methods.

2   Feature Modelling and Classification of Dynamic Texture

As in conventional texture classification, spatiotemporal texture classification process is also contain two phases: feature modeling phase and classification phase. In feature modeling phase, main aim is to build a feature model for the texture portion of each texture class present in training data. To do this a texture analysis method is used which combine motion pattern with the appearance and yield the feature vector to represent the specific DT. The feature vector can be a histogram or statistical distribution or some numeric value. In classification phase, feature vector for unknown sample dynamic texture created using the same texture analysis method. Then this feature vector compared with other feature vector of training textures using some classification method and sample assigned to the class with best match.

Various methods to describe the temporal texture are proposed which can be categorised into following classes:

(a) Discriminative based methods
(b) Model based methods
(c) Flow based methods
(d) Transform based methods

Discriminative based methods do not explicitly model the underlying dynamic system for classification purpose. These methods used the statistical properties of the spatial distribution of gray values and compute the local feature at each pixel in image. There are various studies [5, 7–9] based on the statistical features of grey level values. Among the discriminative based method, local binary pattern (LBP) [10] is most widely used texture operator that is inspired by gray level differences approach [11]. Zhao et al. [7] introduced volume local binary pattern (VLBP) method to extend LBP to spatiotemporal domain. In VLBP, each frame consists of P equally spaced pixels on a circle of radius R. Similar to LBP, volume center pixel is used to threshold the neighbouring pixel along the helix in the cylinder constructed with the neighbouring frame. However, VLBP texture features are insensitive to rotation and monotonic gray-scale changes but the size of feature descriptor (i.e. LBP histogram) increases rapidly as number of neighbouring pixel increases. Due to this exponential growth, it is hard to extend VLBP to a large neighborhood, and this limits its applicability. To reduce the dimensionality of VLBP feature descriptor Zhao et al. [5] proposed the local binary pattern in three orthogonal plane (LBP-TOP) and Ren et al. [9] introduced optimized LBP structure method. Ren et al. [8], also proposed enhanced LBP features to deal with the reliability issues and reduce the dimension of LBP histogram. Ghanem et al. [12] introduced DL-PEGASOS that uses Pyramid of Histogram of Oriented Gradients (PHOG) with LBP to represent the spatial texture information and demonstrate that spatial information contributes more in DT recognition. Some rotation invariant version of LBP [13, 14] in spatiotemporal domain also proposed for classification purpose. Zhao et al. [13] present a method to compute rotation-invariant features from histograms of local non-invariant patterns. Rotationally invariant features are computed by applying discrete Fourier transform (DFT) on histogram. Tiwari et al. [15] proposed a completed VLBP framework to include the contrast and center pixel structure information to improve the classification accuracies of DT. CVLBP also demonstrate robustness to noise. Apart from LBP, fractal methods also used statistical feature for DT recognition. Mandelbrot [16] first introduced the concept of fractal to describe images. Fractal is a mathematical set that exhibits a statistical self-similar repeating pattern that displays at every scale and characterized by so-called fractal dimension, which is correlated to the perceived roughness of image texture. Derpanis et al. [17] developed a particular space-time filtering formulation for measuring spatiotemporal oriented energy and is used for recognizing dynamic textures based on fractal analysis. Xu et al. [18] used fractal geometry method and proposed dynamic fractal system (DFS) which encodes sturdy discriminative information regarding multi-scale self-similarities existing in DT. To capture natural fractals Xu et al. [19] used multi fractal spectrum [20] to extract the power law behaviour of the local SIFT-like feature descriptor, 3D-OTF, estimated over multiple scale. The feature descriptor 3D-OFT shows strong robustness to both geometric and photometric variations due to the combination of SIFT and MFS approaches. Ji et al. [21] also make use of MFS to recognise DT. To achieve robustness against many environmental changes scale normalization and multiorientation image averaging are also used in WMFS feature descriptor. Recently Xu et al. [22] also published a an extension of study [18] on DT recognition using fractal analysis. In [22], principal curvature is included to measure the shape information of the local surface of image sequence.

Model based methods portray primary texture process by making parametric model to show observed intensity distribution. Impressive results have recently been achieved in DT recognition using the framework based on state-space model [23, 24] and beg of words (BoW) model [25–27]. State-space based approach to characterise DT is first introduced by Doretto et al. [23]. The autoregressive moving average (ARMA) based linear dynamical system (LDS) model used to describe the DT. LDS model used the system identification theory to estimate model parameters. Saisan et al. [28] applied the LDS model [23] to classify different temporal textures. Fujita et al. [29] modified the approach [28] and used impulse responses of state variables to represent texture and identify model. Various other studies [26, 30–33] also used state space model to describe DT due to its ability to capture appearance and motion precisely. However, LDS model fails to decompose visual processes consisting of multiple, co-occurring dynamic textures into different region of homogenous and distinct dynamics. To overcome this limitation Chan et al. [34, 35] layered dynamic texture model. The LDT is also a state-space model and it modelled a video as a collection of stochastic layers of different appearance and dynamics. Each layer is represented as a temporal texture sampled from a different linear dynamical system. Apart from LDS model, Szummer et al. [24] used the spatiotemporal autoregressive model (STAR) to capture dynamic texture. STAR is a three dimensional extension of autoregressive models, which code each pixel as a linear combination of the neighboring pixels lagged both in space and time. However, spatial causality constraint in the STAR model severely restricts its ability to capture complex dynamical textures with rotation and acceleration. Recently Qiao et al. [36] used Hidden Markov Model(HMM) to describe and classify DTs. HMM encodes the encodes the appearance information of the dynamic texture with the observed variables, and the dynamic properties over time with the hidden states. Experimental results show that HMM model results in superior performance than LDS model.

Along with LDS model some studies also used BoW model to classify temporal texture. A typical BoW model usually includes following three steps: feature description, codebook generation and mapping of image to codebook. Ravichandran et al. [26] used LDS as texture descriptor and extends Bag of features (BoF) (a variant of BoW) to classify DT. In [27], approach [26] is modified to handle view point and scale changes. In [25], used chaotic features to extract pixel intensity series information from a video. Chaotic features vector matrix is formed by concatenating chaotic features and used to model DTs. The feature vector matrixes are clustered and codebook is learned by assigning a codeword to each chaotic feature vector. Lastly each feature vector matrix is represented by a histogram based on chaotic feature vectors. Recently [37] also clubbed BoW approach with deep neural network to create histograms of high level feature which is used to classify DTs.

Transform based methods analyze the frequency content of the image. Relatively few studies [38–42] used transform based method to recognize DTs. Smith et al. [43] first used Spatiotemporal wavelets for indexing dynamic video content. Spatiotemporal wavelets used to decompose motion into local and global at different spatiotemporal scales. The wavelet energy used to captures the texture patterns simultaneously along the spatial and temporal dimensions. Feature vector constructed from wavelet energy is used for classification purpose. Later on Gonçalves et al. [38] used well known Gabor transform to recognize DTs. Feature vector is constructed by calculating the energy statics of spatiotemporal Gabor wavelet response. Qiao et al. [40] used dual tree complex wavelet transform (DT-CWT) to extract the dynamic texture feature at different scale and then use CGGD to model the complex wavelet coefficients. The model parameters serve as the texture feature for the dynamic texture classification. Qiao et al. [41] also used gumbel transform with DT-CWT to classify DTs. Results in [41] shows better classification performance than the existing energy-based feature. Recently Dubois et al. [39] attempts to use 2D+T curvelet transform for extracting descriptors to the purpose of dynamic texture recognition.

Last category is **optic flow based methods** which attempt to capture temporal texture information in form of sequence of apparent motion pattern viewed as static texture. Numerous studies [44–55] have been conducted on the use of optical flow for temporal texture classification. Polana et al. [3, 4] first attempted to use optical flow to derived certain statistical and temporal features having invariant properties and can be used to classify motion pattern such as ripples on water, windblown trees and chaotic fluid flow etc. The features [3, 4] are gray-scale and color invariant but fails to capture structural features properly and unable to accurately estimate the optical flow for unstable temporal texture. To deal with this issue Otsuka et al. [44] proposed to extract dynamic texture features based on the tangent plane representation of motion trajectory. From the distribution of tangent planes a dominant optical is estimated. Nelson and Polana's work influenced many studies [45–47, 49, 50] to use optic flow to define DT features. Bouthemy and Fablet used optical flow to characterize DT and published series of paper [46, 47, 49] to classify DTs. In [49] motion features extracted from temporal cooccurrence matrices and then feature vector is defined by extracting the global motion feature, such as homogeneity, acceleration or complexity, from temporal co-occurrence matrices. Later on the authors enhanced the approach [49] and introduced a more sophisticated approach [47] using the Gibbs model. This approach also does not look appropriate to large database of dynamic textures or to non-segmented videos [6]. Peh et al. [56] extends [3] to include more dynamic and spatial information and define temporal texture by mapping the magnitudes and directions of the normal flow. Peteri et al. [48, 50] attempts to use normal flow along with texture regularity to describe motion magnitude, directionality and periodicity. Their features are robust to rotation variation but does not support multiscale analysis of DTs. Lu et al. [51] introduced a novel way to characterize dynamic textures by spatio-temporal multiresolution histogram based on velocity and acceleration fields. Structure tensor method is used to estimate the velocity and acceleration fields of dissimilar spatiotemporal resolution DTs. The multiresoution helps to provide promising results but the method lacks in processing large volume of spatiotemporal data. Chetverikov et al. [53] gives a solution to this problem in their study. Their method used optical flow along with SVD based algorithm to approximate periodicity. Method is rotation and scaling invariant in the image plane but does not provide affine invariance.

3 Conclusion

Dynamic texture recognition is a new area with limited literature. Static texture recognition provides the basic ground for DT recognition methods and inspired various studies. In 1992, Polana and Nelson published first paper on dynamic texture analysis using the optical flow/normal flow method. Their work inspired the many studies and optical flow became the most popular tool to characterize and classify the DT. For next 15 years, various studies tried to combine normal flow, complete flow with multiresolution histograms, co-occurrence, adaptive-SVD and other known tool for classification purpose. Although, normal flow gives more information for point-wise motion but it is a tough task to reliably estimate normal flow field for chaotic motions. In addition to this, normal flow is good for only small number and low quality DTs. For large database with high quality images it is appropriate to use complete flow to describe the features of DTs as complete flow will take a more effort to compute and the additional effort of computing the complete flow for better evaluation of image dynamics does not pay off on small datasets [52, 54]. Apart from optical flow methods some studies also began to use model based approach to describe and classify DTs in late nineties. Despite the success of model based approaches their suitability to DT recognition is questionable for several reasons: they are computationally more challenging and time-consuming due to the fact that they applied directly to pixel intensities instead of more concise representations; it assumes stationary DTs well-segmented in space and time, and the accuracy drops drastically if they are not [6]; limited type of motion can be model due to various constraint.

In early years of 21st century, some studies extended discriminative methods from static domain to spatiotemporal domain. Discriminative methods outperform model based method approaches due to their robustness to environmental changes, view changes and computational efficiency. However, the existing discriminative methods do not work well with complex DT as they unable of reliably capturing inherent stochastic stationary properties of complex DTs. Though transform based approaches are very popular for conventional texture analysis only few studies used transform based approach for DT classification. This may be due to the unavailability of transform method in spatiotemporal domain.

The survey is concluded with remark that dynamic texture recognition is a novel, challenging and growing research area. Previous studies have already set up some trend in this area but still a lot of new approaches are to be introduced. Various aspects of dynamic texture recognition are still underworked such as noise tolerance, robustness to blurring and a lot of work is still to be done in this area.

References

1. Gibson, J.J.: The perception of the visual world, Boston, MA, Houghton Mifflin, (1950).
2. Julesz, B.: Visual pattern discrimination. IRE Trans. Info Theory, IT-8 (1962).
3. Nelson, R.C. Polana R.: Qualitative recognition of motion using temporal texture. CVGIP: Image Understanding vol. 56, pp. 78–89 (1992).
4. Nelson, R.C. Polana R.: Temporal texture and activity recognition. Motion-Based Recognition, pp. 87–115 (1997).
5. Zhao G., Pietikinen M.: Dynamic texture recognition using local binary patterns with an application to facial expressions. IEEE Trans. PAMI, vol. 29, no. 6, pp. 915–928 (2007).
6. Chetverikov D., P´eteri R.: A Brief Survey of Dynamic Texture Description and Recognition. Proc. Int'l Conf. Computer Recognition Systems, pp. 17–26 (2005).
7. Zhao G., Pietikinen M.: Dynamic Texture Recognition Using Volume Local Binary Patterns. Proc. Workshop on Dynamical Vision WDV 2005/2006, LNCS, pp. 165–177 (2007).
8. Ren, J., Jiang, X., Yuan, J.: Dynamic texture recognition using enhanced LBP features. In IEEE ICASSP pp. 2400–2404 (2013).
9. Ren J., Jiang X., Yuan J., Wang G.: Optimizing LBP Structure For Visual Recognition Using Binary Quadratic Programming. IEEE trans. SPL, vol. 21, no. 11, pp. 1346–1350 (2014).
10. Ojala T., Pietikainen M., Maenpaa T.: Multiresolution Gray Scale and Rotation Invariant Texture Analysis with Local Binary Patterns. IEEE Trans. PAMI, vol. 24, no. 7, pp. 971–987 (2002).
11. Weszka J., Dyer C., Rosenfeld A.: A comparative study of texture measures for terrain classification, IEEE Trans. Syst. Man. Cybernet. vol. 6, pp. 269–285 (1976).
12. Ghanem B., N. Ahuja N., Daniilidis K., Maragos P., Paragios N.: Maximum margin distance learning for dynamic texture recognition. In *Computer Vision–ECCV 2010*, ser. LNCS Berlin, Germany: Springer, vol. 6312, pp. 223–236 (2010).
13. Zhao, G., Ahonen, T., Matas, J., Pietikäinen, M.: Rotation-invariant image and video description with local binary pattern features. IEEE Trans. on Image Processing., vol 21, no 4, pp. 1465–1467 (2012).
14. Zhao G., Pietikinen M.: Improving rotation invariance of the volume local binary pattern operator. In *Proc. IAPR Conf. Mach. Vis. Appl.*, pp. 327–330 (2007).
15. Tiwari D., Tyagi V.: Dynamic texture recognition based on completed volume local binary pattern. Multidimensional Sys. and Signal Processing, doi:10.1007/s11045-015-0319-6.
16. Mandelbrot B.: Fractal Geometry of Nature, W.H. Freeman and Co., New York (1983).
17. Derpanis, K. G., Wildes, R. P.: Dynamic texture recognition based on distributions of spacetime oriented structure" In CVPR, pp. 191–198 (2010).
18. Xu, Y., Quan, Y., Ling, H., & Ji, H.: Dynamic texture classification using dynamic fractal analysis. In *IEEE ICCV*-2011 pp. 1219–1226 (2011).
19. Xu, Y, Huang S., H. Ji, Fermüller C.: Scale-space texture description on SIFT-like textons. Comp. Vision and Image Understanding vol. 116, no. 9, pp. 999–1013, (2012).

20. Xu Y., Ji H., Fermuller C. : Viewpoint invariant texture description using fractal analysis, IJCV, vol. 83, no. 1, pp. 85–100 (2009).
21. Ji H., Yang X., Ling H., Xu Y.: Wavelet domain multifractal analysis for static and dynamic texture classication. IEEE Trans on Img Process., vol 22, no. 1, pp. 286–299 (2013).
22. Xu, Y., Quan, Y., Zhang, Z., Ling, H., & Ji, H.: Classifying dynamic textures via spatiotemporal fractal analysis, Pattern Recognition, doi:10.1016/j.patcog.2015.04.015.
23. Doretto G., Chiuso A., Soatto S., Wu Y.N.: Dynamic Textures. Int. Journal of Computer Vision, vol. 51, no. 2, pp. 91–109, (2003).
24. Szummer, M., Picard, R.W.: Temporal Texture Modeling. Proc. of Int. Conference on Image Processing, Vol. 3, pp. 823–826 (1996).
25. Wang Y., Hu S.: Chaotic features for dynamic textures recognition Soft Computing, doi:10. 1007/s00500-015-1618-4, Feb, 2015.
26. Ravichandran A., Chaudhry R., Vidal R.: View-invariant dynamic texture recognition using a bag of dynamical systems," in Proc. CVPR, pp. 1–6,(2009).
27. Ravichandran A., Chaudhry R., Vidal R.: Categorizing dynamic textures using a bag of dynamical systems. IEEE Trans. PAMI, vol 35, no. 2, pp. 342–353 (2013).
28. Saisan P., Doretto G., Wu Y. N., Soatto S. Dynamic texture recognition. In Proc. CVPR, vol. 2, pp. 58–63, Kauai, Hawaii, (2001).
29. Fujita K., Nayar S.K.: Recognition of dynamic textures using impulse responses of state variables. In Proc. Int. Workshop on Texture Analysis and Synthesis, pp. 31–36 (2003).
30. Woolfe F., Fitzgibbon A.: Shift-invariant dynamic texture recognition," in Proc. ECCV, pp. 549–562 (2006).
31. Vidal R., Favaro P.: Dynamicboost: Boosting time series generated by dynamical systems. In Proc. ICCV, (2007).
32. Chan, A., Vasconcelos, N.: Mixture of dynamic textures. In Proc. of ICCV, vol. 1, pp. 641–647, (2005).
33. Chan A., Vasconcelos, N: Classifying video with kernel dynamic textures. in Proc. CVPR, pp. 1–6 (2007).
34. Chan, A., Vasconcelos, N.: Layered dynamic textures. IEEE Trans. on Pattern Analysis and Mach. Intelligence. pp. 1862–1879 (2009).
35. Chan, A., Vasconcelos, N.: Variational layered dynamic textures. In Proc. CVPR, pp. 1063–1069 (2009).
36. Qiao Y., Weng L., "Hidden Markov model based dynamic texture classification" IEEE signal processing lett., vol. 22, no. 4, pp. 509–512(2015).
37. Wang Y., Hu S.: Exploiting high level feature for dynamic textures recognition. Neurocomputing, vol. 154, pp. 217–224 (2015).
38. Gonçalves W.N., Machado B. B.,Bruno O. M.: Spatiotemporal Gabor filters: a new method for dynamic texture recognition CoRR, vol no: abs/1201.3612, 2012.
39. Dubois S., Pteri R., Mnard M., "Characterization and recognition of dynamic textures based on the 2d + t curvelet transform," Signal, Image Video Process., pp. 1–12 (2013).
40. Qiao Y., Wang F., Song C.: Wavelet-based Dynamic Texture Classification Using Complex Generalized Gaussian Distribution. In Proc. AISS vol. 4, no. 4, (2012).
41. Qiao Y., Wang F., Song C.: Wavelet-based dynamic texture classification using gumbel distribution, Mathematical Problems in Engineering, vol. 2013, 7 pages, (2013).
42. Yu-Iong Q., Fu-shan W.: Dynamic Texture Classification Based On Dual-Tree Complex Wavelet Transform". In Proc. of ICIMCCC, pp. 823–826 (2011).
43. Smith J.R., Lin C.Y., Naphade M.:Video texture indexing using spatiotemporalwavelets. In Proc. ICIP, vol. 2, pp 437–440 (2002).
44. Otsuka K., Horikoshi T., Suzuki S., Fujii M. : Feature extraction of temporal texture based on spatiotemporal motion trajectory. In Proc. ICPR. Vol. 2,pp1047–1051(1998).
45. Peh C.H., Cheong L.-F.: Exploring video content in extended spatiotemporal textures. In Proc. 1st European workshop on Content-Based Multimedia Indexing, pp. 147–153 (1999).
46. Fablet R., Bouthemy P.: Motion recognition using spatio-temporal random walks in sequence of 2D motion-related measurements. In Proc. ICIP, pp 652–655 (2001).

47. Fablet R., Bouthemy P.: Motion recognition using nonparametric image motion models estimated from temporal and multiscale co-occurrence statistics. IEEE Trans. PAMI, vol. 25, pp. 1619–1624, (2003).
48. Peteri R., Chetverikov D.: Qualitative characterization of dynamic textures for video retrieval. In Proc. Int. Conf. on Comp. Vision and Graphics (ICCVG), (2004).
49. Bouthemy P., Fablet R.: Motion characterization from temporal cooccurrences of local motion-based measures for video indexing. In Proc. ICPR, vol. 1, pp 905–908, (1998).
50. Peteri R., Chetverikov D.: Dynamic texture recognition using normal flow and texture regularity. In Proc. IbPRIA, Portugal (2005).
51. Lu Z., Xie W., Pei J., Huang J.: Dynamic texture recognition by spatiotemporal multiresolution histogram. In Proc. IEEE WACV/MOTION'05 (2005).
52. Fazekas S., Chetverikov D.: Normal versus complete flow in dynamic texture recognition: A comparative study. In Int. Workshop on Texture Analysis and Synth., pp 37–42(2005).
53. Chetverikov D., Fazekas S.: On motion periodicity of dynamic textures, in BMVC, vol. 1, pp. 167–177 (2006),
54. Fazekas S., Chetverikov D.: Analysis and performance evaluation of optical flow features for dynamic texture recognition. Sig. Process.:Img Comm., vol. 22, no.7–8, pp 680–691(2007).
55. Fazekas S., Chetverikov D.: Dynamic texture recognition using optical flow features and temporal periodicity," Int. Workshop of Content-Based Multi. Index., pp 25–32 (2007).
56. Peh C.H., Cheong L.-F.: Synergizing spatial and temporal texture. IEEE Trans. on Image Processing, vol 11, pp. 1179–1191 (2002).

Performance Analysis of Fully Depleted Ultra Thin-Body (FD UTB) SOI MOSFET Based CMOS Inverter Circuit for Low Power Digital Applications

Vimal Kumar Mishra and R.K. Chauhan

Abstract This paper demonstrates the integration of fully depleted ultra thin-body Silicon on Insulator MOSFET (FD UTB SOI n and p-MOSFET) into CMOS inverter circuit. The proposed MOS device shows the better Ion to Ioff ratio, lower subthreshold slope and low threshold voltage at 50 nm gate length. The proposed CMOS circuit shows the good inverter VTC curve, and minimum delay has been obtained at 50 nm gate length. The proposed structures were designed and simulated using Sentaurus device simulator.

Keywords CMOS inverter · FD UTB SOI MOSFET · VTC curve · Transient delay

1 Introduction

Silicon on Insulator (SOI) technology has been regarded as another mainstream technology for realizing sub-100 nm CMOS very large scale integrated (VLSI) circuit because of its intrinsic properties such as low parasitic resistance, low sub threshold current, threshold voltage roll off etc. The SOI MOSFETs are classified into two parts such as PD-SOI and FD-SOI MOSFET. The PD-SOI MOSFET has several advantages in low power digital application but, it has certain disadvantages associated, like its physical limits for scalability in MOSFETs. It has been shown by researchers that fully depleted SOI (FD-SOI) is helpful for ultra-low power digital circuits. Leakage and power consumptions are drastically reduced and random fluctuation in threshold voltage is also better controlled in the FD-SOI technology.

V.K. Mishra (✉) · R.K. Chauhan
Department of Electronics and Communication Engineering,
Madan Mohan Malaviya University of Technology, Gorakhpur, India
e-mail: vimal.mishra34@gmail.com

R.K. Chauhan
e-mail: rkchauhan27@gmail.com

© Springer India 2016
S.C. Satapathy et al. (eds.), *Information Systems Design and Intelligent Applications*, Advances in Intelligent Systems and Computing 434,
DOI 10.1007/978-81-322-2752-6_37

Because of all such advantages researchers are presently using ultra-thin film FD-SOI MOSFETs as potential ways to drastically cut power consumption and leakages while preserving high performance and minimizing short channel effects. Thin Film FD-SOI has been an attractive technology for deep sub micron CMOS low power and high speed application. Complementary thin-film fully depleted Silicon on Insulator MOSFET (FD-SOI MOSFET) circuitry consisting of inverters is the building block for realizing integrated logic circuits on large-area. During the past few years, SOI MOSFETs with attractive properties such as high Ion current, low off state current, low subthreshold slope, low power dissipation and low-temperature process compatibility have become potential candidates for the implementation of large-area electronics [1–4]. Although, the PD-and FD-SOI MOSFET have been reported previously by many researchers, most of them were based on n-channel MOSFET (NMOS) technology [1–4] and some are p-channel SOI MOSFFTs [5, 6]. Only a few PD and FD-SOI based CMOS inverters have been demonstrated [6–8].

In this paper, first, we developed n and p-type FD UTB SOI MOSFET with relatively-steep subthreshold slopes and sufficiently large Ion/Ioff current ratios at 50 nm gate length. Next, the n and p-channel FD UTB SOI MOSFET are integrated with matched threshold voltages to demonstrate fully depleted SOI based CMOS inverters. The proposed CMOS inverters the dc characteristics and switching behaviour are studied. Finally, we take a step forward of connecting two proposed inverter circuit back to back to see the performance of proposed inverter circuit.

2 Device Structure and Specifications

The device design and meshing of n- and p-type FD UTB SOI MOSFET at 50 nm gate length are shown in Figs. 1 and 2. This is a 2-Dimensional model formed on the Sentaurus 2d device simulator. Instead of bulk MOSFET, a SOI layer is taken below the fully depleted MOSFET. In proposed FD UTB SOI MOSFET the Si layer thickness is 15 nm, which is about one-fourth of gate length. The insulator thickness of SiO_2 is 10 nm, which are shown in Table 1. The work function of gate

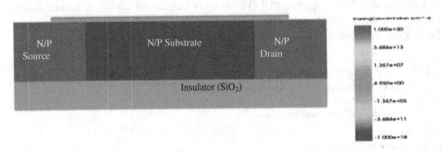

Fig. 1 Device design of n- and p-FD UTB SOI MOSFET

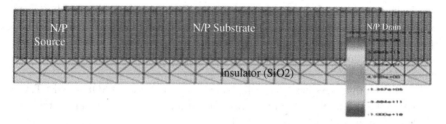

Fig. 2 Mesh analysis of n- and p-type FD SOI MOSFET

Parameters	n-MOSFET	p-MOSFET
Gate length	50 nm	50 nm
Tox	2 nm	2 nm
Doping density in substrate	1e16	1e16
Doping density in source/drain	1e20	1e20
Work function	4.2	5.2
Silicon thickness	15 nm	15 nm
Insulator thickness	10 nm	10 nm

Table 1 Device parameter of FD UTB SOI MOSFET based inverter circuit

n- and p-type metal is 4.2 and 5.2 respectively. The gate oxide thickness is 2 nm and the material used Hfo2 is high–k dielectric to avoid the leakage current from substrate to gate. The source and drain was formed by phosphorous ion implantation and the substrate was implanted by boron ion implantation.

SOI transistors can be distinguished into two parts:

(1) Partially Depleted (PD) SOI MOSFET—if the silicon film thickness is approx 100 nm or more above the buried oxide layer is thicker than the depletion region thickness which is formed below the gate oxide.

(2) Fully Depleted (FD) SOI—If the silicon film thickness is one fourth of the gate length or less and the doping concentration of the substrate is low enough to be fully depleted. FD SOI MOSFET have several advantages over PD SOI MOSFET in terms of extremely low sub-threshold slope less, no channel length modulation effects, Kink effects variation at low threshold voltage. These effects reduce the power consumption in SOI MOS device.

3 Device Concepts and Simulation

Structure of SOI MOSFET is very different from bulk conventional MOSFET, but its operation shows better electrical performance. SOI MOSFET reduces the leakage current by using insulator region at the bottom of the silicon layer. Due to low leakage current, I_{off} current reduces. In SOI MOSFET threshold voltage reduces, which results in high I_{on} current compared to bulk conventional MOSFET. The channel length modulation is found to be negligible in SOI MOSFET compared to bulk conventional MOSFET, because of which its working in saturation region enhances.

The Silicon thickness has been reduced in the proposed UTB SOI MOSFET. On simulating the proposed structure, it was noticed that the silicon thickness reduced till the depletion width of the channel. The depletion width can be computed from Eq. 1 [9].

$$Xdm = \sqrt{\frac{2\varepsilon si.|2\Phi F|}{qNA}} \tag{1}$$

where,

$$\Phi Fp = \frac{kT}{q} \ln \frac{ni}{NA} \quad \text{and} \quad \Phi Fn = \frac{kT}{q} \ln \frac{ND}{ni}$$

The device simulations were performed using Sentaurus TCAD device simulator [10] coupled with some analytical calculation. A Sentaurus TCAD device simulator which is based on continuity equation, drift diffusion model, and physical device model in TCAD includes effects such as mobility, Shockley-Hall-Read and Augur recombination. The channel current obtained from this FD UTB SOI n-MOSFET structure is near to the reported results [1, 11–14], when simulated using Sentaurus TCAD device simulator.

4   Results and Discussions

The input characteristics of FD UTB SOI n-MOSFET are shown in Fig. 3. The I_{off} current obtained from the simulation was 8 pA and I_{on} current as 1 mA. The ratio of I_{on}/I_{off} ratio is therefore 10^9. The proposed n-MOSFET shows a minimum subthreshold slope of 57 mV/decade at room temperature with negligible DIBL. In addition, it exhibits a higher drive current as compared to bulk conventional MOSFET at 50 nm gate length, which help to minimum delay in the inverter circuits. The SOI n-MOSFET has a low threshold voltage of 0.17 V.

Proposed MOSFET based inverter circuit is simulated in this work. Figure 4 shows the voltage transfer characteristics of UTB SOI MOSFET based CMOS inverter circuit. In this circuit the input voltage is varying from 0.1 to 0.3 V the corresponding output voltage is high nearly equal to 1.5 V, shows that n-mos transistor is in cut-off and p-mos is in linear region. As the input voltage increases from 0.3 to 1 V the output voltage decreases from 1.5 to 0.3 V that show the both n and p-mos are in saturation region. On further increasing input voltage the output

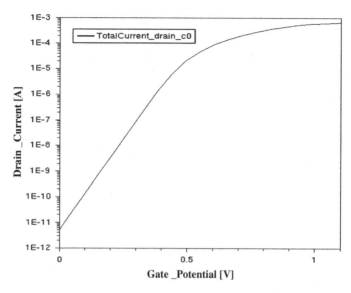

Fig. 3 Input characteristic of proposed SOI n-MOSFET

voltages decreases from 0.3 to 0 V in which the n-mos is in linear region and p-mos in cut-off region. The threshold voltage found in inverter circuit is 0.75 V which is nearly equal to the ideal CMOS inverter circuit.

The input and output voltage waveforms of a proposed inverter circuit are shown in Fig. 5. The voltage in the graph shows the input supply voltage and the voltage 1 shows the output voltage. The propagation delay terms, *Tphl* and *Tplh* and the average propagation delay *Tp* of inverter can be written as:

$$Tphl = \frac{Cload}{Kn(Vdd - Vtn)}\left[\frac{2Vtn}{(Vdd - Vtn)} + \ln\left(\frac{4(Vdd - Vtn)}{Vdd} - 1\right)\right] \quad (2)$$

$$Tplh = \frac{Cload}{Kp|Vtp|}\left[\frac{2(Vdd - |Vtp| - VoL)}{|Vtp|} + \ln\left(\frac{2|Vtp| - (Vdd - V50\%)}{Vdd - V50\%}\right)\right] \quad (3)$$

$$Tp = \frac{(Tphl + Tplh)}{2} \quad (4)$$

where,

"*Tphl*" is the time delay between the fifty percent change of the rising input voltage and the fifty percent change of falling output voltage.

Tplh is defined as the time delay between the fifty percent change of the falling input voltage and the fifty percent change of rising output voltage.

Tp characterizes the average time required for the input signal to propagate through inverter.

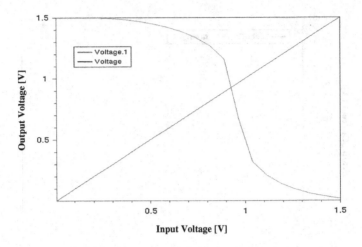

Fig. 4 VTC curve of FD UTB SOI MOSFET based CMOS inverter circuits

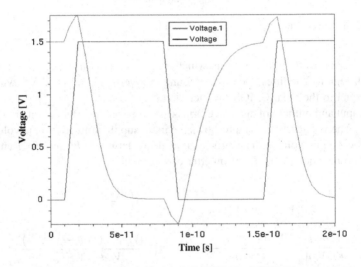

Fig. 5 Output waveforms of the FD UTB SOI MOSFET based CMOS inverter circuit

When input signal switches from low to high voltage, n-MOS device turns on and p-MOS device turns off. Similarly, when input signals switches from high to low voltage, p-MOS device turns on and n-MOS device is turns off.

The transient analysis of proposed inverter circuit is shown in Fig. 5. From figure one finds that the Tphl is 7 ns and Tplh is 10 ns, which is too small to affect the performance of proposed inverter. Analytical calculation shows that the Tphl is 6 ns and Tplh is 12 ns which is quite near to the simulated results.

To see the better performance of proposed inverter circuits, we take a step forward of connecting two proposed inverter circuit back to back. Output

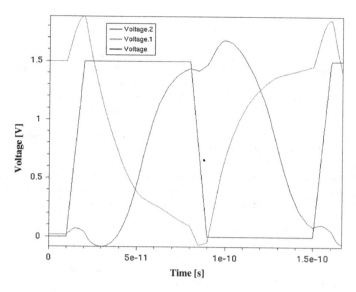

Fig. 6 Output waveforms of the two proposed inverter circuits connected back to back

waveforms of the two proposed inverter circuits connected back to back are shown in Fig. 6. The voltage in the graph shows the input supply voltage and the voltage 1 shows the output voltage of Ist inverter and voltage 2 shows the output voltage of IInd inverter. From Fig. 6 it can be infer that the minimum delay in proposed inverter is not much affect the transient characteristics when two inverters circuits are connected back to back, moreover figure shows that appreciable noise in the proposed inverter circuit.

5 Conclusions

In this work an n and p type FD UTB SOI MOSFET based inverter structures was designed and simulated using Sentaurus device simulator. The proposed structure of SOI n-MOSFET exhibits lower sub-threshold slope, low I_{off} current and high I_{on} to I_{off} ratio. The SOI MOSFET based inverter circuit at 50 nm gate length was simulated and analysed. It was found that the circuit shows low transient delay, desired threshold voltage (Vm). The other digital circuit application of proposed inverter was also analysed by considering two inverter circuit connected back to back, that give better performance with appreciable noise.

Acknowledgments This work is supported by the AICTE under research promotion scheme (RPS-60). Authors would also like to thank incubation cell of IIT Kanpur to provide Sentaurus tool in VLSI/EDA lab.

References

1. J. Chen, J. Luo, Q. Wu, Z. Chai, T. Yu, Y. Dong, and X. Wang, A Tunnel Diode Body Contact Structure to Suppress the Floating-Body Effect in Partially Depleted SOI MOSFETs, *IET, Electron Device Lett.*, vol. 32 no. 10, pp. 1346–1348, Oct. 2011.
2. Y. Wang, X.-W. He, and C. Shan, A Simulation Study of SoI-Like Bulk Silicon MOSFET With Improved Performance, *IEEE Trans. Electron Devices*, vol. 61 no. 9, pp. 3339–3344, Sept. 2014.
3. A. Ohata, Y. Bae, C. Fenouillet-Beranger, and S. Cristoloveanu, Mobility Enhancement by Back-Gate Biasing in Ultrathin SOI MOSFETs With Thin BOX, *IEEE Electron Device Lett.*, vol. 33, no. 3, pp. 348–350, March. 2012.
4. M. Miura-mattausch, U. Feldmann, Y. Fukunaga, M. Miyake, H. Kikuchihara, F. Ueno, H. J. Mattausch, S. Member, and I. Paper, Compact Modeling of SOI MOSFETs With Ultrathin Silicon and BOX Layers, *IEEE Trans. Electron Devices*, vol. 61, no. 2, pp. 255–265, Feb. 2014.
5. J. Luo, J. Chen, Q. Wu, Z. Chai, T. Yu, and X. Wang, TDBC SOI technology to suppress floating body effect in PD SOI p-MOSFETs, *IEEE Electronics Lett.*, vol. 48, no. 11, pp. 652–653, May. 2012.
6. V. P. Trivedi, S. Member, and J. G. Fossum, Scaling Fully Depleted SOI CMOS, *IEEE Trans. Electron Devices* vol. 50 no. 10, pp. 2095–2103, Oct. 2003.
7. S. Zhang, R. Han, X. Lin, X. Wu, and M. Chan, A Stacked CMOS Technology on SOI Substrate, *IEEE Electron Device Lett.*, vol. 25 no. 9 pp. 661–663, Sept. 2004.
8. J. B. Kuo, W. C. Lee, and J. Sim, Back-Gate Bias Effect on the Subthreshold Behavior and the Switching Performance in an Ultrathin SO1 CMOS Inverter Operating at 77 and 300 K *IEEE Transactions on Electron Devices*, vol. 39, no. 12, pp. 2781–2790, Dec 1992.
9. S. M. Sze. *Physics of Semiconductor Devices, 2nd edition.*. New York: John Wiley & Sons, 1981.
10. *Sentaurus Device User Guide*,Mountain View, CA: Synopsys, Inc., 2010.
11. J. Singh and C. Sahu, Device and circuit performance analysis of double gate junctionless transistors at Lg = 18 nm, pp. 1–6, Feb. 2014.
12. Y. Khatami and K. Banerjee, Steep Subthreshold Slope n- and p-Type Tunnel-FET Devices for Low-Power and Energy-Efficient Digital Circuits, *IEEE Trans. Electron Devices*, vol. 56, no. 11, pp. 2752–2761, Nov. 2009.
13. V. Nagavarapu, S. Member, R. Jhaveri, and J. C. S. Woo, The Tunnel Source (PNPN) n-MOSFET : A Novel High Performance Transistor, *IEEE Transactions on Electron Devices*, vol.55, no. 4, pp. 1013–1019, April 2008.
14. H. Zhongfang, Ru. Guoping, and R. Gang, Analysis of The Subthreshold Characteristics of Vertical Tunneling Field Effect Transistors, *Journal of Semiconductors*, Vol. 34, no. 1, pp. 1–7, Jan. 2013.

Mathematical Vector Quantity Modulated Three Phase Four Leg Inverter

Bhaskar Bhattacharya and Ajoy Kumar Chakraborty

Abstract In this paper it has been shown that any unbalanced and distorted three-phase currents can be efficiently generated as output from a four-leg (4L) inverter with mathematical implementation of three dimensional (3D) space vector pulse width modulation (SVPWM) in $\alpha-\beta-\gamma$ frame. The 3D SVPWM technique in $\alpha-\beta-\gamma$ frame is a computation intensive method. It depends on correct mapping of the reference signal vector within a predefined tetrahedron space that is defined with three concurrent non-zero (NZ) switching state vectors (SSV). A SSV represents a valid switching state (SS) of the inverter under modulation. The on/off timings of individual switches are worked out analytically in vector domain and gate pulses for corresponding timings are sent out in time domain. A fast mathematical execution is necessary for proper functioning of 3D SVPWM. A MATLAB simulation of a 4L voltage source inverter working on the proposed method has been presented to validate the proposition.

Keywords A–β–γ frame · Four-leg voltage source inverter · 3D SVPWM · Switching state vectors · Tetrahedron

1 Introduction

Pulse width modulation is a method which generates pulses of uniform amplitude but of varying widths. Width of each pulse is proportional to the instantaneous magnitude of the signal. For that a control signal and a carrier signal are required. These signals are physical quantities and can be measured and displayed. In SVPWM, on the contrary, only one control signal is required instead of two and that is not a physical quantity but is a mathematical entity. It cannot be measured or

B. Bhattacharya (✉) · A.K. Chakraborty
Department of Electrical Engineering, National Institute of Technology, Agartala 799046, India
e-mail: bhaskarohmm@gmail.com

© Springer India 2016
S.C. Satapathy et al. (eds.), *Information Systems Design and Intelligent Applications*, Advances in Intelligent Systems and Computing 434,
DOI 10.1007/978-81-322-2752-6_38

383

displayed. The output pulses from the modulated converter determine ON time durations within a sampling time period of all converter switches. The inverter power output follows the control signal by shape. Control signal generation is thus significant in designing an inverter. In SVPWM it is generated from a three phase signal of the shape that the inverter power output is desired to replicate. The control signal thus generated cannot be compared with any physical line current since it is a single quantity for all three phases and is not any physical current that can be measured or displayed. With a high sampling frequency (10 kHz) compared to power frequency (50 Hz) the input changes from sample to sample. Hence no feedback can be applied and the inverter works on feed forward method. During each sampling duration all the required activities are done starting from measuring the sample waveform at that corresponding instant, generating the control signal vector, resolving it along applicable SS vectors, computing duty cycles and finally applying pulses at the gates of the converter switches. The output waveform of the inverter, generated through control signal, is compared with the sample wave to verify the correctness of the control signal. SVPWM method is a fast computation intensive process. Hence a complete mathematical implementation of the process with an appropriate processor can yield best results.

1.1 Review of Literature on 3D SVPWM

In 1997 Zhang et al. [1] first three dimensional space vector pulse width modulation (3D SVPWM) in $\alpha-\beta-\gamma$ frame for four leg (4L) converters. Since its introduction 3D SVPWM implementation in $\alpha\beta\gamma$ frame has been found difficult and was mentioned in [2]. Look up tables were used and few decisions were taken based on voltage polarities of non-zero switching state vectors (NZSV). But for specific advantages of direct access of the fourth wire by the converter's fourth leg and for better utilization of the converter dc voltage, there have been researches for simpler alternative implementations. Studies reported so far in the allied field have presented different algorithms and their applications in four wire (4 W) three phase inverters [2–7]. These applications have not applied SVPWM proper. In those reports mapping of active tetrahedron, determination of active vectors, determining the matrix to compute duty cycles have not been decided mathematically but had been done intuitively with the help of look up tables [2, 6]. A recent work on four-leg inverters has been reported to deliver active power to the balanced or unbalanced and linear or non-linear loads using a cost function to minimize errors between the reference currents and the output currents to generate gate pulses but it has not used SVPWM [5]. The 4L applications with 3D SVPWM and $\alpha-\beta-\gamma$ frame have been reported in [8, 9] but these are on Active Shunt Filter (ASF).

Park [10] and Clarke [11] presented d–q–0 and $\alpha-\beta-\gamma$ transformation theories which deal with the concept of decomposition of the three phase ac to two space vectors. Since then the method for converting a balanced three phase system into a two dimensional (2D) vector has been well established. Those basics are:

(i) developing an active space in $\alpha-\beta$ frame by valid switching states (SS) of the converter, (ii) developing a vector for reference signal at an instant, (iii) mapping the reference signal in the active space to find the two adjacent SS vectors, (iv) resolving the control vector upon the two SS vectors found, and (v) to compute from the resolutions the duty cycles of all switches of the converter and to generate desired output pulses applicable for that instant.

On the contrary, there is no generally accepted implementation method in 3D SVPWM to apply on four leg converters for unbalanced three phase four wire systems. This has still been in a state of research and different methods have been reported as alternatives for ease in understanding and for less mathematical complicacy.

A literature identified the $\alpha-\beta-\gamma$ frame of coordinates as the source of difficulties and complexities in applying 3D SVPWM. To overcome the difficulties, the *abc* frame of coordinates has been proposed for 3D vector analysis in 2003 [12]. Since then a number of literatures using *abc* frame have been reported [13–16]. But as *abc* frame has all its axes on a common 2D plane, mathematical analysis of 3D vectors is not possible in *abc* frame. The essential mathematical condition for a frame of three axes to represent a three dimensional vector is, that for any combination of axes, the third axis must lie outside the 2D plane formed by the rest two axes of the frame. Literatures that reported the use of $\alpha-\beta-\gamma$ frame in implementing 3D SVPWM on different applications have not explored a total mathematical solution. Instead, one or other method has been applied to bypass mathematics and have applied intuitive decision making and use of look up tables to map the active tetrahedron or to determine active vectors or to determine the matrix to compute duty cycles [2, 6]. A generalized 3D SVPWM in $\alpha-\beta-\gamma$ frame for 4Leg, 3Level inverter application has been reported in 2006. It has used four adjacent vectors for duty cycle computations against the use of three vectors as per vector mathematics [9]. Recent works have reported different other techniques and topologies and have not explored any mathematical solution for 3D SVPWM for a 4L inverter. Those have been: (i) use of effective time concepts presented to avoid high computational burden of conventional SVPWM [17], (ii) application of predictive current control strategy which is a different strategy that predicts the switching state in the (k + 1)th cycle that minimizes the error between the predicted currents and their references computed at the (k)th cycle [3], (iii) applying per-phase vector control strategy instead of the conventional unified approach for all three phases in SVPWM [18], (iv) using a combination of SVPWM with the DC-side voltage control with a converter having 4 switches in 2 legs instead of six switches in three legs [19], and (v) using a three-leg inverter for a four-wire active filter [20]. First mathematical implementation of 3D SVPWM has been reported in [21] but the inverter application was not for unbalanced three phase alternating current system.

1.2 The Motivation

The motivation of this work has been to verify performance of a fully mathematical implementation of 3D SVPWM on an application using a 4-leg bridge circuit. There would be no reference to any lookup table nor there be any decision making based on any polarity change or any other similarity or dissimilarity. The inverter output would follow a control signal which is a mathematical quantity representing the desired waveform. Successful performance of the inverter would verify: (1) that the reference signal vector, the mathematical entity, generated from measured values of any three phase sample system, does truly represent the source; and (2) that the analytical method applied to modulate the bridge circuit as per the proposed method has been correct.

To develop the proposed inverter the first step is to generate a control signal from a sample three phase system and in the second stage to make the control signal drive the inverter to generate the output. The sample three phase currents are produced by connecting a three phase four wire load to a three phase balanced source. Measured instantaneous values of line currents are the inputs for control signal generation. The second step is comprised of the four leg inverter, its controller to implement 3D SVPWM and different meters. A MATLAB/Simulink simulation model in discreet mode has to be designed accordingly. During each sampling interval a finite set of activities is to be executed at the second part of the model with the measured values of phase currents flowing in the first part of the model. The input measured values are fed to the controller. All analytical processes in vector domain, required as per 3D SVPWM, are performed at controller. The outputs of the controller are time domain pulses for IGBT switches of the inverter. Inverter output terminals are connected to an four wire load which is exactly equal to the load connected in the sample current generating circuit in the first part. Finally comparison of inverter output currents with the sample currents from which the control signal was generated would be made.

The paper has been organized in different sections. In Sect. 2, a general presentation of 3D SVPWM has been made. Section 3 presents the Matlab/Simulink model for a four-leg voltage source inverter working as proposed. Simulation results and analyses have been presented in Sect. 4. Section 5 presents the conclusion of the work.

2   3D SVPWM: A General Introduction

In SVPWM a three phase time varying quantity is represented as a single vector in space domain in $\alpha-\beta-\gamma$ frame. Cartesian frame of coordinates and $\alpha-\beta-\gamma$ frame are similar. For unbalanced three phase system the representative vector is a 3D vector in $\alpha-\beta-\gamma$ frame. For unbalanced systems the time varying symmetrical sequence components i.e. positive and negative sequence components are

Fig. 1 4-leg inverter circuit

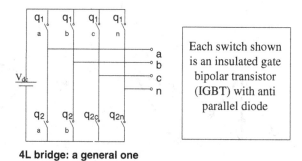

4L bridge: a general one

Each switch shown is an insulated gate bipolar transistor (IGBT) with anti parallel diode

represented by a 2D vector on the $\alpha-\beta$ plane and the zero sequence component is mapped along the γ axis.

In 3D SVPWM the reference vector generated is mapped into a 3D space, called the active space. This active space is built around all valid SS vectors of the converter that has to be modulated. For a 4L inverter as shown in Fig. 1 there are fourteen concurrent non-zero SS vectors and the overall active space is a hexagonal right cylindrical space as shown in Fig. 2.

The common point of all SS vectors is the origin (0, 0, 0). With the mapping of the control signal vector in the active space of the hexagonal right cylinder, three SS vectors which are immediately adjacent to the control signal are identified. To facilitate this task, the overall active space is divided in twenty four tetrahedron spaces each of which is a space defined by three SS vectors. These tetrahedrons are obtained by dividing the overall cylindrical space into six vertical prisms and then subdividing each prism into four tetrahedrons. This has been illustrated in Fig. 3 by

Fig. 2 3D hexagonal right cylindrical active space defined by switching state vectors

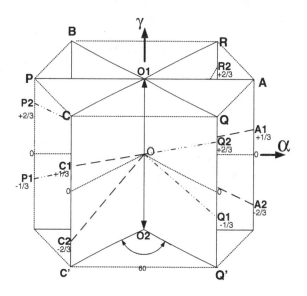

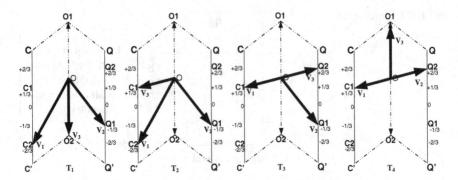

Fig. 3 Four different tetrahedrons in a prism

showing all four tetrahedrons (T_1, T_2, T_3, T_4) of the prism bound by triangular planes CO_1Q on top and $C'O_2Q'$ at the bottom as shown in Fig. 2.

With successful mapping of the control signal into the correct tetrahedron, the three switching state vectors which comprise the tetrahedron are identified as the required set of three SS vectors along which the control signal vector is to be resolved. Let $\vec{I}_{ref}$ be the reference vector and $\vec{I_1}$, $\vec{I_2}$ and $\vec{I_3}$ be the SS vectors comprising the active tetrahedron. Then the resolutions of $\vec{I}_{ref}$ along $\vec{I_1}$, $\vec{I_2}$ and $\vec{I_3}$ can be expressed as ($d_1 \cdots \vec{I_1}$), ($d_2 \cdots \vec{I_2}$), and ($d_3 \cdots \vec{I_3}$), where d_1, d_2 and d_3 are fractional numbers. Expressed mathematically,

$$\vec{I}ref = (d_1.\vec{I_1}) + (d_2.\vec{I_2}) + (d_3.\vec{I_3}) \tag{1}$$

Equation (1) implies that $\vec{I}ref$ during a sampling time T_s would be produced by keeping I_1 switching state on for ($d_1 \times T_s$) time, I_2 switching state on for ($d_2 \times T_s$) time and I_3 switching state on for ($d_3 \times T_s$) time. Fractional numbers d_1, d_2 and d_3 are the duty cycles (ratio of on-time to the sampling time) computed for the three SS vectors identified. It is required to add null values during a sampling period when ($d_1 + d_2 + d_3$) < 1 when null value switching state is to be kept on. The duration of the null state is ($d_Z \times T_s$) where d_Z is obtained from:

$$d_Z = 1 - (d_1 + d_2 + d_3) \tag{2}$$

With the knowledge of all duty cycles the gate pulses for individual switches are computed and are sent out as output.

3 MATLAB/Simulink Simulation for a Four-Leg Inverter

A Matlab/Simulink model, as shown in Fig. 4, has been developed as per the stated objective. The block-I of Fig. 4 shows a standard three phase source feeding two three phase loads through individual circuit breakers, no. 1 and no. 2. All line currents are measured. In block-II of Fig. 4 the measured signals have been processed to prepare them as input for the generation of the reference signal vector and for further computations. It is apparent that from the initial input of measured values of line currents up to the output stage the entire process is analytical. To do justice to the process and to obtain the best performance a program code has been developed to translate all of the steps mathematically and has been applied in the present simulation as an embedded function in the "MATLAB Fcn" block shown in block-II of Fig. 4. The block executes the program on the input data and finally sends out 8-bit digital words as output. These output words are fed as gate inputs to the four leg bridge circuit of the inverter shown next in block-III of Fig. 4. The inverter output is filtered by the filter block and is connected next to two three phase loads through circuit breakers no. 3 and no. 4 in block-IV. These loads are the same as those in the block-I of Fig. 4. The sampling frequency of the simulation is 10 kHz.

The proposed model is developed in MATLAB-8.1.0.604 (R2013a) language on a personal laptop with an Intel (R) core (TM) 2 Duo CPU with 2.00 GHz with 2-GB RAM. The signal is directly fed to gate drivers of the 4-Leg VSI block

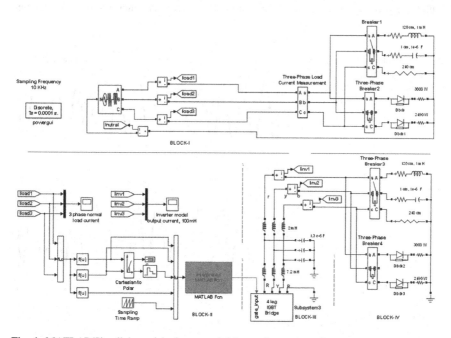

Fig. 4 MATLAB/Simulink model of proposed 4-Leg voltage source inverter

("subsystem3" of the model). The output from the VSI has been shown to pass through a filter block subsystem. Values of different circuit elements have been shown in the drawing. The objective of the simulation has been to verify whether the inverter output currents I_{inv1}, I_{inv2}, and I_{inv3} are same as the currents I_{load1}, I_{load2}, and I_{load3} drawn by the same load from a 3-phase power source. Circuit breaker no. 3 is closed when circuit breaker no. 1 is closed with other circuit breakers open. Similarly when even numbered breakers are closed then the odd numbered breakers are open. Two scopes to display these currents are shown in Fig. 4. With this arrangement the inverter operation has been tested for two distinctly different loads, both by type of loads as well as by magnitudes.

4 Results and Analyses

Results obtained have been presented in Figs. 5 and 6. Figure 5 shows that the current delivered by the proposed inverter to a three phase four wire unbalanced load is same by value and shape as the load current supplied by a normal voltage source to an exactly identical three phase four wire unbalanced load. Both sample currents and inverter currents are shown separately in Fig. 5. There is a minor phase delay in the inverter output currents. This can be attributed to the response time of the inverter as it follows the control signal that is generated from the load currents which are to be replicated. In Fig. 6 the proposed inverter has been tested on a highly non-linear load. The result obtained is highly encouraging about the

Fig. 5 Sample wave and the inverter output wave for load circuit breakers 1 and 3 on

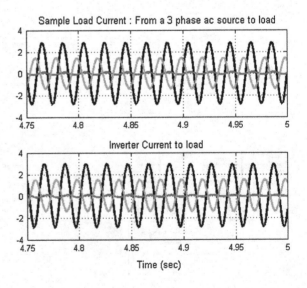

Fig. 6 Sample wave and the inverter output wave for load circuit breakers 2 and 4 on

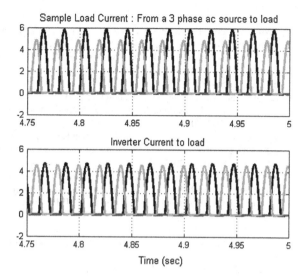

potential of proposed method. Results obtained also validate the correctness of the mathematical program code applied to modulate the inverter with 3D SVPWM in $\alpha-\beta-\gamma$ frame.

5 Conclusion

In this work it has been shown that an unbalanced three phase system can be converted to a single mathematical vector quantity in $\alpha-\beta-\gamma$ frame. The paper also has established that with proper analytical implementation of 3D SVPWM, a 4-leg converter can be successfully modulated with such a control signal. A physical control signal generation process needs to be compared with physical inputs to eliminate errors. To match phases the phase locked loops (PLL) are also required. Control signal generation in the proposed method does not have error as proper mathematical computations does not produce errors. This leads to another benefit of the proposed implementation that neither feedback loop nor PLL is required.

References

1. Zhang, R., Boroyevich, D., Prasad, H., Mao, H., Lee F.C., Dubovsky, S. C.: A three-phase inverter with a neutral leg with space vector modulation. In: Proc. APEC'97, Atlanta, GA, pp. 857–63, (1997).
2. Zhang, R., Prasad, H., Boroyevich, D., Lee F.C.: Three-Dimensional Space Vector Modulation for Four-Leg Voltage-Source Converters. IEEE Trans. Power Electronics, 17 (3), 314–326, (2002).

3. Yaramasu, V., Rivera, M., Wu, B., Rodriguez, J.: Model Predictive Current Control of Two-Level Four-Leg Inverters-Part I: Concept, Algorithm, and Simulation Analysis. IEEE Transactions on Power Electronics, 28, 3459–3468, (2013).
4. Li, X., Deng, Z., Chen, Z., Fei, Q.: Analysis and Simplification of Three-Dimensional Space Vector PWM for Three-Phase Four-Leg Inverters. IEEE Trans on Industrial Electronics, vol. 58(2), pp. 450–464, (2011).
5. Karthikeyan, R., Pandian, S.C.: An efficient multilevel inverter system for reducing THD with Space Vector Modulation. International Journal of Computer Applications, vol. 23(2), pp. 0975–8887, (2011).
6. Zhou, J., Wu, X., Geng, Y., Dai, P.: Simulation Research on a SVPWM Control Algorithm for a Four-Leg Active Power Filter, Journal of china university of mining and technology, vol. 17 (4), pp. 590–94, (2007).
7. Szczepankowski, P., Nieznanski, J., Sleszynski, W.: A New Three-Dimensional Space Vector Modulation for Multilevel Four-Leg Converters Based on the Shape Functions of Tetrahedral Element. In: 39th Annual Conference of the IEEE Industrial Electronics Society (IECON 2013), pp. 3264–3269, (2013).
8. Chaghi, A., Guetta, A., Benoudjit, A.: Four-Legged Active Power Filter Compensation for a Utility Distribution System, *Journal of electrical engineering*, vol. 55, (1–2), pp. 31-35, (2004).
9. Dai, N.Y., Wong, M.C., Han, Y.D.: Application of a Three-level NPC Inverter as a Three-Phase Four-Wire Power Quality Compensator by Generalized 3DSVM, *IEEE Transaction on Power Electronics*, Vol. 21(2), pp. 440–449, (2006).
10. Park, R.H.: Two reaction theory of synchronous machines, *AIEE Transactions,* vol. 48, pp.716–730, (1929).
11. Duesterhoeft, W.C., Schulz, M.W., Clarke, E.: Determination of Instantaneous Currents and Voltages by Means of Alpha, Beta, and Zero Components, Transactions of the American Institute of Electrical Engineers, 70(2),1248–1255 (1951).
12. Perales, M.A., Prats, M.M., Portillo, R., Franquelo, L.G.: Three dimensional space vector modulation in *abc* coordinates for four-leg voltage source converters, IEEE Power Electron. lett., vol. 1(4), pp. 104–09, (2003).
13. Franquelo, L.G., Prats, A.M., Portillo, R.C., Galvan, J.I.L., Perales, M.A., Carrasco, J.M., Diez, E.G., Jimenez, J.L.M.: Three-Dimensional Space-Vector Modulation Algorithm for Four-Leg Multilevel Inverters using abc Coordinates, IEEE Trans. Industrial Electronics, vol. 53(2), pp. 458–466, (2006).
14. Prats, M.M., Franquelo, L.G., León, J.I., Portillo, R., Galván, E., Carrasco, J.M.: A 3-D space vector modulation generalized algorithm for multilevel converters, IEEE Power Electron. Lett, vol. 1(4), pp. 110–14, (2003).
15. Chang, G.W., Chen, S.K., Chu, M.: An efficient *a–b–c* reference frame-based compensation strategy for three-phase active power filter control, Int. J. Electric Power System Research, vol. 60(3), pp. 161–166, (2002).
16. Garcesa, A., Molinasa, M., Rodriguez, P.: A generalized compensation theory for active filters based on mathematical optimization in ABC frame, Int. J. Electric Power System Research, vol. 90, pp. 1–10, (2012).
17. Varaprasad, O.V.S.R., Sarma, D.V.S.S.S.: An improved SVPWM based shunt active power filter for compensation of power system harmonics. In: 16 th International Conference on Harmonics and Quality of Power (ICHQP), 2014 IEEE, Bucharest, pp: 571–575, (2014).
18. Ninad, N.A., Lopes, L.: Per-phase vector control strategy for a four-leg voltage source inverter operating with highly unbalanced loads in stand-alone hybrid systems, Electrical Power and Energy Systems, vol. 55, pp: 449–459, (2014).
19. Wang, W., Luo, A., Xu, X., Fang, L., Chau, T.M., Li, Z.: Space vector pulse-width modulation algorithm and DC-side voltage control strategy of three-phase four-switch active power filters, IET Power Electronics, vol. 6(1), pp: 125–135, (2013).

20. Villalva, M.G., deOlivera, M.E., Filho, E.R.: Detailed Implementation of a Current Controller with 3D Space Vectors for Four Wire Active Filters, In 5th International conference on Power Electronics and Drive Systems, pp. 536–541 (2003).
21. Chakraborty, A.K., Bhattacharya, B.: A new three-dimensional space-vector modulation algorithm for four-leg inverter, In: Proc. APSCOM, Honkong, pp. 1–6, (2012).

21st C New Technologies a Pathway to Inclusive and Sustainable Growth

Pradnya Chitrao

Abstract The 21st C began with many profound technological, economic and social transformations. Improved quantity and quality of information are resulting in rapid advances in science and engineering. Today a megabyte of semiconductor memory is very affordable for the common man. By 2020, one desktop computer will equal all the computers currently in Silicon Valley. The 21st C highly competitive IT sector is making things *faster*, *cheaper* and *smaller*. Information technologies are connecting every part of the world, and also enabling development of major new technologies like automated knowledge work tools, advanced robotics and 3D printing. IT progress will revolutionize production, transportation, energy, commerce, education and health. While this will certainly result in more environmentally sustainable products through a less wasteful production process, there will also be a serious impact on the demographic equilibrium of the world, if effective countermeasures are not put in place by business leaders and policy makers. It will affect the way business is done. Intermediaries are now fast disappearing, and businesses can pass on the benefits to the customers. Tomorrow's Information Technologies will radically improve the capacity to communicate and simulate. It will lead to learning by "doing", joint experimental research and moving at one's own pace for every "wired" person. But increasing technology has increased the risk of natural and man made threats. It is also difficult to adopt new attitudes, and accept alternative approaches to risk management. It requires concerted efforts on part of individuals, businesses and policy makers. The paper analyses the 21st C scenario and the various measures that need to be taken to counter the risks and threats. The author concludes that public opinion will have to ensure that emerging new technologies, which by their very nature vest in the hands of corporate giants, are harnessed for the common good and for sustainable development. This research which is primarily based on secondary sources can be enriched by a primary investigation of the methods employed by the corporate, the governments and the non-government bodies to face the challenges and the risks arising out of the technologies emerging in the 21st century.

P. Chitrao (✉)
Symbiosis Institute of Management Science (SIMS), Range Hills Road, Opp. EME
Workshop, Kirkee Cantt., Pune 411020, India
e-mail: pradnyac@sims.edu; drpradnya.sims@gmail.com

© Springer India 2016
S.C. Satapathy et al. (eds.), *Information Systems Design and Intelligent Applications*, Advances in Intelligent Systems and Computing 434,
DOI 10.1007/978-81-322-2752-6_39

Keywords IT progress · Faster · Cheaper · Smaller · Risks · New business models · Continuous responsibility · Creativity · Transparency

1 Introduction

A plethora of newly emerging advanced technologies has already begun to affect very significantly the 21st C workplace as also the social fabric in general. It has put in forward fast the process of globalization. Companies of all sizes today do business with customers all over the world. Businesses can establish satellite offices in practically any country as long as there is Internet access [1]. Computers and information technology have become integral to our lives. The question is how far these technologies and their consequent repercussions will help develop and retain sustainable businesses and communities.

2 Literature Review/Background

The primary benefit of technology is efficiency. Businesses today provide products and services at a faster, more efficient rate resulting in higher profits [2]. Some technologies can alter people's life and workstyles, rearrange value pools, and lead to entirely new products and services [3]. Business leaders must assess whether technologies might bring them new customers or force them to defend their existing bases or whether there is a need to invent new strategies. Companies prefer customized programs offered by functional and technology consultants.

Policy makers and societies must invest in new forms of education and infrastructure, and foresee how disruptive economic change will affect their present comparative advantages [4]. Lawmakers and regulators must manage new biological capabilities and protect the rights and privacy of citizens.

Different forces like demographic shifts, labour force expansion, urbanization, or new patterns in capital formation can bring about large-scale changes in economies and societies. At the same time, technology also renders old skills and strategies irrelevant [5]. There are thus both opportunities and threats. Leaders can seize these opportunities if they start preparing now.

3 Objectives

The paper attempts to have an overview of the changes that the 21st C technologies are bringing in different aspects and what needs to be done to use them for improvement of human existence rather than letting them spell doom for mankind.

4 Research Methodology

The paper is an analytical survey of secondary sources that may at a later date be enriched by a primary investigation of the methods employed by corporate to face the changes and the risks of the 21st C IT progress.

5 Material Welfare an Outcome of New Technologies

The 21st C essentially represents the third phase of the digital revolution that began in the post-World War II period. The first phase was the development of computing machines. The second phase comprised revolution in the technology of information transfer with the introduction of the Internet. The third phase is pervasion of IT enabled technologies in the fields of manufacturing, service sector, knowledge work, curative medicine and agriculture [6].

The collective impact of these three phases of evolution presents a fascinating picture.

a. *Data Processing* As pointed out by McKinsey Global Institute. The cost of recording, processing and storage of data has come down dramatically, with a good chunk of the benefits getting passed on to the final customer.
   Whatever the size or speed of a computer, all of them are now controlled through "chips" or transistors of various materials other than silicon. The overall trend is towards the ever smaller. A variety of computer technologies are already wearable. Even later technologies beyond optical computing are being developed including bio-computers and quantum computers.

b. *Data Transfer* Rapid advances in the technology have resulted in extremely rapid systems of data transfer, and on line management of businesses in the far flung corners of the world. Internet technology provides free and open access to a very powerful tool namely a common standard. The strengthening of Internet and the felicity of rapid data transfer will accelerate the formation of the global village [7]. Global competition for every job opening will accentuate.

c. *Manufacturing* The advanced Robot of 21st century is light weight, capable of a much larger variety of actions and can easily perform a number of tasks currently performed by humans. They are now replacing humans [8].

d. *Service Sector* With the increasing sophistication building up in IT enabled services, routine and relatively non-complicated service sector jobs will increasingly be replaced by robots that cost far less than the humans they will be replacing. In sum, a whole range of service sector jobs earlier performed by humans are liable to be taken over by IT enabled equipment [9].

e. *Knowledge work* Work like translation from one language to another is being taken over by computers [10]. IT enabled equipment is also being increasingly deployed for graphics design. Computers are being increasingly deployed for activities like preparing medical prescriptions and even for preparation of legal submissions.

f. *Curative Medicine* Performance of medical tests and laboratory analysis of samples, even diagnosis of diseases and determination of curator care are likely to pass to computers. In the 21st C, breakthroughs are expected in understanding genetic and biological processes. Technological advances and well thought out changes to the regulatory systems that protect the present patterns of control of health-related information, could turn patients from passive consumers to active controllers [11].

g. *Agriculture* Coupled with large scale mechanization already evident in the agricultural 'sector, 21st century is likely to witness a two-fold attack on employment in this sector-both by reducing the skill levels required as well as the head count required for delivery.

h. *Exploration and exploitation of natural resources* IT enabled equipment is also spreading its wings in this sector, through highly sophisticated exploration techniques as also highly mechanized means of exploitation of the resources identified.

i. *Disintegration of individual organization's learning conventions* The rapid changes brought about by 21st C technologies has brought in the breakdown of the individual organization's time tested internal consensus on what and how its members have to learn in order to cope with new and radically different environmental conditions.

6 Threats and Challenges Arising Out of the New Technologies

a. *Potential of widespread unemployment* A significant characteristic of the emerging technologies in the 21st C is that IT enabled technologies such as advanced robotics have the potential to (and are already in the process of) eliminating a wide spectrum of low to medium skill jobs in the manufacturing industry. As a result of introduction of advanced technologies, unemployment in the world is on the rise. 2013 saw the highest number recorded for unemployed in the world, at 202 Million [12].

b. *Widening gap between the rich and the poor* All over the world, there is a sharp and growing increase in the income disparities between the privileged few and the vast multitude of ordinary people. According to research findings by OXFAM, the United States is the country where the income disparities are widening at the fastest pace [13]. This is hardly surprising, since the 21st century technologies are orientating in the developed world.

c. *Increase in the financial power of the multinational corporations* The development of these technologies is almost completely controlled now by MNC's, who have the sole discretion and control for the use and dissemination of these technologies (Pavitt, K. and P. Patel [14]. The avowed objective of these corporations is to maximize the earnings for their shareholders.

d. *Lack of a safety net for those less gifted* The advent of the 21st century technologies makes it clear that now there is less and less of safety net for those who are not gifted [15]. This trend is likely to get accentuated if the penchant for less government continues.

7 Corrective Actions Available

In conclusion, the collective impact of the new technologies emerging in the 21st century is highly beneficial. Possible corrective actions are summarized below:

a. *Rapid and thoroughgoing changes in the educational system* These digital tools are necessary to more effectively predict and plan for educational needs. The digital age has expanded data as also unpredictability. Today, it is a priority to *develop an educational system that thrives on such change* and eliminates the problem of "change shock". Public schools, which generally do not have any or adequate computing and network resources, must realize the disadvantages that their students will face in such a world.

b. *Need to teach collaboration skills through networking under guided supervision* Due to the emergence of fast computer networking, the *team approach* has emerged to faster processing of a task [16]. The Internet and the formation of teams using computer networks will significantly improve the rate of interaction and the depth of interaction in learning. Educators view this knowledge, which is critical to current economic growth, as frivolous or dangerous for school productivity [17]. At the same time, the need for this interaction knowledge is increasing exponentially.

c. *Need to teach questioning and higher order thinking skills* Information is transforming the 21st century, and is giving rise to a series of "information induced gaps that create serious challenges for society, educators and students in classrooms" [18]. Information management consists of storing, communicating, and computing. We therefore need more students to use higher order thinking skills for optimally utilizing this data. We also need students to learn to express ideas through a wide variety of Web based media which are two-way systems of communication and storage than the traditional systems like books, newspapers and journals. Schools need to emphasize questioning and higher order thinking skills so that students will have foundational skills that are relevant to the digital world [19].

d. *Control of the operations of MNC's through Public Policy* One key effect of globalization is the growing importance of MNCs who are today responsible for a major chunk of R&D in the 21st C. They disseminate knowledge directly through their operations in foreign countries Developing nations now have to attract and use FDI effectively [20]. Economies that cannot offer intrinsic advantages like attractive profit opportunities through supply to those local markets or through their use as export platforms for other markets will have to find alternative ways of accessing relevant foreign knowledge (e.g. copying, reverse technology, and arm's length transactions).

e. *Effective Management of associated cultural change* The assimilation of computer technology into almost every aspect of our agricultural, industrial and cyberspace tools is constantly increasing. This means that our cultural systems will rapidly incorporate computer technology into our thinking, philosophy and psychology, just as was the case with the introduction of writing into human culture [8].

f. *Creation of a safety net for the underprivileged* Another challenge of the 21st C technologies is that of delivering knowledge to those who need it the most. For most nations, the acquisition of existing knowledge yields higher productivity. Easy communication allows access to technical information in printed or electronic form, especially including what can be accessed through the internet. It requires a strong regulatory environment that facilitates the enforcement of contracts. Thus new technologies require a well-developed economic and institutional regime for maximum benefit [21].

The information revolution and advances in science are accelerating the creation and dissemination of knowledge. So it is important that tapping into the ever growing stock of global knowledge by developing countries, is not impeded [22]. Advanced countries want a legal license to protect their advantage through a strict enforcement of Intellectual Property Rights (IPR). Developing countries will have no option but create their own centres of excellence to ensure for themselves a place in the sun.

g. *Creating systems with inbuilt protection mechanisms* Designers need to build systems that expect imperfection and do a better job of handling it. Software failures due to easily avoidable coding errors have cost companies and governments dearly [1]. Computer systems used in dangerous situations must include fail safe interrupts that provide protection until the computer can be fixed or restarted.

h. *Need to switch over to innovative and cost effective production processes* The digitalization of information and the consequent reduction in transportation and communication costs has resulted in the physical disintegration of production. So it is now also necessary to tap in to global supply chains. The end result is that production of labour intensive goods is shifting to developing countries with lower labour costs. Developed countries try to remain ahead in the competition by concentrating on the skill and technology intensive sectors.

i. *Creating a vision for the organization* Visions are goals shared by persons within an organization. Visions are necessary for providing orientation, co-ordination and for motivation.

8 Conclusions

Overview of the changes in the pipeline Information Technology has begun to enable development of other technologies that could extend higher benefits to the final consumer, but at the same time could potentially exert an adverse impact on employment and accelerate the process of heightening social inequities. For sustainable development therefore, it is not just adequate to accept change but also be willing to learn something new in order to stay adaptive. The new technologies are eliminating jobs at the bottom of the skill ladder whilst creating new and better paying jobs at the top of the ladder. This upward movement to better jobs and better lives sustains efforts to change.

Communication technologies have reduced transportation costs, expanded and globalized trade, and created a "real time world" Developing nations must develop greater capabilities in order to respond quickly to new threats and opportunities.

The Internet has created the need for new business models. Online stores are already affecting standard business models. Due to knowledge boom created by the 21st C technologies, everything gets cheaper, and this factor has to be taken into account in all business plans. In today's economy, it is essential to create new products and services at a very rapid rate.

9 The Possible Way Forward

The conclusion is clear that while the technological advances of the 21st century have the potential to enhance the spectrum of material comfort available to mankind, these are also likely to accentuate inequities, create widespread unemployment and finally lead to problems that will endanger the survival of the less gifted, whether individual, groups or even nations. Intervention of policy makers is definitely required, both for retraining of the work force and also for ensuring a minimum safety net for those who are less gifted. Apart from the Government, all those who are working for the public good, including education institutions, charitable and Non-Government organizations, will need to dedicate themselves to the ideal of creating a safety net for those who are less privileged. There is a need for co-operation and empathy and collective decision making. There is an urgent need for the introduction of new mechanisms that are capable of furnishing reliable and inexpensive information. Rapid technological development and diffusion are unlikely to take place without creativity, spontaneity and transparency on the part of employees and consumers. In the absence of these checks and balances, the 21st century technology may well turn out to be a monster devouring the human race.

References

1. Levinson, Meredith, Let's Stop Wasting 78 Billion a Year, (Oct. 15, 2001).
2. Singh, N., Services-Led Industrialization in India: Prospects and Challenges, (2007).
3. Toffler, A. & Toffler, H., Revolutionary Wealth. New York: Knopf, (2006).
4. Dahlman, C. J. and A. Utz, India and the Knowledge Economy: Leveraging Strengths and Opportunities, World Bank, Washington, D.C., (2005).
5. Akrich, M., "Beyond Social Construction of Technology: The Shaping of People and Things in the Innovation Process" in M. Dierkes and U. Hoffmann (eds.), *New Technology at the Outset: Social Forces in the Shaping of Technological Innovations*, Campus, Frankfurt am Main, pp. 173–190, (1992).
6. Hausmann, R. and B. Klinger (2006), Structural transformation and patterns of comparative advantage in the product space, mimeo, Harvard University, (2010, April 14).
7. Baldwin, R., Globalization: The Great Unbundling(s), paper contributed to event on Globalization Challenges to Europe and Finland organized by the Secretariat of the Economic Council, Prime Minister's Office (June), (2006).
8. Toffler, A. & Toffler, H., Future Shock. New York: Random House, (1970).
9. Appelbaum, E. and R. Schettkat, "The End of Full Employment? On Economic Development in Industrialised Countries", *Intereconomics,* May/June, pp. 122–130, (1994).
10. Evans, B., Global CIO: Google CEO Eric Schmidt's Top 10 Reasons Why Mobile Is #1.
11. Staff of Science, Introduction to special issue: Challenges and opportunities. 331(6018), 692–693. doi:10.1126/science.331.6018.692, (2011, February 11).
12. Economist, The, The New Titans: A Survey of the World Economy, Special supplement in September 16th edition, (2006).
13. Kniivilä, M., Industrial Development and Economic Growth: Implications for Poverty Reduction and Income Inequality, (2007).
14. Pavitt, K. and P. Patel, "Global Corporations and National Systems of Innovation: Who Dominates Whom?" in J. Howells and J. Mitchie (eds.), *National Systems of Innovation or the Globalisation of Technology,* Cambridge University Press, Cambridge, (1998).
15. Toffler, A. & Toffler, H, The Third Wave. New York: Random House, (1980).
16. Smyth G., Wireless Technologies: Bridging the Digital Divide in Education, Anil Varma (Ed), "Information and Communication Technology in Education", First edition, Icfai University Press, Hyderabad, p. 179, (2008).
17. Nooriafshar M., The Role of Technology based Approaches in Globalizing Education, Anil Varma (Ed), "Information and Communication Technology in Education", First edition, Icfai University Press, Hyderabad, p. 53, (2008).
18. Houghton, R.S., The knowledge society: How can teachers surf its tsunamis in data storage, communication and processing, (2011).
19. Oliver R., The Role of ICT in Higher Education for the 21st Century, (2008).
20. Lall, S., Foreign Direct Investment, Technology Development and Competitiveness: Issues and Evidence, in Competitiveness, FDI and Technological Activity in East Asia, S. Lall and S. Urata, eds., Edgward Elgar, Northampton, (2003).
21. Toffler, A. & Toffler, H., Powershift: Knowledge, Wealth and Violence at the Edge of the 21st Century. New York: Bantam Book, (1990).
22. Berger, S., How We Compete: What Companies around the World are Doing to Make it in Today's Global Economy, Random House, New York, (2006).

Dual Image Based Reversible Data Hiding Scheme Using Three Pixel Value Difference Expansion

Giri Debasis, Jana Biswapati and Mondal Shyamal Kumar

Abstract In this paper, we proposed dual image based reversible data hiding scheme using three pixel value difference and difference expansion. We take consecutive three pixels from original image and embed 13 bits secret data by using three pixel value difference (TPVD) and difference expansion (DE) method. Using TPVD, we get modified three pixel values and using DE, we get another three modified pixel values after embedding secret data. These two sets of three stego pixel values are stored on dual images to achieve reversibility. We extract the secret data successfully using TPVD and DE method using two sets of three pixel values from dual images. The classical PVD and TPVD was not reversible data hiding scheme, we achieve reversibility using DE and dual image that means, we recover original cover image from dual stego images without any distortion. Finally, we compare our scheme with other state-of-the-art methods and obtain reasonably better performance in terms of data embedding capacity.

Keywords Reversible data hiding (RDH) · Pixel value difference (PVD) · Difference expansion (DE) · Image steganography · Dual image

G. Debasis (✉)
Department of Computer Science and Engineering, Haldia Institute
of Technology, Haldia 721657, West Bengal, India
e-mail: debasis_giri@hotmail.com

J. Biswapati
Department of Computer Science, Vidyasagar University,
Midnapore 721102, West Bengal, India
e-mail: biswapati.jana@mail.vidyasagar.ac.in

M.S. Kumar
Department of Applied Mathematics with Oceanology and Computer Programming,
Vidyasagar University, Midnapore 721102, West Bengal, India
e-mail: shyamal_260180@yahoo.com

© Springer India 2016   403
S.C. Satapathy et al. (eds.), *Information Systems Design and Intelligent
Applications*, Advances in Intelligent Systems and Computing 434,
DOI 10.1007/978-81-322-2752-6_40

1 Introduction

Secret writing is commonly used to protect multimedia data. It is classified into two categories, irreversible and reversible. In irreversible techniques [1–3], secret data can be embedded and extracted successfully but it cannot recover the original image. On the other hand, reversible data hiding schemes [4–9] can extract the secret message and recover the original image. In recent years, many researchers [10–15] are working to improve the embedding capacity with minimum distortion. Wu and Tsai [16] proposed a data hiding method based on PVD. Tian [8] proposed a difference expansion (DE) data hiding approach to conceal the secret data into the difference of a pair consecutive pixel values with high payload size. Reversible data hiding technique has been proposed by Ni et al. [7] which are based on histogram shifting with zero or minimum change of the pixel gray values. Being reversible, both the original data and the embedded data can be completely restored. Dual image based data hiding become important research issue in reversible data hiding schemes and applies into medical science, military and other many areas. Chang et al. [17] proposed dual-image based data hiding technique using exploiting modification direction (EMD) method. Lee and Huang [18] converted secret data into quinary-based secret symbols and combined every two secret symbols as a set for embedding. Lu et al. [11] used the least-significant-bit (LSB) matching method for embedding. Lee et al. [19] embed secret data using directions to achieve high image quality, but the embedding capacity could only reach 0.75 bits per pixel (bpp). Chang et al. [17] embed secret data through the modulus function matrix to achieve a higher capacity that is 1.00 bpp, but image quality was inferior to that using the method by Lee et al. Lu et al. [12] could achieve good quality stegos and their Peak Signal to Noise Ratio (PSNR) is 49.20 dB but still capacity belongs in 1.00 bpp. Thus, the challenge to enhance embedding capacity while maintaining high image quality through the use of dual-image techniques is still an important issue.

1.1 Motivation

In this paper, dual-image based reversible data hiding scheme has been introduced through three PVD and DE (TPVDE):

• Our motivation is to enhance the embedding capacity and achieve reversibility in data hiding. Data embedding using TPVD was not reversible.
• Another motivation is security. We distribute stego pixels among dual image. Finally, original image can be recovered from dual image without any distortion.

The rest of the paper is organized as follows. Proposed data hiding scheme in detail has been discussed in Sect. 2. Experimental results with comparisons are discussed in Sect. 3. Finally, some conclusions are given in Sect. 4.

2 Propose TPVDE Method

In this paper, a new reversible data hiding scheme has been proposed combining Three Pixel Value Difference (TPVD) with Difference Expansion (DE) on dual image. We select three consecutive pixels P_1, P_b and P_2 from cover image I shown in Fig. 1. Replace three bits secret data by the last three bits of P_b. Then we calculate two pixel value differences d_1 and d_2 between the pixel pair (P_1, P_b) and (P_b, P_2) that is $d_1 = |P_1 - P_b|$ and $d_2 = |P_b - P_2|$. Number of secret bits which will be embedded in the cover image is determined by the range table, which have equal sub range $[lb, ub]$ having length w that is $w = ub - lb + 1$. In this scheme we set w as 8 and the contiguous sub-ranges are $\{0-8, 8-16, 16-24, …, 248-255\}$ which has capability to embed two 4 secret bits within each pixel pair in cover image. After embed 4 bits between two pixel pairs we get three stego pixels P'_1, P'_b and P'_2. Then we apply DE on the four differences as shown in Fig. 1. We calculate $t_1 = |P'_1 - P'_b|$; $t_2 = |P'_b - P'_2|$; and $h_1 = |P_1 - P_b|$; $h_2 = |P_b - P_2|$. After that, we calculate $f_1 = |h_1 - t_1|$; $f_2 = |h_2 - t_2|$. Now, one bit is embed on $P''_1 = |P_b +/- h'_1|$; $P''_2 = |P_b +/- h'_2|$, where $h'_1 = f(f_1)$ and $h'_2 = f(f_2)$, where $f(f_i) = 2 \times f(i) + w$ and $w = 1$ bit secret data and $i = 1, 2$. Finally, three pixels, P''_1, $P''_b = P_b$ and P''_2 are stored on stego image SA and P'_1, P'_b and P'_2 are stored on SM. The detailed schematic diagram of our proposed data embedding is shown in Fig. 1. The numerical example is shown in Fig. 2. At the receiver end, the secret data extraction and original image reconstruction has been performed by taking three consecutive pixels from both the stego images SM and SA. The extraction process and numerical example is shown in the Figs. 3 and 4 respectively.

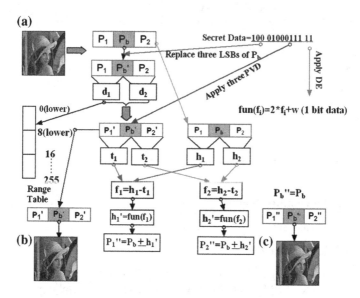

Fig. 1 Block diagram of embedding process of our proposed method. **a** Original image I(M × N). **b** Stego image(SM). **c** Stego image(SA)

(3 x 3) block

96	162	150
100	120	134
112	116	145

Data=10101111001110

P_1	P_b	P_2
96	162	150

Three Pixels

$(162)_{10}=(10100010)_2$

Three bits data (101) is replace in least three bits of P_b; $P_b'=(10100101)_2=(165)_{10}$

$d_1=|P_1-P_b'|=|96-165|=69$; d_1 belongs to the range[64,255] ,so 6 bits data is embedded

$d_2=|P_2-P_b'|=|150-165|=15$; d_2 belongs to the range[8,15] ,so 3 bits data is embedded

$d_1'=30$ (011110)+64 (lower bound) = 94; $d_2'=3$ (011)+8 = 11;

$P_1'=P_b'-d_1'=165-94=71$; $P_2'=P_b'-d_2'=165-11=154$;

$P_1'=P_b'+d_1'=165+94=259$ $P_2'=P_b'+d_2'=165+11=176$;

Finally, $P_1'=71$ (Since,|96-71|<|96-259|) $P_2'=154$ (Since,|150-154|<|150-176|)

After embedding 11 bits within three pixels,
we get stego pixels P_1', P_b' and P_2' are

P_1'	P_b'	P_2'
71	165	154

Apply DE:

Calculate $h_1=|P_1-P_b|=|96-162|=66$; $h_2=|P_2-P_b|=|150-162|=12$;

$t_1=|P_1'-P_b'|=|71-165|=94$; $t_2=|P_2'-P_b'|=|154-165|=11$;

$f_1=t_1-h_1=94-66=28$; $f_2=t_2-h_2=11-12=-1$;

$h_1'=2*f_1+w=2*28+1=57$ $h_2'=2*f_2+w=2*(-1)+0=-2$

$P_1''=P_b-h_1'=162-57=105$ $P_2''=P_b-h_2'=162-(-2)=164$

(Since, $P_b>P_1$)   (Since,$P_b>P_2$)

$P_b''=P_b$ (162);

After embedding 2 bits within three pixels,
we get another set of stego pixels P_1'', P_b'' and P_2'' are

P_1''	P_b''	P_2''
105	162	164

We store two sets of stego pixels within dual image

P_1' P_b' P_2'

71	165	154
100	120	134
112	116	145

SM

P_1'' P_b'' P_2''

105	162	164
100	120	134
112	116	145

SA

Fig. 2 Example of data embedding in proposed method

Example of Data Embedding

Consider an original image I (256 × 256).

Example of Data Extraction

Consider two stego images SM and SA of size (256 × 256) each.
    A (3 × 3) block of SM and SA has been considered for simplicity.

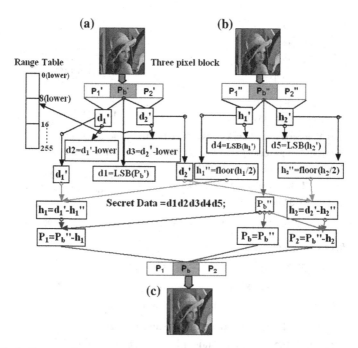

Fig. 3 Block diagram of extraction process of proposed method. **a** Stego image(SM). **b** Stego image(SA). **c** Original image I(M × N)

3 Experimental Results and Comparison

In this section, our proposed method has been verified and tested using gray scale image of size (256 × 256) pixels. The original image and after embedding the secret messages, dual stego images, SM and SA have been generated which are shown in Fig. 5. Our developed embedding and extraction algorithms are implemented in MATLAB Version 7.6.0.324 (R2008a). Here, distortion is measured by means of two parameters namely, Mean Square Error (*MSE*) and Peak Signal to Noise Ratio (*PSNR*). The *MSE* is calculated as follows:

$$MSE = \frac{\sum_{i=1}^{M} \sum_{j=1}^{N} [X(i,j) - Y(i,j)]^2}{M \times N},$$ (1)

where M and N denote the total number of pixels in the horizontal and the vertical dimensions of the image. $X(i, j)$ represents the pixels in the cover image and $Y(i, j)$ represents the pixels of the stego image.

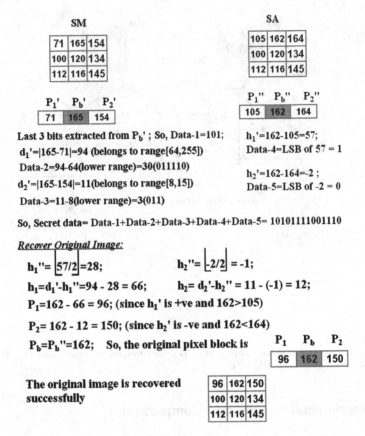

Fig. 4 Example of data extraction in proposed method

$$PSNR = 10 \log_{10} \frac{I_{\max}^2}{MSE},\qquad(2)$$

After that the (*PSNR*) has been calculated by Eq. (2). Where $I_{\max}$ is the maximum intensity value of image. Higher the values of *PSNR* can be the better image quality. The analysis in terms of *PSNR* of cover image with stego images has given good results which are shown in Table 1. The average *PSNR* of our proposed scheme varies 35.86–38.95 dB when data embedding capacity varies from 40,000 bits to 1,61,992 bits. To calculate payload in terms of bits per pixel (bpp), we calculate total bits $B = 256 \times 256/3 \times 13 = 851,968$ bits. We have used gray scale dual image of size (256×256). So, Payload (bpp) = 851,968/$(2 \times 256 \times 256)$ = 2.166.

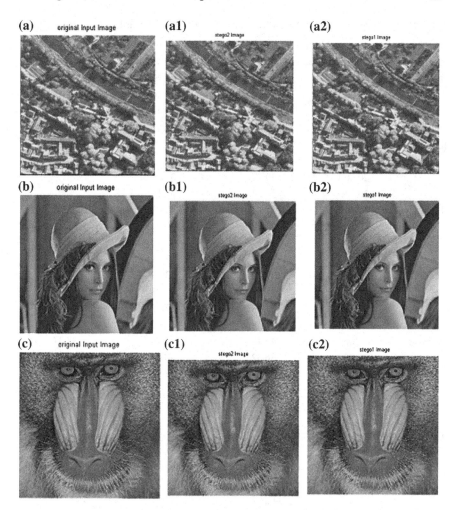

Fig. 5 Standard test images with (256 × 256) pixel collected from [21] and dual stego images, SM and SA with (256 × 256) pixel. **a** Original aerial 1256 × 256. **a1** Stego SM 256 × 256. **a2** Stego SA 256 × 256. **b** Original Lena 256 × 256. **b1** Stego SM 256 × 256. **b2** Stego SA 256 × 256. **c** Mandrill 256 × 256. **c1** Stego SM 256 × 256. **c2**Stego SA 256 × 256

The average PSNR of the stego images of the proposed method is 37.186 dB which is lower than the method proposed by Qin et al.'s [20], Lu et al.'s [11, 12], Chang et al.'s [17] and Lee et al.'s [19] scheme. But the average PSNR of proposed method is higher than the method proposed by Lee et al.'s [5] and Zeng et al.'s [10] scheme. The embedding capacity of the proposed method is 2.166 bpp which is higher than the other existing schemes shown in Table 2.

Table 1 Experimental result: PSNR of S1 with I and PSNR of S2 with I and average PSNR

PSNR (dB) wirh Data embedding capacity (in bytes)

Image I	Capacity (bytes)	PSNR (I and SM)	PSNR (I and SA)	Avg PSNR
Jetplane	5000	47.07	43.29	37.88
	10,000	37.18	39.36	
	20,000	31.65	36.48	
	20,249	31.62	36.41	
Lena	5000	40.31	43.78	36.93
	10,000	35.31	40.19	
	20,000	30.77	37.28	
	20,249	30.67	37.18	
Living room	5000	38.93	43.47	36.70
	10,000	34.18	40.02	
	20,000	31.37	37.19	
	20,249	31.31	37.11	
Pirates	5000	39.79	43.75	37.05
	10,000	35.29	40.28	
	20,000	31.58	37.15	
	20,249	31.58	37.09	
Woman	5000	40.44	43.75	37.37
	10,000	36.32	40.20	
	20,000	32.00	37.19	
	20,249	31.92	37.12	

Table 2 Comparison of average PSNR and payload (bpp) with existing schemes

Scheme	Average PSNR (dB)	Capacity (bpp)
Chang et al. [17]	45.1225	1.00
Chang et al. [19]	48.14	1.00
Lee et al. [5]	34.38	0.91
Zeng et al. [10]	32.74	1.04
Lee and Huang [21]	49.6110	1.07
Qin et al. [17]	52.11	1.16
Lu et al. [13]	49.20	1.00
Proposed Scheme	37.186	2.166

4 Conclusions

In this paper, on the basis of three pixel value difference and difference expansion a dual image based reversible data hiding scheme has been introduced. Here, the reference table has been used for data embedding. We keep the difference value of a sub-range from reference table which helps to recover the original image from stego images. In our proposed method, PVD method achieves reversibility which demands the originality of our work. Also, the scheme still maintains low relative entropy. In addition, it gains good PSNRs and higher payload than other existing methods.

References

1. C. K. Chan and L. M. Cheng, Hiding data in images by simple LSB substitution, Pattern Recognition, vol. 37, no. 3, pp. 469–474, (2004).
2. T. D. Kieu and C. C. Chang, A steganographic scheme by fully exploiting modification directions, Expert Systems with Applications, vol. 38, pp. 10648–10657, (2011).
3. H. J. Kim, C. Kim, Y. Choi, S. Wang, and X. Zhang, Improved modification direction methods, Computers and Mathematics with Applications, vol. 60, no. 2, pp. 319–325, (2010).
4. C. F. Lee and H. L. Chen, Adjustable prediction-based reversible data hiding, Digital Signal Processing, vol. 22, no. 6, pp. 941–953, (2012).
5. C. F. Lee, H. L. Chen, and H. K. Tso, Embedding capacity raising in reversible data hiding based on prediction of difference expansion, Journal of Systems and Software, vol. 83, no. 10, pp. 1864–1872, (2010).
6. C. F. Lee and Y. L. Huang, Reversible data hiding scheme based on dual stegano-images using orientation combinations, Telecommunication Systems, pp. 1–11, (2011).
7. Z. Ni, Y. Q. Shi, N. Ansari, and W. Su, Reversible data hiding, IEEE Trans. Circuits and Systems for Video Technology, vol. 16, no. 3, pp. 354–362, (2006).
8. J. Tian, Reversible data embedding using a difference expansion, IEEE Trans. Circuits and Systems for Video Technology, vol. 13, no. 8, pp. 890–896, (2003).
9. H. W. Tseng and C. P. Hsieh, Prediction-based reversible data hiding, Information Sciences, vol. 179, no. 14, pp. 2460–2469, (2009).
10. Zeng XT, Li Z, Ping LD, Reversible data hiding scheme using reference pixel and multi-layer embedding, AEU Int J Electron Communication, vol. 66, no. 7, pp. 532–539, (2012).
11. T.C. Lu, C.Y. Tseng, and J.H. Wu, Dual Imaging-based Reversible Hiding Technique Using LSB Matching, Signal Processing, vol. 108, pp. 77–89, (2015).
12. Lu, Tzu-Chuen, Jhih-Huei Wu, and Chun-Chih Huang. Dual-image-based reversible data hiding method using center folding strategy. Signal Processing 115, pp. 195–213, (2015).
13. S.k.Lee, Y.H Suh, Y.S. Ho, Lossless data hiding based on histogram moification of difference images, Pacific Rim conference on multimedia, Springer-Verlag, vol.3333, pp.340–347, (2004).
14. D. M. Thodi, J. J Rodriguez, Expansion embedding techniques for reversible watermarking, IEEE Translation on image processing, vol. 16, no. 3, pp. 1057–7149, March (2007).
15. C-C. Lee, H-C Wu, C-S Tsai, Y-P Chu, Lossless Steganographic scheme with Centralized Difference Expansion, Pattern Recognition, vol. 41, pp. 2097–2106, (2008).
16. D. Wu, W. Tsai, A steganographic method for images by pixel-value differencing. Pattern Recognition Letters, vol. 24, pp. 1613–1626, (2003).
17. Chang, C. C., Kieu, Duc, Chou Y. C. Reversible data hiding scheme using two steganographic images. In TENCON IEEE region 10 conference, (2007).
18. Chin-Feng Lee, Yu-Lin Huang, Reversible data hiding scheme based on dual stegano-images using orientation combinations, Telecommun Syst, vol. 52, pp. 2237–2247, (2013).
19. Lee, C. F., Wang, K. H., Chang, C. C., and Huang, Y. L., A reversible data hiding scheme based on dual steganographic images, In Proceedings of the third international conference on ubiquitous information management and communication, pp. 228–237, (2009).

20. Qin C, Chang CC, Hsu TJ, Reversible data hiding scheme based on exploiting modification direction with two steganographic images, Multimed Tools Appl, doi: 10.1007/s11042-014-1894-5, (2014).
21. University of Southern California, The USC-SIPI Image Database, http://sipi.usc.edu/database/database.php.

Design of Adaptive Filter Using Vedic Multiplier for Low Power

Ch. Pratyusha Chowdari and J. Beatrice Seventline

Abstract This paper deals with an architectural approach of designing an adaptive filter (AF) with Vedic Multiplier (VM) and is an efficient method in achieving less power consumption without altering the filter performance-called as Low Power Adaptive Filter with Vedic Multiplier (LPAFVM). AF consists a variable filter (VF) and an algorithm which updates the coefficients of filter. Generally, Filters plays the major role in effecting power in an adaptive system; Power will be significantly reduced by cancelling number of unwanted multiplications, based on the filter coefficients and amplitude of data at input. In less number of steps, VM performs multiplication. LMSA-Least Mean Square algorithm is used for designing the FIR filter. Adaptation process takes place by performing convergence of output computed by the VF to a desirable output of an LMS algorithm is used. The Xilinx ISE 14.6 is used to simulate and synthesize the proposed architecture. Power is calculated on Xpower Analyzer in Xilinx ISE suit.

Keywords Reconfigurable design · Low power filter · LMS algorithm · Adaptive filtering · Vedic multiplier

1 Introduction

The main aim of a digital Signal Processor (DSP) System is to visualize, analyze and convert the analog signal which has information to useful form of signal which can be applied to real time scenarios. There are many technical advancements happening now a days in Digital signal processing such as echo cancellation, noise cancellation, voice prediction etc [1–7]. People are always looking forward to get

Ch. Pratyusha Chowdari (✉)
ECE Department, GRIET, Hyderabad, India
e-mail: pratyushachowdarich@gmail.com

J. Beatrice Seventline
ECE Department, GITAM University, Visakhapatnam, India
e-mail: samsandra2003@gmail.com

© Springer India 2016
S.C. Satapathy et al. (eds.), *Information Systems Design and Intelligent Applications*, Advances in Intelligent Systems and Computing 434,
DOI 10.1007/978-81-322-2752-6_41

acceptable and quick solutions for the problems. So, an Adaptive filtering technique is implemented than standard DSP technique. Any way filtering is most widely used operation in DSP. A filter that extracts desired signal from noise signal, whereas adaptive filter is a system which has a linear filter that include a transfer function controlled by variable parameters and it has an optimization algorithm which adjust the parameters accordingly. Filtering and adaptation are the two processes involved in the operation of AF. Fir filter is used for normal filtering operation, and the filter requires multiplier. In the place of standard multiplier a Vedic multiplier has been used to reduce number of steps for computation and speed of operation be improved. The adaptation algorithm is used to auto-update the filter coefficients. Whole system performance and power consumption is mostly based on the functionality of filter. There by design of low power filter has become an important issue. The proposed architecture LPAFVM has high speed of operation with less power consumption as it is implemented using Adaptive filter and Vedic Multiplier.

2 Introduction and Block Diagram of Vedic Multiplier

In the early 20th century the ancient Indian mathematics system is rediscovered and it is based on 16 Sutras (principles or word-formulae). These sutras focus a light in the field DSP, i.e., a digital multiplier can be implemented using vedic multiplication algorithm. The VM architecture is implemented using the Urdhva Triyakbhyam Sutra (Vertically and Cross wise). For performing two decimal numbers multiplication in less time the Sutra was used by ancient India. The same concept is adapted to the binary number (radix 2) system, which gives an ease for designing the digital hardware.

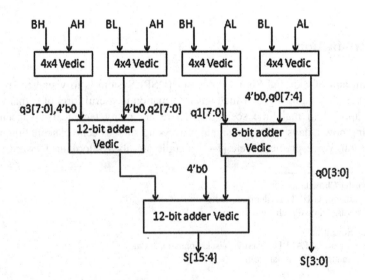

Fig. 1 Block diagram of 8 × 8 bit multiplier

VM Algorithm

1. Take two 8-bit numbers. ex: a = a7a6a5a4aa3a2a1a0 and b = b7b6b5b4b3 b2b1b0.
2. Divide two 8-bit numbers into two parts, so that each part contains two 4-bit numbers.

   ex: {AL = a3a2a1a0, AH = a7a6a5a4,
   BL = b2b3b1b0, BH = b7b6b5b4}.

3. Assign these four parts as input to the four 4 × 4 multiplier blocks as per circuit.

   a3a2a1a0 and b3b2b1b0 as inputs to the first 4 × 4 multiplier.
   a3a2a1a0 and b7b6b5b4 as inputs to the second 4 × 4 multiplier.
   a7a6a5a4 and b3b2b1b0 as inputs to the third 4 × 4 multiplier.
   a7a6a5a4 and b7b6b5b4 as inputs to the fourth 4 × 4 multiplier.

4. From the output of first 4 × 4 multiplier assign four LSB bits to the LSB bits of final result.
5. The output of second multiplier and the remaining bits of first multiplier's output by appending four zeros as MSB(to form a 8-bit number) as inputs to the 8-bit adder. i.e.,{q1[7:0],{00,q0[7:4]}}.
6. The output of 3rd and 4th multipliers are of 8-bit each, we are appending four zeros as MSB bits to 3rd multiplier and as LSB bits to 4th multiplier.
7. These two are the inputs to first 12-bit adder. i.e., {{q3[7:0],0000},{0000,q2 [7:0]}}.
8. Now the output of 1st 12-bit adder and the output of 8-bit adder are the inputs to the 2nd 12-bit adder. i.e.,{12-bit sum value,{0000,8-bit sum value}}. Assign second 12-bit adder sum value and 4 LSB bits of 1st 4 × 4 multiplier to the result. i.e., {sum value of 12-bit adder, q0[3:0]}.

To design 8 × 8 bit VM requires 4 × 4 VM and 4 × 4 bit VM requires 2 × 2 bit VM. Here in this context the block diagrams for 2 × 2 bit VM, 4 × 4 VM and 8 × 8 VM are given in the below Figures. Basic multiplier any way requires simple adders like half adder and full adder (Figs. 1, 2, 3).

3 Architecture of the Adaptive Filter

The architecture of AF is such that it has an auto-adjusting characteristic which is a property of digital filter. The impulse response of the filter is adjusted so that eliminates correlated signal from input. And sometimes the filter does not need prior knowledge on characteristics of signal and noise, but they do require desired response of the signal that is correlated. Under the non-stationary conditions the AF have the capability of tracking the signal. The block diagram of an AF is given in Fig. 4.

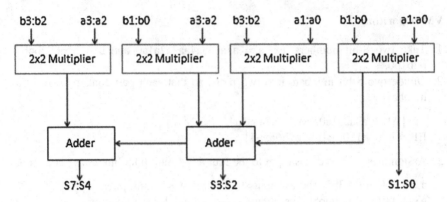

Fig. 2 Representation of 4 × 4 bit vedic multiplier

Fig. 3 Representation of
2 × 2 bit vedic multiplier

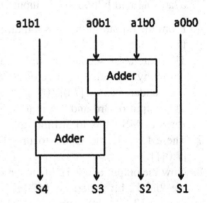

Fig. 4 General block
diagram of adaptive filter

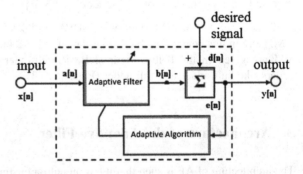

AF has three basic modules: basic filter, performance verifier and the adaptive algorithm. The filter structure specifies the amount of output from the input samples. As stability is also a measure, FIR filter is preferred than the IIR filter. Secondly, the performance verifier is chosen according to the application. There are three methods for doing performance verification are: (1) mean least squares, (2) squared error and (3) weighted least squares. The performance verifier drives

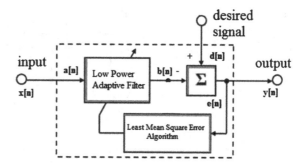

Fig. 5 Block diagram of reconfigurable adaptive filter

adaptive algorithm to improve performance. The adaptive algorithm is used to correct filter coefficient automatically.

The low power adaptive filter is given in Fig. 5. This filter is a reconfigurable FIR filter and has very low power consumption as the VM has been implemented. LMS algorithm (LMSA) is used for adaptation process. Here LMSA self-update filter coefficients based on performance verifier leads to mean square error value.

4 Architecture of Reconfigurable FIR Filter

FIR filter is a digital filter, which has a finite impulse response. FIR filters are the simplest filters and operate only on present and past. They are also called as non recursive filters. Time-invariance and Linearity are the two important properties of FIR filter.

Mathematically,

$$h(n) = \begin{cases} 0, & n \geq n_1 \\ 0, & n \geq n_2 \end{cases}$$

where the range of n_1 and n_2 is $(-\infty, \infty)$, $h(n)$ is the impulse response of digital filter, n represents discrete time index and n_1 and n_2 represents range, which is constant.

The discrete time equivalent of a continuous time differential equation is called as difference equation. The difference equation of a FIR digital filter is

$$y(n) = \sum_{r=0}^{n-1} b_r x(n - r)$$

Where the output at different values of n is represented as y(n), rth feed forward tap is denoted by br. Σ denotes the addition from r = 0 to 'n − 1', x(n − r) represents the input at filter delayed by r samples and n is the number of feed forward taps. Filter output is said to be finite, because if only single impulse is present in the

Fig. 6 Architecture of conventional FIR filter

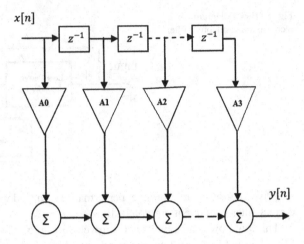

input, then output zero after a finite time. The number of filter coefficients stands for time required by the FIR filter output to become finite almost zero. The FIR filter can also be known as non-recursive filter as the filter behavior is described only in terms of past and present inputs.

The proposed FIR filter structure eliminates unwanted multiplication. By accounting filter coefficients and data samples amplitude the order of the FIR filter can be dynamically changed. Performance is described by the quantization error, obtained by doing multiplication for data sample and filter coefficient which is very small (Fig. 6).

Figure 7 represents the reconfigurable low power FIR filter. It includes three modules. Amplitude Monitor (AM), Multiplier Control Decision window (MCDW) and Self-Restraint Network (SRN). By eliminating the unwanted multiplications, the proposed method reduces the power consumption. More specifically, as switching activity is directionally proportional dynamic power can be reduced. SRN generates the required control signals.

Depending on the filter coefficients and amplitude of data samples the multipliers are automatically turned off and hence power can be saved. So the stated reconfigurable low power architecture dynamically gets updated. If the amplitude of product of filter coefficient and data sample is less than some predetermined value called as threshold, then the corresponding multiplier is turned off. The output becomes logic-1, when the amplitude data sample of incoming signal is lesser than the threshold (x_{th}) value. There is a problem occurs by using this method is that, when data samples changes rapidly for every clock cycle that leads to continuous switching of multiplier, results in increase in dynamic power. To avoid this switching activity, MCDW is used. Where in AM is used for each filter to stabilize amplitude of filter coefficient. The AM compares each filter co-efficient with the

Fig. 7 Block diagram of reconfigurable low power FIR filter

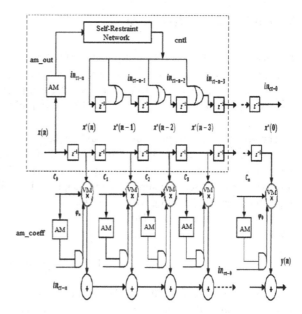

amplitude of data sample w.r.t. a predetermined threshold (c_{th}) value. If it is very small, then the output (am_coeff) becomes logic 1. MCDW is used to monitor the switching activity and reduce the frequency of occurrence. The self-restraints network is responsible for this by generating control signals.

The self-restraint network has counter internally, till the condition $[n] < x_{th}$ is fulfilled the counter counts the number of samples at input. The counter counts up for every output of the amplitude monitor(am_out) when set to logic 1. When the number of counts i.e., size of MCDW reaches a value say m, then the self-resistant network is set to logic 1. Therefore, a logic-1 value at *ctrl* signal indicates there are *m* number of small inputs that are monitored, and now the multipliers are ready to turn off.

The additional signal *inct−n* is controlled by signal *ctrl*. This additional signal accompanies with the input data to indicate that $[n] < x_{th}$. The multiplications get cancelled when the filter coefficient amplitude is lesser than c_{th}, results in adding a delay element just before the first tap to synchronize the signals *inct−n* and $x*$ (*n*). When the amplitude of the data sample at input and filter coefficient is lesser than the thresholds, the signal βn is set to logic 1 and the multiplier is turned off forcing the output to zero. The values of thresholds (x_{th}, c_{th}) are chosen such that power consumption and filter performance are impassive. If the threshold values are significantly large, it gives rise to low power consumption but at the cost of performance of the filter leading to a tradeoff between them.

5 Adaptation Algorithm

The basic adaptive filter is illustrated in Fig. 4, operating in the discrete-time domain. X(n) is the input signal and y(n) is the output signal. Consider input to the adaptive filter is a[n], output from adaptive filter is b[n] and the desired signal (noise component) is represented as d(n). The error signal can be represented as $e(n) = d(n) - b(n)$.

The adaption algorithm is used to update the filter coefficient vector w(n), with the help of error signal according to its performance criteria. Generally, this entire adaptation process aims to get minimum amount of the error signal leads to approximate the reference signal by the adaptive filter. There are many adaptation algorithms meant according to their performance.

The equation for Wiener-Hopf which is a way out for optimal filter weights is given below:

$$w = R - 1 \ P = Woptimum$$

LMS in the limit, the Wiener-Hopf equations is a recursively used stochastic approximation to the process of steepest descent. LMSA is a noisy approximation of "steepest descent algorithm" and by considering a step in the direction of negative gradient, it auto updates the coefficient vector

$$w(n+n0) = w(r) - \frac{\lambda}{2} \frac{\delta Jw}{\delta w(n)}$$

where λ, represents the step size that controls the convergence speed and stability, n0 is equal to 1 in this case. LMSA uses an intent function to find value of the gradient, that consists of an estimated value of the mean square error, $Jw = e^2(n)$ resulting $\delta Jw/\delta w(n) = -2e(n) * x(n)$.

LMSA:

For each value of n
{
y(n) = conv $(w^T(n), x(n))$
error = d(n) − y(n)
w(n + n0) = w(n) + 2λ error x(n)
}

For tracking desire output the LMSA update the filter coefficient for using the error signal (e(n)). Output of filter y(n) is measured as $w^T(n)$. The transposed form of filter coefficients $w^T(n)$, x(n) is the filter samples at input, error signal e(n) are used to update the filter coefficients and step size is represented as λ, also known as learning factor.

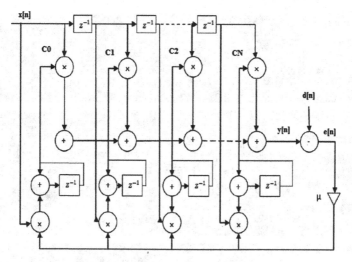

Fig. 8 Conventional LMS adaptive filter structure

General direct form of LMS adaptive filter structure is shown in Fig. 8. This structure is useful for updating filter tap weights. The equation required to update the filter tap weights is given as $w(n + n_0) = w(n) + \lambda e(n) \times (n)$.

6 Simulation Results

The analysis of Vedic Multiplier and Low Power Adaptive Filter is done with the help of Verilog, which is later simulated and synthesized on Xilinx ISE 14.6. The input data samples and the coefficients used to describe the filter are 16-bit. The results are compared against the conventional FIR filter. Verified the filters response using Modelsim also, the simulation results are shown. Figure 9 which represents the snapshots of VM, Fig. 10 shows the conventional FIR filter and Fig. 11 shows the AF. Xpower analysis in Xilinx ISE suit is used for power analysis depicted in Fig. 12. Comparison is graph is provides in Fig. 13.

7 Conclusion

The proposed LPAFVM is a combination of a reconfigurable adaptive FIR filter with LMSA. LMSA is a low complexity algorithm that updates the weights of a reconfigurable AF. The synthesis reports show significant reduction in power. The power that is used by the conventional FIR filter is 6.122 W, whereas LPAFVM consumes only 1.499 W. The LPAFVM is way above superior to the conventional FIR filter by 76 %.

422   Ch. Pratyusha Chowdari and J. Beatrice Seventline

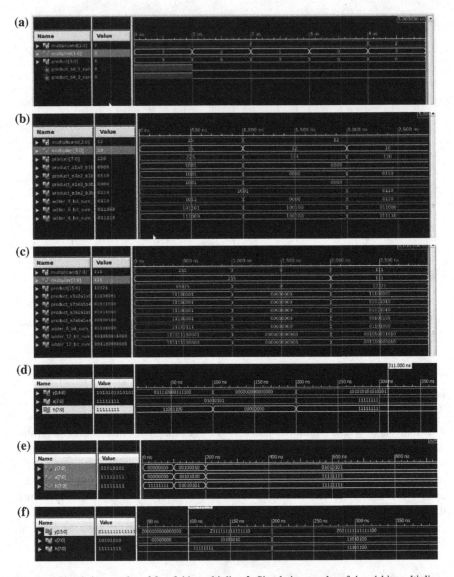

Fig. 9 **a** Simulation results of 2 × 2 bit multiplier. **b** Simulation results of 4 × 4 bit multiplier. **c** Simulation results of 8 × 8 bit multiplier. **d** Simulation results of linear convolution. **e** Simulation results of circular convolution. **f** Simulation results of correlation

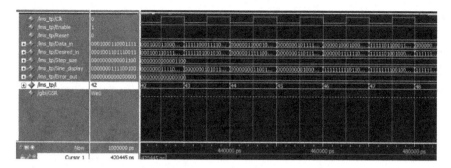

Fig. 10 Simulation results of conventional FIR filter

Fig. 11 Simulation results of low power adaptive filter

(a)

Name	Power (W)	Used	Total Available	Utilization (%)
Clocks	0.515	1	—	—
Logic	0.061	550	9312	5.9
Signals	0.468	1217	—	—
IOs	4.793	99	232	42.7
MULTs	0.002	17	20	85.0
Total Quiescent Power	0.188			
Total Dynamic Power	5.934			
Total Power	6.122			

(b)

Name	Power (W)	Used	Total Available	Utilization (%)
Clocks	0.887	1	—	—
Logic	0.070	4634	9312	49.8
Signals	0.286	7295	—	—
IOs	0.043	130	232	56.0
MULTs	0.002	20	20	100.0
Total Quiescent Power	0.105			
Total Dynamic Power	1.395			
Total Power	1.499			

Fig. 12 a Power report of conventional FIR filter. **b** Power report of low power adaptive filter

Fig. 13 Comparision graph from the report generated in Xpower analysis

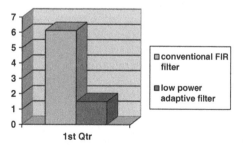

8   Future Work

The LPAFVM can be implemented by using the neural network concept. The proposed LPAFVM architecture is used in the applications like noise cancellation, echo cancellation and identification etc.

References

1. Haykin, Simon, and Bernard Widrow. *Least-mean-square adaptive filters*. Vol. 31. John Wiley & Sons, 2003.
2. Proakis, John G. "Digital signal processing: principles, algorithms, and application-3/E." (1996).
3. Tan, Li, and Jean Jiang. *Digital signal processing: fundamentals and applications*. Academic Press, 2013.
4. Kadu, Mr Pravin Y., and Ku Shubhangi Dhengre. "High Speed and Low Power FIR Filter Implementation Using Optimized Adder And Multiplier Based On Xilinx FPGA." *IORD Journal of Science & Technology E-ISSN*: 2348-0831.
5. Dhillon, Harpreet Singh, and Abhijit Mitra. "A Digital Multiplier Architecture using UrdhvaTiryakbhyam Sutra of Vedic Mathematics." *Department of Electronics and Communication Engineering, Indian Institute of Technology, Guwahati* 781 (2008): 039.
6. Kumar, G. Ganesh, and V. Charishma. "Design of high speed vedic multiplier using vedic mathematics techniques." *International Journal of Scientific and Research Publications* 2.3 (2012): 1.
7. Kumar, Ch Harish. "Implementation and analysis of power, area and delay of Array, Urdhva, Nikhilam Vedic Multipliers." *International Journal of Scientific and Research Publications, ISSN* (2013): 2250–3153.

Issues and Approaches to Design of a Range Image Face Database

Suranjan Ganguly, Debotosh Bhattacharjee and Mita Nasipuri

Abstract Development of new databases contributing much among researchers for solving many challenging tasks that might have an important role during the implementation of efficient algorithms to handle all difficulties for an automatic system. In this paper, authors have introduced the issues and approaches that have been considered during image acquisition procedure during designing of own face database. This database consists of almost all the challenges in the domain of computer vision especially face recognition. Acquisition of database's images are done in our own institute's laboratory with variations of facial actions (i.e. movement of facial units, expression), illumination, occlusion, as well as a pose. Along with the 3D face images, corresponding 2D face images have also been captured using Structured Light Scanner (SLS). Particularly, this image acquisition technique is not harmful as laser scanner does. Moreover, authors have made the visualization of practical representation of laboratory setup within this article that would again be helpful to the researchers for better understanding the image acquisition procedure in detail. In this databases, authors have accomplished the X, Y planes along with range face image and corresponding 2D image of human face.

Keywords Face recognition · Image acquisition · Structured light scanner · Database variations · Range image

S. Ganguly (✉) · D. Bhattacharjee · M. Nasipuri
Department of Computer Science and Engineering, Jadavpur University,
Kolkata, India
e-mail: suranjanganguly@gmail.com

D. Bhattacharjee
e-mail: debotoshb@hotmail.com

M. Nasipuri
e-mail: mnasipuri@cse.jdvu.ac.in

© Springer India 2016
S.C. Satapathy et al. (eds.), *Information Systems Design and Intelligent Applications*, Advances in Intelligent Systems and Computing 434,
DOI 10.1007/978-81-322-2752-6_42

1   Introduction

Face recognition that is considered to be an efficient approach for a biometric based security application continues to be the most interesting and ever increasing research area in the domain of computer vision and pattern recognition. Now along with the design and development of face recognition algorithms, many research groups are also evolving various face image database. Sometimes, the newly designed databases are focused on a particular domain specific application.

Again, based on the image acquisition procedure, captured face images could be grouped into three categories, namely: 2D visual image, 2D thermal image, and 3D image. Here, authors have briefed the development of a new 3D face database especially range face image database. 3D face images have many advantages [1] compared to other available types of the face image based on the different face recognition challenges, like expression, illumination, pose, occlusion, etc. Additionally, 3D images can be represented into four categories [2], such as raw data (i.e. point cloud, range image, etc.), surfaces (i.e. mesh etc.), solids (i.e. octree, BSP tree etc.) and high-level structures (i.e. scene graph etc.). Here, authors have focused to represent the 3D human faces from range face image. In the following section authors have explained more about the range face image.

In brief, this paper describes the detail about image acquisition, its' issues and approaches towards development of new 3D face database by structured light scanner which is being created in our institution's laboratory. Moreover, this database also consists of two multi-sensor spectral, namely: 3D face image and corresponding 2D visual image. The features from this database have been summarized below:

1. This database contains the images with intra-class variations. It has individuals from two different genders.
2. This database consists of frontal face images with different facial actions or expressions, different illumination variations, different occlusions along with pose variations. Moreover, it also consists of face images with mixture of pose and expression as well as pose and occlusion.
3. Structured lights are projected from the optical lamp from projector equipment or sensor of SLS system. The projected light is not harmful to the volunteers. Particularly, the source of the projected light is optical i.e. like fluorescent bulb in nature.
4. This database contains more than 60 range face images and the corresponding 2D images of each volunteer.
5. Along with the depth data preserved in range face image, corresponding X–Y planes are also contained in the database.

The main contribution of the authors to this paper are thus four folds:

- The database is currently in developing stage. Once it is developed, the database will be freely available to the research community.
- The step-by-step process of system calibration process is also detailed here.

- Consistent 2D-3D multi-sensor face images have been accomplished within this database.
- Moreover, X–Y planes are also contained with the database for implementing various 3D volume analysis from range images, such as curvature computation etc.

The paper has been organized as follows. Section 2 describes the background details of the research analysis. In Sect. 3, detailing about the image acquisition and database design have been presented. Conclusion with a future scope has been discussed in Sect. 4.

2 Background

Here, authors have studied about two aspects of the database design. One of them is literature study of the available 3D face databases and another one is introduction of range face image.

2.1 Database Survey

Creation of face database is an important step towards the investigation of human face images. Particularly, many databases have already been created based on the requirement of the purpose. For discussing the better representation of the strengths and limitations of the existing databases, here authors have carried out a comparative analysis with the new developing database based on the different properties. 3D RMA database [3] consists of pose variant 3D face images, each of which having approximately 4,000 points. University of York 1 [4] and University of York 2 [5] collected the 3D face images in terms of range face image with the pose, expression and occlusion variations. GavabDB database [6] contains 3D face images by accomplishing data points with pose and expression variations. Frav3D database [7] is another database where illumination, pose, expression variations could be located. Bosphorus database [8], one of the challenging database has been accomplished by pose, expression as well as occlusion variations. In BJUT-3D Chinese face database [9], the face images are of neutral expression without accessories. Casia 3D database [10] again consists of 3D face images by preserving depth data in range images. There are a pose, expression, illumination variations as well as their combination in the same database. Other this Texas 3D database [11] also contains range face images of frontal face images with varying illumination with neutral and expression. There are face images with expression and occluded 3D face images from UMB-DB [12] and expression, disgust and other variations in ND 2006 [13] databases. Moreover, expression and expression with small pose variations are comprised of BU3D FE database. FRGC v.2 database [14] also consists of varying expression face database. Other than this, little expression variations could be brought from 3D facial expression database [15]. There is also a twins database, 3D Tec database [16], from where neutral as well as smiling expression face images could be collected. Moreover, BFM database [17] also consists of 3D face images with 2,000 points that are pose and illumination variant.

In comparison with the literature study about these databases, authors' own database contains 3D facial images having various facial actions, such as: lower facial unit, upper facial unit, combination of both, different expressions, illumination changes by different light sources as well partially occluded face images by various options, like: hand, palm, glass, hair, handkerchief etc. Additionally, pose variations are also there in the database. Furthermore, this database is focused on creating range face images. Therefore, rather than 3D face images, its equivalent depth image is contained with the database along with corresponding X–Y planes. Although Texas 3D face database already belonging with range face images but unavailability of X–Y plane for each depth image makes it difficult to implement various 3D volume analysis.

2.2 Range Face Image

Range face image contains the normalized depth data from 3D image. Actually, during 3D scanning of a face image, nose region, typically nose tip or pronasal is the closest point to the scanner. Therefore, it is having minimum depth data and thus whole face image is scanned and corresponding detail data is preserved in 3D image. Now, the data is in reverse direction and to represent and visualize it, it has been normalized. Range face images are also known as 2.5D image for preserving valuable facial depth data. As because of range images does not preserve very little information like 2D intensity images and as much as information like 3D data, it is termed as in between them as 2.5D. There is another property of such images that being the closest point to the scanner, maximum depth data is always accomplished in pronasal, especially for frontal range face image. In Sect. 3.3, authors have detailed the process for creating the range face images for own database.

3 Image Acquisition and Database Design

Image acquisition is a major concern for accomplishing face images as a biometric property. Acquisition of face images by optical cameras preserve visual face image. In case of a thermal camera, thermal face images are captured. Now, for 3D face images different 3D scanners are also available. Again, among different 3D scanners, there are laser scanner as well as structured light scanner based image acquisition system. Among them laser scanner belongs with some disadvantages, such as: much higher cost, incorrect geometry from laser beam, effect of laser spikes on human face etc. Therefore, during the development of own database, authors have used DAVID-SLS 1 scanner. Structured light scanners basically deals with the digitization process of the optical light sources, especially black and white structured strip patterns, with various undergoing stages.

Here, authors are going to describe the most detailed procedure for image acquisition during the development process of the database. This stage is again divided into two sub-stages, namely: camera calibration and 3D scanning.

3.1 Camera Calibration

As the name suggest, the calibration process of the camera measures the degree of adjustment of the instruments for accurate image acquisition purpose. In SLS-1 instrument, there are particularly two instruments, like 3D scanner (DAVID 3-M camera) and a projector. This projector is a normal, ordinary projector that used for projection. Now, during SLS-1 based image acquisition procedure, it is useful to project various structured patterns during image acquisition time. Now, during the calibration process, authors have selected 'V3 Pattern' with the scale of 240 mm. Other than V3 Pattern there is 'Old Pattern', but currently the available system does not recognize it. Additionally, there are various measurements, like 120, 60 and 30 that are labeled in the vertical direction of the pattern. Based on the object size, this scale is being selected. As because authors are capturing human face image, researchers have selected 240 as scale size from the pattern. In Fig. 1, authors have depicted a DAVID SLS-1 along with V3 Pattern. There is a typical measurement of V3 Pattern. From the Fig. 1b it could be observed that there are six big circles along with other 64 small black dots in the pattern spreaded equally over two halves that form a 90° angle between them.

Now, two groups of circles on two halves, each of them having three circles are also arranged in such a way that they there is an angle of 90° between them.

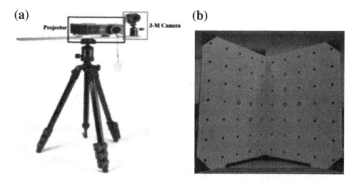

(a)   **(b)**

Projector   3-M Camera

Fig. 1 Structured light scanner with the V3 pattern. **a** 3-M camera and projector **b** V3 pattern with 70 marks from our lab

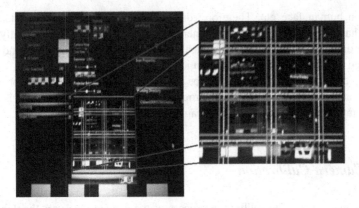

Fig. 2 Detected marks from V3 pattern after the successful calibration

Because such arrangement for successful calibration is to procure exact image by discarding other objects that might be there in and around during 3D scanning. Here, the scanning object is human face image. Now, if the human face is properly positioned between the two half, then only the face image would be scanned by discarding the background portion. It is a crucial feature for calibrated 3D scanning that is not possible for 2D visual images. There is a significant contribution of the points (black circles and dots) of the V3 Pattern for accurate calibration that leads towards the successful acquisition of face image. During the calibration process, these six circles must be appeared in the window of the 3D camera along with other dots along rows or column of the corresponding circles. Now, in this situation if the camera detects these circles and points, it will be presented by green dots in the acquisition system as illustrated in Fig. 2. These green dots represent the successful detection of various points from V3 pattern and the localized area is the window for image acquisition.

If the calibration process is successful, it will describe the detected camera and projector setup along with different parameter set up, such as vertical triangulation angle, horizontal triangulation angle, and total triangulation angle. Otherwise, it will show the error message regarding the failure of detection of measured points and model parameter. Typically during acquisition of the 3D face images for the developing database the values of vertical, horizontal and total triangulation angle are 3.90°, 9.90° and 10.65°. In general triangulation, angle determines the position of the points from the fixed baseline. Thus, the face within the calibrated window is perfectly recognized for measuring the depth details and omits the outliers. SLS is basically a optical 3D scanning system where the triangulation principle is considered for determining the dimensions and geometry structure of the object. Basically, the distance between the sensors i.e. camera and projector along with the point on the object's surface defines a triangle in spatial domain. Thus, the 3D coordinate points are calculated from the triangular relations. Other than camera calibration, it is also notable that there are 24 different structured light patterns,

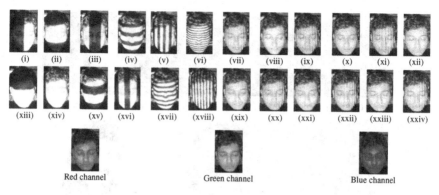

Fig. 3 Various structured patterns along with R-G-B projections

shown in Fig. 3, during image acquisition process. These projections are basically non coherent lights and it work like a video projector. Typically during the acquisition of the 3D face image for the database it is 24fps. Again, the projector also emits three colored light, such as red, green blue to capture corresponding 2D texture or visual image of the 3D face image.

Among 24 various structured patterns, there are exactly 12 horizontal and 12 vertical patterns. From Fig. 3, it is also clearly observable. In Fig. 3, an odd number of images are of vertical and remaining images are of the horizontal projection of structured light on the human face. There are certain advantages about these stripes or structured patterns during 3D scanning. The shift of each stripes from one another along with its variations in terms of width, determines the 3D coordinates. Variations in strip's width actually demonstrates the slope of the object's surface that is being considered for generating 3D coordinate points. It has further been observed that SLS based scanners take average of 12.42 s to capture a 3D face image photo for the database.

It has further been studied that there is a particular importance of such calibration for image acquisition, especially the 3D image. During the calibration process, not only the instruments are fixed with their different parameter setup but the distance between scanner and object is also fixed. Therefore, the camera will consider the human face from a particular distance and will focus on certain calibrated window. Thus, outside the calibrated window everything is just ignored. In Fig. 4, this phenomenon is illustrated. Here two type of visual face images have been considered, in Fig. 4a, visual face image from SLS-1 and in Fig. 4b visual face image from a normal digital camera with 16.1 megapixels. Both have been captured from the same distance.

Now, from Fig. 4a it is clearly visible that the background of the face image is almost neglected with proper calibration, whereas in Fig. 4b, face image along with the other background details has been captured. Additionally, the conversion of the gray image and gradually implementation of a binary image with threshold value 0.2 exhibits more about the differences between image acquisition procedure. The

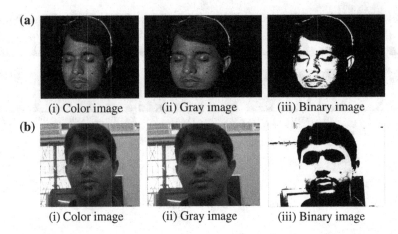

(a)

    (i) Color image   (ii) Gray image   (iii) Binary image

(b)

    (i) Color image   (ii) Gray image   (iii) Binary image

Fig. 4 The requirement of the calibration process. **a** SLS based captured 2D images. **b** Digitally captured images

RoI of the 2D face images i.e. only the face region of the input images from the database has been easily detected by the implementation of the largest connected component. However, for the normal digitized image, this mechanism fails to isolate face image. Thus, the importance of system calibration during 3D image acquisition has been observed.

But there is certain limitations of SLS. First of all, generation of valid 3D points is dependent on projection patterns and width of the structured light (i.e. strip's width). Moreover, it is also dependent on the quality of optical lamp of the projector as well wavelength of the light source. Additionally, the surfaces with transparent as well as reflectance characteristics cause difficulties for 3D scanning. As because of the projector with optical lamp source is used during 3D scanning, the stripe patterns are not clearly detectable by the camera.

3.2 Database Design and Its Characteristic

The very next step after the successful calibration process, various parameters values are setup, like: camera format, exposer, brightness of the projector, etc. Typically, during the development of the database different parameters details that have been preserved is shown in Table 1.

Additionally, there is another fixed measurement that has also been followed. There is approximately the 1.5 ft distance between the SLS and human face. In Fig. 5, authors have depicted the camera set up in the laboratory.

During the development of the database, the main focus is to capture 3D face images under varying facial actions, illumination, occlusion as well as a pose and

Instrument	Parameter name	Parameter values
Camera format	Frame rate	25.099
	Color space/compression	Y800
	Output size	1280 × 960
Projector format	Exposer	1/60 s
	Brightness	188
	Gain	34

Table 1 Different parameter set up during image acquisition procedure

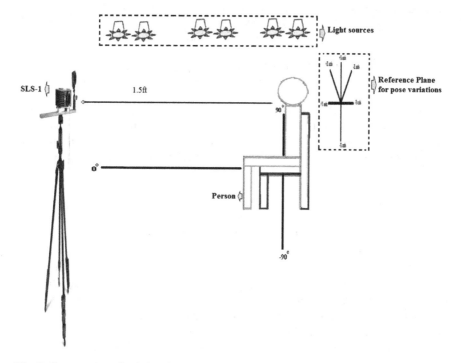

Fig. 5 Camera set up for during development of own database

mixture of them. For considering these issues, the lamps are set about 11ft height from the ground and placed over in the image acquisition environment. In Table 2, authors have described and presented the available images by categorizing with various face recognition challenges.

It has further been observed that there is no effect of variation of the external light sources in terms of illumination change in the captured 2D image. As because of the face is only illuminated with the projected light sources, there is no effect of the external light source during image acquisition. 3D images are inherently free from illumination variations due to its accomplishment of depth data rather than

Table 2 Detailing about database design

Categorization	Variations
Frontal face images	Neutral face image along with movement of lower face units i.e. mouth and lips, movement of upper face units i.e. eyebrow and forehead, combination of lower and upper face unit, various expressions i.e. sad, happiness, laugh, angry, fearful, disgusted and surprised
Illuminated face images	Three different types of illumination variations from three light sources. Again, in each variation there are three more various facial images, such as frontal positioned neutral expression, random pose change along any direction and random occlusion
Occluded face images	Partial occlusion of face images by either hand, palm, hair, glass, and handkerchief
Pose variation	Facial pose changes in each +ve and −ve directions along X, Y and Z axes
Mixture of pose, expression and occlusion	Authors have introduced new face recognition challenges with mixture of pose and expression, pose and occlusion, etc

intensity values. Moreover, 2D images procured from SLS are again free from this effect. Therefore, SLS based image acquisition technique would be effective to capture illumination invariant face images for further processing of human face images.

3.3 Range Face Image from the Database

DAVID SLS-1 preserves the 3D image in '.obj' format that contains 3D data points along with color, texture and reflection map of individual face images which have been scanned. Due to the scanning error or other acquisition factor, 3D images may contain spikes and (or) holes that causes zero depth values in the face image. Now, to remove these outliers authors have applied 3 × 3 window based max filter. Therefore, the gap with zero depth value is used to fill by neighbor's maximum depth values. This process has been repeated three times. Based on the size and shape of the outliers the number of repetitions of filtering window have been changed. In Fig. 6, range face image during preprocessing stages have been presented.

Other than subjective analysis, authors have carried out an quantitative measurement by computing number of the depth data points from face image. It has been observed that in an random selected range image, as shown in Fig. 6, consists of 7,281 valid depth data points where as in preprocessed image it has been increased to 8,299, i.e. approximately 13.9816 % depth data have been regenerated by removing spikes and holes that would be definitely significant for further processing of human faces, such as: face detection, face normalization, hybridization, feature extraction, face recognition etc.

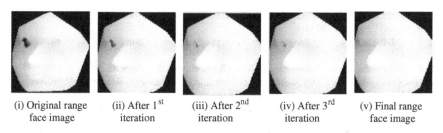

(i) Original range	(ii) After 1st	(iii) After 2nd	(iv) After 3rd	(v) Final range
face image	iteration	iteration	iteration	face image

Fig. 6 Preprocessed range face image from own database

4 Conclusion

This paper describes the design and development of the range face database in details. In this paper, authors have also explained the description of the image acquisition technique, particulars about the calibration mechanism, along with its various equipment setup as well as characteristics of the database. A survey of the database along with a comparison with the developing database has also been done in this literature. Additionally, authors have also explained the preprocessing task that have been carried out on range face images to remove spikes and holes for further processing of human faces. Further, authors have also aimed to create a synthesized range face image dataset from the developing database with common distortion, such as: Gaussian noise and Gaussian blur. The main challenges during the development of this database are the availability of volunteers, equipment set up (light, camera) and make the volunteers habituated with these variations and structured light scanner. Availability of volunteers for the development of any database is always a challenge. Additionally, the structured light scanner is very much different from other types of cameras. To cope with this mechanism is again a challenging task to the authors.

Acknowledgments Authors are thankful to a project supported by DeitY (Letter No.: 12(12)/ 2012-ESD), MCIT, Govt. of India, at Department of Computer Science and Engineering, Jadavpur University, India for providing the necessary infrastructure for this work.

References

1. Ganguly, S., Bhattacharjee, D., and Nasipuri, M.: 3D Face Recognition from Range Images Based on Curvature Analysis. In: ICTACT Journal on Image and Video Processing, Vol. 04, Issue. 03, pp. 748–753 (2014).
2. Funkhouser, T.: Overview of 3D Object Representations. Princeton University, COS 597D (2003).
3. 3D RMA face database, http://www.sic.rma.ac.be/~beumier/DB/3d_rma.html. Retrieved 26th July, 2014.
4. Univ. of York 1 database, http://www-users.cs.york.ac.uk/~tomh/3DfaceDatabase.html. Retrieved 22nd July, 2014.

5. Univ. of York 2 database, http://www-users.cs.york.ac.uk/~tomh/3DfaceDatabase.html. Retrieved 22nd July, 2014.
6. GavabDB database, http://www.gavab.etsii.urjc.es/recursos_en.html. Retrieved 16th August, 2014.
7. Frav3D database, http://archive.today/B1WeX. Retrieved 16th August, 2014.
8. Bosphorus database, http://bosphorus.ee.boun.edu.tr/Home.aspx. Retrieved 16th August, 2014.
9. BJUT-3D Chinese face database, http://www.bjut.edu.cn/sci/multimedia/mul-lab/3dface/facedatabase.htm. Retrieved 24th July, 2014.
10. CASIA-3D FaceV1, http://www.idealtest.org/dbDetailForUser.do?id=8. Retrieved 24th July, 2014.
11. Texas 3D face database, http://live.ece.utexas.edu/research/texas3dfr/. Retrieved 23rd July, 2014.
12. UMB-DB face database, http://www.ivl.disco.unimib.it/umbdb/. Retrieved 22nd July, 2014.
13. ND2006 3D face database, http://www3.nd.edu/~cvrl/CVRL/Data_Sets.html. Retrieved 22nd July, 2014.
14. FRGC v.2 face database, http://www3.nd.edu/~cvrl/CVRL/Data_Sets.html. Retrieved 22nd July, 2014.
15. 3D Facial Expression Database, http://pdf.aminer.org/000/262/410/d_geometric_databases_for_mechanical_engineering.pdf. Retrieved 16th July, 2014.
16. 3D Tec database of 3D Twins Expression, http://www3.nd.edu/~cvrl/CVRL/Data_Sets.html. Retrieved 22nd July, 2014.
17. BFM database, http://faces.cs.unibas.ch/bfm/main.php?nav=1-0&id=basel_face_model. Retrieved 1st August, 2014.

A Secure Homomorphic Routing Technique in Wireless Mesh Network (HRT for WMN)

Geetanjali Rathee and Hemraj Saini

Abstract As Wireless Mesh Network (WMN) is deliberated as a key technology in today's networking era, security during designing of such system plays a significant role. A number of techniques (i.e. PANA, LHAP) have been proposed by several researchers in order to provide the security, but leads to certain vulnerability i.e. computational overhead, network congestion and encryption/decryption timing delay. In order to overcome against such drawbacks, this manuscript gives a novel technique based on homomorphic encryption (HE). Homomorphic encryption is a technique employed to heighten the security using algebraic operations without increasing the computational overhead. The suggested technique is evaluated against encryption/decryption parameter and proves the efficiency in comparison of existing protocols.

Keywords Wireless mesh network · Homomorphic encryption · Routing protocol · Authentication · Security

1 Introduction

WMN [1], a combination of Ad hoc and mesh networks have two sorts of nodes i.e. Mesh Router (MR) and Mesh Client (MC). Mesh Routers are those through which network services are accessed while mesh clients utilizes the network services accessed from mesh routers. There exist certain benefits of WMN (i) self-configured, self-healing, self-organizing characteristic (which does not require a system administration to tell how to get the message) and multi hop nature (in which a message can be conveyed through multiple nodes). The architecture of WMN is

G. Rathee (✉) · H. Saini
Computer Science Department, Jaypee University of Information Technology,
Waknaghat, Solan, Himachal Pradesh, India
e-mail: geetanjali.rathee123@gmail.com

H. Saini
e-mail: hemraj1977@yahoo.co.in

© Springer India 2016
S.C. Satapathy et al. (eds.), *Information Systems Design and Intelligent Applications*, Advances in Intelligent Systems and Computing 434,
DOI 10.1007/978-81-322-2752-6_43

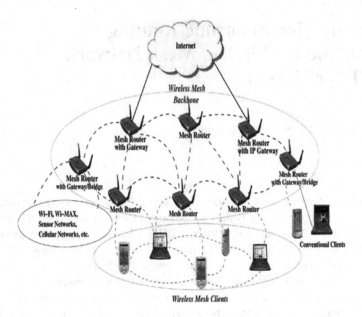

Fig. 1 Key distribution and inter cluster communication

basically divided into three types, i.e. (i) client WMN, (ii) infrastructure WMN and (iii) hybrid WMN. The depicted Fig. 1 shows the hybrid WMN architecture.

Due to the dynamic nature of WMN, security [2, 3] is taken to be an important parameter. A malicious node may get the unauthorized access to the network services by forging the legitimate node address and proving itself as a valid node. In general, security is the state of being free from hazard or an attack. Because of multi hop environment and wireless media, a number of attacks can be performed by an attacker during the transmission. In order to reduce the attacks and computational overhead, communication must be carried out with minimum number of steps. An algorithm taking less computation and communication steps may increase the security and reduces the possibility of attacks by malicious node.

The paper is divided into five sections. Section 2 discussed the related work. The proposed approach is explained in Sect. 3. The comparative study of the suggested approach is evaluated against existing technique in Sect. 4. Finally Sect. 5 concludes the paper.

2 Related Work

This section discusses various existing routing protocols [4, 5]. Researchers have proposed numerous solutions [6, 7] in order to enhance the security in WMN. A brief summary of existing approach is presented in Table 1. In Sect. 2.1, one of the existing approach i.e. TAODV routing protocol is discussed in detail.

Table 1 Comparative analysis of previous proposed protocols

Protocol	Technique	Drawback	Possible attack
PANA [8]	Authenticate clients based on IP protocols	Increase network congestion	Active attack
LHAP [9]	Authenticate hop by hop	Increased computational overhead	Spoofing attack, eavesdrop attack
AKES [10]	Polynomial based authentication	Increase communication steps	Spoofing attack
TAODV [11]	Authentication using third party participation	Asymmetric coding, third party involvement	Eavesdrop attacks

2.1 TAODV Routing Protocol

Initially, each Mesh Router (MR) and Mesh Client (MC) needs to register itself to a certificate authority (CA) from which they may get the cryptographic details and generate their own public or private keys. If source node 'S' wants to communicate with destination node 'D', Source 'S' will contact to its Authentication Server (AS). Authentication Server identifies the client by checking its certificate and issuing a ticket for the source node 'S'. After that node 'S' will send the message by attaching its ticket after encrypting with its private key. Each intermediate node will verify the certificate by detaching the previous ticket and forward the message by attaching its own ticket. As the message is reached to a destination node, it will read the message by detaching its previous node ticket.

The major drawback of this protocol is computational overhead (as each time MR or MC will contact with AS to get the certificate), length of the message is very high. In order to reduce the listed drawbacks, several techniques may be used (i.e. ID Based cryptography, homomorphic encryption, etc.). In our proposed approach, homomorphic encryption technique is considered for reducing the communication overhead during message transmission.

3 Proposed System

Homomorphic encryption [12, 13] is into and onto mapping of algebraic operations performed over plaintext. It is used during the message encryption in order to ensure the security.

3.1 Homomorphic Technique

In this manuscript, we have used a gray code converter which is used to cipher text a binary coded file. The process of cipher text generation is shown in Fig. 2. For

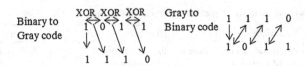

Fig. 2 Homomorphic conversion

encrypting the message, binary coded file will be converted into gray code so that the original file may not be accessible. At the destination node, a reverse process will be followed to get the original plaintext.

3.2 Homomorphic Routing Technique

The entire network is divided into several domains. Each domain has its own domain server in order to reduce the computational overhead. Domain server has a direct contact with main server for accessing the network services. Now, the proposed approach is explained by an example in order to understand its proper working.

Let us suppose a Mesh Client MC 1 wants to communicate with Mesh Client MC 2 (as depicted in Fig. 3). The message can be transmitted by applying any shortest path algorithm. The goal of this manuscript is to reduce the computation overhead.

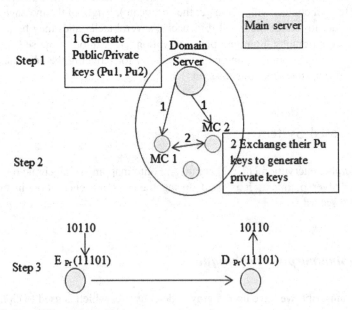

Fig. 3 Homomorphic routing technique

Initially both MC1 and MC2 will contact to its domain server in order to get their public keys. After exchanging their public keys, both the clients will generate their private keys. Further in order to guard its message from forging, MC1 will encrypt it with the GRAY code converter in which the transmitted message will be converted into binary form and then into gray code message using GRAY code converter technique. Finally, the gray code message will be sent to MC2 after encrypting with its private key. After reaching the message to the MC2, it will follow the reverse process in order to get the original message again. The steps of above text is described as follows:

1. Mesh Client MC 1 will contact with its domain server.
2. A Domain server will generate pu/pr keys for both MC 1 and MC 2 and correspondingly send the Pu keys to both the clients.
3. By getting the public key form domain server. MC 1 and MC 2 exchange their public keys to generate their private keys.
4. The message will be converted into binary form by using type cast operation (i.e. character to decimal, decimal to binary).
5. Further the binary form of the message will convert into the gray code through gray code converter.
6. Mesh Client MC 1 will send the cipher text message by attaching its private key to MC 2. The diagrammatic step of proposed approach is shown in Fig. 3.

By using the above six steps, the message M will reach to the destination node in encryption form, then destination node will follow its reverse order to get the original plaintext. As the approach has its own computational overhead because the whole text message is initially converted into binary form and then into gray code format but computational overhead is much less as compared to existing technique i.e. TAODV.

4 Performance Evaluation

The proposed approach is evaluated against encryption, decryption time parameter and proved the efficiency against previous approach.

4.1 Encryption Time

It is defined as how much time an algorithm takes to encrypt the message. The process of encryption is depicted in Fig. 4.

The depicted Fig. 5 shows that the encryption time of the proposed approach is increased at a constant rate while TADOV encryption time is increased at the exponential rate as the size of the file increases.

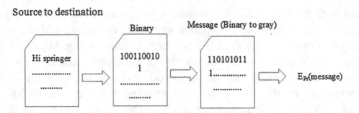

Fig. 4 Encryption process

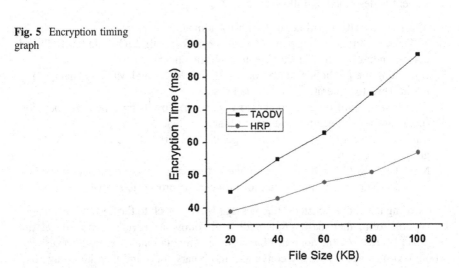

Fig. 5 Encryption timing graph

4.2 Decryption Time

It is measured at the destination node. The time taken by decryption is always less than an encryption file as there is no need to write the entire file into an array and do the typecasting. The comparisons chart of decryption is shown in Fig. 6.

4.3 Security Analysis

Let encrypted message is forged by an attacker during the transmission from source node to destination. As the message is encoded in GRAY format, even if the text is forged by an attacker, it is unaware of the format used to encrypt the file (an attacker will never be able to understand that transmitted file is an original file or encoded in some other formats); it will interpret the file in some form and will never be able to get the original text.

Fig. 6 Decryption timing graph

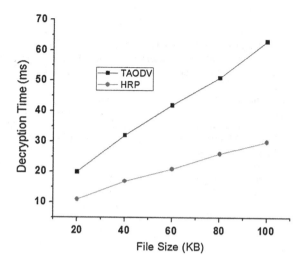

5   Conclusion

In order to enhance the security with reduced computational overhead, a new algebraic technique is encountered i.e. homomorphic encryption in routing. To prove the authenticity, the proposed approach is evaluated over encryption, decryption parameters and shows the comparison chart against computational overhead network parameter with TADOV routing protocol.

References

1. Akyildiz, Ian F., and Xudong Wang. A survey on wireless mesh networks. In; IEEE conference on Communications Magazine, 43(9); 2005.
2. A. A. Franklin and C. S. R. Murthy. An introduction to wireless mesh networks. Security in Wireless Mesh Networks (book chapter), CRC Press, USA; 2007.
3. Ben Salem, N. & Hubaux, J.-P.: Securing Wireless Mesh Networks. In:, IEEE Wireless Communication, 13(2), pp. 50–55, 2006.
4. Campista, Miguel Elias M., et al. "Routing metrics and protocols for wireless mesh networks." Network, IEEE 22.1 (2008): 6–12.
5. Siddiqui, Muhammad Shoaib, and Choong Seon Hong. "Security issues in wireless mesh networks." Multimedia and Ubiquitous Engineering, 2007. MUE'07. International Conference on. IEEE, 2007.
6. J. Sen, N. Funabiki et al. Secure routing in wireless mesh networks. Wireless Mesh Networks (book chapter), INTECH, Croatia; 2011.
7. Cheikhrouhou, O.; Maknavicius, M. & Chaouchi, H.: Security Architecture in a Multi-Hop Mesh Network. In:, Proceedings of the 5th Conference on Security Architecture Research (SAR), Seignosse-Landes, 2006.
8. Parthasarathy, Mohan. "Protocol for carrying authentication and network access (PANA) threat analysis and security requirements." 2005.

9. Zhu, Sencun, et al. "LHAP: a lightweight hop-by-hop authentication protocol for ad-hoc networks." Distributed Computing Systems Workshops, 2003. Proceedings. 23rd International Conference on. IEEE, 2003.
10. He, Bing, et al. "An efficient authenticated key establishment scheme for wireless mesh networks." Global Telecommunications Conference (GLOBECOM 2010), 2010 IEEE. IEEE, 2010.
11. Uddin, Mueen, et al. "Improving performance of mobile Ad Hoc networks using efficient tactical On demand distance vector (TAODV) routing algorithm." International Journal of Innovative Computing, Information and Control (IJICIC), pp. 4375–4389, 8(6) 2012.
12. Ogburn, Monique, Claude Turner, and Pushkar Dahal. "Homomorphic encryption." Procedia Computer Science 20, pp. 502–509, 2013.
13. Singh, Vineet Kumar, and Maitreyee Dutta. "Secure Cloud Network using Partial Homomorphic Algorithms." International Journal of Advanced Research in Computer Science 5.5, 2014.

Fuzzy Based Fault Location Estimation During Unearthed Open Conductor Faults in Double Circuit Transmission Line

Aleena Swetapadma and Anamika Yadav

Abstract In this paper fuzzy inference system (FIS) has been designed for faulty phase identification and fault location estimation during unearthed open conductor faults in double circuit transmission line. Inputs given to the proposed relay are the fundamental component of current of both circuits of double circuit line measured at one end of the line only. Output of the proposed method will be in terms of fault location in each phase circuit-1: LA1, LB1, LC1 and circuit-2: LA2, LB2 and LC2 in time domain from which fault location and faulty phase(s) will be estimated. Proposed method is tested for varying fault type, fault location, fault inception angle, fault resistance and power flow angle. Faulty phase identification accuracy is very high and percentage error in fault location estimation lies within 1 % for large number of test fault cases.

Keywords Open conductor faults · Fundamental component of current · Fuzzy logic · Fault location estimation · Fault phase identification

1 Introduction

Power system relaying involves monitoring of real time power system information; power systems fault analysis and removal, maintenance of components of power system and repair if any damage occurs. For fault analysis SCADA (supervisory control and data acquisition system) and DFR (digital fault recorder system) are used. Power transmission networks are of large physical dimension and consist of complex components. Different types of fault that occur in transmission lines are

A. Swetapadma (✉) · A. Yadav
Department of Electrical Engineering, National Institute of Technology,
Raipur 492010, C.G., India
e-mail: aleena.swetapadma@gmail.com

A. Yadav
e-mail: ayadav.ele@nitrr.ac.in

© Springer India 2016
S.C. Satapathy et al. (eds.), *Information Systems Design and Intelligent Applications*, Advances in Intelligent Systems and Computing 434,
DOI 10.1007/978-81-322-2752-6_44

classified into two category viz. shunt faults (short circuit faults) and series faults (open conductor faults). Unlike shunt faults or short circuit faults which are characterised by substantial increase in current and decrease in voltage profiles, the series faults or open conductor faults are characterised by low magnitude of current and almost no change in voltage profiles. Thus the schemes which are developed for shunt faults are not appropriate for detection, classification and location of open conductor fault. In particular the distance protection scheme is not able to locate the open conductor faults as there is very low magnitude of current flow which leads the impedance seen by the relay much higher than the protective zone impedance. A lot of research work has been reported on protection of transmission line against shunt faults using different soft computing techniques like ANN [1, 2], Fuzzy [3–5], combined wavelet-fuzzy [6], wavelet-ANN-fuzzy [7] and wavelet-SVM [8]. Recently protection against shunt faults using artificial neural network have been addressed by authors in [2]. However there is little research done considering series faults or open conductor faults.

An open phase conductor detector system is described consisting of one or more transmitters and a receiver [9]. The transmitter(s) detects the open phase conductor by monitoring the phase conductor voltage using redundant inputs. Broken conductor protection schemes are also described using carrier communication [10]. A method for open conductor fault calculation in four parallel transmission lines using twelve sequence component methods is discussed in [11]. ANN based technique have been used in [12] for enhancement of distance relay performance against open-conductor fault in HV transmission lines. Digital distance relaying scheme which takes care of a simultaneous open conductor and ground fault occurring co-incidentally on the same phase at the same point on a series-compensated double-circuit line is proposed in [13].

In this paper, a method is proposed which identifies the open conductor faulty phase and locates the fault distance using fuzzy inference system in time domain. Section 2 describe the procedure of designing of fuzzy inference system for fault phase identification and fault location estimation, Sect. 3 shows the results of the proposed method, Sect. 4 summarizes the advantages of the proposed method and Sect. 5 concludes the findings of the proposed research work.

2 Design of Fuzzy Based Fault Phase Identification and Location Estimation Method

Fuzzy based faulty phase and location estimation during open conductor faults has been designed as per flow diagram shown in Fig. 1.

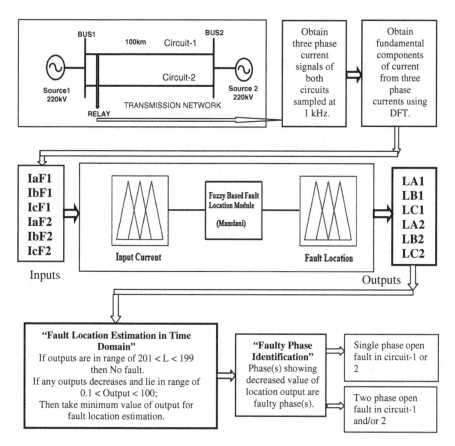

Fig. 1 Flow diagram of the proposed fuzzy based fault phase identification and fault location method

2.1 Power System Network

Transmission line network is shown in Fig. 1 consists of double circuit line of 100 km length fed from 220 kV, 50 Hz sources at both the ends. The double circuit line is modelled considering the zero sequence mutual coupling effect [14] using distributed parameter line block of Simpowersystem library of Matlab 2012 soft ware version. Various open conductor fault simulation studies have been carried out considering variation in different fault parameters such as fault type, fault location and fault inception angle.

2.2 Signal Processing

Following an open conductor fault, the three phase currents undergo changes i.e. the magnitude decreases to very low value while the voltage remains same [14]. Thus only the three phase currents are utilized for detection of open conductor faults. The three phase current signals of both the circuit-1 and 2 of the transmission network shown in Fig. 1 are measured at one end of the double circuit line and sampled at a sampling frequency of 1 kHz.

To detect open conductor fault, signal processing is necessary. By employing the discrete Fourier transform (DFT) block of simulink toolbox of MATLAB, the fundamental components of three phase currents signals of ckt-1 (IaF1, IbF1, IcF1) and 2 (IaF2, IbF2, IcF2) are obtained in time domain. The DFT estimates the fundamental values of current at discrete interval of time depending upon the sampling frequency for the complete simulation time. With each new sample available the DFT block takes one previous cycle samples to calculate the next sample.

2.3 Design of Fuzzy Inference System

Mamdani type fuzzy inference system (FIS) [15] is used in this work which identifies the faulty phase(s) and estimate the location of faults. Inputs given to the FIS are the fundamental component of current signal only measured at one end of the double circuit line. One fuzzy inference system (FIS) is designed of one phase say "A1" of circuit-1 which takes corresponding fundamental phase current (IA1) as input and gives the fault location of A1 phase "LA1" as output. Six outputs of different phases are represented as LA1, LB1, LC1, LA2, LB2 and LC2. This FIS developed for one phase; works for other phases also but one thing is to make certain that the input given to other phase FIS is it's corresponding fundamental phase current. The input variable (magnitude of fundamental current) is divided into 50 ranges using triangular member function corresponding to different fault locations in step of 2 km. Similarly the output also contains 50 ranges of triangular member function. For each 2 km one rule is designed, and one rule for no-fault situation, thus total fifty one rules have been formed for faulty phase identification and fault location estimation task. Then the outputs obtained for each rule are combined into a single fuzzy set, using a fuzzy aggregation operator which in is maximum in this case. The fuzzy set is then transformed into a single numerical value using the 'centroid' defuzzification method [16] which returns the centre of the area under the fuzzy set.

2.3.1 Fault Location Estimation

The FIS developed takes the fundamental current of each phase as input. The total number of outputs are six representing the fault location LA1, LB1, LC1, LA2, LB2 and LC2 in time domain of each phases of circuit-1: A1, B1, C1 and circuit-2: A2, B2 and C2 as shown in Fig. 1. The output of FIS with respect to time should be '199–201' if there is no fault (representing the fault location value outside the protection zone boundary). On the other hand if there is a fault in any phase; the FIS gives the 'Location of fault' as output in corresponding faulty phase which lies between range 0.1 < location < 100. As the fault location is obtained for each phase, the presence as well as the type of open conductor fault at different locations can be easily identified.

2.3.2 Faulty Phase Identification

From the time domain fault location outputs of the developed FIS, faulty phases can also be identified as shown in Fig. 1. During no fault; the fault location value of all the phases will show value between 199 and 201 km. Following a fault; the fault location output of corresponding faulty phase(s) decreases and get settled to a value which lies between 0.1 and 100 km. The output phase(s) whose fault location value lies between 0.1 and 100 km is/are faulty phase(s). If only one phase fault location value decreases then it is single phase open conductor fault, while if two phase fault location values decreases at same time then it is two phase open conductor fault. Further as all the six fault location outputs are in time domain the response time to detect, locate and classify the open conductor fault can be calculated.

3 Results and Discussions

The proposed fuzzy based open conductor faulty phase identification and fault location method is tested for various fault cases. The test fault cases considers the variations of different parameters: fault type, fault location (including fault near boundaries), fault inception angle. Also the effect of power swing, noise and variation in power transfer angle has been investigated and results are discussed in detail in the following. The percentage error in estimated fault location is calculated using (1).

$$\%\text{Error} = [(\text{Actual Location} - \text{Estimated location})/\text{Line length}] * 100 \quad (1)$$

3.1 Variation in Fault Type

Different types of open conductor fault cases have been simulated and tested to check the performance of the proposed scheme. Types of open conductor fault considered are single phase open and two phase open faults. Figure 2a shows the three phase currents of circuit 1 during one phase open conductor fault at 45 km in phase B1 of circuit 1 at 60 ms time. Figure 2b shows the corresponding fundamental component of currents of circuit 1. Figure 2c shows the outputs of fuzzy based faulty phase identification and fault location estimation module. Up to 60 ms time; all the outputs shows '201' indicating that there is no fault in the system. After 60 ms; the healthy phases outputs remains constant except the B1 phase output fault location "LB1" starts decreasing and settles down to 44.95 km value after 88 ms time. Thus B1 phase is identified as faulty phase and the fault type is one phase open conductor fault. Response time of the proposed method is obtained by subtracting $88 - 60 = 28$ ms. Error in estimated fault location is calculated according to (1) which comes out to be 0.7 %. Further test results of different open conductor faults in terms of response time and %error in fault location estimation are shown in Table 1. Estimated locations of different one conductor open and two conductor open faults are given. %Error of the tested fault cases are within 1 % for all the tested fault cases. Time required to locate these faults are given as response time in

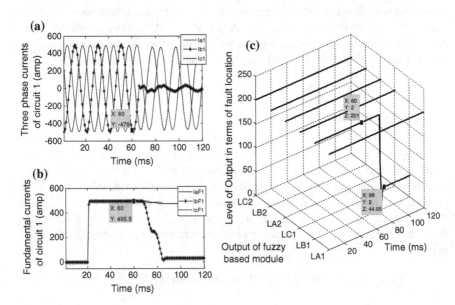

Fig. 2 During one phase open conductor fault in phase B1 of circuit 1 at 45 km at 60 ms time. **a** Instantaneous three phase current signals of circuit 1. **b** Fundamental component of currents of circuit 1. **c** Outputs of fuzzy based fault phase identification and location module

Table 1 Fault location estimation and % error analysis for different open conductor faults

Fault type	Actual fault location (km)	Estimated fault location (km)	%Error	Response time (ms)
A1	15	15.02	−0.02	30
B2	21	21.83	−0.83	27
C1	25	24.84	0.16	24
A1 and B1	31	31.02	−0.02	27
B2 and C2	35	35.25	−0.25	24
C1 and A1	41	41.04	−0.04	24
A2 and B2	45	44.98	0.02	24
A1 and B2	51	51.21	−0.21	28
A1 and C2	55	55.34	−0.34	24
B1 and A2	61	61.12	−0.12	27
B1 and C2	65	64.91	0.09	25
C1 and A2	75	74.85	0.15	25
C1 and B2	85	84.92	0.08	25

Table 1. It is clear from this Table 1 that, proposed scheme correctly identifies the faulty phase and locate the fault accurately with very low % error.

3.1.1 Open Conductor Faults Occurring in Different Phases of Different Circuits at Same Location

In this section, two phase open conductor faults that occur in same location in different phases of different circuits at same time are considered and test results are given in Table 1. The maximum time taken by the two phase open fault to settle down to the final fault location value is found to be 28 ms as per Table 1.

3.1.2 Multi-location Faults in Different Phases at Same Time

Open conductor faults that occur in different locations in different phases at same time are also known as multi-location faults. In this section, multi-location fault occurring in different phases at same time are discussed. Different multi-location faults are tested and some results are given in Table 2 in terms of estimated location and time required to estimate the location. Percentage error in location is within 1 % for all the tested fault cases. Therefore it can be concluded that the proposed fuzzy based scheme can accurately detect, classify and locate the multi-location open conductor faults.

Table 2 Response time, fault location estimation and % error analysis for open conductor faults

Fault 1 (km)	Estimated fault location (km)	Fault 2 (km)	Estimated fault location (km)	Response time (ms)
A1 at 11	10.75	B1 at 39	38.48	27
B2 at 36	35.83	C2 at 67	66.67	24
C1 at 47	46.38	A1 at 83	86.75	23
A1 at 56	55.31	C1 at 85	84.51	24
B2 at 78	77.44	A2 at 89	88.29	24
C1 at 83	82.77	B1 at 93	92.47	24

3.2 Performance in Case of Change in Power Transfer Angle

Proposed Fuzzy based method is tested for different open conductor fault at varying locations with power transfer angle 45° and −45°. Test results for different power transfer/flow angles are given in Table 3. Response time is within 30 ms for all the tested fault cases and % error in fault location is within 1 %. Figure 3a, b shows the outputs of fuzzy based scheme during different power transfer angle 45° and −45° during A1 fault at 81 km at 60 ms time. The response time is 30 ms and fault location error is 0.888 % during 45° power transfer angle. Figure 3b shows outputs of fuzzy based scheme for 45° power flow angle. Response time is 24 ms in this case and fault location error is 0.173 %.

Table 3 Response time, fault location estimation and % error analysis for different power flow angle

Power flow angle (°)	Actual fault location (km)	Estimated fault location (km)	%Error	Response time (ms)
+45°	11	10.83	0.17	30
	31	30.77	0.23	29
	51	50.48	0.52	29
	71	70.52	0.48	30
	91	90.76	0.24	30
−45°	11	10.62	0.38	24
	31	30.95	0.05	23
	51	50.94	0.06	24
	71	70.93	0.07	24
	91	91.21	−0.21	23

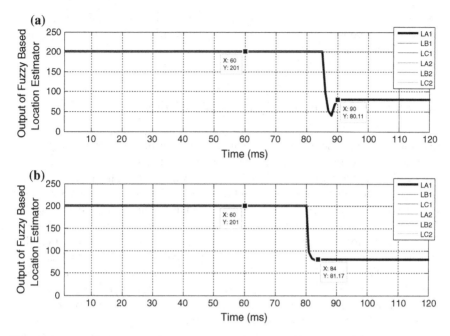

Fig. 3 Outputs of fuzzy based faulty phase identification and location module during open conductor fault occur at 81 km in phase A1 at 60 ms tim, **a** with 45° power flow angle, **b** with −45° power flow angle

3.3 Performance in Varying Fault Inception Angle

Proposed method is tested for different fault inception angle; as fault may occur at any instant of time. Some of the test results for different inception angle with $\phi = 0°$, 45°, 90°, 180°, 225°, 270°, 315° and 360° are shown in Table 4. From table it can be observed that the error in fault location estimation is within 0.2 %. Response

Table 4 Response time, fault location estimation and % error analysis for different phase faults

Fault inception angle (°)	Actual fault location (km)	Estimated fault location (km)	%Error	Response time (ms)
0	7	6.12	0.88	23
45	17	16.3	0.70	30
90	27	26.37	0.63	29
135	37	36.54	0.46	26.5
180	47	46.71	0.29	25
225	57	56.90	0.10	29.5
270	67	67.17	−0.17	28
315	77	77.20	−0.20	26.5
360	87	87.15	−0.15	24

time to locate the faults is within 30 ms time for all the tested fault cases. From Table 4 it can be observed that performance of the relay is not affected by variation in fault inception angle.

3.4 Overall Performance of Proposed Method

Performance of the proposed method evaluated in terms of accuracy in faulty phase identification. Accuracy of faulty phase identification is 99.99 % for all the tested fault cases. Location error distribution is also analysed to check the performance of the proposed method which is shown in Fig. 4a for all fault cases. Percentage errors are divided into ranges and for each range, percentage of fault cases are calculated. Fault location error is in the range of ±0.01 for 71 % cases, ±0.3 for 17 %, ±0.6 for 10 % and ±0.9 for 2 % cases. It can be seen that for most of the fault cases, the estimated location error lies below 0.01 %. Response time for different fault cases is also analysed and plotted in Fig. 4b. Response time for all the tested fault cases are within 30 ms time (1 and ½ cycles). Advantage of using fuzzy logic to estimate the location of open conductor faults over ANN [1, 2] is that it does not require extra training module. For ANN based scheme, fault classification is pre-requisite criteria to locate the faults and for each type one location module is required. But in fuzzy based scheme simultaneously faulty phases are identified and fault locations are estimated.

4 Advantages

 i. Fault classification or faulty phase identification is not required for beginning the proposed fault location estimation algorithm.
 ii. Optimization of resources—Fault phase identification and fault location estimation from a single fuzzy module in time domain.

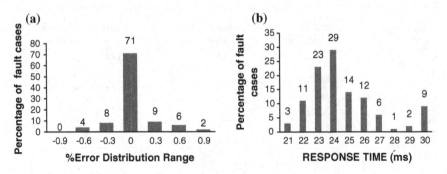

Fig. 4 a Overall %error range of total fault cases tested and **b** response time for different fault cases

iii. Less computation complexity—No training phase in fuzzy inference system, so less complex computation work as compared to other training based method like ANN, SVM and DT etc.
iv. High accuracy of fault phase identification.
 v. Very less percentage error in fault location estimation—within 1 % for all the cases.
vi. Response Time is within 1 and ½ cycles.
vii. Improved Reach Setting—Proposed method can provide protection up to 98 % of line length.

5 Conclusion

Open conductor faults are not discussed by many researchers as occurrence of this type of fault is rare. So in this work, fuzzy based faulty phase identification and fault location estimation technique is proposed for open conductor faults in double circuit line. Inputs to the fuzzy network are the fundamental component of currents of both the circuits. Proposed fuzzy based open conductor faulty phase identification and fault location estimation scheme is tested for wide variation in fault parameters including multi-location faults, pre-fault power flow angle. Total numbers of fault cases tested are 7630. The proposed faulty phase identification scheme has accuracy of 99.99 %. The percentage error in fault location estimation lies within 1 % for all the tested fault cases. The proposed method determines the fault location without knowing the fault phase using single terminal data; this is additional advantage over other techniques which requires fault type to be known first. Response time for the proposed method is within 30 ms for all the fault cases.

References

1. Koley, E., Yadav, A. and Thoke, A. S.: A new single-ended artificial neural network-based protection scheme for shunt faults in six-phase transmission line. Int. Trans. on Elec. Energy Systems, (2014).
2. Yadav, A., Swetapadma, A.: Improved first zone reach setting of ANN based directional relay for protection of double circuit transmission lines. IET Gen. Trans. Dist. 8, 373–388 (2014).
3. Samantaray, S.R., Dash, P.K., Panda, G.: A combined S-transform and fuzzy expert system for phase selection in digital relaying. Euro. Trans. on Elec. Power 18, 448–460 (2008).
4. Saradarzadeh, M., Sanaye-Pasand, M.: An accurate fuzzy logic-based fault classification algorithm using voltage and current phase sequence components. Int. Trans. on Elec. Energy Systems, doi:10.1002/etep. 1960, (2014).
5. Mansour, M.M., Wahab, M.A.A., Soliman, W.M.: Fault diagnosis system for large power generation station and its transmission lines based on fuzzy relations. Int. Trans. on Elec. Energy Systems, doi:10.1002/etep.1782, (2014).
6. Reddy, M.J., Mohanta, D.K.: A wavelet-fuzzy combined approach for classification and location of transmission line faults. Int. J. Elect. Power Energy Syst. 29, 669–678 (2007).

7. Jung, C.K., Kim, K.H., Lee, J.B., Klockl, B.: Wavelet and neuro-fuzzy based fault location for combined transmission system. Int. J. Electr. Power Energy Syst. 29, 445–454 (2007).
8. Yusuff, A.A., Jimoh, A.A., Munda, J.L.: Fault location in transmission lines based on stationary wavelet transform, determinant function feature and support vector regression. Elec. Po. Sys. Res. 110, 73–83 (2007).
9. Westrom, A.C., Sakis Meliopoulos, A.P., Cokkinides, G.J., Ayoub A.H.: Open Conductor Detector System. IEEE Trans. on Power Del. 7, 1643–1651 (1992).
10. Senger, E.C.: Broken Conductor Protection System Using Carrier Communication. IEEE Trans. on Power Del. 15, 525–530 (2000).
11. Xu, P., Wang, G., Li, H., Liang, Y., Zhang, P.: A New Method for Open Conductors Fault Calculation of Four-parallel Transmission Lines. Po. and En. Eng. Con. Asia-Pacific 1–4 (2010).
12. Gilany, M., Al-Kandari, A., Hassan, B.: ANN based technique for enhancement of distance relay performance against open-conductor in HV Transmission Lines. Comp. and Auto. Eng. 5, 50–54 (2010).
13. Makwana, V.H., Bhalja, B.R.: A new digital distance relaying scheme for series-compensated double-circuit line during open conductor and ground fault. IEEE Trans. on Power Del. 27, 910–917 (2012).
14. Anderson, P.M.: Power System Protection. IEEE Press: New York, (1999).
15. Mamdani, E.H.: Applications of fuzzy logic to approximate reasoning using linguistic synthesis. IEEE Trans. on Comp. 26, 1182–1191 (1977).
16. MATLAB® 2012, "Fuzzy Logic Toolbox".

Optimization in Round Robin Process Scheduling Algorithm

Anurag Upadhyay and Hitesh Hasija

Abstract Round Robin (RR) scheduling algorithm certainly is one of the most popular algorithms. In this algorithm, a static time quantum is given to each process. However it suffers from certain problems which are mainly related to the size of time quantum. Larger the time quantum, larger is the response and waiting time of processes. Similarly if the time quantum is too small then the overhead of CPU increases because CPU has to perform greater number of context switches. This paper focuses on the optimization techniques in Round Robin algorithm. Several algorithms have been proposed which use a dynamic time quantum, rather than a static one. The concept of mean, median, dispersion and others are used to calculate time quantum for processes in ready queue based on their remaining burst time. An approach based on multiple time quanta has also been proposed in this paper. Finally it has been shown through implementation and results that these algorithms are able to solve the problems of conventional Round Robin algorithm. A better turnaround time, response time and waiting time has been achieved through the implementation of these algorithms.

Keywords Round robin algorithm · Dynamic and multiple time quantum · Means · Median · Dispersion

1 Introduction

Process is defined as a program in execution [1]. In the modern time sharing operating systems, process is often considered as a unit of work. Based on the type of processes, the task of operating system is twofold [2]. First it must facilitate the

A. Upadhyay (✉) · H. Hasija
Delhi Technological University, Delhi, India
e-mail: anuragupadhyay35@gmail.com

H. Hasija
e-mail: hitoo.hasija@gmail.com

© Springer India 2016
S.C. Satapathy et al. (eds.), *Information Systems Design and Intelligent Applications*, Advances in Intelligent Systems and Computing 434,
DOI 10.1007/978-81-322-2752-6_45

457

working of system processes which execute system code. At the same time, an operating system must also expedite execution of user level processes. Theoretically all these processes can be executed at the same time. However practically, it's not possible due to resource crunch. So a CPU tries to switch between various processes based on the given policy and working environment. CPU executes a process until it requests for an I/O operation. This causes the process to go into waiting condition. During this time, rather than sitting idle, operating system directs CPU to undertake another process. This pattern goes on. The selection of a process to execute depends on the scheduling algorithm. The short term scheduler (also called CPU scheduler) selects a process from ready queue and gives it to the CPU to execute [2]. Various scheduling algorithms are available. Among these algorithms, Round robin is used quite extensively by various operating systems. In this algorithm, a fixed amount of time slice, called time quantum, is assigned to each process in the ready queue in a circular fashion. One major advantage of round robin is that it treats all processes as equal which is similar to FCFS. However unlike FCFS, it has relatively lesser average waiting time. It is especially important if a particular mix of processes contains both very long burst processes and several short burst processes.

Even though Round Robin scheduling algorithm remains one of the most popular CPU scheduling algorithms, it encounters several problems [3] which mostly pertain to the size of the time quantum. The dilemma is to choose an optimal size for time quantum because the performance of round robin algorithm, by and large, depends on the chosen time quantum size. If the chosen time quantum is too large in size, then there is a great chance that the waiting and response time of processes will also become large. On the other hand, if the chosen time quantum is very small in size, then CPU will undergo a large number of context switches, which will substantially increase the CPU overhead. These problems give us a fairly good intuition that solution lies in having a dynamic time quantum, rather than a static one. Thus motivation for this paper is to rectify these problems of round robin algorithm by using measures like mean, median and dispersion along with use of multiple time quanta. Another motivation for this paper has come from the fact that good number of research papers have dwelt into these problems. This large literature thus provides opportunities for both learning as well as for exploring and innovating new solutions. The proposed Algorithms are implemented and compared with the round robin algorithm and its variants. It has been shown that proposed algorithms perform better than existing algorithms in terms of waiting and turnaround time.

The whole paper has been organized into five sections, including the Sect. 1. The rest of the sections focus on following aspects: Sect. 2 is Literature survey. Section 3 contains the various proposed algorithms and approaches which have been proposed in this paper. Section 4 provides the results from implemented algorithms. Section 5 provides the conclusion and other remarks, as well as future work.

2   Literature Survey

Samih M. Mostafa, S.Z. Rida and Safwat H. Hamad have used integer programming approach so as to improve round-robin-algorithm [4]. Their aim has been to minimize the context switches in the round-robin-algorithm by selecting a not too big and not too small time quantum. A similar approach has also been used by Praveen Kumar, T. Sreenivasula Reddy and A. Yugandhar Reddy [5]. Mahesh Kumar MR, Renuka Rajendra B, Sreenatha M and Niranjan CK have used linear programming approach to solve the problem of dynamic time quantum [6]. They have shown that their approach is successful in minimizing the context switches and performs better than conventional round-robin-scheduling algorithm. Shahram Saeidi and Hakimeh Alemi Baktash have developed a non-linear-mathematical-model to improve the round-robin-algorithm [7]. Their aims have been to minimize the average waiting time. Helmy and Dekdouk have developed a proportional-sharing scheduling algorithm [8]. It is a variant of round-robin-algorithm which gives more preference to shorter jobs. Thus to some extent it resembles as a combination of Shortest-job-first and round-robin. The merit of this method is that the processes which are about to get completed, will be removed from the ready queue quickly. The average waiting time is reduced and throughput is increased. Bisht, Ahad and Sharma have also used a similar approach [9]. Their focus is also on completing the execution of those processes which are near their completion. Behera, Mohanty and Nayak have proposed an approach which uses calculation of median value for determining optimal size of time quantum [10]. They have named it as Dynamic-Quantum-with-Readjusted-Round-Robin (DQRRR) algorithm. Alam, Doja and Biswas have used fuzzy inference system and fuzzy logic to determine a good value for time quantum [11]. Rajput and Gupta have tried to integrate priority-scheduling-algorithm in the existing algorithms [12]. Each process is assigned a specific priority. Based on the assigned priority, they are taken up by round robin scheduling algorithm. Thus their approach is not based on dynamic time quantum, but rather focusses on appropriate ordering of processes. Noon, Kalakech and Kadry have given a new algorithm called AN algorithm [13]. This algorithm calculates size of time quantum by using mean value, based on the remaining burst time of processes. Siregar has used genetic algorithm to improve round-robin-scheduling algorithm [14]. He uses inverse of average waiting time as a good fitness function so as to determine best time quantum size. Roulette wheel method, one point crossover method and flip mutation have been used by the authors. Trivedi and Sajja have used neuro fuzzy approach to optimize round-robin in a multitasking environment [15]. Their algorithm was successful in increasing throughput and decreasing the average waiting time. Yaashuwanth and Ramesh have coined the term 'intelligent time slice' for round-robin-algorithm for scheduling tasks in real time operating systems [16]. The time slice for each process is calculated based on priority, shortest burst time and time to avoid a context switch.

3 Proposed Algorithms

3.1 Round Robin Using Knowledge of Mean Median and Dispersion

The basic idea of this modified round-robin approach is that the value of time quantum is repeatedly changed based on the value of remaining time of processes in the ready queue. Here time quantum is calculated at the beginning of each cycle in the following way

$$TQ = (\text{Mean} + \text{Median} + \text{Min value} + \text{Max values})/4$$

3.2 Modified Round-Robin Using Arithmetic and Harmonic Mean

This algorithm is also very similar to above algorithm as it too calculates the size of time quantum dynamically. Here at every cycle of round-robin, we calculate the arithmetic mean and the harmonic mean of the processes based on their remaining times. Thus we obtain a new time quantum in every cycle as follows

$$TQ = (\text{AM} + \text{HM})/2$$

where AM = Arithmetic mean and HM = harmonic mean.

3.3 Incremental Round-Robin Algorithm

The basic idea of this algorithm is to start with a very low value for the size of time quantum. Initially it can be set as 1 or it can also be set equal to the size of burst time of the smallest process in the ready queue. This time quantum is applied to each process in the ready queue. In every next cycle the size of time quantum is incremented by one. However once it reaches a threshold maximum value, the size is again set to the initial value.

3.4 Multiple Time Quanta Round-Robin Algorithm

The basic idea here is to use more than one time quantum for different process. The basic intuition is that we should use relatively larger time quantum for smaller processes so that they end up quickly. On the other hand, we should use relatively smaller time quantum for longer processes, so as to make them gradually a process with smaller remaining burst time. Once it becomes sufficiently small, it can be ended quickly by using relatively bigger time quantum used in first case. The two time quanta are calculated as follows

$$TQ1 = (\text{Mean} + \text{Median} + \text{Min\_Val} + \text{Max\_Val})/4$$
$$TQ2 = (\text{Mean} + \text{Median} + \text{Min\_Val} + \text{Max\_Val})/8$$

4 Results and Comparison

The implementation of the proposed algorithms provides a simple user interface which is command line based. However it is user friendly to some extent too as it allows users to enter the burst time and arrival time of each process. The user can input total number of processes in the beginning along with the value for time quantum if needed. Then the program automatically produces the output so as to show average turnaround time and average waiting time (Fig. 1).

This tool has following features:

1. Interface is very easy and simple to use. A user can easily input the data about burst time and arrival time of process.

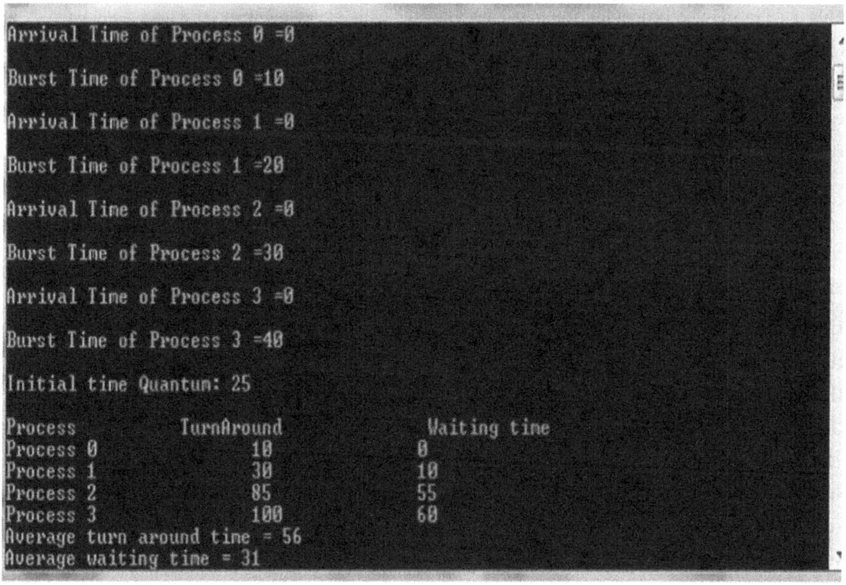

Fig. 1 Screenshot example

2. Output is produced in a step wise fashion. This is helpful as it facilitates the error detection process.
3. The coding has been done in C++. An IDE like Code block or Dev C++ should be used.
4. The tool will run in any operating system higher than or equal to windows XP.
5. RAM should be more than 128 MB.

In this section, the results from the implemented programs are presented. In order to validate the proposed algorithms, two major criteria were:

1. Reduction in average turnaround time.
2. Reduction in average waiting time.

Now the sample data for the process can vary a lot. So as to accommodate variations in the data in terms of arrival and burst time, four different cases have been considered here. A further testing of these algorithms has also been performed whose results are presented later in this chapter in the form of graphs. In the below mentioned cases, following terminology has been used:

New RR1 = Round Robin using mean median and dispersion.
New RR2 = Round robin using Arithmetic and harmonic mean.
New RR3 = Incremental Round robin algorithm.
New RR4 = Round robin using multiple time quanta.

Case 1: Based on the values of arrival and burst time of processes, following values were obtained for various algorithms (Tables 1 and 2).

Table 1 Timings of processes for first case

Process no.	Arrival time	Burst time
P1	0	20
P2	0	40
P3	0	60
P4	0	80

Table 2 Results for first case

Algorithm	Average waiting time	Average turnaround time
RR with TQ = 10 ms	85	135
RR with TQ = 20 ms	70	120
RR with TQ = 30 ms	72	122
New RR1	62	112
New RR2	61	111
New RR3	74	124
New RR4	56	106

Table 3 Timings of processes for second case

Process no.	Arrival time	Burst time
P1	0	10
P2	0	14
P3	0	70
P4	0	120

Case 2: Based on the values of arrival and burst time of processes, following values were obtained for various algorithms (Tables 3 and 4).

Case 3: Based on the values of arrival and burst time of processes, following values were obtained for various algorithms (Table 5 and 6).

Case 4: Based on the values of arrival and burst time of processes, following values were obtained for various algorithms (Table 7 and 8).

These results can be summarized in the following graphs (Figs. 2 and 3).

From the above comparisons, it is clear that dynamic calculation of the size of time quantum is more efficient than conventional method of using static time quantum. Conditions under which a particular heuristic is best suited are as follows:

Case 1: In case, all the processes are relatively very large, then first two algorithms will have a very large time quantum too. So they will act similar to an FCFS algorithm. So, average waiting time will be high. The last two algorithms will then perform better.

Table 4 Results for second case

Algorithm	Average waiting time	Average turnaround time
RR with TQ = 10 ms	52	105
RR with TQ = 20 ms	47	100
RR with TQ = 30 ms	47	100
New RR1	46	99
New RR2	41	94
New RR3	40	93
New RR4	39	92

Table 5 Timings of processes for Third case

Process no.	Arrival time	Burst time
P1	0	18
P2	4	70
P3	8	74
P4	16	80

Table 6 Results for third case

Algorithm	Average waiting time	Average turnaround time
RR with TQ = 10 ms	115	175
RR with TQ = 20 ms	105	165
RR with TQ = 30 ms	105	165
New RR1	102	163
New RR2	97	158
New RR3	89	149
New RR4	81	141

Table 7 Timings of processes for fourth case

Process no.	Arrival time	Burst time
P1	0	10
P2	6	14
P3	13	70
P4	21	120

Table 8 Results for fourth case

Algorithm	Average waiting time	Average turnaround time
RR with TQ = 10 ms	42	95
RR with TQ = 20 ms	37	90
RR with TQ = 30 ms	37	90
New RR1	36	89
New RR2	31	84
New RR3	30	83
New RR4	29	82

Fig. 2 Average turnaround time of algorithms

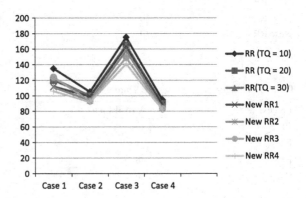

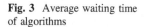

Fig. 3 Average waiting time of algorithms

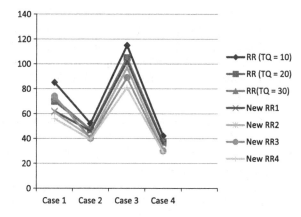

Case 2: In case, most of the processes are small and few processes are relatively very large, then in Algorithm 4, there will be larger number of context switches. This is because smaller TQ2 will be applied again and again to trim down the larger processes.

5 Conclusion and Future Work

In light of the effectiveness and the efficiency of the Round robin process scheduling algorithm and the need to find an optimal time quantum, this paper implemented four algorithms. They provide a fairly good way to calculate time quantum, without user intervention, based on the process mix and thus perform better than algorithms using a fixed time quantum. On the basis of simulation and analysis, it was found that all the four proposed algorithms perform better than the conventional Round robin algorithm. Further we can draw following conclusions from the results:

- All the four proposed algorithms were able to reduce the Average waiting and average turnaround time of the processes.
- The reduction in the number of context switches was less consistent. There were some occasional increases in the number of context switches.
- The additional complexity of these algorithms comes from the fact that they calculate time quantum dynamically. However the calculation of mean, median and dispersion; or use of multiple time quanta or calculation of harmonic mean etc. can be done in linear or constant time complexity. Thus there is only a nominal increase in complexity of the round robin algorithms in these modified forms.

Thus it can be concluded that all four proposed algorithms give better performance than round robin algorithm. There is a good scope for improvement in the existing methods which calculate the size of time quantum dynamically. The earlier works have tended to use concepts of mean and median to identify time quantum size. This paper tries to combine other measures with the concept of mean and median so as to calculate time quantum size. Some other ways to improve round robin scheduling algorithm have also been explored in this paper. Overall the paper involves following:

- Use of measures like dispersion and harmonic mean and combining them in the existing methods to calculate size of time quantum dynamically.
- Exploring the concept of multiple time quanta. It is an extension of the concept of dynamic time quantum.
- Showing that the incremental increase in the size of time quantum effects and improves the average turnaround and average waiting time of given processes.

References

1. Stallings, W: Operating Systems, Internals and Design Principles. Prentice Hall, Englewood Cliffs (2001).
2. Silberschatz, Galvin and Gagne: Operating systems concepts, 9th edition, Wiley, (2012).
3. Andrew Tanenbaum, "Modern operating systems", 3rd edition, Pearson education international, (2007).
4. Samih M. Mostafa, S. Z. Rida and Safwat H. Hamad, "Finding Time Quantum Of Round Robin Cpu Scheduling Algorithm In General Computing Systems Using Integer Programming", International Journal of Research and Reviews in Applied Sciences (IJRRAS), Vol 5, Issue 1, (2010).
5. D Praveen Kumar, T. Sreenivasula Reddy, A. Yugandhar Reddy, "Finding Best Time Quantum for Round Robin Scheduling Algorithm to avoid Frequent Context Switch", International Journal of Computer Science and Information Technologies, Vol. 5 (5), (2014).
6. Mahesh Kumar M R, Renuka Rajendra B and Sreenatha M, Niranjan C K, "An Improved approach to minimize context switching in rounr orbin scheduling algorithm using optimization techniques", International Journal of Research in Engineering and Technology Volume: 03 Issue: 04, (2014).
7. Shahram Saeidi and Hakimeh Alemi Baktash, "Determining the Optimum Time Quantum Value in Round Robin Process Scheduling Method", I.J. Information Technology and Computer Science, (2012).
8. Tarek Helmy and Abdelkader Dekdouk, "Burst Round Robin as a Proportional-Share Scheduling Algorithm", In Proceedings of The fourth IEEE-GCC Conference on Towards Techno - Industrial Innovations, pp. 424-428, Bahrain, (2007).
9. Aashna Bisht, Mohd Abdul Ahad and Sielvie Sharma, "Calculating Dynamic Time Quantum for Round Robin Process Scheduling Algorithm", International Journal of Computer Applications (0975 – 8887) Volume 98 – No. 21, July (2014).
10. Rakesh Mohanty, H. S. Behera and Debashree Nayak, "A New Proposed Dynamic Quantum with Re-Adjusted Round Robin Scheduling Algorithm and Its Performance Analysis", International Journal of Computer Applications (0975–8887), Volume 5– No. 5, August (2010).
11. Bashir Alam, M. N. Doja and R. Biswas, "Finding Time Quantum of Round Robin CPU Scheduling Algorithm Using Fuzzy Logic," International Conference on Computer and Electrical Engineering, (2008).

12. Ishwari Singh Rajput and Deepa Gupta, "A Priority based Round Robin CPU Scheduling Algorithm for Real Time Systems", International Journal of Innovations in Engineering and Technology (IJIET) ISSN: 2319 – 1058 Vol. 1 Issue 3 Oct (2012).
13. Abbas Noon, Ali Kalakech and Seifedine Kadry, "A New Round Robin Based Scheduling Algorithm for Operating Systems: Dynamic Quantum Using the Mean Average", IJCSI International Journal of Computer Science Issues, Vol. 8, Issue 3, No. 1, May (2011).
14. Siregar, "A New Approach to CPU Scheduling Algorithm: Genetic Round Robin", International Journal of Computer Applications, Vol. 47, No. 19, (2012).
15. Jeegar A Trivedi and Priti Srinivas Sajja, "Improving efficiency of round robin scheduling using neuro fuzzy approach", IJRRCS vol. 2, No. 2, April (2011).
16. C. Yaashuwanth, R. Ramesh, (2010) "Intelligent Time Slice for Round Robin in Real Time Operating Systems", IJRRAS, 2(2):126–131.

Frequency Dependent Lumped Model of Twin Band MIMO Antenna

Vilas V. Mapare and G.G. Sarate

Abstract Latest communication systems support interactive multimedia, voice, video, wireless internet, other broadband services with very high speed, high capacity and low cost/bit. To compounds these with the desire of mobile operators to expand their band allocation with compact devices as smart phones or similar devices and what result is a difficult design arena. These advanced features can be confined within compact devices by the development of smaller and admissible MIMO antenna. The simulated and the measured results are in a good agreement. The proposed structure minimizes the frequency dependent lumped component with proper arrangement of array element and covers the 3G/4G range of 2.1–2.29 GHz.

Keywords Multi band · Rectangular microstrip antenna · Gap coupled rectangular microstrip antenna · Multiple resonators · Parasitic resonators · Multi band MIMO antenna

1 Introduction

Multiple-Input Multiple-Output (MIMO) technology made a great breakthrough by satisfying the demand of high speed, high fidelity mobile communication services in today's fast paced word without using any additional radio resources [1, 2]. MIMO (Multiple Input Multiple Output) systems are proved to achieve higher data rates by deploying multiple antennas at both the transmitter and receiver instead of a monopole antenna at the respective locations without demanding additional

V.V. Mapare (✉)
Sant Gadge Baba Amravati University, Amravati, Maharashtra, India
e-mail: mapare.vilas@gmail.com

G.G. Sarate
Government Polytechnic Amravati, Amravati, Maharashtra, India
e-mail: ggsanshu@gmail.com

© Springer India 2016
S.C. Satapathy et al. (eds.), *Information Systems Design and Intelligent Applications*, Advances in Intelligent Systems and Computing 434,
DOI 10.1007/978-81-322-2752-6_46

469

bandwidth or an increase in the respective power. The huge potential of MIMO technique is evidenced by a rapid adoption into the wireless standards, such as WLAN, LTE (Long Term Evolution), and WIMAX. However, when multiple antennas are involved at closer spacing the technical challenges are more pronounced compared to a SISO (Single Input Single Output) system. Hence, the basic aim of MIMO antenna design is to reduce the correlation between the multiple signals. The parameter that describes the correlation between the received signals in highly diversified environments is mutual coupling, which deteriorates the performance of the communication system [2]. Higher mutual coupling reduces the antenna efficiency and thus minimizes the system channel capacity, radiation efficiency.

Some of the solutions have been reported in [3–9] to minimize the mutual coupling between antenna elements which are in closed proximity. In [3–6], the mutual coupling was effectively reduced by using defected ground structures (DGS). Antenna with low mutual coupling, operating at 2.45 GHz is reported in [3], and the improvement is achieved by putting two λ/2 slot on the ground plane. A combine effect of rectangular slot ring and inverted T-shaped slot stub has been shown in [3] to reduce the mutual coupling between two quad band antenna. Ground plane with N-section resonator slot [4], protruded T-shaped ground plane [5], and the tree-like structure on ground plane [6] were implemented to improve port-to-port isolation. In [7] transmission characteristics of different DGS and their equivalent circuit models were presented. Electromagnetic band gap (EBG) structures, neutralization techniques, and lumped circuit network are also attractable solutions to produce high isolation. Some of the EBG structures were reported in [8] to optimize the mutual coupling by reducing the surface waves. However, EBG structure requires an intricate fabrication process and also a large area. Some of the neutralization techniques have been suggested in [8] to improve isolation by utilizing a field cancelling concept. In [6], antennas are designed for the DCS and UMTS bands and studied the mutual coupling between them by considering the feeding strip facing and shorting strip facing cases. They used a single suspended line of different length cancelling the field. The lumped components circuit networks presented are used to provide good isolation between two antenna elements. But the lumped components induce the reactive losses, which in turn strongly affect the total antenna efficiencies. The resonant frequency increases as the number of elements increases.

2 Antenna Design and Configuration

For an Rectangular Microstrip Antennas (RMSA) having resonant frequency f_r, with length L, and width W, the effective dielectric constant reduces because of fringing fields. This effective dielectric constant, ε_e can be calculated as

$$\varepsilon_e = \frac{\varepsilon_r + 1}{2} + \frac{\varepsilon_r - 1}{2}\left[1 + \frac{10h}{W}\right]^{-\frac{1}{2}} \tag{1}$$

To calculate ε_e, width W has to be known. It can be calculated using the following formula,

$$W = \frac{c}{2f_r\sqrt{\frac{\varepsilon_r + 1}{2}}} \tag{2}$$

Also, as mentioned the effective length of the RMSA, L_e, is slightly more than the actual length. The effective length can be given as,

$$L_e = L + \Delta L \tag{3}$$

Where ΔL is the extension in length due to fringing fields. This ΔL can be calculated as,

$$\Delta L = \frac{h}{\sqrt{\varepsilon_e}} \tag{4}$$

Thus by calculating ΔL, length L can be easily calculated for a given resonant frequency f_r, using (3). The initial antenna structure consists of single microstrip which are located on top of the substrate FR4 (PCB of mobile device with $\varepsilon_r = 4.3$ and $\tan\delta = 0.02$). The dimension of the microstrip is chosen as 46.08 mm × 35 mm by using the fundamental formula.

3 Parametric Study and Analysis

The rectangular antenna dimensions and coordinates are displayed in Fig. 1. Normally, the patch length L is between $\lambda_0/3$ and $\lambda_0/2$, and its width W is smaller than λ_0 while the substrate thickness d is very small. To be a resonant antenna, the length L should be around half of the wavelength. In this case, the antenna can be considered a $\lambda/2$ microstrip transmission line resonant cavity with two open ends where the fringing fields from the patch to the ground are exposed to the upper half space ($z > 0$) and are responsible for the radiation. This radiation mechanism is the same as the slot line, thus there are two radiating slots on a patch antenna, as indicated in Fig. 1. This is why the microstrip antenna can be considered an aperture-type antenna. The radiating fringing fields at the ends are separated by $\lambda/2$, which means that they are 180° out of phase but are equal in magnitude. Viewed from the top of the antenna, both fields are actually in phase for the x components, which leads to a broadside radiation pattern with a main lobe in the z direction (Figs. 2 and 3).

Fig. 1 Single patch antenna

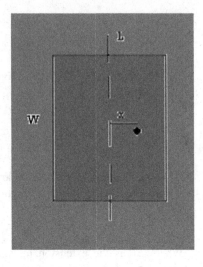

Fig. 2 S11 plot of single
patch antenna

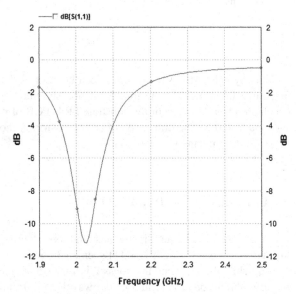

3.1 Effect of Distributed Parameter on Single Patch RMSA Configurations

The width of single patch w = 46.08 mm, length is L = 35 mm, dielectric constant
of FR4 is $\varepsilon_r = 4.3$, height of patch is h = 1.59 mm and loss tangent of the material is
$\tan \delta = 0.02$ as shown in Fig. 1 (Fig. 4).

In single patch antenna, at resonance, the capacitive and inductive reactance are
equal. Thus we expect to see the apparent or measured L increases near resonance,

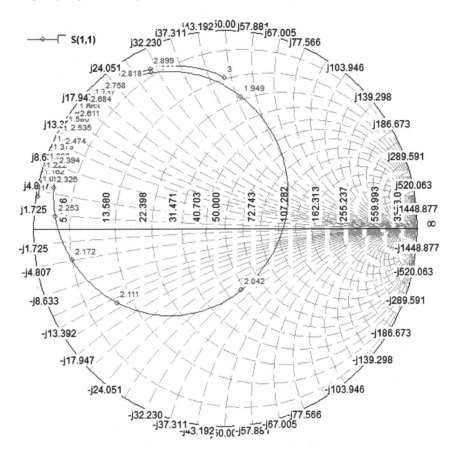

Fig. 3 Smith chart of single patch antenna

Fig. 4 Two strip configuration

then reach zero at resonance and then switch to a high negative value just above resonance frequency 2.02 GHz and then dropping with increasing frequency as shown in Table 1 and Fig. 3.

In the Smith chart, a resonant circuit shown up as a circle. The larger the circle, the stronger the coupling. The radius of the resonance circle is larger than 0.25.

Table 1 Lumped model of single patch antenna

Freq (GHz)	Q	L (nH)	R (Ohm)
1.98e		4.45e+00	6.74e+001
2.00e	3.75e−001	2.67e+000	8.98e+001
2.01e	−2.33e−001	−1.69e−001	9.18e+001
2.11e	−1.38e+000	−1.07e+000	1.03e+001
2.03e	4.75e−001	3.67e+000	9.98e+001
2.04e	7.25e−001	4.45e+00	8.74e+001

Hence the centre of the chart lies inside the circle. At critical coupling i.e. 2.02 GHz the resonance circle touches the centre of the Smith chart as shown in Fig. 3.

3.2 Effect of Distributed Parameter on Two Patch RMSA Configuration

For two patch configuration the number of resonant frequencies obtained is two, and a single loop is observed in the input impedance loci on the Smith Chart which represent resonance. The size of the loops and hence the frequencies ratios depend on coupling between the two strips, which primarily depends on the gap. At resonance, the capacitive and inductive reactance are equal. Thus we expect to see the apparent or measured L increases near resonance frequency 2.053, 2.158 GHz and then reach zero at resonance and then switch to a high negative value just above resonance, dropping with increasing frequency as shown in Table 3 and in Fig. 5. As the width of the gap increases, the coupling decreases and the resonant frequencies ratios also decreases, reducing the loop size as can be seen from Fig. 5.

In the Smith chart, a resonant circuit shown up as a circle. The larger the circle, the stronger the coupling. In two patch the radius of resonance circle is smaller. Hence the centre of the chart lies outside the circle. At critical coupling i.e. 2.05 and 2.15 GHz the resonance circle touches the centre of the Smith chart as shown in Fig. 5. If tan δ is 0.001, displacement current is increases, there is a strong coupling. In the Smith chart this is shown by larger circle. If tan δ is 0.02, displacement current is decreases, there is a not-strong coupling. In the Smith chart this is shown by smaller circle. Whenever there is a critical coupling the circle touches the centre of the chart as shown in Fig. 5. Effect of feed point location on the behaviour of two patch RMSA configuration for two different values of feed point, i.e. x = 6 mm and x = 9 mm is interpreted (Figs. 6, 7, 8 and Tables 2, 3, 4).

When the feed point location is at 6 mm, the centre of the chart lies outside the circle under critical coupling When the feed point location is 9 mm, the circle touches the unity circle which is critical coupling. as shown in Fig. 9.

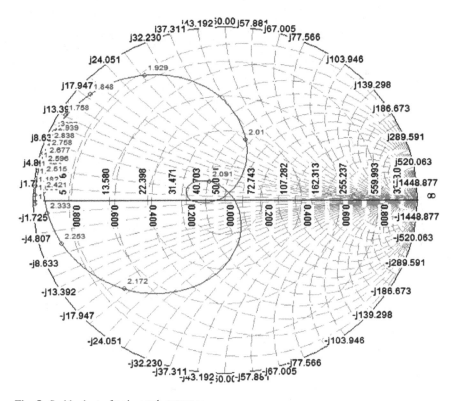

Fig. 5 Smith chart of twin patch antenna

Fig. 6 Effect of gap on the behaviour of two patch RMSA configuration, for two different values of gap, i.e. S = 0.2 mm and S = 2 mm, VSWR plot

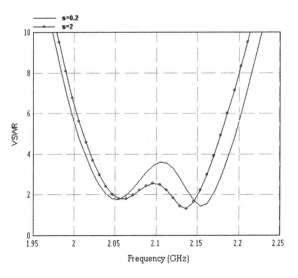

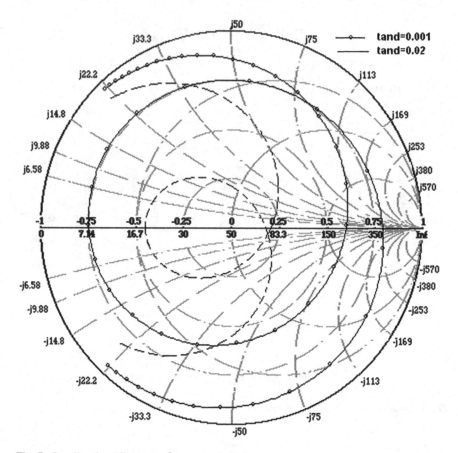

Fig. 7 Coupling for different tan δ

3.3 Effect of Distributed Parameter on Three Patch Configuration

As the number of coupled elements in a patch configuration go on increasing the flexibility to obtained the higher frequency ratio is high, as the individual Twin radiating elements can be optimized. The simulation results for three patch configuration, though does not yield as high bandwidth as two patch configuration, however it is found that it can be very effectively used for Twin and triple frequency operation.

For three patch configuration, it is observed, that the separation between two resonance frequencies is considerably more than that observed for two patch configuration. This is due to increased coupling as explained before. For same gap of $S_1 = 0.2$ mm, the two resonance frequencies for two patch configuration are, $fr_1 = 2.05$ GHz and $fr_2 = 2.15$ GHz, while for three patch configuration these are,

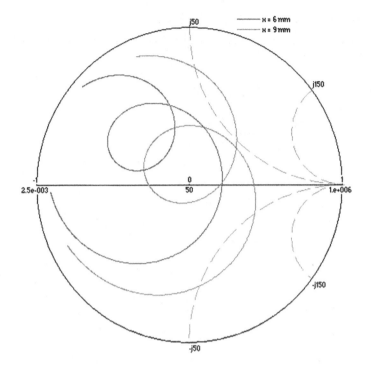

Fig. 8 Coupling for different feed point

Table 2 Twin patch configurations (w = 23.04, L = 35 mm, ε_r = 4.3, h = 1.59 mm, tan δ = 0.01)

No. of elements	Gap width, S (mm)	Feed point, x (mm)	Fr_1 (GHz)	Fr_2 (GHz)	Frequency ratio
2	0.2	8	2.053	2.158	1.05
	2	8	2.053	2.152	1.03

Table 3 Lumped model of twin patch antenna

Freq (GHz)	Q	L (nH)	R (Ohm)
2.03	1.04e−001	4.29e−001	5.27e+001
2.05	−3.61e−002	−1.08e−001	103e+001
2.07	1.15e−001	2.90e−001	3.27e+001
2.15	−1.59e+000	−1.14e+000	1.82e+001
2.16	−2.07e+000	−9.63e+000	4.07e+001
2.18	−2.38e+000	−1.26e+000	7.26e+000
2.19	−2.53e+000	−2.88e−001	5.39e+000

Table 4 Effect of distributed parameter on input impedance

No. of elements	Gap width, S (mm)	Feed point, x (mm)	tan δ	Coupling
2	0.2	8	0.001	Strong
2	0.2	8	0.02	Weak

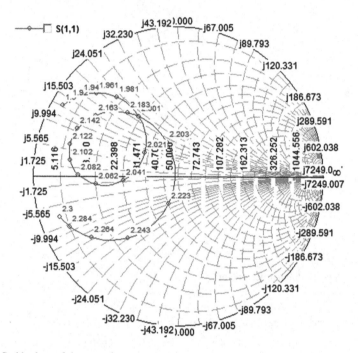

Fig. 9 Smith chart of three patch antenna

fr_1 = 2.05 and fr_2 = 2.22 GHz as can be seen from Fig. 9. So, using three patch configuration, much higher frequency ratio can be obtained for Twin frequency operation, as compared to the two patch configuration (Table 5).

In Smith chart, a resonant circuit shown up as a circle. The larger circle, the stronger the coupling. In three patch the radius of resonance circle is smaller. Hence the centre of the chart lies outside the circle. At critical coupling i.e. 2.05 and 2.22 GHz the resonance circle touches the centre of Smith chart as shown in Fig. 9 (Table 6).

Table 5 Twin frequency response of three patch configuration using patches with different lengths (W = 15.36 mm, ε_r = 4.3, h = 1.59 mm, tan δ = 0.01)

L_1 (mm)	L_2 (mm)	L_3 (mm)	S (mm)	X (mm)	fr_1 (GHz)	fr_2 (GHz)	BW_1 (MHz)	BW_2 (MHz)	fr_2/fr_1
35	35	35	0.2	5.2	2.05	2.22	17	13	1.09

Table 6 Lumped model of single patch antenna

Freq (GHz)	Q	L (nH)	R (Ohm)
2.00e	3.51e−001	8.20e−001	2.96e+001
2.02e	2.43e−001	5.27e−001	2.93e+001
2.03e	1.47e−001	3.27e+001	2.82e+001
2.05e	−1.14e−001	−1.82e−001	1.84e+001
2.20e	3.14−001	9.87e+001	4.35e+001
2.21e	1.39−001	4.73e+001	4.70e+001
2.22e	−1.03e+001	−1.53e+001	1.76e+001
2.25e	−1.05e+001	−9.83e−001	6.13e+001

Table 7 Twin frequency response of the designed four patch configuration (W = 11.52 mm, ε_r = 4.3, h = 1.59 mm, tan δ = 0.01)

L_1 (mm)	L_2 (mm)	L_3 (mm)	L_4 (mm)	S (mm)	X (mm)	Fr_1 (GHz)	BW_1 (MHz)	fr_2 (GHz)	BW_2 (MHz)	fr_2/fr_1
34	35	36	35	S1 = 1.7 S2 = 2.8	10	2.010	45	2.184	41	1.12

3.4 Effect of Distributed Parameter on Four Patch Configuration

Resonance frequencies of patches 1 and 4 and patches 2 and 3 are quite close to each other as equal lengths elements radiate at the same resonant frequency. So by adjusting the lengths of these elements and suitably providing the matching requirements, Twin frequency operation is possible. The length of first element is reduced to 34 mm to increase its resonant frequency, while the length of the third element is increased to 36 mm to reduce it's resonant frequency and by properly adjusting the gaps and feed point, Twin frequency operation has been obtained. The two resonance frequencies, corresponding to minimum return loss are, fr_1 = 2.010 GHz and fr_2 = 2.184 GHz, with a bandwidth of around 40 MHz at each frequency band. Table 7 gives the details of various parameters. The field is broadside at the lower frequency while it becomes narrower at the second frequency.

In the Smith chart, a resonant circuit shown up as a circle. The larger the circle, the stronger the coupling. In four patch the radius of resonance circle is smaller. Hence the centre of the chart lies outside the circle. At critical coupling i.e. 2.010 and 2.184 GHz the resonance circle touches the centre of the Smith chart (Fig. 10 and Table 8).

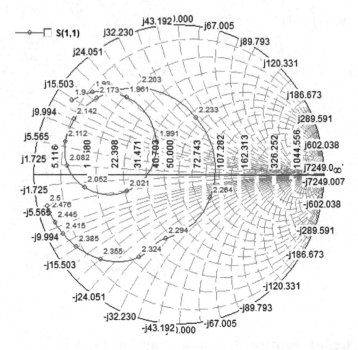

Fig. 10 Smith chart of three patch antenna

Table 8 Lumped model of four patch antenna

Freq (GHz)	Q	L (nH)	R (Ohm)
1.97e+000	7.85e−001	3.10e+000	4.98e+001
1.99e+000	1.01e−001	5.41e−001	6.71e+001
2.01e+000	−7.10e−001	−1.064e+000	1.23e+001
2.15e+000	3.17e−001	6.27e−001	2.67e+001
2.17e+000	5.29e−002	1.62e−001	4.174e+001
2.18e+000	−1.17e+000	−1.67e+000	1.31e+001
2.25e+000	−2.16e+000	−1.52e+000	9.94e+000

3.5 Effect of Distributed Parameter on Five Patch Configuration

The parasitic elements with shorter lengths produce more resonance frequency as the length is inversely proportional to resonant frequency. The elements L1 and L5 have same length so their resonant frequencies are also nearly same which add up to produce one of the frequency bands. The feed element will produce the second frequency band. The return loss for L2 and L4 are high due to longer length which

Fig. 11 S11 chart of five patch antenna

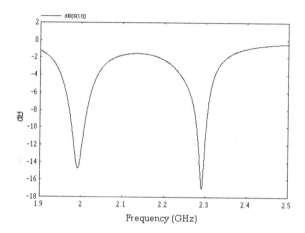

Table 9 Twin frequency response of the designed five patch configuration (W = 9.21 mm, ε_r = 4.3, h = 1.59 mm, tan δ = 0.01)

L_1 (mm)	L_2 (mm)	L_3 (mm)	L_4 (mm)	L_4 (mm)	S (mm)	x (mm)	Fr_1 (GHz)	BW_1 (MHz)	fr_2 (GHz)	BW_2 (MHz)	fr_2/fr_1
33	36	35	36	33	S1 = 0.5 S2 = 2.8	7.5	1.993	34	2.29	26	1.15

Table 10 Lumped model of five patch antenna

Freq (GHz)	Q	L (nH)	R (Ohm)
1.90e+000	4.300e+000	1.32e+000	3.68e+000
1.94e+000	2.27e+000	1.8376e+000	9.88e+000
1.99e+000	−4.66e−002	−1.14e−001	1.08e+001
2.10e+000	1.08e+000	4.42e−001	5.38e+000
2.18e+000	2.32e+000	1.63e+000	9.68e+000
2.29e+000	−1.82e+000	−1.95e+000	1.06e+001
2.389e+000	−3.72e+000	−1.02e+000	4.13e+000

provides the separation between the two frequency bands. Figure 11 shows the return loss for five patch configuration.

Thus it can be concluded that five patch configuration offers greater flexibility for Twin frequency operation, as compared to other configurations. Significantly higher frequency ratio is possible using this configuration, and the radiation pattern is also in broadside direction in the two frequency bands (Tables 9 and 10).

In the Smith chart, a resonant circuit shown up as a circle. The larger the circle, the stronger the coupling. In three patch the radius of resonance circle is smaller. Hence the centre of the chart lies outside the circle. At critical coupling i.e. 1.99 and 2.29 GHz the resonance circle touches the centre of the Smith chart as shown in Fig. 12.

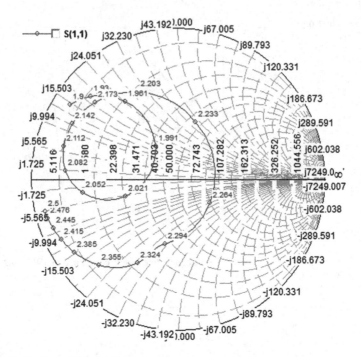

Fig. 12 Smith chart of five patch antenna

4 Conclusion

In this paper, a compact, Twin bands MIMO antenna with low mutual coupling operating over the range of 2.1–2.29 GHz. is proposed. A single resonator along the width is divided so that width of each element is equal, then the substrate parameter is varied. The mutual coupling between the elements is changes due to displacement current that yield multiple frequencies. But there is limitation on the number of elements due to mutual coupling that combines the indivi Twin frequency bands result into overcritical and under critical coupling. The proposed configurations have Twin band with VSWR ≤ 2 are in the range of 2.1–2.29 GHz. Lumped model equivalent circuit is just a fitted model of S parameters. The values of the circuit elements may or may not have or match any physical meaning. This concept can be used to design multiband antenna, which can be used in multiple Input Multiple Output (MIMO) technology.

References

1. K. P. Ray. "Gap coupled rectangular microstrip antennas for dual and triple frequency operation", Microwave and Optical Technology Letters, 06/2007.
2. K. P. Ray. "Compact broadband gap-coupled rectangular microstrip antennas", Microwave and Optical Technology Letters, 12/2006.
3. Guofu Zhang. "Research on Comprehensive Evaluation of Agent Coalition Based on D-S Evidential Reasoning", ourth International Conference on Fuzzy Systems and Knowledge Discovery (FSKD 2007), 08/2007.
4. Atif Jamil. "Design and performance evaluation of multiband MIMO antennas", 2011 National Postgraduate Conference, 09/2011.
5. Sheng-Lyang Jang. "A CMOS LC-tank frequency divider with 3D helical inductors", Microwave and Optical Technology Letters, 06/2007.
6. K. P. Ray. "Broadband gap-coupled half hexagonal microstrip antennas", Microwave and Optical Technology Letters, 02/2008.
7. XueMing Ling, and RongLin Li. "A Novel Dual-Band MIMO Antenna Array With Low Mutual Coupling for Portable Wireless Devices", IEEE Antennas and Wireless Propagation Letters, 2011. 26.
8. Qinghao Zeng. "Tetraband Small-Size Printed Strip MIMO Antenna for Mobile Handset Application", International Journal of Antennas and Propagation, 2012.
9. T. S. P. and Z. N. Chen. "Diversity performance of a dual-linear polarisation suspended gap-coupled microstrip antenna", IEE Proceedings - Communications, 2006.

A Novel Framework for Integrating Data Mining Techniques to Software Development Phases

B.V. Ajay Prakash, D.V. Ashoka and V.N. Manjunath Aradhya

Abstract In software development process, phases such as development effort estimation, code optimization, source code defect detection and software reuse are very important in order to improve the productivity and quality of the software. Software repository data produced in each phases have increased as component of software development process and new data analysis techniques have emerged in order to optimize the software development process. There is a gap between the software project management practices and the need of valuable data from software repository. To overcome this gap, a novel integrated framework is proposed, which integrates data mining techniques to extract valuable information from software repository and software metrics are used in different phases of software development process. Integrated framework can be used by software development project managers to improve quality of software and reduce time in predicting effort estimation, optimizing source code, defect detection and classification.

Keywords Software effort estimation · Code optimization · Defect prediction and classification · Software reuse · Software development process

1 Introduction

Increase in demand for the software made Information technology (IT) industries to deliver quality software product efficiently and effectively, but the failure in software have not been reduced. Failure of software may be because, software product

B.V. Ajay Prakash (✉)
Department of Computer Science and Engineering, SJBIT, Bengaluru, India
e-mail: ajayprakas@gmail.com

D.V. Ashoka
Department of Computer Science and Engineering, JSSATE, Bengaluru, India

V.N. Manjunath Aradhya
Department of MCA, Sri Jayachamarajendra College of Engineering, Mysuru, India

© Springer India 2016
S.C. Satapathy et al. (eds.), *Information Systems Design and Intelligent Applications*, Advances in Intelligent Systems and Computing 434,
DOI 10.1007/978-81-322-2752-6_47

fail to meet the need of software requirement specification due to defects in coding or logical error which makes software to produce unpredicted results, may be effort and budget estimated to deliver the software may exceed estimated effort and budget. This situation is a 'software crisis' which made IT people to think about the software development methodology and practices which can improve the quality of software, deliver software product on estimated effort and reduce the defects. A software development methodology (SDM) is defined as documented collection of procedure, process and policies used by the software development people to develop the software product and give the complete IT solutions to the customers [1, 2]. The important aspect of software product is its software development life cycle, as it goes into number phases to accomplish to final software product. The software development tasks which are followed are feasibility study, requirement specification analysis, project planning, design, implementation and testing. In requirement specification analysis phase, customer describes the features and its functionality which need to be included in the software product. Most important activity in project planning for effective development of software is effort estimation. If effort estimation is not accurately estimated software project may lead to failure and budget may exceed. In implementation phase and testing, defects should be identified. Defects identification is to find the code rule violations while developing code, review algorithm implementation, etc. Defects classification is an integral part of development and testing where defects are classified based on the severity to give scale of impact on quality of software product. In order to reduce development time and effort software reuse task place an important role in development. Using existing versions or features software productivity and quality can be improved. There is need of method which can extract useful software component from software repository to reduce the development time.

To solve these problems, an integrated framework is proposed which integrates data mining techniques and software metrics to achieve the following research goals:

- Accurately estimate the software development effort
- Optimize the source code
- Identify defects and classify the defects based on severity
- Effectively reuse the existing domain knowledge while development code
- To improve the quality of the software product

This paper is organized as follows: Sect. 2 gives detailed survey of different data mining techniques applied to various aspects of software development phases while Sect. 3 brief on proposed integrated framework, Sect. 4 presents the results we have obtained so far. Finally, in conclusion and future work, concluded the paper and present our goals for future research.

2 Related Works

This section summarizes research works which have been carried out by researchers to accurately estimate software development effort, defect prediction and classification, to improve quality and productivity and framework for software development process.

Many research works has been made on developing efficient effort estimation using data mining techniques such as classification, clustering and fuzzy logic [3, 4]. Kelly et al. [5] provides methodology to explored neural networks, genetic algorithms for software cost estimation. Artificial neural network are effectively used in estimating effort due its capability to train from previous data [6, 7]. Idri et al. [8], proposed fuzzy logic rules and neural network on COCOMO 81 dataset to estimate the cost. Parag [9] applied a Probabilistic Neural Networks (PNN) model for predicting effort values of software development parameters (either software size or effort) and concluded that the probability of actual value parameter will be less than its predictable value. The idea of inferring code refactoring is proposed by Demeyer et al. [10], by comparing two program, versions based on software metrics such as LOC, method calls within a method. Zou and Gofrey [11] coined the term origin analysis, for refactoring and reconstruction by measuring similarity in code elements using multiple criteria such as names, metric value, signatures, callers and callees. Zou and Godfrey conclude merge, split, and rename refactoring. Godfrey et al. [12], proposed a techniques for identifying merging and splitting of source files when design changes. Shepherd et al. [13], proposed a mechanism which automates feature mining technique for exploiting the Program Dependence Graphs (PDG) and Abstract Syntax Tree (AST) representations of a program. Fluri et al. [14], presents an approach to extract the fine grained changes that occur across different versions.

Kim et al. [15] classified change request as buggy to clean using new techniques. Jalbert and Weimer [16] identified duplicate bug reports automatically from software bug repositories. Cotroneo et al. [17], performed defect failure analysis of java virtual machine. Guo et al. [18] explored factors which are affecting in fixing the bugs for Windows Vista, and Windows 7 systems. Davor et al. [19], text classification method is proposed for automatic bug tracking. They used supervised Bayesian learning algorithm for bug assignment to developer. Evaluation of their method can predict 30 % of the report correctly to report assignment to developers. Several work has being done in creating platforms for machine learning and on software engineering based on software reusable components.

3 Proposed Framework

From the previous research works it is revealed there is a need of applying data mining techniques to various phases of software development life cycle. Software development processes are complex and related tasks often generates large number variety of software artifacts makes well suited to apply data mining techniques.

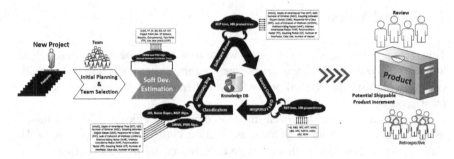

Fig. 1 Integrated framework for applying data mining techniques and software metrics

Our proposed integrated framework, integrated data mining techniques in various phases such as software effort estimation, code optimization defect detection and classification, software reuse. The proposed framework is as shown in Fig. 1:

Proposed methodology by integrating data mining techniques

Step 1 **Collecting software requirement specification**
In this step, software Features are analyzed, find how to solve the context scenarios, risk factors analysis, cost and effort estimation, initial planning, team selection.
Software metric used: Source Lines of Code (SLOC), Function Point (FP), External Inputs (EI), External Outputs (EO), External Enquires (EQ), Input Logical Files (ILF), External Interface files (EIF), Object Point (No. Of Screens, Reports and Components), Test Point (TP), Use case points (UCP), software design.
Data Mining Techniques applied: probabilistic neural networks (PNN) and generalized regression neural networks (GRNN) model.

Step 2 **Applying Iterative Method**
Here source code is developed and refactored several times to optimize the code. Defects are predicted and classified based on severity. Based on bugs severity proposed method is divided into three layers: core, abstraction and application layer. Core layer contains the core functionality of the system, the defects in this layer affects critical functionality of the system and leads to failure of the project. An abstraction layer defect affects major functionality or major data of the system. Application layer contains user oriented functionality responsible for managing user interaction with the system defects in this layer affects minor functionality of the system.
Software metric used: Cyclomatic Complexity (CC), LOC Total, LOC Executable (LOCE), LOC Code and Comments, LOC Comments (LOCC), LOC Blank (LOCB), Node Count (NC), Edge Count (EC), Weighted Methods per Class (WMC), Coupling

between Object Classes (CBO), Number of Lines (NOC), Response for a Class (RFC), Lack of Cohesion of Methods (LCOM1), Method Hiding Factor (MHF), Depth of Inheritance Tree (DIT), Method Inheritance Factor (MIF), Polymorphism Factor (PF), Number of interfaces, Class size, Number of classes, Coupling Factor (CF). Data Mining Techniques applied: REP tree, J48 pruned tree, Naive Bayes, Multilayer Perceptron, GRNN, PNN models.

Step 3 **Using domain knowledge as Software reuse component**
Identifying software components, locating software modules, adapting existing software components and retrieval of existing API from software repository.
Software metrics used: Weight Method per Class (WMC), Depth of Inheritance Tree (DIT), Number of Childers (NOC), Coupling Between Classes (CBO), Response for Class (RFC), Lack of Cohesion Method (LOC), Coupling Afferent (CE), Number of public method (NPM).
Data Mining Techniques applied: REP tree, J48 pruned tree.

Step 4 **Review potentially working software product increment**
Continuously check the quality of software increment in each iteration by unit integration and load testing. Meet the customer change in requirement specification.

4 Results

To evaluate the proposed framework, software repository is created by downloading the publicly accessible datasets for research work from PROMISE repository (http://promisedata.org/data) for applying data mining techniques to various phases of software development. Nowadays neural network (NN) models are widely used as information processing model that is simulated by brain. NN models are configured to apply on specific application such as image processing, pattern recognition, data classification etc. First for software effort estimation phase probabilistic neural networks (PNN) [20] and generalized regression neural networks (GRNN) [21] models are applied for accurately estimating the effort. GRNN is type of supervised learning model based on radial basis function (RBF) [22] which can be used for regression, classification and time series predictions. GRNN consists of four layers, which are named as input layer, pattern layer, summation layer and output layer. PNN is mainly a classifier map to any input variables to a number of classes. PNN works on supervised multilayered feed forward network with similar four layers as GRNN. MATLAB 11 is used for implementation of algorithms GRNN and PNN for effort estimation. Here output is effort in terms of

Table 1 Comparative results of actual and estimated effort with MRE using china datasets

Project ID	Actual effort	Software development effort estimated using		MRE using	
		GRNN	PNN	GRNN	PNN
371	89	102	118	14.60	32.58
75	139	98	142	24.03	10.07
449	170	123	210	27.64	23.52
48	204	154	168	24.50	17.64
53	281	340	382	20.99	35.94
460	374	325	424	13.10	13.36
93	481	402	248	16.42	11.01
325	752	702	812	6.64	7.97
391	1210	998	829	17.52	31.48
395	1741	1554	2008	10.74	15.33
208	2520	2138	2245	15.15	10.91
5	2994	2558	2694	14.56	10.02
320	3877	3109	3228	19.80	16.73
378	15,039	13,821	12,884	8.09	14.32
326	29,399	25,482	26,700	13.32	9.18
435	54,620	48,553	49,324	11.10	9.69

required person per month. 15 variables are considered as independent attributes and 1 dependent attribute that is actual effort variable. Table 1 shows the comparative result.

Secondly, for defect classification, publicly available NASA MDP data sets are used. GRNN, PNN, J48 and Multilayer Perceptron (MLP) classification techniques are applied on various data sets like CM1, JM1, KC1, KC3, MC1, MC2, MW1, PC1, PC2 and PC3. Weka tool is used to apply data mining techniques on publicly available NASA MDP data sets In our proposed method source code defects are predicted and classified based on severity. Severity is categorized into three layers namely core, abstraction and application. Core layer termed as critical defects which can cause software failure and damage the human life of the customers. Abstraction layer defects are the software functionality defects which are major defects.

Application layer defects are minor defects such as conditions, logical files defects. Here in our proposed method data mining algorithms such as GRNN, PNN. Table 2 shows the results obtained from various classifiers after applying MDP datasets.

The benefits of the proposed framework are as follows:

- By considering risk factor prioritization of work is done do implement in the earlier iterations.

Table 2 Performance of various classifiers

Data sets	Classifiers	TP rate	FP rate	Precision	Recall	F-measure
CM1	J48	0.817	0.717	0.801	0.817	0.808
	Naive Bayes	0.792	0.66	0.804	0.792	0.798
	MLP	0.847	0.794	0.797	0.847	0.816
JM1	J48	0.764	0.628	0.725	0.764	0.736
	Naive Bayes	0.782	0.651	0.739	0.782	0.742
	MLP	0.79	0.711	0.75	0.79	0.727
KC1	J48	0.748	0.504	0.725	0.748	0.729
	Naive Bayes	0.727	0.538	0.699	0.727	0.706
	MLP	0.758	0.582	0.733	0.758	0.714
KC3	J48	0.794	0.562	0.776	0.794	0.783
	Naive Bayes	0.789	0.52	0.782	0.789	0.785
	MLP	0.768	0.611	0.75	0.768	0.758
MC1	J48	0.974	0.892	0.963	0.974	0.967
	Naive Bayes	0.889	0.661	0.962	0.889	0.922
	MLP	0.974	0.913	0.962	0.974	0.967
MC2	J48	0.712	0.374	0.705	0.712	0.707
	Naive Bayes	0.72	0.432	0.712	0.72	0.696
	MLP	0.68	0.423	0.669	0.68	0.672
MW1	J48	0.87	0.76	0.838	0.87	0.85
	Naive Bayes	0.814	0.414	0.874	0.814	0.837
	MLP	0.866	0.668	0.852	0.866	0.858
PC1	J48	0.901	0.722	0.88	0.90	0.88
	Naive Bayes	0.881	0.59	0.886	0.881	0.883
	MLP	0.921	0.616	0.907	0.921	0.911
PC2	J48	0.972	0.979	0.957	0.972	0.965
	Naive Bayes	0.907	0.858	0.959	0.907	0.932
	MLP	0.976	0.979	0.957	0.976	0.967
PC3	J48	0.85	0.655	0.831	0.85	0.839
	Naive Bayes	0.357	0.162	0.859	0.357	0.409
	MLP	0.863	0.762	0.821	0.863	0.834

- Accurately estimation of software development effort by using data mining techniques makes less chances of software failure.
- Continuous verification of quality by effectively classifying defects.
- Usage of knowledge data base makes developers to optimize the source code.
- Responsibility of each module is clearly defined which leads to improved traceability.
- Software evolution made easier to add new functionality and adopt to the change in requirements.

- Software reuse module reduces the development time and improves the productivity of the software product.
- Cost of features implementation can be reduced by avoiding cost of modifying many modules to implement.

5 Conclusion

In view of reducing development effort and also improve the quality of software product. This paper presents a novel framework which integrates data mining techniques and software metrics in various software development phases. The phases such as project planning, effort estimation, code optimization, defect detection and software reuse are important in software development methodology. In each phases data mining techniques such as neural network models are applied to get better results. The proposed framework can be used as guidelines for decision making by project managers in various phases of software development process. Future work will include the implementation of automation tool which includes data mining techniques and software metrics in framework.

References

1. J. Iivari, R. Hirschheim H.M. Klein, A dynamic framework for classifying information systems development methodologies and approaches Journal of Management Information Systems, Vol. 17 (3), 2000, pp. 179–218.
2. T. Ravichanran, A. Rai, Total quality management in information systems development, Journal of Management Information Systems, Vol. 16 (3), 1999, pp. 119–155.
3. J. Ryder, PhD thesis, Fuzzy COCOMO: Software Cost Estimation., Binghamton University, 1995.
4. A. C. Hodgkinson and P. W. Garratt, "A neuro-fuzzy cost estimator," in Proceedings of the Third Conference on Software Engineering and Applications,1999, pp. 401–406.
5. Kelly, Michael A. A methodology for software cost estimation using machine learning techniques. NAVAL POSTGRADUATE SCHOOL MONTEREY CA, 1993.
6. Albrecht, Allan J., and John E. Gaffney Jr. "Software function, source lines of code, and development effort prediction: a software science validation." Software Engineering, IEEE Transactions, Vol SE. 9, (6), 1983, pp. 639–648.
7. Matson, Jack E., Bruce E. Barrett, and Joseph M. Mellichamp. "Software development cost estimation using function points." Software Engineering, IEEE Transactions on Vol. 20, (4), 1994, pp. 275–287.
8. Idri, Ali, Taghi M. Khoshgoftaar, and Alain Abran. "Can neural networks be easily interpreted in software cost estimation?." Fuzzy Systems, 2002. FUZZ-IEEE'02. Proceedings of the 2002 IEEE International Conference on. Vol. (2), 2002, pp. 1162–1167.
9. Pendharkar, Parag C. "Probabilistic estimation of software size and effort." Expert Systems with Applications Vol. 37, (6), 2010, pp. 4435–4440.
10. M. Fowler : Refactoring:-Improving the Design of Existing Code, 2000.

11. Demeyer, S., Ducasse, S., Nierstrasz, O., Object-Oriented Reengineering Patterns. Morgan Kaufman, 2003.
12. Godfrey, Michael W., and Lijie Zou. "Using origin analysis to detect merging and splitting of source code entities." Software Engineering, IEEE Transactions on Vol. 31, (2), 2005, pp. 166–181.
13. Shepherd, David, Emily Gibson, and Lori L. Pollock. "Design and Evaluation of an Automated Aspect Mining Tool." Software Engineering Research and Practice, 2004, pp. 601–607.
14. Fluri, Beat, et al. "Change distilling: Tree differencing for fine-grained source code change extraction." Software Engineering, IEEE Transactions on Vol. 33, (11), 2007, pp. 725–743.
15. S. Kim, E. J. Whitehead Jr. Y. Zhang, (2008) "Classifying Software Changes: Clean or Buggy?" IEEE Transactions on Software Engineering, 34(2):181–196.
16. Jalbert, Nicholas, and Westley Weimer. "Automated duplicate detection for bug tracking systems." Dependable Systems and Networks with FTCS and DCC, 2008. DSN 2008. IEEE International Conference on. IEEE, 2008, pp. 52–61.
17. Cotroneo, Domenico, Salvatore Orlando, and Stefano Russo. "Failure classification and analysis of the java virtual machine." Distributed Computing Systems, 2006. ICDCS 2006. 26th IEEE International Conference on. IEEE, 2006, pp. 1–10.
18. Guo, Philip J., et al. "Characterizing and predicting which bugs get fixed: an empirical study of Microsoft Windows." Software Engineering, 2010 ACM/IEEE 32nd International Conference on. Vol. 1, 2010, pp. 495–504.
19. Čubranić, Davor. "Automatic bug triage using text categorization." In SEKE 2004: Proceedings of the Sixteenth International Conference on Software Engineering & Knowledge Engineering, 2004.
20. Donald F Specht. (1990), Probabilistic Neural Networks. Neural Networks, Vol. 3, pp. 109–118.
21. Patterson. D.W. artificial neural networks. Prentice Hall, 1995.
22. Masters, T., Advanced Algorithms for Neural Networks. Wiley, New York, 1995.

Survey of Improved k-means Clustering Algorithms: Improvements, Shortcomings and Scope for Further Enhancement and Scalability

Anand Khandare and A.S. Alvi

Abstract Clustering algorithms are popular algorithms used in various fields of science and engineering and technologies. The k-means is example unsupervised clustering algorithm used in various applications such as medical images clustering, gene data clustering etc. There is huge research work done on basic k-means clustering algorithm for its enhancement. But researchers focused only on some of the limitations of k-means. This paper studied some of literatures on improved k-means algorithms, summarized their shortcomings and identified scope for further enhancement to make it more scalable and efficient for large data. From the literatures this paper studied distance, validity and stability measures, algorithms for initial centroids selection and algorithms to decide value of k. Then proposing objectives and guidelines for enhanced scalable clustering algorithm. Also suggesting method to avoid outliers using concept of semantic analysis and AI.

Keywords Clustering · Unsupervised · k-means · Large data · Semantic analysis

1 Introduction

Data mining is a technique for finding, summarizing, analyzing and visualizing knowledge from data set. Clustering is one of the data mining techniques. Clustering is the process of forming group or clusters of similar data items. Clustering is unsupervised technique because it deals with unlabeled data. There are various clustering algorithms are available such as k-means, k-medoids etc. This paper is focusing on enhancement of k-means clustering algorithm. The k-means is

A. Khandare (✉)
Department of CSE, SGB Amravati University, Amravati, India
e-mail: anand.khandare1983@gmail.com

A.S. Alvi
Department of CSE, PRMIT&R, Badnera, Amravati, India
e-mail: abrar_alvi@rediffmail.com

© Springer India 2016
S.C. Satapathy et al. (eds.), *Information Systems Design and Intelligent Applications*, Advances in Intelligent Systems and Computing 434,
DOI 10.1007/978-81-322-2752-6_48

Fig. 1 Working of standard
k-means algorithm

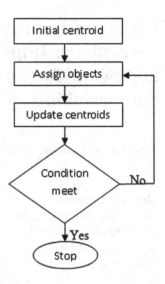

widely used in many applications [1–3]. The basic k-means algorithm is divided into three steps. The first step is randomly select the initial mean or centroids from data objects. Then calculate distance between centroids and data objects. Then assigning data objects into clusters based on minimum distance calculated. The second and third steps are repeated until mean not change. Outline of basic k-means algorithm is given as follows and also shown in Fig. 1.

1. Randomly select data objects as initial mean or centroids from data set.
2. Assign data object into cluster which has the minimum distance.
3. Update centroids for new cluster and repeat step 2–3 until condition meet.

Because of randomly selecting the initial centroids, k-means has shortcomings such as k-means may produce empty clusters, cluster quality is not good. More calculations involved in finding centroids and comparing it with all data objects in each step. It is inefficient and computationally complex when used for big data. k-means also not work well for outliers. To overcome these problems, various algorithms are proposed in following literatures. This paper studied these literatures, summarized and find scope for enhancement for making correct cluster efficiency. Organization of this paper is as follows. Section 1 given introduction and working of k-means, Sect. 2 explains the surveys of improved in k-means, Sect. 3 proposing methodologies objectives, guidelines, parameter for new algorithm and sample algorithm for cluster formation. Experimental results of basic k-means are shown in Sect. 4. Conclusion is presented in Sect. 5.

2 Related Work

Huge research work related clustering algorithms is carried out by academic and industries and research to improve efficiency, accuracy, quality of clusters, and methods for initial centroid selection, cluster validity, to decide number of clusters automatically. Following section explained some of literatures and their scope for further improvements. Initialization method [1] for the k-means algorithm is proposed to improve basic k-means clustering. This method selects proper initial centroids which are well separated and having the potential to form high quality clusters. But this work is only focusing on to improve the quality of clusters. Also this method may take more time to search centroids. The authors [4] proposing circular k-means algorithm for the cluster vectors containing information in a circular shift invariant manner. And also fourier domain representation this measure is given to reduce computational complexity of algorithm. And then splitting and merging technique to reduce local minima and to estimate correct number of clusters is proposed. The proposed approach is focusing only on computational complexity and estimating correct number of clusters but not focusing on quality of clusters. Paper [5] proposing index for stability i.e. sum of the pair wise individual and ensemble stabilities. This index was used to identify number of clusters. Paper [6] proposing spectra analysis of eigen values of the data set to estimate the number of clusters. But similarity measure used in this work is not appropriate. Symmetry clustering algorithm [7] is proposed for determining the number of clusters and proper partitioning of data set. In this work, assignments of data objects to different clusters are done based on point symmetry. This work only concentrating on one form of symmetry i.e. point symmetry not on line symmetry. Assignment order algorithm [8] is proposed which is based on the combination of clustering algorithms. Accuracy of proposed algorithm may be increased again.

Authors in paper [9] proposed the cluster validity indices such as davies bouldin, dunn, generalized dunn, point symmetry, I, xie beni, fs, k, and sv indices. These indices are used to indicate both appropriate numbers of clusters and as well as partitioning of data set also. The cluster validity index in paper [10] is based on the decision theory rough set. This index determines required number of clusters. Paper [11] introducing three algorithms fuzzy moving, adaptive moving and adaptive fuzzy moving k-means algorithms. These algorithms are fast and accurate. But the accuracy of proposed algorithms can further be improved. Paper [12] presented an improved k-means algorithm with the help of noise data filter with density based detection methods based on characteristics of noise data. This algorithm is suitable for small data sets and but time will increase when dealing with large data sets. Paper [13] presented kernel based learning k-means clustering based on statistical learning theory. This is a fast and effective algorithm. Paper [14] is proposing multiple centers technique to represent single cluster. This algorithm consists of the three sub algorithms fast global fuzzy k-means, best m-plot, and grouping multi-center algorithms. The quad tree based k-means algorithm [15] is proposed for finding initial cluster centroids. In this algorithm gain is used to determine the quality of clusters. TW k-means algorithm [16] is the extension of the standard k-means clustering algorithm. The efficiency of this algorithm is high large dimensional data. To improve the clustering quality [17] authors are proposing variable neighbourhood search and also proposed the vsail and pvsail algorithms. These algorithms help to improve the clustering quality with minimum computations, but there is scope for improvement in efficiency. The competitive k-means algorithm [18] to minimize problems of scalability and inconsistency are proposed. These algorithms improve accuracy and decreases variance. The double selection based semi supervised cluster ensemble framework proposed in [19] for improvement in performance of clustering using the normalized cut algorithm. Heuristic method is presented in [20] for pruning unnecessary computations in k-means algorithms. This is a simple and a novel method for avoiding unnecessary computations. Centroid ratio [21] is introduced to compare clustering results to find unstable and incorrectly located centroids. The proposed index [22] focuses on the geometric structure of the dataset and consists of two parts i.e. Compactness and separation. The new algorithm [23] is the extended version of k-means to find weight of dimension in each cluster and then using this weight to find the subsets of dimensions. This algorithm can be used for large data also. Grid density based clustering algorithm [24] is proposed which is based on an artificial immune network for making proper clusters. The proposed algorithm is maintaining higher efficiency and stability also. In [25], validity index is presented to find the required number of clusters. This new index is using one of a measure for multiple cluster overlap and a separation data point. The paper [26] proposing a new method for initialization of centroids based on three parameters such as centrality, sparsity, and isotropy and producing good clusters and improving speed. In this paper, other parameters are not considered. An improved version of the moving k-means algorithm called enhanced moving k-means algorithm presented in [27]. This algorithm is able to generate good quality clusters and less cluster variance and also less depends on the initialization of initial centroids. The paper [28] introduced two new, faster algorithms, i.e. Geometric progressive fuzzy, c-means and random fuzzy, c-means to find the subsample size.

3 Proposed Methodologies

3.1 Objectives

From the above literature survey, it is observed that there is huge scope for improvements in k-means. This paper is focusing on to make k-means clustering algorithm optimized and modern so that it will work efficiently with big and different types data for different applications. This work merging AI techniques with k-means clustering for enhancement. From survey this paper formed the some objectives which are listed given.

Table 1 Parameters for improvements

S. no.	Parameter	Characteristics of algorithm
1	Correct number of clusters	Algorithm should find required number of clusters from data set
2	Semantic cluster	Cluster elements should be semantically closer to each other. This will done by using semantic analysis and AI
3	Correct initial centroids	Algorithm should find required proper initial centroids
4	Quality of clusters	Data items must be in respective cluster only
5	Efficiency of clusters	Algorithm should produce clusters in minimum time
6	Scalability of algorithm	Algorithm should work for large and different types of data
7	Stability	For same data algorithm should produce similar clusters for each run
8	Validity	Validity is the goodness of clustering
9	Optimum path within clusters	Within clusters there must be some optimum path or relation

- To design the improved algorithm for finding the correct value of a number of clusters and initial centroid.
- To improve the quality and efficiency of the cluster using AI.
- To make algorithm scalable for large and different types of data.
- To design the algorithm for optimized distance, similarity and validity indices.
- To design algorithm for improving and finding correct path within clusters.
- To design algorithm for avoiding outliers using semantic analysis.

For the enhancement of algorithms this paper is emphasizing on various parameters. These parameters are listed in Table 1.

3.2 Guidelines

To enhance performance of k-means algorithm, this paper is focused on various aspects such as determining the correct number of clusters, initial centroids selection, optimizing calculations of data objects and cluster centroids and achieves more matching with clustering goals. Guidelines of proposed methodology are given as follows.

a. **Correct cluster and centroids selection**. Select centroids in better way rather than random selection. For centroids selection, this work will propose new algorithms which will minimize chances of empty clusters problem and produce high quality clusters. The desired value of k is estimated using new methods.

b. **Cluster formation**. To form the correct clusters this algorithm will use three criteria.
   Cluster with minimum computation. Form the cluster by avoiding unnecessary calculations and correct centroids selected using new methods.
   Cluster with minimum distance. Form the cluster using new distance measures based on point symmetry And line symmetry which will also work better for outliers in data also and will be used for all types of data.
   Cluster based on semantic analysis. Form the cluster not only on minimum distance of data objects but also the semantic relationship of among the data objects. This step will improve quality of clusters and minimize outliers in clusters.
c. **Finding and improving correct path within clusters**. After formation of clusters this step will check correctness of clusters by comparing clusters whether data objects are placed in proper cluster or not if not improvements will done in clusters.
d. **Cluster fine tuning**. The aim of fine tuning is to make clusters correct based on some parameters like quality, compactness, and separation in cluster.
e. **Updation**. Form the final clusters by updating centroids and redistributing data objects to the final clusters with quality and minimum computations.
f. **Validation**. Evaluation of clustering results is the cluster validation. Using validation measures validate clusters so that cluster will be globally optimized.
g. **Termination**. Stop execution of algorithms based on criterion.

3.3 Sample Algorithm for Cluster Formation

Following is tentative algorithm based on above objective and guidelines.

1. Input data objects apply knowledge representation techniques of AI and organize data objects by their properties, relationship, and category.

   1.1 Represent data object with their value and category and relationship.

2. Then select one data objects from each category of data objects as initial centroids.
3. Then calculate the distance between data objects and centroids.
4. Assign data objects into clusters by using minimum distance, category and relationship. This will improve the quality of clusters.
5. Handle outliers using

   5.1 Using deduction and reasoning techniques, calculate probability handle the outlier.
   5.2 Using semantic analysis handle outliers

6. After first iteration algorithm will select only boundary elements from category of data objects for comparison and checks if objects are in their proper clusters then ignore these in next iterations to minimize computations. Repeat steps from 2 to 6.
7. Stop when criteria meet.

4 Result

To demonstrate working and problems of k-means algorithm, this paper written program for k-means clustering. And set numbers of required clusters (k) are three. For data set{8, 13, 8, 5, 7, 6, 15, 10, 9, 4, 4, 3}. Then randomly select initial centroids are $k_1 = 8$, $k_2 = 13$ and $k_3 = 5$. Result of k-means is shown in Table 2.

From the above example it is observed that value of k is 3 but actually formed clusters are 2. Cluster 1 is not having single data. Such cluster is called as empty cluster. Again there are some objects in one cluster which can belong to other clusters. Therefore quality of clusters is not good because of random centroids. In each step, centroids selected and compared with all data objects. This procedure may increase computations and decreasing efficiency. Also for data set {5, 6, 7, 8, 15, 20}, k = 3 initial centroids are $c_1 = 5$ and $c_2 = 7$ then it is happen that 7 can belongs to 5 or 6 centroid cluster. To address these reasons methodology is proposed.

Tentative result of enhanced k-means using AI

Consider the data set {8, 13, 8, 5, 7, 6, 15, 10, 9, 4, 4, 3}. Organize data objects in category 1: even{8, 8, 6, 10, 4, 4} and Relation (+2) {4, 6, 8, 8, 10} and category 2: odd{13, 5, 7, 15, 9, 3} and Relation (+2) {3, 5, 7, 9}, Relation (+3) {9, 13} and Relation (+2) {13, 15}. There are two categories available, hence the value of k = 2. Then select 8 from category 1 and 13 from category 2 and calculate distance and assign objects based on minimum distance, relation and category. Tentative result is shown in Table 3.

Table 2 Result of basic k-means

Cluster no.	Data objects
1	–
2	1, 3, 4, 4, 5, 6, 9, 10
3	2, 3, 7, 8

Table 3 Result of enhanced k-means

Cluster no.	Data objects
1	8, 8, 6, 4, 4, 5, 7, 9, 10
2	13, 15

5 Conclusion

This paper provided brief study of existing improved k-means algorithms, methods for centroid selection, value of k, and measures for distance, validity and stability, for dealing with outliers and deciding number of clusters in advanced. But still there is scope for improvements. No research focused to remove all deficiencies together and make the single algorithm. This paper summarized shortcomings of algorithms from literature and given the methodology with objectives and guidelines for developing enhanced scalable clustering algorithm for big data. Also highlighted the parameters to make algorithm enhanced and given sample algorithm.

References

1. Robert Harrison, Phang C., Zhong, Gulsah Altun, Tai, and Yi Pan: Improved k-means clustering algo. For exploring protein sequence motifs representing common structural property, ieee trans on nanobio, vol 4, no 3 (2005).
2. V. N. Manjunath Aradhya, M. S. Pavithra: An application of k -means clustering for improving video, text detection, advances in intelligent systems and computing volume 182, 2013, pp 41–47(2012).
3. C. Rajalaxmi, K. P. Soman, S. Padmavathi: Texel identification using k-means clustering method, advances in intelligent systems and computing volume 167, pp 285–294(2012).
4. Dimitrios Charalampidis: A modified k-means algorithm for circular invariant clustering, ieee trans, on patrn anlyss, vol. 27, no. 12 (2005).
5. Kunchev and Dmitry P. Vetrov, Ludmila I.: Evaluation of stability of k-means cluster ensembles with respect to random initialization, ieee tran. On patrn. analys. and machine intelligence, vol. 28, no. 11 (2006).
6. Wenyuan Li, Wee-Keong NG, Ying Liu, Member, and Kok-Leong Ong: Enhancing the effectiveness of clustering with spectra analysis, ieee trans on knowledge and data engineering, vol. 19, no. 7 (2007).
7. Sanghamitra Bandyopadhyay and Sriparna Saha: A point symmetry-based clustering technique for automatic evolution of clusters, ieee transactions on knowledge and data engineering, vol. 20, no. 11 (2008).
8. Yi Hong and Sam Kwong: Learning the assignment order of instances for the constrained k-means clustering algorithm, ieee trans on systems, man, and cybernetics—part b: cybernetics, vol. 39, no. 2(2009).
9. Sanghamitra Bandyopadhyay, and Sriparna Saha: Performance evaluation of some symmetry-based cluster validity indexes, ieee transactions on systems, man, and cybernetics—part c: vol. 39, no. 4 (2009).
10. Pawan Lingras, Min Chen, and Duoqian Miao: Rough cluster quality index based on decision theory, ieee trans. on knowldge and data engineering, vol. 21, no. 7(2009).
11. Nor Ashidi Mat Isa, Samy A. Salamah, Umi Kalthum Ngah: This is an adaptive fuzzy moving algorithm of k-means clustering for image segmentation, ieee trans on consumer electronics, vol. 55, no. 4 (2009).
12. Juntao Wang, Xiaolong Su: An improved k-means clustering algorithm, comm. software and networks ieee 3rd international conference (2011).
13. Kong Dexi, Kong Rui: A fast and effective kernel-based k-means clustering algorithm, ieee conf on intelligent system design and engg. Applications (2013).

14. Jiye Liang, Liang Bai, Chuangyin Dang, and Fuyuan Cao: The k-means-type algorithms versus imbalanced data distributions, ieee trans. on fuzzy systems, vol. 20, no. 4 (2012).
15. Partha Sarathi Bishnu and Vandana Bhattacherjee: Software fault prediction using quad tree-based k-means clustering algorithm, ieee trans on knowg and data engg, vol. 24, no. 6 (2012).
16. Xiaojun Chen, Xiaofei Xu, Joshua Zhexue Huang, and Yunming Ye: TW-k-means: automated two-level variable weighting clustering algorithm for multiview data, ieee trans. on knowlge and data engi, vol. 25, no. 4 (2013).
17. Jie Cao, Zhiang Wu, Junjie Wu, Member, Ieee, and Hui Xiong: Sail: Summation-based incremental learning for information-theoretic text clustering, ieee trans on cybernetics, vol. 43, no. 2 (2013).
18. Rui Máximo Esteves, Thomas Hacker, Chunming Rong: Competitive k-means-a new accurate and distributed k-means algorithm for large datasets, ieee international conference on cloud computing technology and science (2013).
19. Zhiwen Yu, Hongsheng Chen, Jane You, Hau-San Wong, Jiming Liu, Le Li, and Guoqiang Han: Double selection based semi-supervised clustering ensemble for tumor clustering from gene expression profiles, ieee/acm trans on computational biology and bioimrmtcs, vol. 11, no. 4 (2014).
20. Kazuki Ichikawa and Shinichi Morishita: A simple and powerful heuristic method for accelerating k-means clustering of large-scale data in life science, ieee/acm trans compt biology and bioinf, vol. 11, no. 4(2014).
21. Qinpei Zhao and Pasi Fränti: Centroid ratio for a pairwise random swap clustering algorithm, ieee trans on knowld and data engg,. V. 26, no. 5 (2014).
22. Hongyan Cui, Mingzhi Xie, Yunlong Cai, Xu Huang, Yunjie Liu: Cluster validity index for adaptive clustering algorithms, iet commun., vol. 8, iss. 13(2014).
23. Liping Jing, Michael K. Ng, and Joshua Zhexue Huang: An entropy weighting k-means algorithm for subspace clustering of high-dimensional sparse data, ieee trans. on knowledge and data engg, vol. 19, no. 8 (2007).
24. Chuan Ming Chen, Dechang Pi and Zhuoran Fang: Artificial immune k-means grid- density clustering algorithm for real-time monitoring and analysis of urban traffic", electronics letters vol. 49 no. 20 pp. 1272–1273 (2013).
25. Hoel Le Capitaine and Carl Fr´elicot: A cluster-validity index combining an overlap measure and a separation measure based on fuzzy-aggregation operators, ieee trans on fuzzy systems, vol. 19, no. 3 (2011).
26. Pilsung Kang, Sungzoon Cho: k-means clustering seeds initialization based on centrality, sparsity, and isotropy, lncs vol 5788, 2009, pp 109–117(2009).
27. Fasahat Ullah Siddiqui and Nor Ashidi Mat Isa: Enhanced moving k-means algo. for image segmentation", ieee trans. On cons electrcs, vl. 57, no 2 (2011).
28. Jonathon K. Parker, and Lawrence O. Hall: Accelerating fuzzy-c means using an estimated subsample size", ieee trans on fuzzy system, vo. 22, no. 5(2014).

Attack Identification Framework for IoT Devices

Jagan Mohan Reddy Danda and Chittaranjan Hota

Abstract With the emergence of Internet and embedded computing, Internet of Things (IoT) is currently becoming an area of interest amongst researches. IoT enable interconnection of embedded devices capable of running application like smart grid, smart traffic control, remote health monitoring etc. As the IoT devices can be connected virtually as well as physically, cyber attacks are likely to become a major threat. An attacker who have an access to the on-board network connecting IoT devices or appliances can spy on people, can inject malicious code into these embedded devices creating serious security concerns. In this work, we propose a framework to monitor security threats possible on IoT devices. The framework consists of several modules like data capture, anomaly detector and alert generator. The data capture module collects the application level data, transport and network headers of the traffic that goes into the IoT device. The anomaly detector module uses a signature based approach to detect threats. The proposed framework is tested on a testbed comprising of Arduino boards with Wiznet Ethernet shield as the IoT device communicate with Samsung Android smart-phone over a bridge connected through WiFi. We ran SNORT Intrusion Detector on the bridge with rules for generating alerts for intrusion.

Keywords P2P · Detection · Embedded devices · IDS

J.M.R. Danda (✉) · C. Hota
Birla Institute of Technology and Science-Pilani, Hyderabad Campus,
Shameerpet, R.R. District 500078, Telangana, India
e-mail: jagan.reddy507@gmail.com

C. Hota
e-mail: hota@hyderabad.bits-pilani.ac.in

© Springer India 2016
S.C. Satapathy et al. (eds.), *Information Systems Design and Intelligent Applications*, Advances in Intelligent Systems and Computing 434,
DOI 10.1007/978-81-322-2752-6_49

1 Introduction

Internet of Things (IoT) is an emerging area that is playing a central component of a smart world where service Ubiquity and network convergence are the main pillars [1]. Recent advances in mobile wireless networks, cloud and Big data technologies have supported the generation of IoT services. IoT promises to connect various objects (Physical and Virtual) to the Internet and enable a plethora of newer applications. In IoT, devices gather and share information directly with each other leaving security loop holes that can be exploited by smart attackers. The Internet-of-Things (IoT) is an emerging research area with revolutionary applications in a number of environments, including:

- **Health Domain**: The IoT can be used at hospitals and medical centers to collect real-time data on the patients, monitor their health status, help doctors with consistent journaling of patients' treatments, etc. Obviously, privacy, confidentiality, and integrity are of utmost concern in such an area.
- **Agriculture**: The IoT can be deployed at massive scales in farming lands to monitor the conditions at which crops grow, e.g., monitor weather conditions, use of pesticides, spread of infections, existence of parasites, etc.
- **Corporate Infrastructure Management**: The measurement of resource consumption (electricity, gas, water, etc.) can be carried out over an extensive IoT network. The prevention of tampering with the meters setup within an organization, or of external attacks (such as setting a whole corporate office in the dark), will be of critical importance in the future as cyber attacks are becoming common in modern warfare.

The fascinating opportunities presented by the IoT for future applications, can only be realized through the massive deployment of inexpensive miniature devices in the physical world. But along with massive deployment come massive problems. The operation of devices in non-controlled, possibly hostile environments, puts the reliability of IoT systems at stake. More specifically, adversaries may inject malware into the IoT devices, intercept and modify the communication pattern between two/more IoT devices, and clone the devices to instrument the data gathering and overall operation to their interest. OpenDNS provides DNS service and enables reliable Internet connectivity with zero downtime, when we operate these devices remotely anywhere, anytime. The OPenDNS is still under development to protect the devices. In recent times, the OpenDNS Security Lab [2] discovered, the providers hosting malicious domains to damage the IoT infrastructures. They found several vulnerable and incorrectly configured SSL connections.

This paper proposes a testbed for monitoring malicious activities in smart home appliances. The major objective is to set-up an experimental environment where the proposed framework generates alerts when malicious activities are being carried out on an IoT device in real-time setup. For this work we capture the network traces from the IoT devices and analyze the behavioral communication patterns between those devices.

Once the behavior of the IoT device is analyzed, we create SNORT rules that can identify newer attacks in future. To demonstrate a higher degree of penetration (attack possibility), we considered Souliss [3] which is an open source Peer-to-Peer (P2P) application to run on IoT devices in our framework. Souliss allows peers to communicate directly with other peers in a distributed manner, without any intervention of a Central authority. The anomaly detector can learn zero-day attack patterns and can create new signatures to protect IoT devices.

The main contributions in this paper are:

- It proposes a framework for IoT security over IP-based communication.
- It demonstrates the Intrusion Detection alert classification using signatures.

The rest of the paper is organized as follows: Sect. 2 covers the related work on security challenges for IoT devices. Section 3 describes the overview of testbed design and implementation. In Sect. 4, we describe the detailed analysis of network traces. In Sect. 5 we discuss the framework for IoT device attack identification and Sect. 6 concludes the paper and propose the future work.

2 Related Work

In [4], authors proposed challenges with special focus on standard IP security protocols in the context of IoT devices. The secure protocols should be lightweight and able to run on small embedded devices and support end-to-end security and domain-specific protocol. They also addressed various layers of security and requirements needed for IoT security. In their work, they consider three major challenges: (i) heterogeneous Communications-like DoS resistance, protocol translation and End-to-End security, (ii) bootstrapping of security domain-like distributed versus centralized, identity and privacy-aware identification, (iii) operations-like End-to-End security and group membership and security.

In [5], authors described a systematic approach to security in IoT. They considered identification and authentication, trust, privacy, responsibility, autoimmunity, safety and reliability. They also describe these interactions with various application domains.

John Pescatore [6] presents a survey paper on IoT security. Their survey revealed that, almost 90 % of respondents recognized that changes to security controls will be required, with 50 % believing major (if not complete) enhancements and replacements to many controls. Authors also discuss the critical security concerns in hardware and software inventory, vulnerability assessment and configuration management.

3 Testbed Design

3.1 Testbed Design

We set-up a testbed in our Information Security lab with Arduino board, Laptops, Smart-phones, LED and a WiFi router. The tesbed also contains one Ethernet Wiznet W5100 shield. The basic Arduino board and Ethernet Wiznet shield are attached together to communicate between IoT device and Smart-phone. In this design the Laptop acts like a bridge between Smart-phone and the IoT device (Arduino board) and is also used to capture the network traces using Wireshark [7] which is useful for debugging and troubleshooting to know the connectivity of IoT devices. Using the Smart-phone we control the IoT devices over a P2P network. Figure 1 shows our testbed setup.

3.2 Software

Arduino-1.0.6 environment [8] was used to program the embedded devices. We developed a basic C++ program to make the LED ON/OFF on the board and executed this program on Arduino which is attached to the Ethernet board for communicating with Android Smart-phone (via the bridge). To monitor the communication between the Mobile and IoT device, an open source protocol analyzer Wireshark [7] is also installed in the Laptop.

3.3 Hardware

The implementation has brought the Arduino and Ethernet shield together (we refer as IoT device throughout the paper) to perform the desired function. Arduino node

Fig. 1 Testbed for IoT at our InfoSec. Lab

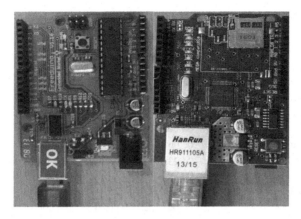

Fig. 2 Arduino board and Wiznet Ethernet Sheild W5100

```
alert udp $EXTERNAL_NET any <> $HOME_NET any (msg:"IoT Packet
Detected"; content:"|0c 0b 17 11 00 25 01 08 0f 00 00 00|";
classtype:IoT_vNet_Packet; nocase; sid:100222; priority:30;)
```

Fig. 3 Ping request from SoulissApp to IoT node as a SNORT signature

is connected to a Laptop through USB interface to provide the power supply and also is used as a serial port to upload the code into the chip-set. In Fig. 2, we show the basic Arduino board and Ethernet shield W5100 that support TCP/IP stack. The detailed data-sheet of this board is described in [9] (Fig. 3).

4 Analysis

In this section, we observe the communication patterns of the IoT device from the network traces. We provide the detailed description of the communication patterns from Android phone to Arduino Node. We installed the SoulissApp on rooted Android phone. Souliss API carries the communication MaCaco over vNet, it builds a flat virtual network for building and routing over different communication media. It also serves event-based and state-less protocol, allows P2P communication between nodes and it saves the battery and offers fast interaction amongst IoT nodes [3]. Using SoulissApp things such as lights, windows, refrigerators, etc. can be handled or monitored remotely in a home environment.

SoulissApp provides different low-level UDP connectivity operations to test whether the IoT device is up or down. Following commands describe different ways of communication from smart phone to IoT device.

- **Ping**: It checks the connectivity of the IoT device. The source SoulissApp uses port number 23,000 and sends a ping message to destination (IoT) on port

Table 1 Commands and signatures that can check the status of IoT

Command		Port number	Signature in Hex decimal	Bytes/Pkt
Ping	Send	23,000	0c 0b 17 11 00 25 01 08 0f 00 00 00	12
	Receive	230	0c 0b 17 25 01 11 00 18 0f 00 00 00	12
Poll	Send	23,000	0c 0b 17 11 00 25 01 27 00 00 00 01	12
	Receiver	230	24 23 17 25 01 11 00 37 00 00 00 18 00 00 00.....	36
TypReq	Send	23,000	0c 0b 17 11 00 25 01 25 00 00 00 01	12
	Receive	230	24 23 17 25 01 11 00 32 00 00 00 18 00 00 00.....	36
Health	Send	23,000	0c 0b 17 11 00 25 01 25 00 00 00 01	12
	Receive	230	0d 0c 17 25 01 11 00 35 00 00 00 01 ff	13

number 230 over UDP. Each sent and received 'ping' packet has 12 Bytes of payload length. The detailed ping commands are shown in Table 1.

- **Poll**: This command requires all Souliss network data without any subscription. This command also works similar to the ping command to obtain the data from the IoT devices. The source SoulissApp uses port number 23,000 and sends a poll message to destination (IoT) on port number 230 over UDP. Each sent 'poll' packet has 12 Bytes and received 'poll' packet has 36 Bytes of payload length as shown in Table 1.
- **TypReq**: Typical request command reads the data changes in IoT device with no subscription. This obtains the data from the IoT devices. The source SoulissApp uses a port number 23,000 and sends a message to destination (IoT) on port number 230 over UDP. Each sent 'TypReq' packet has 12 Bytes and received 'TypReq' packet has 36 Bytes of payload length as given in Table 1.
- **Health**: This command is used to request communication's health to identify healthiness of the IoT device. The source SoulissApp uses a port number 23,000 and sends a message to destination (IoT) on port number 230 over UDP. Each sent 'Health' packet has 12 Bytes and received 'Health' packet has 13 Bytes of payload length as given in Table 1.

Table 1 refers to the communication patterns between SoulissApp and the IoT device. The first and second byte refer to the total frame size. The third byte specifies the protocol identification number for MaCaco that goes over vNet [10]. The fourth and fifth bytes refer the addresses to identify the nodes into the vNet network. The remaining bytes are referred as payload of the frame. Further implementation details of MaCaco and vNet is available at Soluliss-A5.2.1 [10].

5 Framework for IoT Threat Detection

Design of a robust framework for IoT security has gained a lot of importance. To implement a robust security framework, first we need to understand the behavior of these devices. Due to the light-weight protocol design there is not much security and authentication mechanisms on these platforms. But there are several security models for current Internet that has gained enough attention in this field. The objective of our proposed robust framework is to profile the IoT devices that can classify the security threats. Figure 4 describes the proposed framework. The major components of this framework are:

- **Detector**: It identifies presence of IoT devices in the network and the known vulnerabilities within those IoT devices. It constantly inspects the payload of every packet and maps with the existing database signatures. If it finds any suspicious activity, then it raises an alarm.
- **Rule Base Engine**: If any zero-day attack happens, the detector module fails to give alarm. Further, we use a rule base engine, where it creates a new rule for every unknown threat and sends to learning module to form a new signature that updates the database.

Once we extract the signatures from the network captures, we experimented with an open source intrusion detector SNORT. The SNORT alert polices are shown in Fig. 3 from the network traces. In our InfoSec Lab, a dedicated PC is installed to run SNORT which identifies the malicious activities on IoT devices. Once we are ready with alert database, SNORT is able to classify the traffic in real-environment as shown in Fig. 5.

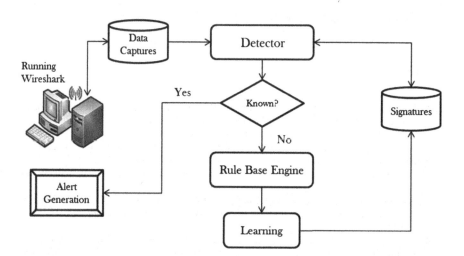

Fig. 4 Framework to identify IoT threats

Fig. 5 SNORT alerts when signature maches with the IoT packet

6 Conclusion and Future Work

In this work we proposed a framework to identify threats on IoT devices based on the traditional signature based techniques. We capture the network traces from the IoT devices and extract the signatures from the payload. After signatures are obtained, we created rules/alerts that are compatible with SNORT rules. SNORT IDS parses the packet payload and maps the existing signature and generates an alert for the administrator. The limitation of our framework is: we check every payload that comes to the IoT device while it communicates with other devices, which is similar to deep packet inspection (DPI) that creates privacy issues. However our framework does this payload check on bridge (not on the IoT device) that has required amount of resources. We will extend this work to identify security threats on IoT devices using Machine Learning techniques that could be more efficient than the DPI work presented in this work. More over these devices are extremely sensitive to hackers to exploit vulnerabilities. The IoT devices need to be monitored closely to provide alarms when an attack is detected.

Acknowledgments This work was supported by grant from Tata Consultancy Services (TCS) under research scholar program, India.

References

1. Y. Berhanu, H. Abie, and M. Hamdi. A testbed for adaptive security for iot in ehealth. In *Proceedings of the International Workshop on Adaptive Security*, page 5. ACM, 2013.
2. O. S. Lab. The 2015 internet of things in the enterprise report: Executive summary. available online at:. http://info.opendns.com/rs/033-OMP-861/images/OpenDNS-2015-IoT-Executive-Summary.pdf/. Accessed on July 2015.
3. Souliss. Souliss. available online at:. https://code.google.com/p/souliss/. Accessed on Jan 2015.
4. T. Heer, O. Garcia-Morchon, R. Hummen, S. L. Keoh, S. S. Kumar, and K. Wehrle. Security challenges in the ip-based internet of things. *Wireless Personal Communications*, 61(3):527–542, 2011.
5. A. Riahi, Y. Challal, E. Natalizio, Z. Chtourou, and A. Bouabdallah. A systemic approach for iot security. In *Distributed Computing in Sensor Systems (DCOSS), 2013 IEEE International Conference on*, pages 351–355. IEEE, 2013.

6. G. S. John Pescatore. Securing the internet of things survey. available online at:. https://www. sans.org/reading-room/whitepapers/analyst/securing-internet-things-survey-34785/. Accessed on Feb 2015.

7. Wireshark. Wireshark. available online at:. https://wireshark.org/, 2015.

8. Arduino. Arduino. available online at:. http://arduino.cc/en/Main/Software/. Accessed on Jan 2015.

9. Wiznet. Wiznet ethernet w5100. available online at:. https://sparkfun.com/datasheets/. Accessed on Jan 2015.

10. S. A5.2.1. Souliss archive. available online at:. https://souliss.googlecode.com/archive/A5.2.1/. Accessed on Jan 2015.

Compact Coalescence Clustering Algorithm (C³A)—A GIS Anchored Approach of Clustering Discrete Points

Anirban Chakraborty, Jyotsna Kumar Mandal, Pallavi Roy and Pratyusha Bhattacharya

Abstract GIS is a subject with multi-disciplinary applications, ranging from military applications, weather forecasting, recognizing biodiversity prone regions to hotspot identification for socio-economic purposes. The main focus of the work is identification of neighboring tourist hot spots based on the Compact Coalescence Clustering Algorithm (C³A). In order to achieve this goal, firstly various tourist spots, along with the major cities and towns are identified (digitized) and based on the proposed clustering algorithm, which actually works in two phases, various existing clusters of neighborhood tourist hot spots are generated. In the first phase of the process using Clustering of Noisy Regions (CNR), several clusters of tourist spots are formed based on certain threshold distance value and accordingly the centroid is computed and updated every time the cluster is expanded. The soft clustering approach Fuzzy C-Means (FCM) is applied on the result produced in order to enhance the compactness of the clusters formed. Furthermore, the location of the tourist spots neighborhood to major cities and towns are displayed graphically on the map.

Keywords GIS · CNR · FCM · Digitized map · Tourist spot · Hotspot · Soft clustering · Hard clustering

A. Chakraborty (✉) · P. Roy · P. Bhattacharya
Department of Computer Science, Barrackpore Rastraguru Surendranath College,
Barrackpore, Kolkata 700 120, West Bengal, India
e-mail: theanirban@rediffmail.com

P. Roy
e-mail: roypallavi8@gmail.com

P. Bhattacharya
e-mail: pratyusha.bhattacharya13@gmail.com

J.K. Mandal
Department of Computer Science and Engineering, University of Kalyani, Kalyani, Nadia,
West Bengal, India
e-mail: jkm.cse@gmail.com

© Springer India 2016
S.C. Satapathy et al. (eds.), *Information Systems Design and Intelligent Applications*, Advances in Intelligent Systems and Computing 434,
DOI 10.1007/978-81-322-2752-6_50

515

1 Introduction

GIS has diverse applications in numerous fields, proving itself as a powerful tool for spatial data analysis, spatial data creation and visualizing the result on a professional quality map. Thus GIS benefits various sectors of almost diverse number of industries. Nowadays there is an increased awareness of the social and economic impact that GIS can have in the society. One such application of GIS includes the identification of hotspots. Hotspot is a statistically significant area which can be distinguished within the whole study area based on certain statistical data analysis which somehow differentiates that area from the rest of the study area. Thus, tourist hotspots are the areas important to the tourists.

In a fast growing economy like India there are several strategies laid out every day to enhance its economic strength, among those strategies tourism is an important sector which has a great impact on India's economy. The proposed work enhances the impact by identifying the tourist hotspots and major towns/cities located near those hotspots, thereby providing a proper vision to the tourists so as to give them an idea about the significant places they can visit anchoring on a specific town or city located nearby it. To achieve desired objective of cluster formation of tourist hotspots, the work has been done in two phases. In the first phase a hard clustering approach CNR (Clustering of Noisy Region) [1] is applied where clustering begins from the northern-most or the western-most tourist spot, depending upon whether the map is of vertical or horizontal type respectively. The point (spot) satisfying threshold distance (in terms of Manhattan distance) is included in the cluster and the centroid is computed. The cluster is expanded by including all the points satisfying the threshold distance from the centroid and inclusion of any point requires re-calculation of the centroid position. If any point, due to not satisfying the above criterion, does not get included in any cluster, is treated as a lonesome point and is forcefully included into its nearest cluster. In order to achieve compactness of each cluster a soft clustering algorithm FCM [2] (Fuzzy C-Means) is applied, thereby the result obtained produces a compact and well separated cluster. Hence, the proposed clustering algorithm is a hybrid of hard and soft clustering algorithm. The major idea behind this hybrid clustering algorithm, is to extract the good features of both CNR and FCM. FCM produces very compact clusters, but at the same time faces two major problems. First, it sometimes produces virtual clusters; clusters associated with no real data points, having no practical impact. Secondly it elapses much iteration before termination with a substantial value of terminating criterion. To increase the compactness of already formed rough clusters by CNR, FCM is applied and hence the above two problems vanishes. In addition, CNR can be regarded as a modified version of DBSCAN, which was mainly targeted for clustering with noisy data and hence CNR has also the same capability of dealing with noise but the task of handling noise was quite a difficult for FCM alone.

The present clustering method can also be extended for the purpose of identification of health centers/nursing homes in the vicinity of a city/town centering around a medicine distributor or diagnostic center for a smooth and timely supply of the necessary aids and in the same way for identification of various schools, serving as examination centers in a specific city/town centering around the controlling sections, enabling easy distribution of examination materials and moreover clustering the schools in such a manner will facilitate the parents with ample choice for selection of a suitable school for their children; similarly clustering of hospitals around a patient's residence helps to choose the nearest one to meet an emergency. The cluster formed by the present method has been compared with K-means [3], MST based [4], Tree based [4], CNR [1] and FCM [2] (Fuzzy C-Means) algorithms and has been found to perform better in most cases, judged on the basis of performance criteria.

Section 2 of this paper deals with the proposed technique. The implemented results are given in Sect. 3. Analysis and comparisons are outlined in Sect. 4 and finally conclusions are drawn in Sect. 5.

2 The Technique

Initially, any raster map [5] is considered and only the tourist spots along with the major towns/cities are digitized [5] by digitation tool, incorporated in the present technique. Figure 1, shows such digitized map, where on digitization major towns are represented by right triangle and tourist spots by circles.

To achieve the objective it is needed to form clusters of neighboring tourist-spots, so that none of the spot is located far than a prescribed threshold, i.e. eps, measured from the cluster center, enabling a tourist to visit those spots under a single trip. Here value of eps is chosen heuristically. A tourist spot, say X is said to be eps *neighborhood* of another tourist spot, say Y, if Manhattan distance between X and Y, say Dist(X, Y), is less than or equal to eps, i.e. Dist(X, Y) <=eps. After forming cluster with respect to eps, a soft clustering algorithm is applied to fulfill the goal of increasing the cluster compactness. For FCM, a parameter "m" is there, when m is large, each data point becomes member of almost every cluster, resulting in a clustering unsuitable for practical application. Generally 'm' needs to be smallest for optimal estimation of maximum number of clusters. However, taking m = 1, degrades FCM into K-Means clustering technique. Although there is no particular selection strategy available for 'm', 'm' should lie between 1.5 and 3 for giving a good result [2].

Fig. 1 Digitized map

Let $T = \{t_1, t_2,\ldots, t_n\}$ are the set of tourist spots. By calculating the Manhattan distance (Eq. 1) between Northern-Southern spot and Eastern-Western spot, it is decided whether the map is spread over horizontally or vertically.

$$\text{DISTANCE} = |X_0 - X_1| + |Y_0 - Y_1| \tag{1}$$

For maps which are vertically spread, the formation of clusters begins from top most point, or else, it begins from the western most point. Initially, this point becomes the sole member of the cluster. If Manhattan distance, between this point and with all other members of the set T is greater than eps then the considered point becomes lonesome member and the step is repeated with next member of the set T, else the point is treated as the core point of the cluster. The point with minimum distance from the centroid and satisfying eps criterion is included in the cluster and the centroid is computed. The cluster is expanded by including all the points satisfying eps criterion from the centroid and inclusion of any point requires re-computation of the centroid position.

When a cluster can't be expanded furthermore, it's accepted as complete and formation of next cluster starts with above procedure and the cluster formation procedure with respect to eps finishes when no more points are there in T to judge. All those points, designated as lonesome point, are finally forcefully included to their respective nearest cluster, although they are not satisfying eps criterion. Thus the cluster with respect to eps will be formed.

To increase the compactness of the clusters formed, FCM is applied with some modification, where the rough clusters obtained so far is the input to the FCM. If X is the number of clusters obtained and n is the total number of points, a membership matrix A ($X \times n$) is formed and initialized in such way that $A_{ij} = 1$ if ith region belong to jth cluster; otherwise $A_{ij} = 0$. At kth step, center vector of the cluster, say $C^{(k)} = [c_i]$, is calculated by the formula given in Eq. 2.

$$c_i = \left(\sum_{j=1}^{n} A_{ij}^m * t_j\right) / \left(\sum_{j=1}^{n} A_{ij}^m\right); \ 1 \leq i \leq X \tag{2}$$

where m is any real number greater than 1

At each iteration, A ($X \times n$) is updated by the Eq. 3.

$$A_{ij} = 1 / \left(\left(\sum_{k=1}^{X} d_{ij}/d_{kj}\right)^{2/m-1}\right); \quad 1 \leq i \leq X; 1 \leq j \leq n$$

$$\text{where } d_{ij} = \sqrt{\left(\sum_{i=1}^{n} (t_i - c_j)\right)} \tag{3}$$

If $\|A(k) - A(k-1)\| < \epsilon$, ($\epsilon$ being the terminating criterion)then the final membership matrix will be formed. Otherwise the procedure is carried out [2].

Here, FCM is related with optimal least square error functional equation $F_m(A,C) = \sum_{i=1}^{n} \sum_{j=1}^{X} \left(A_{ij}^m \right) \left\| t_i - c_j \right\|_w^2$ by providing the mean with the help of the Eqs. 2 and 3 where w is n × n weight matrix. When w becomes identical matrix I, hyper spherical clusters are identified by F_m. Otherwise w form hyper ellipsoidal clusters with axes proportional to the Eigen values of w.

3 Implementation and Results

The implementation results, output and operations are presented in this section. The implementation has been done in Net Beans (Java) [6–8], based on only flat-file system without using databases for increasing portability. Figure 2 shows the appearance of the cluster formed (shown in Fig. 2, bounded by red colored polygons), where the red squares denotes the centroid of each cluster, signifying the fact that to visit all the places belonging to one cluster, its best to stay in a town nearest to the centroid.

Figure 3 depicts the application of finding all the hospitals nearby to the patient's residence, enabling to meet the emergency cases in a easiest way.

4 Analysis and Comparison

For measuring the quality of clusters, two criteria have been used namely Object positioning [9] and Time factor. It is designed so as to measure the quality of cluster sets at lowest level of granularity. Ideally it's needed to generate partitions that have

Fig. 2 Cluster of tourist-spots

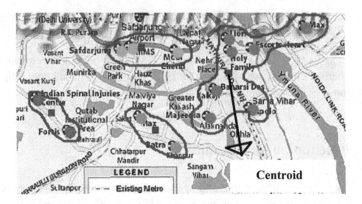

Fig. 3 Cluster of hospitals

compact, well separated clusters. Hence, the criteria used presently combine the two measures to return a value that indicates the quality of the partition thus the value returned is minimized when the partition is judged to consist of compact well separated clusters with the criteria judging different partition as the best one. The second criterion is based on time efficiency.

For comparison purpose, five existing well-known clustering algorithms have chosen, which works efficiently on clustering of discrete points, along with presence of noise points; these are MST (Minimum Spanning Tree) [4], Tree-based [4] and K-means [3], CNR [1] and FCM(Fuzzy C-Means) [2]. Formation of clusters, on execution of these algorithms on the tourist map of West Bengal are depicted in Fig. 4.

To compute the value of object positioning for each test cases (i.e., six maps shown above), the largest distance between two object points O_i and O_k within the same cluster along with smallest distance between that object point O_i with all other object points belonging to all other clusters O_m is computed. The difference between these two gives the value of object positioning, shown in Table 1 for above six maps.

As smaller value of object positioning for any clustering algorithm is considered to be superior so the proposed technique performs better from other five algorithms. However with respect to time efficiency the proposed technique performs better than the tree based algorithms. The proposed algorithm also overcomes the individual demerits of CNR [1] and FCM [2], where, in case of CNR [1], which produces a less compact result and in addition to it the isolated point is not always conveniently positioned and in case of FCM [2] the problem of virtual cluster generation is eradicated here. As the rough clusters are already generated through CNR [1], on which the FCM [2] is applied; so formation of virtual cluster is eradicated but at the same time compactness is increased. This also explains the reason behind increase in time factor for the proposed clustering algorithm, as it works in two phases combining the goodness of CNR [1] and FCM [2] respectively.

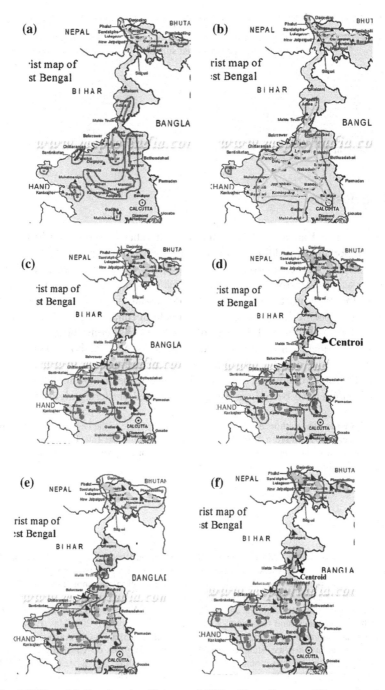

Fig. 4 **a** MST based. **b** Tree based. **c**: K-means. **d** CNR. **e** Fuzzy C means. **f** Proposed technique

Table 1 Comparison of performance of various algorithms based on cluster evaluating matrices

Algorithm	Object positioning (less value is considered superior)	Time factor (ms) (less time is efficient)
MST [4]	2516	2122
Tree-based [4]	833	2098
K-means [3]	1717	90
CNR [1]	813	128
Fuzzy C-means [2]	775	136
Proposed algorithm	525	223

5 Conclusions

From Kashmir to Kanyakumarika, India has a number of spots with historic and religious importance, in addition with natural beauties of the spots attracts a number of local and foreign tourists every year. In another way hotspot identification is an interesting way to identify the significant tourist locations and organize them into a cluster, so that the tourists can have a clear idea about the places they want to visit anchored on a specific city/town. So it would help in enhancement of the economic growth of the country as well. Henceforth, the work has further been expanded by taking into consideration of schools and hospitals. Where the hotspots of schools and hospitals are identified based on the nearby controlling section (for school) and medicine distributor/diagnostic center (for hospital) respectively. In all the cases the cluster formation have been done using the proposed clustering algorithm.

Acknowledgments The authors express a deep sense of gratitude to the Department of Computer Science, Barrackpore Rastraguru Surendranath College, Kolkata-700 120, India and Department of Computer Science and Engineering, University of Kalyani for providing necessary infrastructural support for the work.

References

1. Anirban Chakraborty, J.K. Mandal, Pallavi Roy, Pratyusha Bhattacharya, Clustering of Noisy Regions (CNR)—A GIS Anchored Technique for Clustering on Raster Map, Proceeding 2nd International Conference on Computer and Communication Technologies (IC3T 2015), pp. 511–520, Springer AISC, Volume 381, 2016.
2. James C. Bezdek, Robert Ehrlich, and William Full, "FCM: The Fuzzy c-Means Clustering Algorithm", Computers and Geosciences Vol. 10, No. 2–3, pp. 191–203, May 1984.
3. Anirban Chakraborty, J.K. Mandal, S.B. Chandrabanshi, S. Sarkar, A GIS Anchored system for selection of utility service stations trough k-means Method of Clustering, Conference Proceedings "Second International Conference on Computing and Systems (ICCS-2013)", ISBN 978-9-35-134273-1, McGraw Hill Education (India) Private Limited, pp. 244–251, 21–22 September, 2013.
4. Anirban Chakraborty, J.K. Mandal, A GIS Anchored system for clustering discrete data points —a connected graph based approach, Emerging ICT for Bridging the Future-Volume 1,

Advances in Intelligent Systems and Computing 337, Springer International Publishing, Switzerland, DOI:10.1007/978-3-319-13728-5_5, pp. 43–51, December 2014.

5. Anirban Chakraborty, J.K. Mandal, Arun Kumar Chakraborti, A File base GIS Anchored Information Retrieval Scheme (FBGISIRS) through Vectorization of Raster Map, International Journal of Advanced Research in Computer Science, ISSN No. 0976-5697 volume-2 No. 4, pp. 132–138, July–August 2011.

6. http://www.tutorialspoint.com/java/index.htm, accessed on 14 September, 2013.

7. http://www.zetcode.com/tutorials/javaswingtutorial, accessed on 14 February, 2015.

8. https://netbeans.org/kb/docs/java/javase-intro.html, accessed on 14 February, 2015.

9. Bhavani Raskutti and Christopher Leckie, An Evaluation of Criteria for Measuring the Quality of Clusters, IJCAI Proceedings of the 16th International Joint Conference on Artificial Intelligence, Volume 2, Morgan Kaufman Publishers Inc., San Francisco, pp. 905–910, 1999.

Miniaturized UWB BPF with a Notch Band at 5.8 GHz Using Cascaded Structure of Highpass and Lowpass Filter

Arvind Kumar Pandey, Yatindra Gaurav and R.K. Chauhan

Abstract The objective of this paper is to reduce the size of UWB filter with a notch without any via or DGS to make fabrication easier. The proposed filter is designed by cascading high pass and low pass filter. High pass structure is created by a planar Interdigital structure and low pass is created by a Hairpin line, which make the overall size of the filter much miniaturized. The frequency response of the filter has pass band frequency between 3.1 and 10.6 GHz with wide stop band. There is a notch in pass band at 5.8 GHz with attenuation around 16 dB to avoid interference from the WLAN. The overall size of the filter is 3.36 × 4.132 mm^2 which is much smaller than many previously reported structure of filters.

Keywords UWB · FCC · Interdigital · Hairpin line · Notch band

1 Introduction

In 2002, the Federal Communications Commission (FCC) of the United States fixed the unlicensed frequency range 3.1–10.6 GHz for the Ultra Wideband (UWB) commercial use [1]. Band pass filter is a very essential part of any communication system but for an UWB system a band pass filter with Ultra-Wide pass band bandwidth is required. Design of such wideband filters is definitely a big challenge for traditional filter designers due to some requirements such as Ultra Wide pass band bandwidth, miniaturization, in band frequency rejection notch,

A.K. Pandey (✉) · Y. Gaurav · R.K. Chauhan
Electronics and Communication Engineering Department,
M.M.M.U.T., Gorakhpur, India
e-mail: arvindmknk@gmail.com

Y. Gaurav
e-mail: ygaurav2000@gmail.com

R.K. Chauhan
e-mail: rkchauhan27@gmail.com

© Springer India 2016
S.C. Satapathy et al. (eds.), *Information Systems Design and Intelligent Applications*, Advances in Intelligent Systems and Computing 434,
DOI 10.1007/978-81-322-2752-6_51

525

wide stop band, and low insertion loss over the pass band frequency range. To cope up with these challenges, a number of different approaches have been reported by different authors.

Most of the works in the design of Ultra Wide Band filter are based on the Multi Mode Resonators (MMR), which was initially developed in [2, 3], but the MMR based technique suffers from a narrow upper stop band frequency range due to high order harmonics which rises just after the first pass band.

Some researchers added stub in the center of the MMR structure to raise resonator modes and to improve stop band attenuation of the filter [4, 5].

Furthermore, other structures are added to MMR method to enhance selectivity, out of band rejection, and to reduce the size of the filters [6–8] but suffered from signal error and large size.

Quasi lumped element prototype is another technique which has been used to build UWB band pass filters [9–11].

Although these papers introduce a new technique, but the resulting filters have a large size, complicated in design and difficult in fabricating.

The increasing demand of different wireless technologies in the same spectrum is rapidly increasing radio frequency noise, and threatening the data throughput performance. This problem is more intense especially for UWB, which covers a very broad frequency band and is overlapped with WiMAX (3.5 GHz) in the lower frequency range and with WLAN (5.8 GHz) in the upper circle. The easiest way to get rid of this problem is to create notch band at these frequencies in bandpass range of the UWB filter. To fulfill this requirement many filters with different concepts came into the reality by different researchers [12, 13]. In Refs. [12, 14], resonating structures are used to create notch in the pass band but are suffered from harmonics of resonators and also due to presence of via fabrication is difficult and costly. In Refs. [15] and [13] transversal signal interaction concept is used, but are suffered from larger size. After all, it can be concluded that the of designing UWB bandpass filters with notched band structures is found not to be an easy task and the challenges include strong harmonic response, reduction in frequency utilizing efficiency, large circuit size, or high fabrication cost.

In this paper a microstrip UWB band pass filter with a notch at 5.8 GHz is designed. The structure of the filter is of very small size, compact, planner and there is no via or defected ground structure so, it is very easy to fabricate.

2 Design and Analysis of UWB Filter

The Ultra Wide pass band frequency range of the proposed filter is achieved by the combination of high pass filter and low pass filter. In the pass band range of the filter a notch band at 5.8 is achieved by a stub. So, the structure of the proposed filter can be divided into three parts; (i) High pass filter (ii) Low pass filter and (iii) Stub for notch at 5.8 GHz, as shown in Fig. 1.

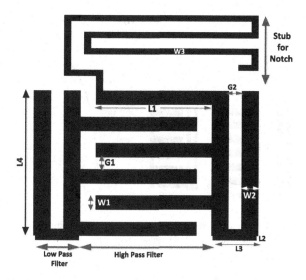

Fig. 1 Configuration of proposed UWB BPF with notch

High pass section of proposed filter is designed by using Interdigital structure. Interdigital structure consists of planar parallel coupled finger lines which is equivalent to combination of some inductor and capacitor, as shown in Fig. 2a. Length of finger line generates effect of inductor L, gap between fingers is equivalent to capacitor C, and the capacitor Cp is equivalent to dielectric material on which structure is designed [16]. For high pass filter the inductor L should be as small as possible to pass higher frequencies. To do this small length of finger is used in the proposed filter.

Low pass section of the proposed filter is designed by a hairpin line structure which generates low pass behavior. Hair pin structure is a folded transmission line in the form of hairpin. Folded section of hairpin generates effect of inductor and gap

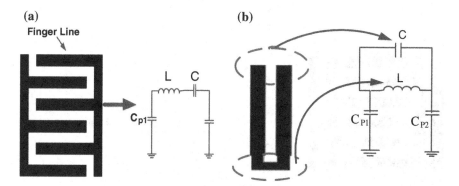

Fig. 2 Equivalent circuit of the high pass and low pass section. **a** Interdigital high pass filter. **b** Hair pin line low pass filter

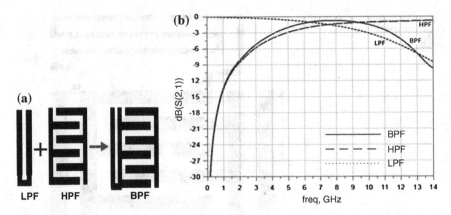

Fig. 3 Composite band pass filter. **a** Cascaded structure of interdigital and hair pin line. **b** Simulated result—S (2, 1) versus frequency

between the lines generates capacitive effect [17]. The dielectric material of the structure generates capacitance Cp1 and Cp2, as shown in Fig. 2b.

The Interdigital high pass filter and hairpin low pass filter shown in Fig. 2 are cascaded to each other to form a band pass filter, as shown in Fig. 3a. The cascaded structure is simulated with respect to frequency. The simulated result shows the graph of S21 in Fig. 3b, which verifies the band pass nature of the structure. The pass band range of this filter is between 4 and 11.2 GHz, which does not cover the UWB range (3.1–10.6 GHz) and attenuation in stop band range for higher frequency is not good.

To cover whole band in Ultra Wide Band range and to improve stop band, one more section of low pass filter is cascaded as shown in Fig. 4a. The simulated result

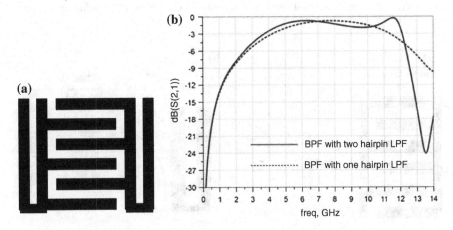

Fig. 4 Composite BPF. **a** Cascaded structure of interdigital and two hair pin line. **b** Simulated result—S (2, 1) versus frequency

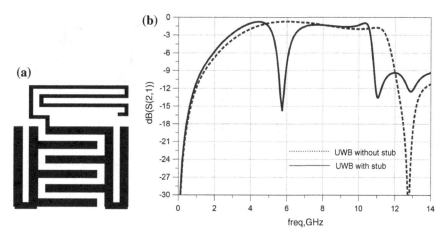

Fig. 5 UWB BPF with notch. **a** Stub loaded UWB BPF structure. **b** Simulated result—S (2, 1) versus frequency

of this structure shows the wider bandwidth which covers UWB range and also attenuation in stop band for high frequency is improved, as shown in Fig. 4b.

A quarter wave length long open stub is loaded on the Interdigital structure of the filter to achieve a notch at 5.8 GHz, as shown in Fig. 5a. To reduce overall size of the filter stub is folded with many folds to confine it into a small area. This structure is simulated and its result in Fig. 5b shows a notch at 5.8 GHz in the pass band range of the filter.

Higher cut off frequency of the filter is dependent on the low pass section of the filter so it can be controlled by folded part of the hairpin structure. The variation in higher cut off frequency with respect to length of folded section, L_f, of the hairpin structure is shown in Fig. 6a. On Increasing Lf the higher cutoff frequency shifted towards lower frequency and vice versa. The higher cut off frequency of proposed filter at 10.6 GHz is achieved for $L_f = 0.7$ mm.

The notch in the pass band is achieved by open stub loaded at the Interdigital structure, so its frequency can be varied by the variation in length of the stub. On varying the length of stub, position of notch and higher cut off frequency both are varied. The variation in notch and higher cut off frequency with respect to variation in stub length, Ls is shown in Fig. 6b.

Folded part of hair pin line controls only the higher cut off frequency of the filter but stub length controls notch as well as higher cut off frequency both. So, by mutual variation in stub length and folded part of hairpin line, the higher cut off frequency and notch frequency of filter are achieved at 10.6 and 5.5 GHz respectively.

Lower cut off frequency of the filter is dependent on the high pass section of the filter, so it is controlled by different dimensions of the Interdigital structure. On the basis of all simulations and analysis the dimensions of the proposed filter are as follows: L1 = 1.73 mm, L2 = 0.05 mm, L3 = 0.6, L4 = 2.66 mm, Ls = 11.01 mm,

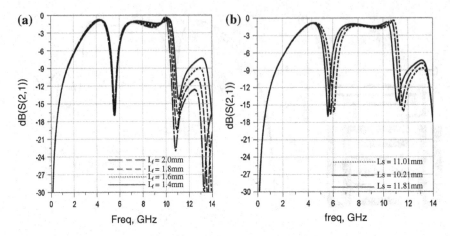

Fig. 6 **a** Variation in S (2, 1) with respect to folded length of hairpin line. **b** Variation in S (2, 1) with respect to stub length of the filter

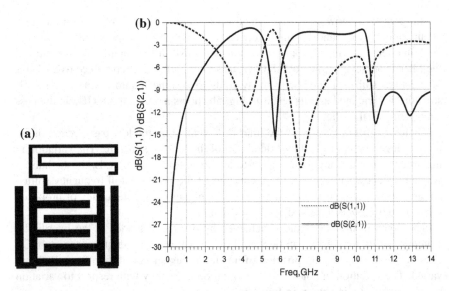

Fig. 7 **a** Proposed structure of UWB BPF with a notch at 5.8 GHz. **b** Simulated result—S (2, 1) and S (1, 1) versus frequency

G1 = 0.23, G2 = 0.2 mm, W1 = 0.25 mm, W2 = 0.25 mm, W3 = 0.1 mm, ε_r = 4.4, h = 1.6 mm.

The final structure of the proposed filter and its simulated result is shown in Fig. 7.

The proposed structure of the filter shows the band pass behavior in whole UWB frequency range with a notch at 5.5 GHz. In design of the filter all capacitors and

Table 1 Comparison of proposed filter with previously reported filters

References	IL/RL (dB)	Size (λg × λg)	ε_r/h (mm)	Notch capability
[12]	1.0/13	0.58 × 0.12	3.5/0.508	Yes
[18]	<1.5/>10	0.752 × 0.456	2.55/0.8	Yes
[14]	<2/>10	0.31 × 0.34	2.2/0.787	Yes
[15]	<0.27/>15	0.511 × 0.259	3.38/0.812	Yes
[13]	0.8/14	0.235 λg × λg	3.55/0.8	Yes
This work	<2/>10	0.145 × 0.177	4.4/1.6	Yes

inductors are implemented by very compact and small size microstrip structures. So, the size of the filter is much smaller than many filters reported in recent years. The comparison of size and performance of this work with different recently reported UWB BPFs is enlisted in Table 1.

3 Conclusions

An UWB BPF with a notch at 5.8 GHz is designed by cascading High Pass and Low Pass filter. This filter is covering whole UWB frequency with good stop band range. The notch at 5.8 GHz enables this filter to reject interference signal from WLAN systems. It is very simple to design and the structures used as High pass and Low pass section are very compact, small and planar without any via or defected ground structure, so its fabrication will be easier and cost effective.

References

1. FCC, Revision of Part 15 of the Commission's rules regarding ultrawideband transmission system, first note and order Federal Communication Commission, ETDocket 98–153, 2002.
2. L. Zhu, S. Sun, and W. Menzel: Ultrawideband (UWB) bandpass filters using multiplemode resonator. IEEE Microw. Wireless Compon. Lett., vol. 15, no. 11, pp. 796–798, Nov. 2005.
3. J. Gao, L. Zhu, W. Menzel, and F. Bogelsack.: Shortcircuited CPW multiplemode resonator for ultrawideband (UWB) bandpass filter. IEEE Microw. Wireless Components Letters, vol. 16, no. 3, pp. 104–106, Mar. 2006.
4. R. Li and L. Zhu.: Compact UWB bandpass filter using stubloaded multiplemode resonator. IEEE Microw. Wireless Compon. Lett., vol. 17, no. 1, pp. 40–42, Jan. 2007.
5. H.W. Deng, Y.J. Zhao, L. Zhang, X.S. Zhang, and S.P. Gao.: Compact quintuplemode stub loaded resonator and UWB filter," IEEE Microwave Wireless Compon. Lett., vol. 20, no. 8, pp. 438–440, Aug. 2010.
6. Q.X. Chu, X.K. Tian.: Design of UWB bandpass filter using steppedimpedance stub loaded resonator. IEEE Microwave Wireless Components Letters, vol. 20, no. 9, pp. 501–503, Sep. 2010.

7. Q.X. Chu, X.H. Wu, and X.K. Tian.: Novel UWB bandpass filter using stubloaded multiplemode resonator. IEEE Microw. Wireless Compon. Lett., vol. 21, no. 8, pp. 403–405, Aug. 2011.

8. A. Taibi, M. Trabelsi, A. Slimane, M.T. Belaroussi.: A Novel Design Method for Compact UWB Bandpass Filters. IEEE microwave and wireless components letters, vol. 25, no. 1, pp. 46, January 2015.

9. T. Kuo, S. Lin, and C. Chen.: Compact ultrawideband bandpass filters using composite microstrip coplanar waveguide structure. IEEE Transactions on Microwave Theory and Techniques, vol. 54, no. 10, pp. 3772–3778, Oct. 2006.

10. Z. Hao and J. Hong.: Ultrawideband bandpass filters using multilayer liquid crystal polymer technology. IEEE Transactions on Microwave Theory and Techniques, vol. 56, no. 9, pp. 2095–2100, Oct. 2008.

11. Z. Hao and J. Hong.: Ultrawideband bandpass filters using embedded stepped impedance resonators on multilayer liquid crystal polymer substrate. IEEE Microwave Wireles Components Letters, vol. 18, no. 9, pp. 581–583, Sep. 2008.

12. Kaida Xu, Yonghong Zhang, Joshua Lewei, William T. Joines, and Qing Huo.: Miniaturized notchband UWB bandpass filters using Interdigital coupled feedline structure. Microwave and Optical Technology Letters, vol. 56, no. 10, pp. 2215–2217. October 2014.

13. Bahman Mohammadi, Arash, Valizade, Javad Nourinia, and Pejman Rezaei.: Design of a compact dual band notch ultrawideband bandpass filter based on wave cancellation method. IET Microw. Antennas Propag., vol. 9, no. 1, pp. 1–9, 2015.

14. Feng Wei, Wen Tao Li, Qiu Lin Huang, and Xiao Wei Shi.: Super Compact UUB BPF with One Narrow Notched Band and Wide Stop band. Microwave and Optical Technology Letters, vol. 57, no. 3, pp. 763–765, March 2015.

15. M. Mirzaee and B.S. Virdee.: UWB bandpass filter with notch band based on transversa signal interaction concepts. Electronics Letters, vol. 49, no. 6, 14th March 2013.

16. Inder Bahl.: Lumped Elements for RF and Microwave Circuits. 2003 ARTECH HOUSE, INC.

17. Ju Hyun Cho and Jong Chul Lee.: Compact Microstrip Stepped impedance Hairpin Resonator Low Pass Filter With Aperture. Microwave And Optical Technology Letters, vol. 46, no. 6, pp. 517–520, September 20 2005.

18. He Zhu and Qing Xin Chu.: UltraWideband Bandpass Filter With a Notch Band Using StubLoaded Ring Resonator. IEEE Microwave and Wireless Components Letters, vol. 23 no. 7, pp. 341–343, July 2013.

Dual-Band Microstrip Fed Monopole Patch Antenna for WiMAX and WLAN Applications

Chandan and B.S. Rai

Abstract A compact dual-band monopole patch antenna is designed for WiMAX and WLAN applications. The proposed antenna operates over two bands a lower band for WiMAX system (3.1–3.9 GHz) and higher band for WLAN system (IEEE 802.11a standard) with resonance at 3.46 and 5.48 GHz respectively. It consists of double C-slots on a rectangular patch which is directly fed by a 50 Ω microstrip line and a truncated ground with two slots cut in it. Wider bandwidth is achieved by cutting slots in the ground plane. The antenna presented is quite small with dimension of $10 \times 15 \times 0.8$ mm^3 and quite wide impedance bandwidth of 22.85 % at resonant frequency 3.46 GHz and 35.42 % at resonant frequency 5.48 GHz. To investigate the designed antenna, various parameters of the antenna are studied such as S-parameter, current distribution, VSWR and radiation pattern.

Keywords VSWR · Monopole antenna · Slots · Defected ground plane and dual-band

1 Introduction

Multiband antennas are becoming more popular in modern wireless communication because by using this, single device can be used for many applications. Mobility of wireless communication devices is necessary for fulfillment of multiple demands which increase the interest for compact and low profile antennas in recent years, but low-profile antennas are now facing some crucial problem like narrow bandwidth, poor impedance matching and low gain. Several methods were presented to achieve wide band operations, such as increasing the substrate thickness, mending the

Chandan (✉) · B.S. Rai
Electronics and Communication Engineering Department, M.M.M.U.T., Gorakhpur, India
e-mail: chandanhcst@gmail.com

B.S. Rai
e-mail: bsr_54@yahoo.co.in

© Springer India 2016
S.C. Satapathy et al. (eds.), *Information Systems Design and Intelligent Applications*, Advances in Intelligent Systems and Computing 434,
DOI 10.1007/978-81-322-2752-6_52

radiator, using a couple feed for wideband impedance matching, and cutting slots or notches on the patch or ground plane [1].

In monopole antenna, keeping pace with the rapid development of wireless communication system, several multiband antennas for dual band WLAN applications have been reported [2–5]. A dual band-notched small monopole antenna is shown for WLAN applications [2] and analysis of ultra-wide band printed planar quasi-monopole antenna is done using only the VSWR [3]. A compact dual band-notched printed monopole antenna and compact polyamide based antennas is proposed for flexile display [4]. Another compact planar plate-type antenna is proposed for WLAN operations [5, 6]. They have some limitations that make them unsuitable for modern multi-band applications such as all these antennas have large size, low impedance bandwidth and low gain.

In this paper, a dual band monopole antenna for WiMAX and WLAN (IEEE 802.11a) band application is realized by using two C slot cut in the patch. Increase in bandwidth is obtained for both bands, for first band impedance bandwidth is 22.85 % (3.1–3.9 GHz) and for second band achieved impedance bandwidth is 35.42 % (4.46–6.38 GHz).

2 Antenna Configuration

The structure of proposed shown in Fig. 1 is of size $10 \times 15 \times 0.8$ mm^3. Size calculated for patch antenna by the formula and optimized by parametric analysis in HFSS software. The antenna comprises a rectangular patch with two C-slots in it, a microstrip line and a defected ground plane with two slots cut in it. The antenna is fed by 50 Ω microstrip line and printed on the low-cost FR4 substrate with relative constant of 4.4, loss tangent 0.02 and thickness of 0.8 mm. Using FR4 as substrate has advantage that it is easily available in the market and its cost is low. The antenna presented here is a very low profile; it is easy to mount it on any surface.

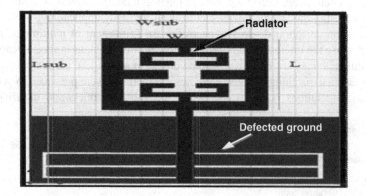

Fig. 1 Geometry of the proposed antenna

Table 1 Comparison of proposed antenna with existing antennas

Antenna	Frequency band covered	Antenna size (mm^2)	Dielectric (FR4)	Substrate thickness (mm)
Ref. [2]	5.2/5.8 GHz Wireless local area networks and 3.5/5.5 GHz WiMAX	12 × 18	4.4	1.6
Ref. [3]	Ultra-wideband (UWB)	40.5 × 45	1.05	1.6
Ref. [4]	3.3–3.8 GHz WiMAX band and 5.1–6 GHz WLAN	20 × 18	4.4	1.0
Ref. [5]	2.45 GHz and 5.2 GHz, WLAN and bluetooth	23 × 25	4.2	1.6
This work	3.1–3.9 GHz WiMAX and 4.46–6.38 GHz WLAN	10 × 15	4.4	0.8

The proposed antenna has ground of length $\lambda_g/4$ and height of strip Ls is 15 mm and width W_f is 1.5 mm. The proposed antenna used two C-slots, first slot has length of 1 mm and width of 7 mm, the second slot has length of 1 mm and width of 7 mm. Dual C-shaped branch is responsible for lower and higher frequency band. The dual-band monopole antenna used two C-slots to generate two bands and optimized to wireless communication system, Worldwide Interoperability for Microwave Access (WiMAX) and Wireless Local Area Networks (WLAN) applications. The parameters of the proposed antenna are studied by changing one parameter at a time and fixing the other. To fully understand the behavior of the antenna structure and to determine the optimum parameters, the antenna was analyzed using ansoft HFSS.

The proposed dual-band antenna has a radiator with an area of 7×2 mm^2, a ground plane of 28×15 mm^2 and overall dimension of ($10 \times 15 \times 0.8$ mm^3). In order to transfer electromagnetic energy from microstrip feed to the radiating element, one end of microstrip feeding line connects to SMA connector, as shown in Fig. 1.

Defected ground plane is responsible to increase the impedance bandwidth of both bands and also to improve the return loss. The proposed antenna is very compact which is a basic need for any device in present time; Table 1 is shown for comparing the dimensions of the antenna with some existing reference antennas.

3 Results and Discussion

The simulated result in Fig. 2 of the proposed antenna shows that the antenna operates for two bands, first band resonates at 3.46 GHz and achieved gain is 4.21 dBi for this band and the second band resonate at 5.48 GHz and gain achieved is 5.41 dBi for this band. This proposed design can be used as indoor base station for WLAN communication. For first band, impedance bandwidth is 22.85 % and

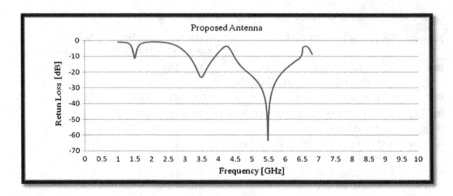

Fig. 2 Simulation result of S_{11} of the proposed antenna

for second band, impedance bandwidth is 35.42 %. The return loss is below −60 dB for the second band which is quite good for communication.

Figure 3a, b shows the 3D diagram of gain of the proposed antenna. For first band, gain is 4.21 dBi at 3.46 GHz resonant frequency and for the second band; gain is 5.41 dBi at 5.48 GHz resonant frequency. The plots clearly indicate that in z-direction, gain is more and in y-direction it is null. Figures 4a, b and 5a, b are shown below which represent the radiation pattern of proposed antenna. Figure 4a, b are radiation pattern display for 3.46 GHz and Fig. 5a, b are radiation pattern display for 5.48 GHz. Figure 6 shows the voltage standing wave ratio (VSWR) of proposed antenna. VSWR ≤ 2 is standard for monopole antenna of the wireless communication systems. VSWR is 1.52 dB for 3.46 GHz resonant frequency and VSWR is 0.98 dB for 5.48 GHz resonant frequency, both are good for dual band monopole antenna of WiMAX and WLAN applications. The current distribution of

(a)   **(b)**

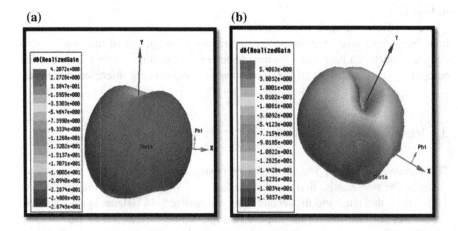

Fig. 3 Gain of the proposed antenna **a** at (3.36 GHz), **b** at (5.48 GHz)

(a)   (b)

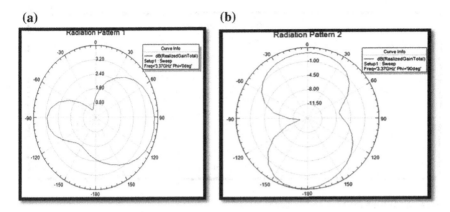

Fig. 4 Radiation pattern of proposed antenna **a** at 3.46 GHz with phi = 0°, **b** at (3.46 GHz with phi = 90°)

(a)   (b)

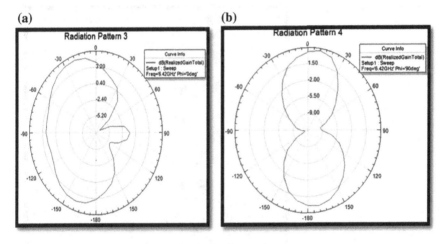

Fig. 5 Radiation pattern of proposed antenna **a** at 5.48 GHz with phi = 0°, **b** at (5.48 GHz with phi = 90°)

the proposed antenna is shown in Fig. 7. Table 2 contains a comparison between antenna designed and some existing antenna. The impedance bandwidth and gain of the proposed antenna is more than most of the existing antenna. Return loss is very important factor of antenna designed; this is also far better than the existing reference antenna. VSWR is less than 2 for the operating bands.

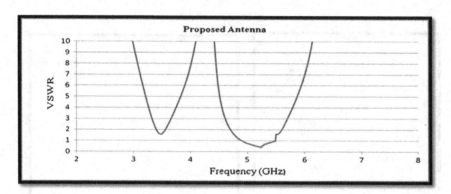

Fig. 6 VSWR of the proposed antenna

Fig. 7 Current distribution of the proposed antenna

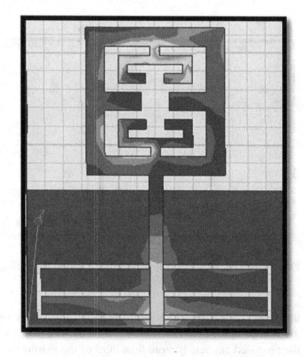

Table 2 Comparisons between proposed antenna and existing antenna

Antenna	Return loss (dB)	Impedance bandwidth (%)	Gain (dBi)	VSWR (dB)
Ref. [2]	–	130	3	0.89
Ref. [3]	–	–	–	1.5
Ref. [4]	–	–	5.5	0.5
Ref. [5]	−23 and −19	5 and 7	−0.6 and 2.4	–
Ref. [6]	−18.8 and −22	16.3 and 16.9	2.1 and 3.1	–
This work	−22 and −60 dB	22.85 and 35.42	4.21 and 5.41	0.98 and 1.52

4 Conclusion

The design of dual band monopole antenna with a compact size of only $10 \times 15 \times 0.8$ mm^3 has been presented for WiMAX and WLAN application. Discussed results show that the antenna has −10 dB wide impedance bandwidth 22.85 % for band1 (3.1–3.9 GHz) and 35.42 % for band2 (4.46–6.38 GHz). Comparison of proposed antenna with previously reported antenna is done on the basis of different parameters in Table 2 and it can be concluded that proposed antenna has more bandwidth and more gain. Good radiation pattern is obtained for phi 0° and 90° of the proposed antenna. Besides it the antenna is very compact which is fully satisfied the present days requirements for communication system. The antenna is applicable for the WLAN and WiMAX communication.

References

1. C.A. Balanis, "Antenna theory: analysis and design," John Wiley and Sons, INC., New York, Third Edition, 2005.
2. Nasser Ojaroudi, "Dual Band-Notched Small Monopole Antenna with Novel W-Shaped Conductor Backed-Plane and Novel T-Shaped Slot for UWB Application," IET Microwave Antenna Propagation, vol. 7, pp. 8–14, 2013.
3. Weixia Wu, "Analysis of Ultra-Wideband Printed Planar Quasi-Monopole Antennas Using the theory of characteristic Modes," IEEE Antenna and Propagation Magazine, vol. 52, No. 7, pp. 67–77, 2010.
4. M. Abdollahvand, "Compact Dual Band-Notched Printed Monopole Antenna for UWB Application," IEEE Antenna and Wireless Propagation Letters, vol. 9, pp. 1148–1151, 2010.
5. Haider R. Khaleel, "Compact Polyimide-Based Antennas for Flexible Display," Journal of Display Technology, vol. 8, No. 2, pp. 91–96, 2012.
6. L. Peng, "A Microstrip Fed Monopole Patch Antenna with Three Stubs for Dual-Band WLAN Applications," Journal of Electromagnetic Wave and Application, vol. 21, No. 15, pp. 2359–2369, 2007.

Script Based Trilingual Handwritten Word Level Multiple Skew Estimation

M. Ravikumar, D.S. Guru, S. Manjunath
and V.N. Manjunath Aradhya

Abstract Skew estimation and correction plays an important role in document analysis. In the present work, we proposed a model to estimate multiple skews present in trilingual such as Devanagari, English, and Kannada handwritten documents at word level with a priori knowledge about the corresponding scripts. The idea of using different skew estimation techniques for different scripts such as Hough transform (HT) for Devanagari words, Gaussian Mixture Models (GMM) and convex hull for Kannada and English words is proposed. The effectiveness of these approaches has been reported by testing on a dataset consisting of 1000 words in each script. Experimental results show that the proposed approaches are effective in estimating and correcting the handwritten skew words.

Keywords Handwritten skew estimation · Skew correction · Gaussian mixture models · Hough transform · Convex hull

M. Ravikumar (✉) · D.S. Guru
Department of Studies in Computer Science, University of Mysore,
Mysore, Karnataka, India
e-mail: ravi2142@yahoo.co.in

D.S. Guru
e-mail: dsg@compsci.uni-mysore.ac.in

S. Manjunath
Department of Computer Science, Central University of Kerala,
Kasaragod, Kerala, India
e-mail: manju_uom@yahoo.co.in

V.N. Manjunath Aradhya
Department of MCA, S.J. College of Engineering, Mysore, Karnataka, India
e-mail: aradhya.mysore@gmail.com

© Springer India 2016   541
S.C. Satapathy et al. (eds.), *Information Systems Design and Intelligent
Applications*, Advances in Intelligent Systems and Computing 434,
DOI 10.1007/978-81-322-2752-6_53

1 Introduction

India is a multilingual country, written communication in government sectors documents and forwarded notices generally contain both printed as well as handwritten
texts in multiple scripts. The technology is being used to set up a paperless office, in
this regard a huge amount of documents need to be digitized. In order to digitize a
document, it is very much essential to develop a multi script OCR for recognition of
all the scripts present in the document. Research community has tackled the problem
with respect to the design of OCR for printed documents. However, a greater
attention has to be given for design and development of multilingual OCR for
handwritten documents. Analysis of multilingual and multiple skewed handwritten
documents pose two important challenges in development of multilingual OCR, i.e.,
identification of scripts and its estimation and correction of skew angle. Normally, to
the given document, separating printed text from the handwritten text followed by
word segmentation, skew correction and later script identification tasks are to be
performed. In case of document containing words with multiple scripts having
multiple skew, order of using skew correction and script identification can be tackled
by using any of the two following ways. First and foremost task is that the script can
be identified and script based skew estimation techniques can be employed. On the
other hand, skew correction followed by script identification. In the later case skew
estimation leads to a problem of selection of appropriate skew estimation technique
as it depends on the script. Hence, the first case of performing script identification
followed by skew estimation is suitable order for such application of developing
OCR for a multilingual handwritten document containing multiple skews.

Script identification can be performed at three levels i.e., script identification at
block level, textline level and word level. Specifically, for multilingual documents
containing multiple skews, word level script recognition is recommended [1, 2]. If
the script of a word is known, then selecting a particular algorithm for estimating the
skew angle will be an easier task. From the literature, it is learnt that the efficiency of
an OCR system mainly depends on the effectiveness of skew estimation and correction. Hence skew estimation in general and at word level in particular plays a vital
role in the field of handwritten document analysis. In this paper, we explore the
approaches to correct the skew of multilingual handwritten document.

The organization of the paper is as follows, in Sect. 2, we present a brief review
only on handwritten skew estimation techniques. In Sect. 3, we present the
description the proposed model and experimental analysis is presented in Sect. 4.
Finally, conclusion is presented in Sect. 5.

2 Related Work

Analysis of any handwritten document image requires that there shall not be any
skew. Producing a document without any skew seems to be inevitable. Hence Skew
estimation and correction are required before the actual document image analysis.

Inaccurate de-skew will significantly deteriorate the subsequent processing stages in document analysis and may lead to incorrect layout analysis, erroneous word or character segmentation, and hence results in misclassification. The overall performance of a document analysis system will subsequently decrease due to the presence of a skew [3, 4].

On the other hand in real time applications, a document may contain multiple skews because, there may be different number of components annotated using different scripts in different direction. Hence an effective algorithm is required for multiple skew detection of multilingual document, which makes OCR system capable of processing. Already a good number of skew estimation algorithms are available in the literature and most of the existing algorithms will work for single language. Hence, estimation of the skew angle for multilingual handwritten documents with multiple skews is still a challenging issue.

Most of the existing skew estimation techniques can be broadly divided into the following categories according to the basic approach they adopt: Projection Profile Analysis, Hough transform, Nearest neighbors clustering, Cross-correlation [5].

A skew angle estimation approach based on the application of a fuzzy directional run length is proposed for complex address images which use a new concept of fuzzy run length, which imitates an extended run length [6]. Skew angle detection of a cursive handwritten Devanagari script at word level is proposed in [7] where each word is fit into a standard frame work and Wigner-Ville distribution is applied to each rotation of a word ranging from $-89°$ to $+89°$ and the skew angle is selected as maximum angle of distribution. In [8], a method for skew detection and correction of entire document using HT and contour detection is proposed.

A method based on mixture models in which expectation maximization algorithm is used for estimation of skew angle in unconstrained Kannada handwritten documents is discussed in [9]. In this work, gaps between the characters are filled using a morphological operation like dilation. Textlines and words are extracted using component extension method and vertical projection profile techniques respectively. Further these words are passed to the GMM to extract the mean vector points. With these mean vector points skew angle is estimated. Some more works related to handwritten skew estimation are reported in the literature [10–12].

Multiple skew estimation in multilingual handwritten documents is discussed in [13]. Each word in a document is segmented using morphological operations and connected component analysis. Skew of each word is estimated by fitting a minimum circumscribing ellipse. The orientation of each word is estimated and then words are clustered using adaptive k-means clustering to identify the multiple blocks present in the document and average orientation of each block is estimated. Estimation and correction of skew angle at word level for Devanagari script is discussed in [14], in which HT is applied on words for skew estimation and rotation transform is used for correction of skew angle.

The methods discussed above concentrates only on skew estimation of handwritten documents containing words written in one script and single skew. In the proposed work, we concentrate mainly on skew estimation of a multilingual document containing multiple skew present in a forwarded official documents comprising three scripts: Devnagari, English and Kannada.

3 Proposed Methodology

The block diagram of the proposed system is shown in Fig. 1. A trilingual document containing printed and handwritten text is subjected to the system. Using connected component analysis the words are segmented. After word segmentation, the words are classified either as printed or handwritten. In order to carry out this we use the work carried out in [15]. Statistical texture features of words such as mean, standard deviation, smoothness, moment, uniformity, entropy and local range including local entropy is extracted. The K-nearest neighbor classifier has been used to classify the words into printed or handwritten.

Normally skew estimation technique depends on the script specifically in case of handwritten. Indeed the knowledge about the script helps us in identifying the skew more accurately. For example, if the document contains Devanagari script we can directly use skew estimation model which works based on line estimation techniques as Devanagari script has Shirorekha feature. In case of document containing Kannada/English then we can use either the application of Gaussian Mixture Models (GMM) or convex hull to estimate the same. Hence, prior to skew detection we apply script identification at word level and we use the work of script identification described in [1]. Once the script of the word is recognized further three different skew estimation techniques are carried out, which are as follows:

1. Hough transform (HT)
2. Gaussian Mixture Models (GMM)
3. Convex hull

From the above three approaches, we estimate candidate points for skew estimation. Once the candidate points are obtained, we use linear regression based line fitting and moments for skew estimation. The brief description about each technique is presented in the following subsections.

Fig. 1 Block diagram of the proposed model

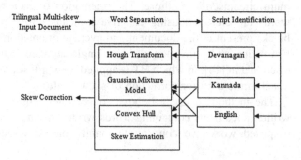

3.1 Hough Transform (HT)

Hough transform is a technique to find lines, circles or any other parametric curves. In the present approach, we use Hough transform to estimate straight line present in the word image and use the same straight line to estimate the skew of the word. Each point in the image plane is transformed to a sinusoidal curve in the parametric space, i.e., $\rho = x \cos\theta + y \sin\theta$ where ρ is the distance of line from the origin, and θ is the angle of ρ with respect to x axis. The parametric space is divided into a number of accumulator cells. Straight line in the image plane will be converted into a number of sinusoidal curve and each sinusoidal curve will crossover at a particular point in the parametric space forming a peak. Once peak points are detected we can find end points of line segments corresponding to peak values. Shirorekha is a longest line present in a Devanagari word. Using Hough transform, we estimate the longest line in the Devanagari word. The problem with direct use of Hough transform as mentioned in case of [14] for our application is that the line will be partitioned and multiple lines will be identified. The estimated lines are treated as candidate lines and in order to identify true Shirorekha we use technique of boundary growing of candidate lines in the direction of the angle obtained by the line. After boundary growing the line which forms longest line will be treated as Shirorekha line. The end points of Shirorekha are candidate points for skew estimation.

3.2 Gaussian Mixture Models (GMM)

In order to provide good approximation to multimodal distributions, it is failed due to its intrinsic unimodal property of Gaussian distributions and also very limited in representing adequate number of distributions. Latent variables also called as hidden or unobserved variables can be used in mixture distribution which allows us to solve the aforesaid issues. Discrete latent variables can also be interpreted as assigning data points towards specific components of mixtures. Expectation Maximization (EM) algorithm is one of the techniques for finding maximum likelihood estimation in latent variable [16]. During learning process, we discover mean centers μ_k and variances Σ of the Gaussian components associated with mixing coefficients which are employed as regulating parameters. Gaussian mixtures used to analyze complex probability distribution and are formulated in terms of discrete latent variables and further defined as the weighted sum of 'K' Gaussian components.

In the present work, the obtained means of k clusters $\mu_k, \forall k = 1, \ldots, K$ is then used for estimating the skew of a word. Selecting K value is a highly subjective in nature. Therefore, we fix the value of K equal to number of connected components in the word image. The centroid of each cluster is treated as candidate points to estimate the skew angle of the word. Detail information regarding GMM can be seen in [9].

3.3 Convex Hull

As words are presented with few connected components, the pixels of connected components are treated as points in two dimensional spaces. A convex polygon is fit to the set of word pixels such that all points of the connected component lie in a polygon. The centroids of each convex hull of the connected components are treated as candidate points for skew estimation [17].

3.4 Skew Estimation

In case of HT based approach, it is recommended to apply only to the script of Devanagari words as Shirorekha. The end points of the line are considered and angle obtained by the end points of the line is used to estimate the skew angle. In case of Kannada and English we recommend either to use convex hull based approach or GMM. Once the candidate points are estimated we use two different techniques to estimate the skew viz., linear regression method and second order moments based approach. In case of linear regression based approach for the candidate points we fit straight line using linear regression and the skew angle of the estimated line is treated as skew angle of the word image. In case of moments based approach, we estimate the skew angle of the candidates based on the statistical properties of the points as mentioned in [14].

4 Experimentation

In order to carry out the experiment, we have created a handwritten word database of trilingual (Devanagari, English and Kannada) scripts. Each script is treated as a class and the each class contains 1000 words. Handwritten documents were collected using flatbed HP scanner at 300 resolution dpi and preprocessing techniques such as binarization and noise removal are also carried out. The words are segmented and used for the experimentation. Sample words of each class are given in Table 1. Experiment was conducted on a desktop computer using Matlab (version 2013), with Windows operating system having 2 GB RAM capacity.

To evaluate the performance of the proposed model, the skew of each word is estimated manually by drawing a line on each word and the orientation of each line is stored. The stored orientation of the line of each word is compared using average relative error obtained as per Eq. 1, with the skew obtained by the proposed model. Class-wise average relative error is reported in Table 2.

$$Average\ Relative\ Error = mean\left(\left|\frac{\theta_{acutal} - \theta_{obtained}}{\theta_{actual}}\right|\right) \tag{1}$$

Table 1 Sample images of trilingual words

Script	Kannada	Devanagari	English
Sample images			

Table 2 Average relative error of the proposed model for different scripts using different methods

Class	Method	Average relative error	
		Linear regression analysis	Second order moments
Devanagari	Hough transform	0.35	0.25
Kannada	Gaussian mixture	0.48	0.43
	Convex hull	0.40	0.38
English	Gaussian mixture	0.21	0.19
	Convex hull	0.20	0.16

From Table 2, it is clear that the convex hull based approach performs better for both Kannada and English words. The reason for high relative error for Gaussian Mixture based model in case of Kannada words is due to the presence of modifiers. When Gaussian mixtures are obtained based on the connected component analysis, the modifiers will also get formed to a cluster and the centroids will make the slope of the line to deviate from its actual slope. Whereas, in case of English it is observed that the Gaussian Mixture and the convex hull based approaches have resulted almost same relative error because in English script there is such modifiers present.

However, most of the researchers have recommended to use accuracy as a performance measure to compare different skew estimation techniques. Hence, we have calculated the accuracy of the proposed model and are tabulated in Table 3.

Table 3 Average accuracy of the proposed model for different scripts using different methods

Class	Method	Average accuracy	
		Linear regression analysis	Second order moment
Devanagari	Hough transform	90.12	91.25
Kannada	Gaussian mixture	89.36	90.14
	Convex hull	92.15	93.48
English	Gaussian mixture	92.81	95.66
	Convex hull	93.58	95.57

From Table 3, it is clear that for Kannada and English scripts, the convex hull based approach performs better than GMM model. Also it should be noted that the Hough transform based model is best suited for Devanagari words.

5 Conclusion

In this paper, a model to detect multiple skew in a multilingual handwritten document image is proposed. As a prerequisite, the input document is expected to be annotated with the associated scripts in the respective locations of the document image. For this purpose, script identification at word level is initially performed using [1] by extracting individual words from the document through the application of connected component analysis. To suggest a skew estimation technique for a word of particular script, experiments were conducted with different techniques such as Hough transform, convex hull and Gaussian mixture models. Through the experimental analysis it was observed that, Hough transform based approach best suits for the skew estimation in Devanagari scripts. Whereas, convex hull based approach gives better performance for skew estimation of both Kannada and English scripts.

References

1. Ravikumar, M., Manjunath, S., Guru, D.,S.: Analysis and Automation of Handwritten Word Level Script Recognition. Proceedings of IJCAISC, Vol. 369, pp. 213–225, (2015).
2. Manjunath, S., Guru, D.,S., Ravikumar, M.: Handwritten script identification: Fusion based approaches. ACEEE proc. of Int Conf on MPCIT, pp. 216–222, (2013).
3. Kasturi, R., Gorman, O., L., Govindaraju, V.: Document image analysis: a primer, Sadhana, Vol. 22, Part I, pp. 3–22, (2002).
4. Pavankumar, M.,N.,S.,S.,K., Jawahar, V.,V.: Information processing from document images. In information technology: principles and Applications(ED) pp. 522–547, (2004).
5. Lin, C., T., Fan, K, W., Yeh, C., M., Pu, C.,H., Wu, F., Y.: High-Accuracy Skew Estimation of Document Images. Int. J. of Fuzzy Systems, Vol. 8, No. 3, pp. 119–126, (2006).
6. Shi, Z., Govindaraju, V.: Skew detection for complex document images using fuzzy run length. ICDAR, Vol. 2. (2002).
7. Kapoor, R., Bagai, D., Kamal, T.,S.: Skew angle detection of a cursive handwritten Devanagariscript. Journal of Indian institute of science, Vol. 82, pp. 161–175.J, (2002).
8. Ramappa, M., H., Krishnamurthy, S.: Skew Detection, Correction and segmentation of Handwritten Kannada Document. Int. J. of Ad. Sci and Tech. Vol. 48, pp. 71–87, (2012).
9. Manjunath Aradhya, V. N., Naveen C., Niranjan S. K.,: Skew estimation for unconstrained handwritten documents. Proceedings of ICACC. pp. 1542–1548, (2011).
10. Mello, C. A. B., Ángel, S., Cavalcanti, G. D. C.: Multiple Line Skew Estimation of Handwritten images of Documents Based on a Visual Perception. Springer, LNCS 6855, pp. 138–145 (2011).
11. Brodić, D., Milivojević, Z.: Estimation of the Handwritten Text Skew Based on Binary Moments. Radio engineering. Vol. 21(1), pp. 162–169 (2012).

12. Kleber, F., Diem, M., Sablatnig, R.: Robust Skew Estimation of Handwritten and Printed Documents based on Grayvalue Images. 22nd ICPR, IEEE, pp. 3020–3025 (2014).

13. Guru. D., S., Ravikumar, M., Manjunath, S., Multiple Skew Estimation in Multilingual Handwritten Documents. IJCSI, Vol. 10, Issue 5, No. 2, pp. 65– 69, (2013).

14. Tripati, A., Jundale, Ravindra, S., Hegadi.: skew detection and correction of Devanagari script using hough transform, procedia computer science 45, pp. 305– 311, (2015).

15. Mallikarjun Hangarge, K.C. Santosh, Srikanth Doddamani and Rajmohan Pardeshi. Statistical Texture Features based Handwritten and Printed Text Classification in South Indian Documents. Proceedings of ICETECIT, Vol. 1, pp. 215–221, (2012).

16. Christopher, M., Bishop.: Pattern recognition and machine learning. Springer, (2006).

17. Manjunath Aradhya V N, Hemantha Kumar G, Shivakumara P. Skew Estimation Technique for Binary Document Images based on Thinning and Moments. Engineering Letters, 14:1, pp. 127–134, (2007).

Deep Convolutional Neural Network Classifier for Handwritten Devanagari Character Recognition

Pratibha Singh, Ajay Verma and Narendra S. Chaudhari

Abstract The performance of two architecture of Neural Networks are compared for handwritten Devanagari character recognition. The first one is the fully connected Feed-forward Neural Network and the second one is deep Convolutional Neural Network. Deep learning is basically a biologically inspired technique based on human brain. A part of brain called neocortex is having layered architecture. The advantage of using CNN is that it does not require complex preprocessing or feature extraction algorithm. Image pixels are the input for the two networks. We obtained the improved result for standard character benchmarking datasets.

Keywords MLP · Features · CNN · Pooling · Convolution

1 Introduction

Multi layer perceptrons are quite efficient in learning complex high dimension problems. It consists of many layers of fully connected hidden layers between input and output layer. However for the patterns having higher dimensions, the learning of MLP becomes costly and time consuming specially for the case of image processing applications with large number of inputs. So to alleviate the difficulty of this over-parameterized NN for high dimensional image processing applications, a model of Neural Network based on Convolutional Neural Network [1] is implemented. The network comes in the category of deep networks. The Convolutional

P. Singh (✉) · A. Verma
IET DAVV Khandwa Road, Indore 452017, India
e-mail: prat_ibh_a@yahoo.com

A. Verma
e-mail: ajayrt@rediffmail.com

N.S. Chaudhari
VNIT, Nagpur, India
e-mail: nsc0183@gmail.com

© Springer India 2016
S.C. Satapathy et al. (eds.), *Information Systems Design and Intelligent Applications*, Advances in Intelligent Systems and Computing 434,
DOI 10.1007/978-81-322-2752-6_54

Neural Network is a system which incorporates the feature extraction and the classification in itself. The arrangement of Convolution and sub-sampling layer builds the feature detector part which is trainable. The experimented scheme is very efficient for the domain of character recognition in the present study applied to datasets of the characters and numerals. The Convolutional Neural Network is applied to character recognition application by some of the researchers [2] on direct image pixel. In this study we compared the performance of Convolutional Neural Network with Neural network using mini-batch learning algorithm. This method uses the shared weight, minimum learning parameter criterion and deep learning concept of Convolutional Neural Network. The main advantage of Convolutional Neural Network is that performance is invariant to basic transformations such as rotation, translation due to the subsampling and pooling operations performed at various layers of the network.

Convolutional Neural Networks (CNN) are inspired from our biological system and it is basically a variant of Multilayer perceptron (MLP). There exists a complex arrangement of cells within the visual cortex [3]. Many such neurally inspired models exists like the NeoCognitron [4], HMAX [5] and LeNet-5 [1]. The standard MLP ignores the topology present in the input, whereas the CNN exploits this topological information. In the model of Convolutional Neural Network the input image is imagined to be made up of small sub-regions called receptive fields. The filter(or kernels) sensitive to these sub-regions of the input space are used for convolution, in such a way so as to cover the entire visual field. These filters are locally connected to the input space therefore more suitable to exploit the local correlation spatially present in natural images. Based on the size of receptive field there exists two type of cells in the CNN namely, simple cell (S) and the complex cell (C). For the edge like patterns in the receptive fields, response is given maximally by simple cell. Complex cells covers larger receptive fields and are locally invariant to the exact position of the stimulus. The major advantage of the CNN is that it requires only few parameters of learning because it is made up of the translated version of same basis function. Hence CNN has capability of producing considerable recognition performance with only few samples. CNN has a capability of building network insensitive to translation invariance. Convolutional Neural Network have been widely used for ample of recognition problems such as in MNIST handwritten digit recognition [6], object recognition [7] and in natural language processing [8]. This paper reports the implementation and performance evaluation of Convolutional Neural Network for handwritten Devanagari characters. The rest of the paper is organized as follows: Sect. 2 gives the introduction of Convolutional Neural Network. Section 3 describes the data preprocessing technique used. Section 4 describes the experimental results and Sect. 5 gives the final conclusion and scope for future development.

Fig. 1 Basic characters of Devanagari script

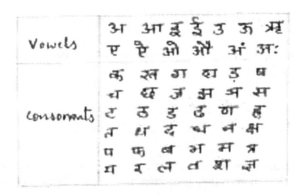

1.1 The Devanagari Script

Digital computers are good at handling problems which are explicitly formulated, but handwritten character recognition is not such a problem. Since the advent of neurocomputing technology, great research effort has been devoted to using it to perform recognition tasks. In this paper the character recognition problem is addressed by analyzing the limitations and advantages of the methodology used. Basic characters in Devanagari script consists of 13 vowels and 36 consonants [9] as shown in Fig. 1. Writing style is from left to write. There is no concept of upper and lower case. Vowels following consonants take a modified shape and known as modified characters. In Devanagari script for all 36 consonants, there exists more than twelve forms of writing them [10], also there exist vowel modifiers whose shape differs, depending on whether the vowel modifier is placed to the left, right, top or bottom of the consonants.

The research efforts in Devanagari OCR is lacking because of the following reasons: lack of data resources like data sets for handwritten words and characters, unavailability of online dictionaries and the scarcity of statistical analysis and evaluation tools. So, the creation of such data resources will augment the research efforts related to Devanagari characters.

2 Data Preprocessing

Data preprocessing is an important step in many deep learning algorithms. It consists of normalization and whitening. The common methods for feature normalization are:

- Simple Rescaling
- Per-example mean subtraction or removal of DC
- Feature standardization (zero-mean and unit variance for each feature across the dataset)

2.1 Simple Rescaling

In simple rescaling, the data vectors are rescaled along each dimension so that the final data vectors lie in the range [0, 1].

2.2 Per-example Mean Subtraction

If the statistics for each data dimension follow the same distribution, then the subtraction of the mean-value from each example (computed per-example) is performed over the samples in this step.

2.3 Feature Standardization

Feature standardization refers to, setting each dimension of the data to have zero-mean and unit-variance. This is the most common method for normalization and is often recommended as a preprocessing step. For this we first calculate the mean of each dimension and subtracts this mean from their respective dimension. Then each dimension is divided with their respective standard deviation.

3 Dataset Experimented

3.1 The MNIST Dataset

The MNIST [2] dataset include the images of handwritten digit samples which are divided in 60,000 samples for the training and 10,000 samples for testing. These images are centered and size normalized to 28 × 28 pixels. In the original dataset, pixels of the image samples are represented by value in range 0–255, where 0 is black, 255 is white and anything in between is represented as shade of grey. An image is rescaled and represented as a 1-dimensional array of 784 (28 × 28) float values between 0 and 1 (0 stands for black, 1 for white). The labels assigned to each digit class are numbers between 0 and 9. When using the dataset, we usually divide it in mini-batches of size 100. The patterns are also underwent preprocessing experimented at the size 32 × 32.

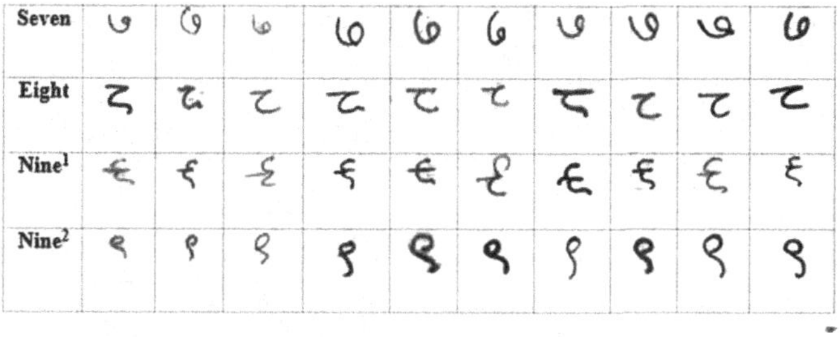

Fig. 2 Image samples of CPAR-2012 handwritten numerals

3.2 The CPAR-2012 Dataset

This dataset is available since the year 2012 [11] to the research community and is developed by Intelligent system group Noida (Centre for Pattern Analysis and Recognition). This is the largest dataset available for the handwritten isolated patterns. It consisting of 35,000 images for numerals (shown in Fig. 2) and 78,400 images of characters. The data is collected from diverse population strata of 2000 writers from various states of India having different religions. The numeral dataset is having 11 classes in this dataset because the pattern corresponding to number '9' can be written in two alternate ways. The character dataset is having 1000 training images and 600 test images of 49 classes each. Table 1 describes the number of pattern for each numeral class.

3.3 CVPR-ISI Dataset

This dataset is available for the global research community since 2009 and is developed by CVPR unit of ISI Kolkata. The Devanagari numeral database includes samples collected from mail pieces and job application forms through specially designed form for data collection. The dataset consists of 22,556 images stored in 'tif' format collected from 1049 writers (shown in Fig. 3). The experiments of CNN are performed for all the three datasets using centroid based pre-processing and size normalization to (i) 28×28 and (ii) 32×32.

Table 1 The number of samples in CPAR-12 numeral dataset

Image	0	1	2	3	4	5	7	8	9^1	9^2
Dataset I	2280	2280	2280	2280	2280	2280	2280	2280	2280	1200
Dataset I	1012	1012	1012	1012	1012	1012	1012	1012	1012	880
Total	3292	3292	3292	3292	3292	3292	3292	3992	3292	2080

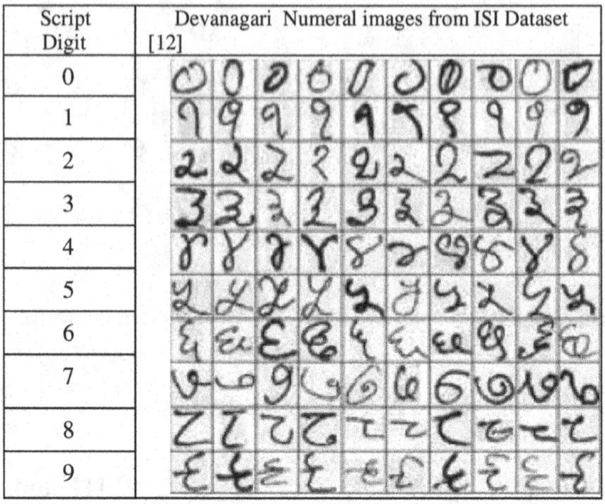

Fig. 3 Sample images from CVPR dataset [12]

Script Digit	Devanagari Numeral images from ISI Dataset [12]
0	
1	
2	
3	
4	
5	
6	
7	
8	
9	

4 The Architecture of Neural Network

Two neural network architectures are implemented for experiments in this study. The first is fully connected architecture of NN and the second one is Convolutional Neural Network based architecture. The Convolutional Neural Network based architecture is based on connecting local region of the previous layer to the next layer. The CNN architecture used is equivalent to Lanet-5 [2]. For both the architectures the experiments are performed on mini-batches of input patterns. Weights used for CNN are sampled randomly from a uniform distribution whose range lies in the interval [−1/fan-in, 1/fan-in], where fan-in is the number of inputs to a hidden unit. For CNNs this is dependent on the number of input feature maps and the size of the receptive fields. Max-pooling is applied as a non-linear down sampling in the next step. Max-pooling is the operation in which the input image is partitioned into a set of non-overlapping sub-regions and for each such sub-region, outputs are calculated as the maximum value of that sub-region. In computer vision Max-pooling operation is quite useful for two reasons: (1) The computational complexity for upper layer is reduced and (2) A form of translational invariance is inherently provided using this. CNN has a property of sharing weights, the number of free parameters in CNN does not increase proportionally with the input dimensions as in the case of standard multi layer Neural networks.

The functions of the main layers of CNN are described as follows:

Convolutional Layer: The function of this layer is to perform 2-D filtering operation on the input image x using the bank of filters and thus producing another set of images h. A connection table, CT is used as a track record for the input-output correspondences. Filter responses from inputs connected to the same output image are linearly combined. The layer performs mapping given by Eq. (1)

$$h_j = \sum_{i,k \in CT_{i,k,j}} x_i * w_k \tag{1}$$

where * indicates the 2D valid convolution. For a particular layer the size of filter w_k is same. It is defined together with the size of the input the size of the output images h_j. After that, a non-linear activation function (e.g. tanh, logistic, etc.) is applied to h similar to standard multi-layer networks.

Pooling Layer: It is used to reduce the dimensionality of the input by a constant factor. It is used for a kind of feature selection at the cost of computational burden. The input images is made up of non overlapping sub-regions out of which one output value is selected. Generally the choice of selection is maxima (called Max-pooling) or average (called Avg-Pooling). Max-Pooling is generally favorable as it introduces small invariance to translation and distortion, leads to faster convergence and better generalization [13]. We have Max-pooling method in this study.

Fully Connected Layer: As in the case of multi-layer network we used fully connected output layer. It performs a linear combination of the product of input vector with a weight matrix. The outputs are normalized with a softmax activation function and therefore approximate posterior class probabilities.

There is an alteration of convolutional layer and max-pooling layer in the CNN as shown in Fig. 4 so as to obtain 1-D feature vector.

Convolution layer has many parameters such as size and the number of feature maps, kernel size, skipping factors and the connection table. In this paper we used the kernel size as 5×5, the number of feature map for the first convolution layer equal to 6 and the second convolution layer equal to 12. The skipping factor specifies the horizontal and vertical number of pixel by which kernel skips between the subsequent convolutions. For each layer if there are M feature maps of size (M_x, M_y), the kernel of size (K_x, K_y), the skipping factors are (S_x, S_y) then the relation of the size of output maps in terms of the abovementioned parameters is given by Eq. (2)

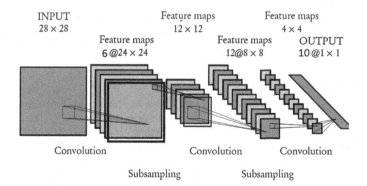

Fig. 4 The Convolutional Neural Network architecture

$$M_{nx} = M_{n-1x} - K_{nx}S_{nx} + 1$$
$$M_{ny} = M_{n-1y} - K_{ny}S_{ny} + 1 \tag{2}$$

where the n represents the index of layer. Each map in layer L_n is connected to at most M_{n-1} maps in layer L_{n-1}. Neurons of a given map share their weights but have different receptive fields.

5 Experimental Results

The experiments are performed on the datasets mentioned in the previous sections. The images of the dataset are undergone some preprocessing steps which are not computationally expensive. All the images are normalized to the size of 28 × 28 pixels. The experiments are performed on the full dataset comprising of thousands of images. Table 2 describes the results of fully connected MLP. The MLP is made up of three layers, and the number of neurons in the first layer equal to the size of image. The hidden layer for the fully connected MLP architecture is experimented with 100 and 200 neurons. The parameter corresponding to L2 weight decay is set to $1 \times e^{-04}$. Experiments are performed on the mini-batches of different sizes. The best results are found for the mini-batch of size 100. The activation function used in hidden layer is hyperbolic tangent and in the output layer is logistic. The momentum is set at value equal to 0.7. Stochastic gradient descent using mini-batch is used as the learning algorithm which is different from ordinary gradient descent [14].

The weight decay is used as regularization method in the experiments performed for fully connected feed-forward neural network. The penalty term is used for penalizing the larger weights, as large weights can cause poor generalization. Excessively large weights in the hidden units of NN can cause the output function to be too rough, pushing it to nearly discontinuities of output. Excessively large weights in the output units can cause wild outputs which may shoot out beyond the range of the data.

For Convolution neural network we used the architecture shown in Fig. 4. The input layer has feature map equal to the size of normalized image of the character.

Table 2 Performance of fully connected feed-forward neural network

Dataset	Neurons in hidden layer	Time for recognition of dataset including training (s)	Percentage accuracy
MNIST handwritten digit	200	7923	97.82
ISI digit	200	2767	97.26
CPAR-2012 character	200	8334	81.4
CPAR-2012 numeral	200	5422	97.47

The second convolution layer consists of 6 feature maps obtained by convolving the kernel of size 5 × 5 with the input feature map. The third layer is sub-sampling layer which is used to down-sample the image. Max-pooling operation is used across 2 × 2 area of the input image for down sampling at this layer. The third layer is the convolution layer which convolves the image from the previous layer with the kernel of size 5 × 5. The number of feature maps obtained at this stage equal to 12. Again in the fourth layer the sub-sampling using 'maxpooling' operation is performed in the area of size 2 × 2. The fifth layer is fully connected layer which drives a feed-forward classifier for giving the output. The layer 1–4 forms the trainable feature extraction part and the next succeeding layer forms the classifier. The results of the Convolutional Neural Network is shown in Tables 3 and 4 for different size feature set. The column 2 represent the recognition rate for number of feature map of 6 and 12 respectively for the layer 2 and four of the architecture. The experiments are also done by varying the number of feature maps stated above to be equal to 12 and 24. The column 4 of Table 3 reports the result corresponding to that. The recognition times reported in the result are the complete time of training and testing.

The result for the available dataset are better in comparison with the previously reported results. The Experiments done on the CPAR digit dataset [15] reported the highest recognition rate of 97.87 % by Rajiv Kumar and Kiran Kumar. While on the character dataset the highest recognition rate reported was 84.03 % with a rejection of patterns. In our experiments the recognition rate comes out to be 84.22 % without any rejection of patterns with proposed architecture of Convolutional Neural Network. The recognition rate for the digit in our results comes out to be 97.66 % which is slightly less than result reported by Kumar Rajiv Kumar and Kiran Kumar. But the results are without any rejection while in the experiments

Table 3 Performance of Convolutional Neural Network with 784 features

Dataset	% accuracy for feature map-6–12	Recognition time for 6–12	% accuracy for feature map-12–24	Recognition time for 12–24
CPAR-2012 char	81.11	20,892	84.21	68,182
CPAR-2012 number	96.95	14,761	97.66	53,251
MNIST digit	98.96	32,193	99.21	104,469
ISI numeral	97.8	10,951	98.11	42,078

Table 4 Performance of Convolutional Neural Network with 1024 features

Dataset	% accuracy for feature map-6–12	Recognition time for 6–12	% accuracy for feature map 12–24	Recognition time for 12–24
CPAR-2012 char	86.48	68,292 s	90.58	243,102
CPAR-2012 number	97.5	48,945 s	97.47	126,306
MNIST	99.11	115,558	99.18	290,491
ISI numeral	97.97	34,476	98.17	94,799

done by Rajiv Kumar and Kiran Kumar results are obtained with the rejection of the degraded patters. Further our method does not involves the complex feature extraction or classifier ensemble approach.

Better results can be obtained if combination of classifier is used in place of a single classifier. Our experiments also reports that the result for ISI-CVPR dataset is better than the proposed approach by Jindal et al. [16]. They reported the recognition percentage of 96.8 % and our results are giving 98.11 % recognition result with CNN.

6 Conclusion

The Convolution Neural network based classifier is implemented for Devanagari character and numeral dataset. The Convolution Neural Network is a deep architecture inspired from the human vision system. It consists of trainable feature detector and classifier embedded in itself. It directly takes the input of the character image in the pixel form. The proposed deep learning architecture is quite effective on the Devanagari character/numeral datasets. However the results can further be improved if classifier fusion is implemented which will be the future direction of our proposed architecture.

Acknowledgments The authors would like to thank Intelligent System Group Noida and CVPR unit of ISI Kolkata for providing the dataset of Devanagari characters/Numerals.

References

1. Y. LeCun, L. Bottou, Y. Bengio, and P. Haffner, "Gradient-based learning applied to Document recognition," in *Proceedings of the IEEE 86*, 1998, pp. 2278–2324.
2. Y. LeCun, C. Cortes, and Christopher J.C. Burges. (1998) The MNIST Dataset of handwritten digits. [Online]. http://yann.lecun.com/exdb/mnist/.
3. D. Hubel and T. Wiesel, "Receptive fields and functional architecture of monkey striate cortex," *Journal of Physiology (London)*, pp. 215–243, 1968.
4. K. Fukushima, "Neocognitron: A self-organizing neural network model for a mechanism of pattern recognition unaffected by shift in position," *Biological Cybernetics*, vol. 193–202, no. 36, 1980.
5. T. Serre, L. Wolf, S. Bileschi, and M. Riesenhuber, "Robust object recognition with cortex-like mechanisms," *IEEE Transactions on Pattern Analysysis and Machine Intelligence*, vol. 29, no. 3, pp. 411–426, 2007.
6. P. Simard, D. Steinkraus, and J. Platt, "Best practices for convolutional neural networks applied to visual document analysis," in *International Conference on Document Analysis and Recognition*, Edinburgh, 2003, pp. 958–963.
7. Y. LeCun, F.J. Huang, and L. Bottou, "Learning methods for generic object recognition with invariance to pose and lighting," in *IEEE International conference on Computer Vision and Pattern Recognition*, Washington, 2004, pp. 97–104.

8. R. Collobert and J. Weston, "A unified architecture for natural language processing: Deep neural networks with multitask learning," in *International Conference on Machine Learning (ICML)*, Helsinki, 2008, pp. 160–167.
9. N Sharma, U Pal, F Kimura, and S. Pal, "Recognition of Off-Line Handwritten Devanagari Characters Using Quadratic Classifier," in *ICVGIP 2006, LNCS 4338*, 2006, pp. 805–816.
10. S. Arora, D. Bhattacharjee, M. Nasipuri, D. K. Basu, and M. Kundu, "Recognition of non-compound handwritten Devanagari characters using a combination of MLP and minimum edit distance," *International Journal of Computer Science and Security*, vol. 4, no. 1, pp. 1–14, 2010.
11. Rajiv Kumar, Amresh Kumar, and Pervez Ahmed, "A Benchmark Dataset for Devanagari Document Recognition Research," in *WSEAS Press*, Lemesos, Cyprus, 2013, pp. 258–263.
12. U. Bhattacharya and B. B. Choudhary, "Handwritten Numeral Databases of Indian Scripts and Multistage Recognition of Mixed Numerals," *IEEE Transaction on Pattern Analysis and Machine Intelligence*, vol. 31, no. 3, pp. 444–457, March 2009.
13. D. Scherer, A. Muller, and S. Behnke, "Evaluation of pooling operations in convolutional architectures for object recognition," in *International Conference on Artificial Neural Networks*, Greece, 2010, pp. 92–101.
14. O. Dekel, R. Gilad Bachrach, O. Shamir, and L. Xia, "Optimal distributed online prediction using mini-batches," *The Journal of Machine Learning Research*, vol. 13, no. 1, pp. 165–202, 2012.
15. Rajiv Kumar and Kiran Kumar Ravulakollu, "On the Performance of Devanagari Handwritten Character Recognition," *World Applied Sciences Journal*, pp. 1012–1019, 2014.
16. T. Jindal and U. Bhattacharya, "Recognition of Offline Handwritten Numerals Using an Ensemble of MLPs Combined by Adaboost," in *ACM Proc. of the 4th International Workshop on Multilingual OCR (MOCR 2013)*, Washington DC, USA, 2013, p. Article No. 18.

A Clustering-Based Generic Interaction Protocol for Multiagent Systems

Dimple Juneja, Rashmi Singh, Aarti Singh and Saurabh Mukherjee

Abstract The paper proposes a clustering based Generic Interaction Protocol for Multiagent Systems (GIPMAS) that exploits clustering methodology for establishing interaction among agents operating in a network of multiagent systems. GIPMAS is a hierarchical protocol that supports intelligent formation of clusters and dynamic election of cluster head and executive cluster head as well. It also describes a recovery mechanism in case cluster head relocates from its respective cluster.

Keywords Mutliagent systems · Agent interaction protocol · Data-aggregation · Agent clustering

1 Introduction

A decentralized multiagent system [1, 2] can be managed by forming clusters of agents with similar interests [3–6] as clustering primarily supports coalition. Very few clustering based protocols dedicated to controlling MAS are available and further none supports intelligent and autonomous clustering of agents. The current work takes the advantage of the void and proposes a novel clustering based protocol

D. Juneja (✉)
Dronacharya Institute of Management and Technology, Kurukshetra, Haryana, India
e-mail: dimplejunejagupta@gmail.com

R. Singh · S. Mukherjee
Banasthali University, Banasthali, Rajasthan, India
e-mail: rashmisi@techmahindra.com

S. Mukherjee
e-mail: mukherjee.saurabh@rediffmail.com

A. Singh
Maharishi Markandeshwar University, Mullana-Ambala, India
e-mail: singh2208@gmail.com

© Springer India 2016
S.C. Satapathy et al. (eds.), *Information Systems Design and Intelligent Applications*, Advances in Intelligent Systems and Computing 434,
DOI 10.1007/978-81-322-2752-6_55

that partitions the wider multiagent system into manageable and self-configurable clusters. Use of clustering is highly advocated in literature [6–9] as it potentially denies broadcast of unwanted messages and other constrained resources such as temporal ability of agents. Complimentary to the advantages this clustering technique would bring, it also had some design challenges. Firstly, multiple agents operating in different cluster but in close proximity might have similar interests and would serve as redundant service providers which in turn would deliver analogous results. Secondly, deciding the cluster size i.e. number of agents that would form a cluster is a big unfolded challenge. It should be controlled either apriori or should be left for the application to decide need to be considered carefully. Third, agents in one cluster may try to relocate, how to manage their mobility is another issue. Another issue that should be dealt with while designing an interaction protocol is deciding the route of interaction or the path to be followed by every agent while transmitting the results. This paper provides GIPMAS as a complete solution addressing the above listed design challenges. GIPMAS executes in three phases to achieve the stated goals. First phase deals with formation of self-organizing clusters, Second is concerned with election of cluster heads in case current cluster head is unavailable and executive cluster heads and third phase describes the rules of interaction of agents (at various levels) in clusters thus formed. Simulation of GIPMAS reveals that it is able to achieve the desired target.

2 Related Works

Researchers have been discussing the challenges while designing the systems based on multiple agents [4, 10–13] and are also putting efforts to resolve issues pertaining to functioning of these systems [14–16] and interaction protocols [11, 17, 18] in particular. While Contract Net Protocol (CNP) [19–21] ensures distributed control and shared responsibility, the commitment based agent interaction protocols [22, 23] emphasize on commitments and allow flexible execution of protocols. The centralized solutions [8] for controlling the distributed agents offer limited autonomy, constrained scalability and more complexity. Few algorithms employing clustering technique to resolve the issues prevailing in a fully decentralized multiagent systems are available in [6, 7, 24]. The survey of work presented above unveils the fact that there are very few researchers who are giving importance to clustering in multiagent systems. To the best of our knowledge and available literature, a generic protocol that facilitates interaction of agents in MAS is lacking. The focus of next section is to propose GIPMAS, a generic protocol where actions taken by agents to produce the desired results are not explicitly specified and depends on the architecture of agent taking the actions. The agents are clustered dynamically and are assigned tasks by the cluster head which in turn is responsible for performing filtering and fusion of information received for efficient communication.

3 Proposed Protocol

GIPMAS is a generic cooperation protocol employing the clustering and it allows agents to execute autonomously and also do not constrain itself to a specific application. It executes in three phases namely Formation of Clusters, Election of Cluster Head (CH) and Executive Cluster Head (ECH), imposing the rules of interaction on all agents (member agents, CH and ECH). In GIPMAS, hierarchy of agents is formed where the agents are allowed to interact with others a particular order. Moreover, agents are allowed to transmit data only in the time slot allotted to them. Each phase is being elucidated in the following sections.

3.1 Formation of Clusters

Autonomous agents operating on a network may participate and organize themselves into local clusters. The agent initiating the process of formation of cluster would be henceforth referred to as cluster head (CH). To initiate the process of forming clusters, an agent (chosen randomly) broadcasts "*Forming_Cluster*" message along with its set of attributes using RCNTEP [16, 18] protocol. All listening agents (LA) with similar attributes (see Table 1) may respond showing their interest to join the particular cluster. The CH reserves all rights to choose its members on the basis of minimum communication delay as well as the agents with largest capability set. The cluster formation is being carried out in accordance to steps depicted in Fig. 1.

The cluster formation algorithm is depicted in Fig. 2. The algorithm was simulated on a system comprising of 84 agents of four different categories deployed randomly across a small network. Figure 3 depicts the initial deployment of agents. As shown in Fig. 4, five clusters containing multiple agents of different categories were formed. A cluster thus formed contains heterogeneous agents and hence becomes generic and application independent. It is obvious that although agents of similar attributes could be clustered in one cluster but these could deliver variant results leading to collisions of results. In order to avoid such a collision, we propose to deploy a cluster head in every cluster. Consequently another issue pertaining to

Table 1 Packet format of message from CH to listening agents

Agent id	Agent timestamp	Announcement	Advertisement	Attributes
Initiator agent$_i$/ listening agent	t_i/t_j	Forming_cluster/interested/welcome/thankyou	Request/response	Communication protocol name of certification agency, parent domain reliability rating

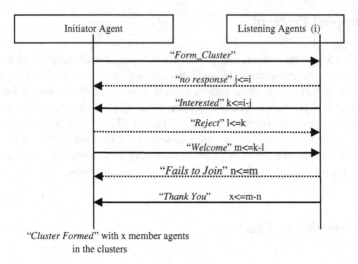

"Cluster Formed" with x member agents
in the clusters

Fig. 1 Interaction diagram depicting interaction of initiator agent and listening agents

```
Form_ClusterAlgo ()
{
cluster_head=agent₀;
time_stamp[0]=t₀;
c=sizeof(cap(Cluster_head));
for (i=1 to n)
        broadcast(Forming_Clusters);
for (j=1;j<=i;j++)
    if (no response)
                goto step 1;
    else
                for( k=1;k<=i-j;k++)
                { if (msg=interested)
                    for ( ∀ capc(LAk) ∈ cap(cluster_head) ∧comm._delay (LAk)< τ )
                        {initial_cluster[i][k]= LAk;
                            Time_stamp[k]:=comm._delay(LAk)
                        }
                    Else reject (LAk);
                for (l=1;l<=k;l++)
                for(m=l+1;l<=k-l+1;k++)
                    if(Time_stamp [m]>Time_stamp [m+1])
                        { temp:= Time_stamp [m];
                            Time_stamp [m]:= Time_stamp [m+1];
                            Time_stamp [m+1]:=temp;
                        }
                for (l=1;l<=k;l++)
                temp_cluster [i][l]=max (capc(LAl));
                            c
                cluster[i] = max(temp_cluster[i][l]) ∧ min(τ)
                          l                          l
```

Fig. 2 Algorithm for formation of clusters

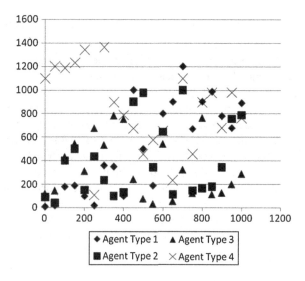

Fig. 3 Random deployment of agents in MAS

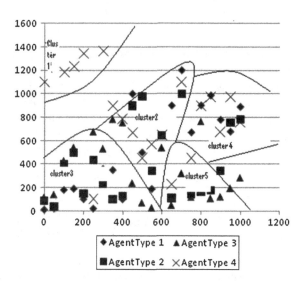

Fig. 4 Clustering due to GIPMAS

cluster-head based multiagent system is the mobility of agents and especially, the mobility of cluster head leading to the election of new cluster head and expansion of cluster, if necessary.

The second phase presents an algorithm for selection of cluster head.

Election of Cluster Head and Executive Cluster Head: Here, the challenge is to choose a new cluster head in the event of relocation of existing cluster head. If there is a gap between the capability set and communication delay of member agents, then the agent with the maximum capabilities and contributing minimum communication delay is straightforward chosen as new cluster head.

$$\text{cluster\_head\_agent}_i = \max_n(\text{cap}(\text{agent}_i)) \wedge \min_n(\text{com\_delay}(\text{agent}_i)) \quad (1)$$

where, $\text{cap}(\text{agent}_i)$ represents capabilities possessed by agent$_i$ and $\text{com\_delay}(\text{agent}_i)$ represents the communication delay induced in transmitting the information from member to cluster head. However, if all members of the cluster are possessed with equal capabilities and contributes to equal communication delay, the challenge is to select such a cluster head that distributes the tasks among all member agents uniformly. Since, cluster head is a more responsible entity and is used more intensively, therefore we propose that each member agent should become cluster head at least once addressing the challenge as posed above. Let us assume there are α member agents in a cluster. Now, in the event of relocation of existing cluster head, the probability $(\pi_{CHi}(t))$ for a member agent to become new cluster head at time (t) is given by Eq. (2). Now, here since all α member agents' have similar capabilities, hence all α agents are the potential candidates to become the cluster head. Therefore, expected number of cluster heads (φ) will be

$$\varphi = \sum_{i=1}^{\alpha} \pi_i(t) * 1 = \alpha \quad (2)$$

If there is a member agent who has the experience of being cluster head in another cluster and also has equivalent capabilities, may become the new cluster head in the particular cluster. According to Eqs. (1) and (2), selection of a cluster head depends on the agents' capabilities and overhead due to communication delays relative to the total capabilities and aggregate communication delay induced by all operational agents in the network. Therefore, probability of being a new cluster head can be expressed as Eq. (3):

$$\pi_{CHi}(t) = \max\left\{ \frac{\text{cap}(\text{agent}_i(t))}{\sum_{i=1}^{\alpha}\text{cap}(\text{agent}_i(t))}\alpha, 1\right\}$$
$$\wedge \min\left\{ \frac{\text{com\_delay}(\text{agent}_i(t))}{\sum_{i=1}^{\alpha}\text{com\_delay}(\text{agent}_i(t))}\alpha, 1\right\} \quad (3)$$

Using this probability functions, agents with higher capability set and lesser communication delay, have more probability of becoming cluster heads. In case of tie, as already mentioned younger agent has to wait and older agent would be chosen as cluster head. The age of the agent is now calculated from the time of it became member of the particular cluster where it is now competing to become cluster head.

If there is η number of clusters thus formed, there would be η cluster heads and all of them are eligible to compete to become executive cluster head (ECH). On the basis of similar arguments as used for selecting a cluster head, an ECH would be elected that would interact with end user. Alike Eqs. (3), (4) computes the probability of selecting the ECH among all η cluster heads:

$$\pi_{ECHi}(t) = \max \left\{ \frac{cap(CH_i(t))}{\sum_{i=1}^{\eta} cap(CH_i(t))} \eta, 1 \right\} \wedge \min \left\{ \frac{com\_delay(CH_i(t)}{\sum_{i=1}^{\eta} com\_delay(CH_i(t)} \eta, 1 \right\}$$

(4)

In order to use these probability functions, all member agents should know about attributes of all other competing agents and here registry agent plays vital role. During the election process, registry agent broadcast the pair of values (both capability set and communication delay offered) to all members of the cluster which can be used by competing agents directly avoiding the delay of one-to-one communication. The values pertaining to capabilities have been taken to be discrete for the current implementation but can be computed using fuzzy logic in future. One must note that while cluster head acts as local coordinator of transmissions within a cluster operating in small network of multiagent system, ECH acts as global coordinator as one ECH can communicate and hence serve many end users belonging to a wider network. Both of the heads are more responsible than member agents as these are required to fuse and filter the relevant information before transmitting to their ancestor. This requirement sometimes may become a bottleneck. Election can also be held if the current head cannot serve as head any more due to some inadvertent failures, it goes non-operational (sleep mode) and remaining members are then required to elect a new CH.

Interaction of ECH: For an end user, ECH acts as the interface agent. However, since various clusters are required to transmit data to ECH and may transmit at the same time leading to inter-cluster interference.

One of the possible solutions is to assign a local id (*lid*) to each CH and a global id (*gid*) to each ECH. For initiating the communication with ECH, the CH sends a request message along with us *lid* showing its interest to send the data packet to the closest ECH. In response, ECH assigns a time-stamp to the request message received and grants the permission to CH to transmit. ECH is required to receive the messages in the order of timestamps and if the messages received are out of order or a message with the assigned timestamp is not received, ECH assumes that some data has been lost during transmission.

Table 2 represents the format of data packet being exchanged between CH and ECH.

Although this policy incurs overhead in terms of communication cost between CH and ECH but while simulating the same very few interferences and loss of data was observed and hence results outperforms the overhead thus incurred. Figure 5

Table 2 Format of data packet exchanged between CH and ECH

Source agent ID	Destination agent Id	Type of message	Time-stamp

Fig. 5 Communication delay
versus number of clusters in
GIPMAS and direct
communication

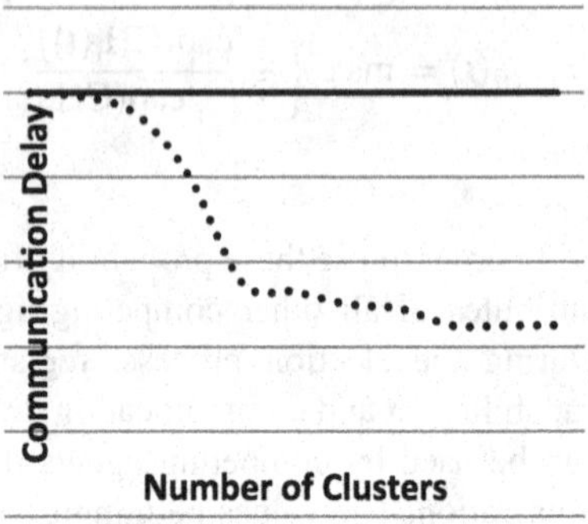

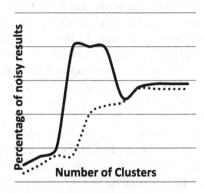

Fig. 6 Noisy results versus
number of clusters in
GIPMAS and direct
communication

shows a clear tradeoff in communication delay compared to direct communication
of agents with end user. The primary reason is that GIPMAS allows agents to
communicate only when they data to send and also since CH aggregates the results
before transferring ahead lead to reduction in noisy information being transmitted
(see Fig. 6).

4 Conclusions

GIPMAS is a clustering based agent interaction protocol that minimizes communication delay and noisy data. It allows the election of CH dynamically and also
supports recovery mechanism in case CH relocates from its respective cluster. CH
has the overhead of gathering information from member nodes, fuse and filter the
results before handing over the same to ECH.

Issues such as security of transmission of data, malicious agents becoming the
head are still untouched and currently beyond the scope of this work.

References

1. Keith D., Sycara K., Williamson M.: Middle-Agents for the Internet. IJCAI, 1 (1997).
2. Wooldridge, M. An Introduction to Multi-Agent Systems. In: John Wiley & Sons (2009).
3. Aknine, S., Pinson S., Shakun F.,M.: An Extended Multi-Agent Negotiation Protocol. *Autonomous Agents and Multi-Agent Systems* 8.1, 5–45 (2004).
4. Jennings, N., R., Faratin, P., Lomuscio, A., R., Parsons, S., Wooldridge, M., J., Sierra, C.: Automated Negotiation: Prospects, Methods and Challenges. Group Decision and Negotiation, 10(2), 199–215(2001).
5. Klusch, M., Gerber, A.: Dynamic Coalition Formation Among Rational Agents. IEEE Intelligent Systems, 3, 42–47(2002).
6. Ogston, E., Overeinder, B., Van Steen, M., Brazier, F.: A Method for Decentralized Clustering in Large Multi-Agent Systems. In: Second International Joint Conference on Autonomous Agents and Multi-Agent Systems, pp. 789–796. ACM Press (2003).
7. Foner, L.,N.: Yenta: A Multi-Agent, Referral-Based Matchmaking System. In: First International Conference on Autonomous Agents (AGENTS'97), pp. 301–307. ACM Press (1997).
8. Jain, A., Murty M., Flynn, P.: Data Clustering: A Review. ACM Computing Surveys, 31, 3, 264–322(1999).
9. Tsang, C. H., Kwong, S. (2005, December): Multi-Agent Intrusion Detection System in Industrial Network using Ant Colony Clustering Approach and Unsupervised Feature Extraction. In: IEEE International Conference on Industrial Technology (ICIT-2005), pp. 51–56. IEEE (2005).
10. Chopra, K.,A., Artikis A., Bentahar J., Colombetti M., Dignum F., Fornara N., Jones I.J.A., Singh P.M., Yolum P.,: Research Directions in Agent Communication. ACM Transactions on Intelligent Systems and Technology (TIST), 4, 2 (2013): 20.
11. Mark D., Kinney D., Luck M.: Interaction Protocols in Agentis. In: IEEE International Conference on Multi Agent Systems. IEEE Press (1998).
12. Jennings, N,. R., Sycara, K., Wooldridge, M. (1998): A Roadmap of Agent Research and Development. Journal Autonomous Agents and Multi-Agent Systems, 1(1), 7–38 (1998).
13. Noriega, P., Padget, J., Verhagen, H., d'Inverno, M.: The Challenge of Artificial Socio-Cognitive Systems. AAMAS (2014). *Online at:* http://homepages.abdn.ac.uk/n.oren/pages/COIN14/papers/p12.Pdf (2014).
14. Stefan, B., Jennings R,.N., Wooldridge M.: Re-Use of Interaction Protocols for Agent-Based Control Applications. *Agent-Oriented Software Engineering III.* Springer Berlin Heidelberg, 73–87 (2003).
15. Marzougui, B., Barkaoui, K. (2013): Interaction Protocols in Multi-Agent Systems based on Agent Petri Nets Model. Interaction, 4(7), 2013.
16. Singh, A., Juneja, D., Sharma. K. A.: Introducing Trust Establishment Protocol in Contract Net Protocol. In: International Conference on Advances in Computer Engineering (ACE-2010), pp. 59–63. IEEE (2010).
17. Aknine, S., Pinson, S., Shakun, M. F.: An Extended Multi-Agent Negotiation Protocol. Autonomous Agents and Multi-Agent Systems, 8(1), 5–45(2004).
18. Singh, A., Juneja, D.: An Improved Design of Contract Net Trust Establishment Protocol. ACEEE Journal on Communications, 4(1), 19–25 (2013).
19. Smith, R.,G.: Communication and Control in Problem Solver. IEEE Transactions on Computers, 29 (1980): 12.
20. Smith, R.,G.: The Contract Net Protocol: High-Level Communication and Control in a Distributed Problem Solver. IEEE Transactions on Computers, 100(29), pp. 1104–1113 (1981).
21. Smith, R.G., Davis, R.: Frameworks for Cooperation in Distributed Problem Solving. IEEE Transactions on Systems, Man and Cybernetics, 11(1), pp. 61–70 (1981).

22. Mallya U.,A., Singh P.M.: A Semantic Approach for Designing Commitment Protocol'. LNAI, 3396, 33—49 (2005).
23. Fornara, N., Colombetti M.: A Commitment-Based Approach to Agent Communication. Applied Artificial Intelligence, 18.9–10, 853–866 (2004).
24. Maturana, F., P., Norrie, D. H.: Multi-Agent Mediator Architecture for Distributed Manufacturing. Journal of Intelligent Manufacturing, 7(4), 257–270(1996).

An Improved Content Based Medical Image Retrieval System Using Integrated Steerable Texture Components and User Interactive Feedback Method

B. Jyothi, Y. Madhavee Latha and P.G. Krishna Mohan

Abstract The advancement in medical technology has resulted in a huge number of medical images saved in a data-base. Content Based Medical Image Retrieval (CBMIR) mechanisms help the radiologist in retrieving the required medical images from an immense database. This paper envisages an effective content based procedure in which the region of the image is taken into account by determining the borders of the image region using gray level gradient method instead of considering the image as a whole. Later, the content within the boundary region of the image is described through the steerable filter in different orientations followed by extracting the second-order statistical components as feature vectors. Medical images correlated to the query image are retrieved by computing the Euclidean distance as a similarity measure between database images and the query image. To enhance the accuracy of the medical retrieval system, Instant Based Relevance Feedback has been used. In this procedure, the user interacts with the system and selects the most relevant image for searching again. The above search procedure is repeated for finding out more precise images by sorting out the first search and the second search similarity distances. Eventually, the corresponding top ranked images are displayed. These results reveal that the proposed algorithm outperforms by of increasing Recall Rate and reducing Rate of Error.

Keywords Content based image retrieval · Boundary recognition · Similarity measure · Relevance feedback

B. Jyothi (✉)
Department of ECE, MRCET, JNTUH, Hyderabad, India
e-mail: bjyothi815@gmail.com

Y. Madhavee Latha
Department of ECE, MRECW, JNTUH, Hyderabad, India
e-mail: madhaveelatha2009@gmail.com

P.G. Krishna Mohan
Department of ECE, IARE, Hyderabad, India
e-mail: pgkmohan@yahoo.com

© Springer India 2016
S.C. Satapathy et al. (eds.), *Information Systems Design and Intelligent Applications*, Advances in Intelligent Systems and Computing 434,
DOI 10.1007/978-81-322-2752-6_56

1 Introduction

The amount of medical images in the database has been in creditably increased due to advancement of technology, Medical images are essential in diagnosing the various diseases in helping the curing process and in supporting the medical decision making process [1]. Precise medical images help the experts in taking effective decision so that the patients can be protected from fatal consequences.

CBIR is an active research area facilitates the process of retrieving the requisite medical images from the bulky data-base [2] with minimum human interface. It is determined as a course of action of searching similar images from the great data-base on the source of their visible features such as shape, color and texture [3].

Medical images are embellished with precise patterns of texture which have detailed information. Texture is very essential and usual surface characteristic, repetitive pixel information regarding the collection and it also provide data with the association of its exterior surroundings. The objects in a medical images distinguished only by their texture sequence. The usual texture features used in CBIR systems can be label into statistical and structural texture features. Haralick et al. [4] utilized Gabor representations which is invariant to scale and rotation. The structural techniques describe the texture patterns formed by repeated arrangement of homogenous gray levels which are described by morphological operations.

The effectiveness of the CBMIR System strongly based on the selection of the set of visible features. These features plays significant role in CBMIR System.

Medical image have a reduced boundaries. Hence there are a small amount of practical troubles at the time of scanning the medical images for the duration of scanning such as a smaller amount illumination; noise and poor contrast results difficulty in identifying the tissue patterns and arrangement of organs are reasons of decrease in the retrieval outcome. A boundary is a line that set apart it from the surroundings. It assist us in Object identification and considering the shape of an entity. Somkath in [5] has made known a technique to defeat the difficulty of boundary recognition for weakly defined, but this has got limitation that the boundary must be closed.

This paper focuses on effective feature extraction procedure. According to this first boundary of the medical image is detected by using intensity gradient and edge map techniques followed by texture feature extraction steerable filer at different orientations followed by extracting statistic texture features using seconds order statistical components at various orientations.

This frame work increases the performance of the retrieval system by using relevance feedback [6]. The Relevance Feedback is a method has been used successfully in human computer interface. This has initially been elaborated for increasing the efficiency of information system. The main objective of relevance feedback for retrieval system is to understand the user requirements. Various relevance techniques have been discussed in [7].

The medical image retrieval system profits preliminary results based on ED as a similarity metric [8]. If the user is not satisfied with retrieved output, he can interact with the system by modifying the query image. The system accordingly analyzes the user feedback and returns superior results. The distances from the first search and second search are sorted in uphill order then the corresponding lower distance images are displayed.

The remaining paper is planned in this fashion. Section 2 focus the CBMIR system. Section 3 describes the Features Extraction. Section 4 gives the Experimental Results with Relevance Feedback. Section 5 concludes the paper.

2 Proposed CBMIR System Architecture

The building block illustration of a typical theoretical content-based retrieval system is illustrated and discussed in Fig. 1. It expressed by three modes: off-line feature extraction mode, online image retrieval mode and feedback mode.

In offline mode, the images saved in the data-base are pre-processed for reducing noise by means of median filter. Later the region of the object is observed by ignoring the background of the image. Next, the visible characteristics of the image are extracted by using statistical components. These components stored in the database as a feature vector. We fallow the suit for input image during online phase.

In online mode the query image is submitted for probing analogous images. In conclusion, the system profits the large amount relevant to the input image by measuring similarity among the feature vectors of the input image and individuals of the data base images.

In feedback mode user interacts with the retrieval system to refine queries representations. It is the process of selecting the most relevant image for searching again and using the information feuded back by the user.

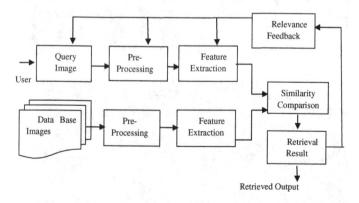

Fig. 1 CBIR system

3 Features Extraction

A feature gives the particulars about the visible property of an image in the vicinity for a little collection of gray values of the entire image. The characteristics of an image traced by means of their features. In the presented approach local feature used despite of extracting total image. The image boundary is identified by using edge following technique and consequent texture features are extricated in various orientations.

3.1 Object Boundary Detection

Medical images recognized with the help of their appropriate boundaries. The design of boundary wrenching out in blurred images is shown in Fig. 2. In this process, the boundaries of the images are recognized using intensity gradient edge subsequent algorithm which is based on magnitudes, directions and edge map [9] of given image f(i, j) is calculated based on the subsequent formulas.

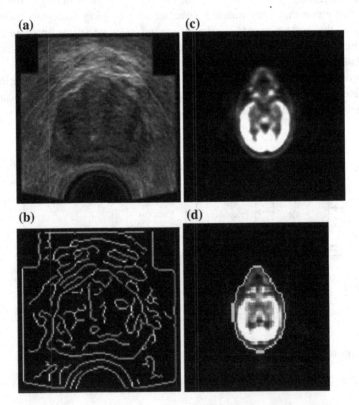

Fig. 2 a Noisy ultrasound image. **b** Edge mapped image. **c** Brain image. **d** Boundary detection

$$H(i,j) = \frac{1}{H_r} \sum_{(i,j) \in N} \sqrt{H_x^2(i,j) + H_y^2(i,j)} \tag{1}$$

$$A(i,j) = \frac{1}{H_r} \sum_{(i,j) \in N} \tan^{-1} \left(\frac{H_y(i,j)}{H_x(i,j)} \right) \tag{2}$$

Hr is the sum of pixels.

The edge map of an of images is determined by convolution with the query image F(i, j) through the texture mask T(x, y) and the ensuing image is R(x, y).

$$TM(x,y) = LM.LM^T \tag{3}$$

$$L = (1,4,6,4,1)^T \tag{4}$$

$$RE(i.j) = F(i,j) * TM(x,y) \tag{5}$$

The boundary of an image obtained by using

$$L_{ij}(i,j) = H_{ij}(i,j) + A_{ij}(i,j) + RE_{ij}(i,j) \tag{6}$$

H (i, j), A (i, j) are the magnitude, angles of gradient and R (i, j) is the map of edges.

3.2 Texture Features Extraction

Medical images are frequently exemplified in gray level; the majority medical image surfaces demonstrate texture. Kriti et al. utilized the capabilities of GLCM for breast cancer classification [10]. We have implemented invariant texture descriptor based on steerable filter rotting explained in [3]. Steerable is synthesized at different orientations and determines the output as a linear combination of basis filters. Steerable oriented filter designed is a quadrature pair to permit adaptive control over phase and orientation. The filter at any orientation θ is a linear combination of basis filters $G_1^{0°}$ & $G_1^{90°}$ and interpolation functions $\cos(\theta)$ and $\sin(\theta)$

$$G_1^{\theta} = \cos(\theta)G_1^{0°} + \sin(\theta)G_1^{90°} \tag{7}$$

$G_1^{0°}$ & $G_1^{90°}$ are set of Basis filters and are derivatives of Gaussian function G(x, y) in x and y directions with scaling and standard deviation set to one.

$$G(x,y) = e^{-(x^2 + y^2)} \tag{8}$$

where $G_1^{0°} = -2xe^{-(x^2+y^2)}$ and $G_1^{90°} = -2ye^{-(x^2+y^2)}$

The steering constraint is

$$F_\theta(m,n) = \sum_{k=1}^{N} b_k(\theta)A_k(m,n) \tag{9}$$

where $b_k(\theta)$ is the interpolation function and $A_k(m, n)$ are the basis filters.

Texture information can be described by appertain second order statistics in a variety of tilting sub-bands using steerable filter. In this paper, we extort the texture features from (b) 10 tilting sub-bands as shown in Fig. 3.

$$SR_i(i,j) = \sum_{i_1} \sum_{j_1} F(i_1,j_1)S_i(i - i_1 j - j_1) \tag{10}$$

where $SR_i(x, y)$ is symbolizing horizontal and S_i symbolize the deviation band pass filters at direction $i = 1, 2, 3, 4, 5, 6...$

The subsequent steps give the depiction of the statistical computation of the statistical elements examination [11].

$$\text{Energy (E)} \quad E = \sum_{x=0}^{N-1} \sum_{y=0}^{M-1} \{F(i,j)\}^2 \tag{11}$$

$$\text{Contrast (C)} \quad C = \sum_{i=0}^{N-1} \sum_{j=0}^{M-1} |i - j|F(i,j) \tag{12}$$

$$\text{Inverse Difference Moment (IDM)} \quad IDM = \sum_{i=0}^{N-1} \sum_{j=0}^{M-1} \frac{1}{1 + (i - j)^2} F(i,j) \tag{13}$$

$$\text{Entropy (E)} \quad E = \sum_{i=0}^{N-1} \sum_{j=0}^{M-1} F(i,j) X \log(F(i,j)) \tag{14}$$

$$\text{Correlation (Cr)} \quad Cr = \sum_{i=0}^{N-1} \sum_{j=0}^{M-1} \frac{\{iXj\}X(F(i,j) - \{\mu_i X \mu_j\})}{\sigma_i X \sigma_j} \tag{15}$$

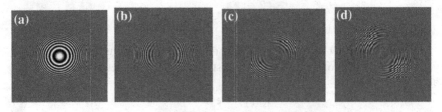

Fig. 3 **a** The input image. **b** Oriented image horizontally. **c** Rotated image with 45°. **d** −45° rotated image

$$\text{Variance (V)} \quad V = \sum_{i=0}^{N-1}\sum_{j=0}^{M-1}(i-\mu)^2 F(i,j) \qquad (16)$$

$$Fv = [ASM, C, IDM, E, Cr, V] \qquad (17)$$

4 Experimental Results with Instance Based Relevance Feedback

To test the performance of proposed approach we make use of the medical images from web accessible international resources called Frederick national laboratory.

The database in our proposed algorithm contains 1000 medical images of the human organs—lungs, brain, abdomen etc. Analogous medical images retrieved by calculating the Euclidian distance [12] between the input image feature vectors and matching feature of the data-base images.

$$D(Fv_1, Fv_2) = \sqrt{\sum_{\forall i}(Fv_1(i) - Fv_2(i))^2} \qquad (18)$$

The feature resemblance measure also plays an important role on the retrieval outcome. We have experimented with a variety of similarity measures to assess the performance as illustrated in Fig. 4. It is found that Euclidian provides better retrieval output when compared with SSIM, MSSIM and NC techniques respectively.

For a given query, the medical image retrieval system returns preliminary results based on ED as a similarity metric. If the user is not satisfied with retrieved output,

Fig. 4 Performance analysis of various distance measures

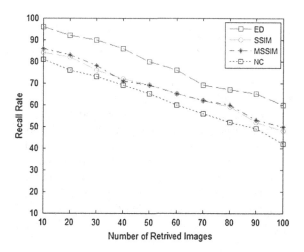

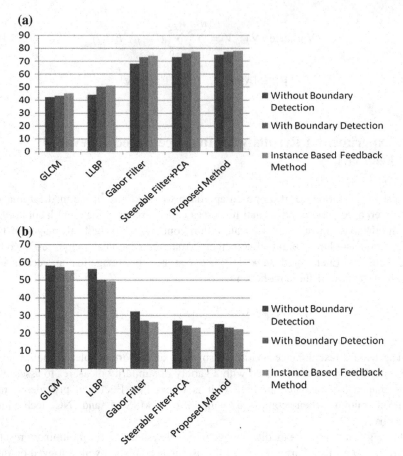

Fig. 5 **a** A comparison of the recall rate with various feature extraction methods. **b** A comparison of error rate with various feature extraction methods

he can interact with the system by modifying the query image. The system accordingly analyzes the user feedback and returns superior results.

In this approach further search will be taken up if the user requirements are not met. The best medical image from the results will be selected by the user and given as input to the medical image retrieval system. The distances from the first search and second search are sorted in ascending order then the corresponding lower distance images are displayed As a result we get superior medical images by sorting the Euclidian distances which have been obtained from first search and second search respectively. On the whole the performance of the presented CBMIR system is evaluated by measuring Recall Rate and Error Rate which have been recruit in [13]. The retrieval output of the presented CBMIR method has been analogize with the following feature extrication techniques such as PCA, LLBP [14, 15].

The existing proposed system has considered the whole image for feature extraction. The existing methods like PCA LLBP giving poor results when the images are corrupted with noise. In our approach we have proposed a more effective approach that enhances the production of the retrieval system. The present approach considers only the object region of the medical image by detecting the boundary of an image. Juxtapose principally above mentioned methods, the presented feature extraction procedure gives superior retrieval showing even in the noisy database images also (Fig. 5a, b).

$$\text{Recall Rate} = \frac{\text{Number of Relvant Images Retrieved}}{\text{Total Number of Relavant Images in Datab}} \qquad (19)$$

$$\text{Error Rate} = \frac{\text{Number of Non Relvant Images Retrieved}}{\text{Total Number of Images Retrieved}} \qquad (20)$$

5 Discussion and Conclusion

In this paper an effective feature extraction method for medical image retrieval has been proposed. This approach utilizes the segmentation method which enormously helped in feature extraction of the noisy medical images. In addition Instance based Relevance Feedback has been integrated with the feature extraction method to enhance the efficiency of CBMIR system. The Proposed approach is limited to time. As the quantity of medical images grows, the computing time increases and retrieval results decreases. To address these limitations, in future multiple features and an integrated with classifiers will be used to enhance accuracy of the result. Thus the CBMIR System will be an effective tool in assisting CAD System.

References

1. Müller H, Michoux N, Bandon D, and Geissbuhler A. A review of content-based image Retrieval systems in medical applications-clinical benefits and future directions. Medical Informatics. 1, 73 (2004).
2. L.A. Khoo, P. Taylor, and R.M. Given-Wilson, "Computer-Aided Detection in the United Kingdom National Breast Screening Programme Prospective Study," Radiology, vol. 237, pp. 444–449, 2005.
3. Young Deok Chun, Nam Chul Kim, Ick Hoon Jang, Content-based image retrieval using multiresolution color and texture features, IEEE Transactions on Multimedia 10 (6) (2008) 1073–1084.
4. Sourav Samanta, SK. Saddam Ahmed, Mohammed Abdul-Megeed, M, Salem, Siddhartha Sankar Nath, Nilanjan Dey, and Sheli Sinha Chowdhury, "Haralick Features Based Automated Glaucoma Classification Using Back Propagation Neural Network." Springer international

publishing Switzerland 2015, vol. 1, Advances in intelligent system and computing, 327, DOI: 10.1007/978-3-319-11933-5-38.

5. Krit Somkantha, Nipon Theera-Umpon, "Boundary Detection in Medical Images Using Edge Following Algorithm Based on Intensity Gradient and Texture Gradient Features," in *Proc. IEEE* transactions on biomedical engineering, vol. 58, no. 3, March 2011, pp. 567–573.

6. Dr. (Mrs) Ananthi Sheshasaayee, Jasmine. C, "Relevance Feedback Techniques Implemented in CBIR: Current Trends and Issues", International Journal of Engineering Trends and Technology (IJETT), Volume 10 Number 4, Apr 2014.

7. Darshana Mistry, "Survey of Relevance Feedback methods in Content Based Image Retrieval", Darshana Mistry/International Journal of Computer Science & Engineering Technology (IJCSET), Vol. 1 No. 2, pp 32–40, ISSN: 2229-3345.

8. Miguel Arevalillo-Herráez, Juan Domingo, Francesc J. Ferri, Combining similarity measures in content-based image retrieval," Pattern Recognition Letters 29 (2008) 2174–2181.

9. Ms. S. Veeralakshmi, Mrs. S. Vanitha Sivagami, Ms. V. Vimala Devi, Ms. R. Udhaya "Boundary Exposure Using Intensity and Texture Gradient Features. IOSR Journal of Computer Engineering (IOSRJCE) ISSN: 2278-0661, ISBN: 2278-8727, Volume 8, Issue 1 (Nov., Dec. 2012), pp. 28–33 www.iosrjournals.org.

10. Kriti, Jitendra Virmani, Nilanjan Dey, Vinod Kumar, PCA-PNN and PCA-SVM Based CAD Systems for Breast Density Classification, Chapter, Applications of Intelligent Optimization in Biology and Medicine Volume 96 of the series Intelligent Systems Reference Library, pp. 159–180.

11. Soaya Cheriguene, Nabiha Azizi, Nawel Zemmal, Nilanjan Dey, Hayet Djellali, Nadir Farah, "Optimized Tumor Breast Cancer Classification Using Combining Random Subspace and Static Classifiers Selection Paradigms", Medicine, Volume 96 of the series Intelligent Systems Reference Library, pp. 289–307.

12. I. El-Naga, Y. Yang, N.P. Galatsanos, R.M. Nishikawa, and M.N. Wernick, "A Similarity Learning Approach to Content-Based Image Retrieval: Application to Digital Mammography," IEEE Trans. Medical Imaging, vol. 23, no. 10, pp. 1233–1244, Oct. 2004.

13. B. Jyothi, Y. Madhavee Latha, P.G. Krishna Mohan, Multidimensional Feature Vector Space for an Effective Content Based Medical Image Retrieval 5th IEEE International Advance Computing Conference (IACC-2015), BMS College of engineering Bangalore, June 12 to 13, 2015.

14. A.S. Syed navaz1, T. Dhevi Sri and Pratap Mazumder, Face Recognition using Principal Component Analysis and neural networks" International Journal of Computer Networking, Wireless and Mobile Communications (IJCNWMC) ISSN 2250-1568 Vol. 3, Issue 1, Mar 2013, 245–256.

15. B. Jyothi, Y. MadhaveeLatha, P.G. Krishna Mohan, Steerable Texture Descriptor for Effective Content Based Medical Image Retrieval System Using PCA. 2nd International conference on Computer & Communication Technologies (IC3T-2015) published by proceedings of IC3T-2015, Springer-Advanced in Intelligent System and Computing Series 11156, vol 379, 380–381.

Improvement of Stability by Optimal Location with Tuning STATCOM Using Particle Swarm Optimization Algorithm

P.K. Dhal

Abstract Now a day's modern power system demands promptly increasing but so many problems are facing by utilities. The problems include power flow violation in the system i.e. voltage dips in the buses, static/dynamic instability and voltage collapse etc. In this paper transient stability is analyzed via optimal location of the Western Science Coordinated Council (WSCC) 9 bus system. The FACT device i.e. STATCOM is introduced properly, when voltage goes down. It is helpful using PSAT software. The system performance has been analyzed by applying three-phase fault. It is assumed the fault time at 1.05 s and clearing time 1.15 s. similarly fault time 3.15 s and clearing time 3.5 s. in this fault condition, the STATCOM is used in the position of optimal location in 9-bus system. The particle swarm optimization algorithm technique is implemented with STATCOM for better improvement of stability.

Keywords STATCOM · Voltage stability · PSO

1 Introduction

The Flexible AC Transmission Systems (FACTS) are played in vital role recent years. FACTS devices are used to inject or absorb the reactive power. FACTS devices are used to improve power system stability and improve the power transform capability. STATCOM is one of the FACTS devices which are used to regulate the system voltage by absorbing and injecting reactive power [1, 2]. It consists of a voltage source inverter which generates controllable ac voltage source behind a transformer leakage reactance and an energy storage capacitor [3]. The optimal location and tuning of STATCOM have an essential role to enhance the stability of

P.K. Dhal (✉)
Department of Electrical and Electronics Engineering,
Vel Tech Dr. RR & Dr. SR Technical University, 42,
Almathi-Vel Tech Road, Avadi, Chennai 600062, India
e-mail: Pradyumna.dhal@rediffmail.com

© Springer India 2016
S.C. Satapathy et al. (eds.), *Information Systems Design and Intelligent Applications*, Advances in Intelligent Systems and Computing 434,
DOI 10.1007/978-81-322-2752-6_57

583

the system. This paper presents the investigation of best location and tuning of STATCOM to enhance the transient stability for three phase fault conditions [4]. The performance of the proposed controller is carried out by the time domain simulation.

2 Transient Stability on STATCOM

The power system is to lose synchronism when it is subjected to severe transient disturbances occur in the system. Here the disturbances might be a fault, loss of generation or loss of a big load. The disturbances affect the rotor angles, bus voltages, power flows and many variables of the system. This paper investigates the effect of fault on bus voltages occurred in the WSCC system [5]. A current injection model of STATCOM has been implemented in PSAT software in MATLAB. The reactive power is

$$Q = i_{sh} * V \tag{1}$$

where i_{sh} = current of the STATCOM which will be injected to the system.

The mathematical model of STATCOM is shown in Fig. 1.

A STACOM can improve the power system performance through the voltage flicker control, transient stability, power oscillation damping, dynamic voltage control and Reactive power control.

3 Particle Swarm Optimization Algorithms on STATCOM

The particle swarm optimization places the particles in the search space with initial velocities. The velocities are assigned to the particles randomly. Each particle in search space will find optimal solution with the help of two parameters. The two

Fig. 1 Mathematical block diagram of STATCOM.
K_r = Regulator gain,
T_r = Regulator time constant

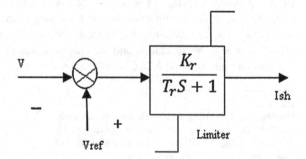

parameters are velocity and position respectively. By the help of the two parameters, the fitness function of the particle has been calculated [6, 7]. Each particle in the problem space would have its best solution. That best value is called as P_{best}. When a particle completes its population, then the best value is global best G_{best}. After finding the two best values, the particle updates its velocity and position according to the following equations

$$V_i^{k+1} = W_k^* V_i^k + C_1^* rand_1^* (P_{besti} - S_i^k) + C_2^* rand_2^* (G_{best} - S_i^k) \qquad (2)$$

$$S_i^{k+1} = S_i^k + V_i^{k+1} \qquad (3)$$

V_i^k = Velocity of agent i at kth iteration, V_i^{k+1} = Velocity of agent i at (k + 1)th iteration, W_k = The inertia weight, C_1 and C_2 = Individual and social acceleration constants (0–3), $rand_1$ and $rand_2$ = random numbers (0–1), S_i^k = Position of agent i at Kth iteration, S_i^{k+1} = Position of agent i at $(K+1)$th iteration, P_{besti} = Particle best of agent i and G_{bestt} = Global best of the group.

3.1 Objective Function on STATCOM

The design of STATCOM parameters has been formulated as an Eigen value based objective function. Here two sub objective functions are used [8, 9]. One is minimization of the real part of the Eigen value and the other one is maximization of the damping ratio. The damping ratio of the ith critical mode is

$$\varsigma_i = \frac{-\sigma_i}{\sqrt{\sigma_i^2 + \omega_i^2}} \qquad (4)$$

where the eigen value $\lambda_i = \sigma_i \pm j\omega_i$. The objective functions are represented as,

$$J_1 = \sum_{i=1}^{n} (\sigma_0 - \sigma_i)^2 \qquad (5)$$

and

$$J_2 = \sum_{i=1}^{n} (\varsigma_0 - \varsigma_i)^2 \qquad (6)$$

where $\sigma i \leq \sigma_0$, $\zeta_i \geq \zeta_0$ for i = 1, 2 ......,n. The combined objective function $J = J_1 + \alpha J_2$ is used to have a closed loop eigen values. The value of α is considered as 7 (Fig. 2).

Fig. 2 Objective function J
of the system

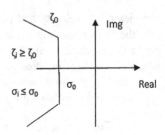

3.2 Algorithm on STATCOM

Step 1: Initialize the parameters K_r, T_r.

Step 2: Calculate the fitness value using the objective function and determine the P_{best}.

Step 3: Determine G_{best} from the P_{best}.

Step 4: Update the parameter values.

Step 5: Check the solution is feasible or not.

Step 6: If the solution is feasible, check the iteration count.

Step 7: If the iteration count reaches the maximum, stop the process.

Step 8: If the iteration count does not reach the maximum, then continue the process from 2 to 7.

3.3 Flow Chart on STATCOM Using Particle Swarm Optimization

The fitness function can be varied depends upon the problem occur in the system. STATCOM parameters K_r, T_r are selected to tune as per algorithmic procedure (Fig. 3).

4 Western Science Coordinated Council (WSCC) 9 Bus System

A 9 bus Western Science Coordinated Council (WSCC) with 6 transmission lines, 3 generators, 3 loads and a local load D is considered. The system performance has been studied by applying a 3 phase fault i.e. (a) Fault time at 1.05 s and clearing time 1.15 s. (b) Fault time at 3.15 s and clearing time 3.50 s. The fault time second case is producing better result than first case. The fault has been applied at the bus 6

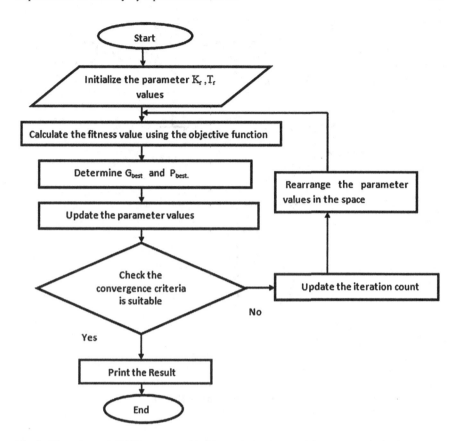

Fig. 3 Flow chart of WSCC system with STATCOM using particle swarm optimization

of the system. In the second case bus 6 has affected severely and the system has low voltage profile because of the fault. According to this case, the optimal location of STATCOM is bus 6. Here, STATCOM has been applied at bus 6 to improve the system stability. (The parameters of untuned STATCOM are—50 for K_r and 0.001 for T_r) (Figs. 4, 5 and 6).

From the Table 1, bus 6 is severely affected to zero value. The STATCOM has been applied at bus 6 of the system. STATCOM is used to improve the stability of the system (Figs. 7 and 8).

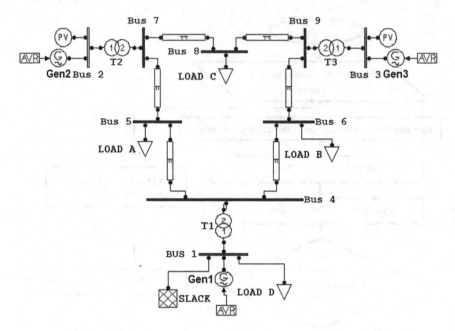

Fig. 4 WSCC 9-Bus systems

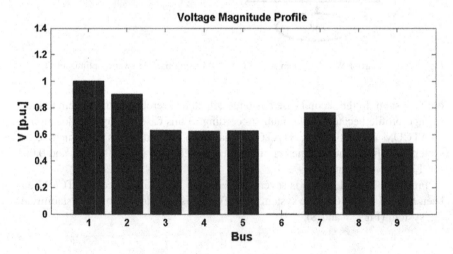

Fig. 5 Voltage profile without STATCOM

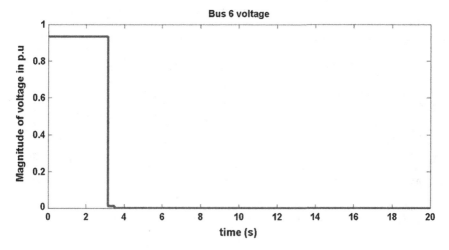

Fig. 6 Bus 6 V without STATCOM

Table 1 WSCC system parameter values without STATCOM

Bus	V (p.u)	phase (rad)	P gen (p.u)	Q ge (p.u)	P load (p.u)	Q load (p.u)
Bus 1	1.00119	1956.4	4.62488	2.01569	1.6	0.65
Bus 2	0.90289	1956.43	1.63	0.62432	0	0
Bus 3	0.63087	1956.46	0.85	0.42558	0	0
Bus 4	0.62573	1956.29	0	0	0	0
Bus 5	0.62698	1956.19	0	0	2	0.9
Bus 6	0	1955.72	0	0	1.6	0.65

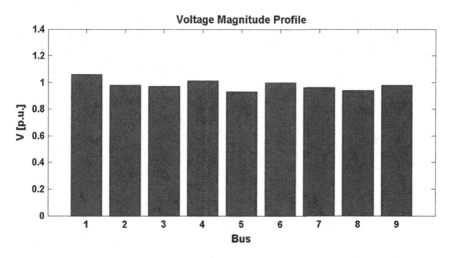

Fig. 7 Voltage profile with STATCOM

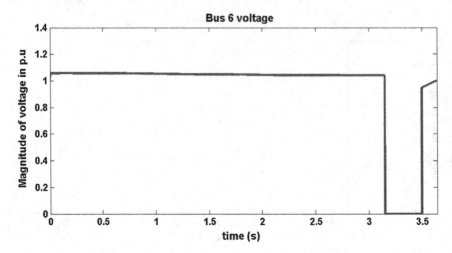

Fig. 8 Bus 6 voltage with STATCOM

Table 2 WSCC system parameter value with STATCOM

Bus	V (p.u)	phase (rad)	P gen (p.u)	Q gen (p.u)	P load (p.u)	Q Load (p.u)
Bus 1	1.06101	−8.3518	4.61375	1.56254	1.6	0.65
Bus 2	0.97774	−8.2964	1.63	0.48664	0	0
Bus 3	0.96985	−8.3813	0.85	0.15559	0	0
Bus 4	1.01259	−8.4716	0	0	0	0
Bus 5	0.9307	−8.5698	0	0	2	0.9
Bus 6	0.99962	−57.779	0	0.65749	1.6	0.65

From Table 2, the bus 6 is improved to 0.99962 p.u with unturned STATCOM. According to PSO tuned system, one STATCOM is applied to bus 6 to improve the transient stability (by using the PSO tuned STATCOM parameters) as shown in Figs. 9, 10, 11 and Tables 3, 4.

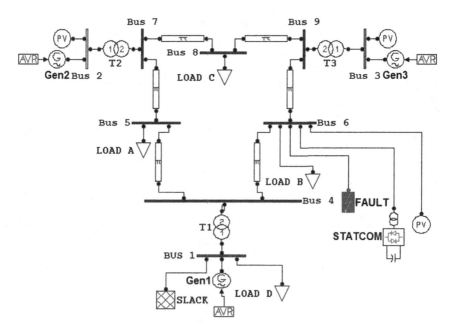

Fig. 9 WSCC systems with tuned STATCOM by PSO

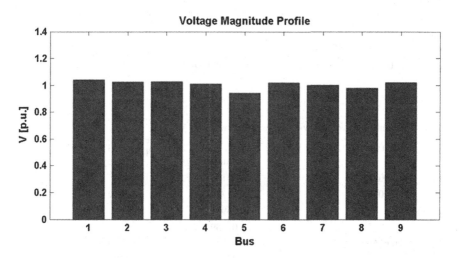

Fig. 10 Voltage profile with STATCOM (tuned by PSO)

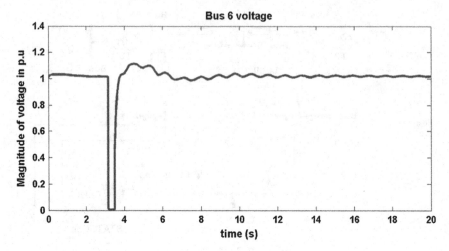

Fig. 11 Bus 6 V with STATCOM (tuned by PSO)

Table 3 Parameter value with tuned STATCOM by PSO

Bus	V (p.u)	phase (rad)	P gen (p.u)	Q gen (p.u)	P load (p.u)	Q bad (p.u)
Bus 1	1.04045	−57.474	4.61375	1.56254	1.6	0.65
Bus 2	1.02449	−57.611	1.63	0.48664	0	0
Bus 3	1.02687	−57.692	0.85	0.15559	0	0
Bus 4	1.01006	−57.643	0	0	0	0
Bus 5	0.94329	−57.779	0	0	2	0.9
Bus 6	1.01894	−57.779	0	0.65749	1.6	0.65

Table 4 Comparison voltage without, with and tuned STATCOM

Bus	Without STATCOM V in p.u	With STATCOM (untuned) V in p.u	With STATCOM (tuned by PSO) V in p.u
Bus 1	1.00119	1.06101	1.04045
Bus 2	0.90289	0.97774	1.02449
Bus 3	0.63087	0.96985	1.02687
Bus 4	0.62573	1.01259	1.01006
Bus 5	0.62698	0.9307	0.94329
Bus 6	0.0	0.99962	1.01894

5 Conclusion

The performance of STATCOM has been investigated through the time domain and power flow simulations. The optimal location and tuning of the STATCOM improves the system's voltage stability. The voltage profile and bus voltage graphs show increase in voltage and transient stability of the system when STATCOM is used. The results are compared with the PSO tuned STATCOM and show the effectiveness of the tuned system. The future work can be carried out using Genetic and BBO algorithm to get better results.

References

1. H. Chen, R. Zhou, Y. Wang, "Analysis of voltage stability enhancement by robust non linear STATCOM control," Proceedings of IEEE PES Summer Meeting, vol. 3, 2000, pp. 1924–1929.
2. Naveen Goel, R.N. Patel, Saji T. Chacko K. "Genetically Tuned STATCOM for Voltage Control and Reactive Power Compensation". International Journal of Computer Theory and Engineering, Vol. 2, No. 3, June, 2010
3. Dian Palupi Rini, Siti Mariyam Shamsuddin, Siti Sophiyati Yuhaniz "Particle Swarm Optimization: Technique, System and Challenges". International Journal of Computer Applications (0975 – 8887) Volume 14–No. 1, January 2011.
4. Rajesh Kr Ahuja, Mukul Chankaya "Transient Stability Analysis of Power System with UPFC Using PSAT". International Journal of Emerging Technology and Advanced Engineering Volume 2, Issue 12, December 2012.
5. Kyaw Myo Lin, Wunna Swe, Pyone Lai Swe "Coordinated Design of PSS and STATCOM for Power System Stability Improvement Using Bacteria Foraging Algorithm". International Journal of Electrical, Computer, Electronics and Communication Engineering Vol:7 No:2, 2013.
6. Haniyeh Marefatjou, Iman Soltani "Optimal Placement of STATCOM to Voltage Stability Improvement and Reduce Power Losses by Using QPSO Algorithm". Journal of Science and Engineering Vol. 2 (2), 2013, 105–119.
7. Ahmed Elsheikh, Yahya Helmy, Yasmine Abouelseoud, Ahmed Elsherif "Optimal Power Flow and Reactive Compensation Using a Particle Swarm Optimization Algorithm". Journal of Electrical Systems 10-1 (2014): 63–77.
8. D.K. Sambariya, R. Prasad "Robust tuning of power system stabilizer for small signal stability enhancement using meta heuristic bat algorithm" Electrical Power and Energy Systems 61 (2014) 229–238.
9. Aribi Fughar, Nwohu, M. N. "Optimal Location of STATCOM in Nigerian 330kv Network using Ant Colony Optimization Meta- Heuristic". Global Journal of Researches in Engineering: F Electrical and Electronics Engineering Volume 14 Issue 3 Version 1.0 Year 2014.

Identifying Crop Specific Named Entities from Agriculture Domain Using Semantic Vector

Ashish Kumar, Payal Biswas and Aditi Sharan

Abstract Named entity extraction is the most primitive task in the field of text mining. This paper is a preliminary attempt to identify domain specific named entities, specifically crop names from text documents in Agriculture domain. The task is challenging as the names of these entities are very generic and hence word level features are not very helpful in differentiating them from routine words. Thus in this paper we have suggested a semantic vector based approach. Two different methods have been suggested, that are based on exploiting the context of the words in order to extract these entities. The methods accept few seed entities, identify their context and then find other words that are sharing the similar context. These words sharing the similar context are expected to be newly identified entities. Considering this as an initial attempt, the results are motivating and inspire us to move further in this direction.

Keywords Semantic vector · Agriculture domain · Entity tagger · Named entity extractor · Co-occurrence matrix

1 Introduction

Named Entity Recognition (NER) is considered to be a fundamental task of Information Extraction. Initially when NER came into picture, the primitive task was to recognize the name of place, person and organization from a given open domain text documents. However if we move towards a specific domain one has to

A. Kumar (✉) · P. Biswas · A. Sharan
Jawaharlal Nehru University, New Delhi 110067, India
e-mail: ashishkumar2912@gmail.com

P. Biswas
e-mail: payal.biswas138@gmail.com

A. Sharan
e-mail: aditisharan@gmail.com

© Springer India 2016

595

S.C. Satapathy et al. (eds.), *Information Systems Design and Intelligent Applications*, Advances in Intelligent Systems and Computing 434,
DOI 10.1007/978-81-322-2752-6_58

tune these classes of entities in accordance with the nature of domain. For example, in case of agriculture domain the appropriate entities would be the name of crops, cereals, fertilizer etc. rather than name of place person and organization. Agriculture is an important domain in Indian context. Extraction of Agricultural named entities is the most primitive task, in order to perform any kind of text mining operations like information retrieval, machine translation, document summarization etc. in agricultural domain.

There are various NER systems available such as Stanford's NER, Python NLTK NER, Learning Based Java1 (LBJ) [1] and many others. But being open domain NER, these NERs can only recognize the name of Place, Person and Organization. Various NER systems have also been developed for biomedical domain such as ABNER [2], BNER, BANNER [3] etc. But since these systems are trained for recognizing biomedical entities they cannot recognize agricultural entities. As no entity recognizer is available in Agriculture domain, we were motivated to work in this direction.

2 Related Work

We are focusing only on related work in domain specific NER, particularly biomedical domain, as this domain is somewhat related to Agriculture domain. Some of the prodigious works in biomedical domain is discussed below.

Lee et al. [4] presented a two-phase namely boundary identification phase and a semantic classification phase for named entity recognizer. It is used to resolve the multi-class problem and unbalanced class distribution problem by employing an ontology based hierarchical classification method. Power Bio NE was developed by Zhou et al. [5] which deals with the special phenomena of naming conventions in the biomedical domain like word formation pattern, morphological pattern, part-of-speech etc. HMM is used to integrate all the features and k-NN to resolve the data sparseness problem. Unlike this Seki and Mostafa [6] presented a hybrid approach without using any NLP tool. The approach completely relied on heuristics and probabilistic model for identifying the names. In recent years, Nested Named Entity has also become popular among various researchers. Alex et al. [7] introduced three techniques namely layering, cascading, and joined label for modeling and recognizing nested entities. They also compared these techniques by means of a conventional sequence tagger. In recent year, Tang et al. [8] exploited a large set of features for Named Entity Recognition. They have investigated three different types of word representation (WR) features for BNER, namely clustering-based representation, distributional representation, and word embedding and improved the F-measure by 3.75 % on the BioCreAtIvE II GM and 1.39 % on JNLPBA corpora.

Researchers have also focused over diverse genres and domains like emails Minkov [9], scientific texts and religious texts Maynard et al. [10]. However no significant work has been done in agriculture NER.

3 Proposed Work

Agriculture domain provides for various significant entities such as: plant names, cereals and crops, fertilizers, pest and pesticides, plant diseases, etc. It is not possible to extract all the possible entities. In this work we have targeted for extraction of cereals and crop names as these can be considered to be most basic and commonly used entity in Agriculture domain. As per our knowledge, till date no significant work has been done in this direction. This is a primitive attempt to identify named entities in agriculture domain. Therefore we have to start from scratch. We faced following challenges in our work:

1. Non applicability of most of the general domain word level features in the Agriculture domain such as word case, digit, punctuations, suffix, prefix, word length etc.
2. Unavailability of annotated corpus for agriculture domain as GENIA corpus is present in case of biomedical domain.
3. Non-existence of a standard full fleshed ontology or knowledgebase for identifying agriculture entities. Though AGROVOC (Multilingual Agricultural Thesaurus) is available as a thesaurus, however, its coverage is low.

As the word level features that are generally used for open domain as well as for biomedical domain do not work properly for Agriculture domain, we have switched over to the word context. Semantic vector can capture the contextual information from the text.

Semantic vector can be understood by defining two terms: Context Pattern and Co-occurrence/Semantic matrix.

Context Pattern Context pattern is a sequence or arrangement of few words/patterns enclosing certain context. Concept of distributional hypothesis says that: words that hang together in similar contexts lean towards having similar meanings [11]. Thus similar entities are expected to share similar context. So it can be assumed that this context helps us in extracting named entities. Context can be found using the window method. Window in concern with text represents a span of words which passes over the whole corpus.

Co-occurrence Matrix For a given document, word-by-word co-occurrence frequency matrix can be generated by using a sliding window of fixed size: all the words present within the window are considered as co-occurring with each other

[12]. Thus each word is represented as semantic vector in word space. We have used a uniform notion of semantic matrix which is neither distance dependent nor direction dependent. Thus it is a symmetric matrix, where information about preceding and following words can be obtained by either row or column.

In this paper we have worked over the semantic vector derived from co-occurrence matrix in order to extract the named entities of agriculture domain. We have proposed two approaches:

SVT (Semantic Vector of Terms)
SV-NV (Semantic Vector of Noun and Verb)

3.1 Proposed Approach 1: SVT

The process in our proposed approach starts with preprocessing the text documents and selection of seed words. Preprocessing includes sentence extraction, POS tagging and noun phrase detection. Seed words are some known preselected entities that help in extracting new entities. Once the data is preprocessed, co-occurrence matrix [12] is constructed using the window based approach. Each word is represented by co-occurrence vector. This is followed by calculating the distance between each pair of words using cosine based distance. As cosine is similarity measure, we used negative of cosine as a score for calculating cosine based distance between two words.

Cosine distance gives the angular cosine distance between vectors u and v as given in Eq. 1. The angle between two vectors u and v is represented as Cosine Similarity and is expressed as given in Eq. 2.

$$\text{Cosine Distance} = (1 - \text{Cosine Similarity}) \tag{1}$$

$$\text{Cosine Similarity} = \frac{u.v}{\|u\|\|v\|} = \frac{\sum_{i=1}^{n} u_i \times v_i}{\sqrt{\sum_{i=1}^{n}(u_i)^2} \times \sqrt{\sum_{i=1}^{n}(v_i)^2}} \tag{2}$$

The cosine distance between each word pair is calculated and this value was stored in the distance matrix. From the distance matrix we extracted all the rows corresponding to seed words. For each seed word we sorted the co-occurring words based on their distance with seed word. The topmost words for each seed word form a similarity list. The similarity list is expected to contain newly discovered entities. The data items used in our approach are presented in Table 1.

The detailed algorithm for approach 1 is shown below:

Table 1 List of variables used in algorithms

Variable name	Variable type	Explanation
SL	Sequence list	List of all words and phrases along with POS tag in sequential order of their occurrence in the corpus
WL	Word list	Dictionary sequence of each term in sequence list with their frequency
CM	Co-occurrence matrix	For each word pair in Word List, CMij represents, total number of times jth term is occurring in context of ith term within a fixed window size
CSM	Co-occurrence sub matrix	This is a sub matrix of co-occurrence matrix, it is constructed only for noun and verbs in Word List. Rows represent nouns and columns represent verbs
SDM	Sorted distance matrix	Each row corresponding to each term in CM represents a vector. Cosine based distance between two terms is measured by computing the one minus cosine of the angle between these two vectors. Distance matrix represents pairwise cosine distance between the terms in CM. DMij stores the cosine based distance between two terms i and j
SWL	Similarity word list	Each row of distance matrix is sorted in increasing order, giving a new matrix sorted distance matrix (SDM), each entry of SD corresponds to a pair of value (word index, distance)
VL	Verb list	We extracted some specific rows of SDM that correspond to the seed entities. For each seed entity we extracted the topmost similar words (below some threshold) from the corresponding rows using SDM. These words were stored in the Similarity Word List SWL of the corresponding seed word. Finally this SWL was used to extract new named entities
EL	Entity list	Verb list contains the verbs that are co-occurring with seed words

```
Input: window size ws, word list WL, sequence list SL, threshold
distance td, seed words (s1, s2, s3…)
Output: similarity Word List SWL for each seed words and Entity List EL
1: Construction of Co-occurrence matrix
        Assign, Total Rows = Total Column = no. of words in WL(say wwl)
        Initialize CM(i,j) = 0 (where i,j = 1 to wwl)

        a: for each word x in SL
        b:      Get the index xᵢ of x in WL
        c:      for all following words yof x within ws in SL
        d:          Get the index yᵢ of y in WL
        e:              CM(xᵢ,yᵢ) = CM(xᵢ,yᵢ)+1;
        f: CM = CM +transpose(CM)
2: Construction of Distance Matrix
        a: for each row i of CM
        b:      Calculate the cosine-distance with each row j of CM
        c:              Distance matrix DM(i,j) = cosine-distance(i,j)
3: Sort each row of DM in ascending order, newly sorted matrix is sorted
distance matrix SDM
4: Add seed words in an entity list EL
5: Extract entities based on seed entities
        For each row i of SDM
                for each column j of SDM corresponding to seeds (s1, s2,
                s3…)
                        if SDM(i,j).distance < td
                                get the corresponding word w
                                addSWL(i,j) = word;
                Find the top k similar words from SWL for i
                Add these k words in to the EL, if not present
6: Return similarity Word Lists SWL,Entity List EL
```

3.2   Proposed Approach 2: SV-NV

In approach 1 we have used all the terms of the document to construct the co-occurrence matrix. However we observed that all the terms do not play significant role for extracting the named entities. In our case nouns and verbs are more important. Here nouns represent the (entities), whereas verbs represent the context pattern for these entities. Hence in this approach we have focused only on the nouns and verbs present in the corpus. We have then used the similar notion of co-occurrence matrix as in approach 1, however in the newly proposed matrix rows of the matrix correspond to the nouns (entities) and columns correspond to the verbs (context).

The detailed algorithm is given below:

```
Input: window size ws, word list WL, sequence list SL, seed words (s1,
s2, s3…)
Output: Entity List EL
1: Construction of Co-occurrence matrix
Assign, Total Rows = Total nouns in WL and Total Columns = Total verbs
in WL
Initialize CSM(i,j) = 0
a: for each noun x in SL
b:       Get the index xi of x in WL
c:       for all following verbs yof x within ws in SL
d:             Get the index yi of y in WL
e:             CSM(xi,yi) = CSM(xi,yi)+1;
2: Get most frequently co-occurring verbs corresponding to seed nouns
for each seed word x from (s1, s2, s3…)
          Find the verbs (v1, v2, v3…) corresponding to x from CSM
          Add these verbs to verb list VL
3: Extract entities using verbs in VL
for each verb v from verb list (v1, v2, v3…)
          Find the corresponding nouns (e1, e2, e3…) of v from CSM
          Add these noun to output entity list EL
4: Return the entity list EL
```

4   Experiment and Result

In the absence of any standard benchmark dataset for our problem, we have designed our own dataset. In order to perform the experiments we have crawled the data from the web. Most of data have been crawled from Agro Products [13]. There are 2206 sentences and 5137 distinct words. We also have a list of cereals and crops which contains 324 cereals and crops names, from different online sources, which is used as benchmark for evaluating our result. In order to assign the part of speech tags to the text Stanford's POS tagger was used.

We observed that the selection of seeds may affect the quality of result. Therefore, we performed our experiments by selecting the seeds randomly and by selecting the seeds manually which we think are more common and are expected to

Table 2 Result of experiment 1 using random seed selection

Number of seeds	10	20	30	40	50
Precision	43.3962264	37.1134021	33.0769231	34.751773	28.1553398
Recall	24.2105263	37.8947368	45.2631579	51.5789474	61.0526316
F-measure	31.0810811	37.5	38.2222222	41.5254237	38.538206

Table 3 Result of experiment 1 using manual seed selection

Number of seeds	10	20	30	40	50
Precision	62.962963	46.1538462	43.3333333	41.0714286	37.5
Recall	17.8947368	31.5789474	41.0526316	48.4210526	56.8421053
F-measure	27.8688525	37.5	42.1621622	44.4444444	45.1882845

provide better result. In each experiment we gradually increased the number of seeds to check the performance on different numbers of seeds.

Experiment 1

This experiment is done using SVB approach. After the construction of co-occurence matrix followed by distance matrix, we selected topmost m similar entities corresponding to given seed entities and we can get a list of newly extracted entities. The result of experiment using random seed selection are presented in Table 2. The parameter values considered are: number of seed entities (variable as shown in Table 2), similarity threshold (td) $\geq$ 0.5, m = 10.

It can be observed from Table 2 that on increasing number of seeds precision degrades but recall increases.

The result of experiment 1 corresponding to manual seed selection are presented in Table 3. The parameter values are same as for random seed selection.

Here also, on increasing number of seeds precision degrades but recall increases. It can be observed that precision is better for manual seed selection whereas random seed selection is giving better recall, F-measure is varying for different number of seeds. As the data sample and number of entities are not very large, we cannot say whether manual selection works better or not. However we can comment that the method of seed selection and number of seeds are important factors in determining the quality of result.

Experiment 2

This experiment is done using NV-SVB approach. The result of experiment using random seed selection are presented in Table 4. The parameter values considered are: number of seed entities (variable as shown in Table 4), similarity threshold (td) $\geq$ 0.5 and m = 10.

Again the experiment was performed by selecting the seeds randomly and manually and number of seeds values was varied to see their effect on quality of

Table 4 Result of experiment 2 using random seed selection

Number of seeds	10	20	30	40	50
Precision	9.20245399	8.46394984	8.54816825	7.85463072	8.14371257
Recall	63.1578947	56.8421053	66.3157895	70.5263158	71.5789474
F-measure	16.064257	14.73397	15.1442308	14.1350211	14.6236559

Table 5 Result of experiment 2 using manual seed selection

Number of seeds	10	20	30	40	50
Precision	7.53424658	7.45614035	7.36728061	7.24174654	7.21102863
Recall	69.4736842	71.5789474	71.5789474	71.5789474	71.5789474
F-measure	13.5942327	13.5054618	13.3595285	13.1528046	13.1021195

result. The results using random seed and manual seed selection are presented in Tables 4 and 5 respectively.

In this experiment random selection and general seed selection does not have much impact on the result. Precision and recall values are almost similar for both types of seed selection.

It can also be observed that second experiment is good for very little number of seeds. Any increase in number of seeds does not make too much increase in performance. Thus requiring a very few number of seed words in an important advantage of this approach.

As compared to first experiment this experiment covers more entities because recall is high but we achieved this performance at the cost of precision loss. In spite of low precision value in comparison to first experiment we cannot say that experiment 2 is less accurate than experiment 1. Firstly as the recall is higher therefore definitely this experiment is covering more entities in comparison to experiment 1. Secondly the decline in the precision is mainly because the length of list containing the resultant entity is very large in comparison to the list obtained in experiment 1. It may not be justifiable to compare the precision when the length of result is different. Obviously the list with longer length is expected to be less precise.

It can be emphasized that recall achieved is quite motivating and we are able to cover up to 75 % entities of our interest.

5 Conclusion and Future Work

This work is a preliminary attempt to identify named entities from Agriculture domain text data. We have confined our work to extract only names of cereals and crops. The work is challenging as these entities are of very general nature, word specific features are not very helpful in identifying these entities. We have suggested a semantic vector based approach that captures the context of entities in order to identify them. The results have been analyzed using recall, precision, F-measure and confusion matrix. Considering as a preliminary attempt, the results are quite motivating. Specially recall is high, thus in future we can think of filtering the false positive entities in order to improve the precision. Moreover as there are lot of entities in Agriculture domain, there is a large scope of pursuing work in this direction.

References

1. Rizzolo, N., and Roth, D. Learning Based Java for Rapid Development of NLP Systems. Language Resources and Evaluation, 957–964, (2010).
2. Leaman, R., and Gonzalez, G. BANNER: an executable survey of advances in biomedical named entity recognition. Pacific Symposium on Biocomputing, 13, 652–663, (2008).
3. Settles, Burr. ABNER: an open source tool for automatically tagging genes, proteins and other entity names in text. Bioinformatics 21(14), 3191–3192, (2005).
4. Lee, K. J., Hwang, Y. S., Kim, S., and Rim, H. C. Biomedical Named Entity Recognition using Two-phase Model Based on SVMs. Journal of Biomedical Informatics, 37(6), 436–447, (2004).
5. Zhou, G., Zhang, J., Su, J., Shen, D., and Tan, C. Recognizing Names in Biomedical Texts: A Machine Learning Approach. Bioinformatics, 20(7), 1178–1190, (2004).
6. Seki, K., and Mostafa, J. A Hybrid Approach to Protein Name Identification in Biomedical Texts. Information Processing & Management, 41(4), 723–743, (2005).
7. Alex, B., Haddow, B., and Grover, C. Recognising Nested Named Entities in Biomedical Text. In Proceedings of the Workshop on BioNLP 2007: Biological, Translational, and Clinical Language Processing, 65–72, (2007).
8. Tang, B., Cao, H., Wang, X., Chen, Q., and Xu, H. Evaluating Word Representation Features in Biomedical Named Entity Recognition Tasks. Biomedical Research International, 1–6, (2014).
9. Minkov, E., Wang, R. C., and Cohen, W. W. Extracting Personal Names from Email: Applying Named Entity Recognition to Informal Text. In Proceedings of the Conference on Human Language Technology and Empirical Methods in Natural Language Processing, 443–450, (2005).
10. Maynard, D., Tablan, V., Ursu, C., Cunningham, H., and Wilks, Y. Named Entity Recognition from Diverse Text Types. In Proceedings of Conference on Recent Advances in Natural Language Processing, 257–274, (2001).
11. Deerwester, S. C., Dumais, S. T., Landauer, T. K., Furnas, G. W., and Harshman, R. A. Indexing by Latent Semantic Analysis. Journal of the Association for Information Science and Technology (JAsIs), 41(6), 391–407, (1990).
12. Lund, K., and Burgess, C. Producing High-Dimensional Semantic Spaces from Lexical Co-occurrence. Behavior Research Methods, Instruments, & Computers, 28(2), 203–208, (1996).
13. Agro Products, Internet: http://www.agriculturalproductsindia.com/ [December 29, 2014].

MED-HYREC: A Recommendation System for Medical Domain

Venkata A. Paruchuri

Abstract Artificial intelligence is widely used in identifying human diseases and their treatments. Every day, millions of people get sick and receive treatments. However, there is no system that can accept the information related to the symptoms, diseases, timelines, medical procedures, and medications experienced by various people and use this information to recommend treatments and to predict possible future diseases to other similar people. The purpose of this research is to develop such a system. Case-based reasoning, which is a subfield of artificial intelligence, is used for this purpose. An algorithm is developed for this purpose. The system has been evaluated to determine its adaptability for change of trend in the field.

Keywords Health care · Artificial intelligence · Medical procedure · Treatment · Case-based reasoning

1 Introduction

The goal of Artificial Intelligence (AI) is to develop systems that can exhibit intelligence similar to human beings and AI is applied to many fields [1]. Case-Based Reasoning (CBR) is an area of AI [2]. CBR solves problems by using the experience gained from solving similar problems in the past. For example, doctors use their past experience in diagnosing or treating similar diseases. The life cycle of CBR looks as in Fig. 1.

CBR works as follows: Once a new problem is submitted to CBR, the problem is treated as a new case that needs to be solved. An old case that has been solved and closely matches the new case will be retrieved from the case-base. The old case will then be modified to solve the new case. The modified case will be tested to make

V.A. Paruchuri (✉)
Florida A&M University, Tallahassee, FL 32307, USA
e-mail: venkatap11@gmail.com

© Springer India 2016
S.C. Satapathy et al. (eds.), *Information Systems Design and Intelligent Applications*, Advances in Intelligent Systems and Computing 434,
DOI 10.1007/978-81-322-2752-6_59

605

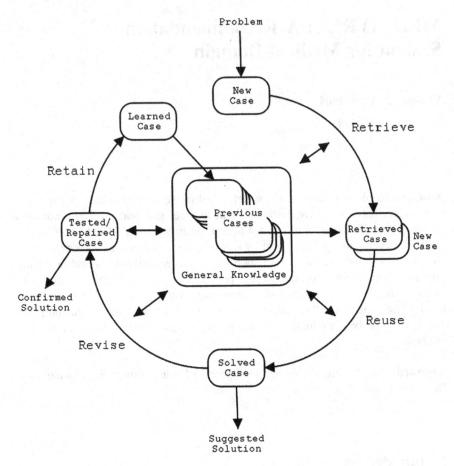

Fig. 1 CBR cycle according to Aamodt and Plaza [14]

sure it is solving the new case and will be repaired if it is failing to solve the new case. Once the modified case successfully solves the new case, the new case (along with its newly calculated solution) will be stored (or retained) into the case-base with proper indexes. In this way, CBR systems keep on adding new cases to their case-bases.

One of the earliest AI systems used for medical purposes was MYCIN [3]. MYCIN is used to identify bacteria that caused some serious infections. The system developed by Jaulent et al. [4] is used for diagnosing histopathology in the breast cancer domain. A system presented by Perner [5] was used to detect degenerative brain diseases and in particular, Alzheimer disease in a patient's CT image. The research conducted by Nilsson et al. [6] was focused on the domain of psychophysiological dysfunctions.

The research conducted by Montani et al. [7] focused on applying CBR for hemodialysis treatments for end stage renal disease. A summary of CBR research in the medical domain as of 2005 is available in Holt et al. [8].

CBR is also suitable for developing medical applications to support clinical processes [9]. A CBR system for complex medical diagnosis, specifically for diseases those involve multiple domains in medicine, is presented by Chattopadhyay et al. [10]. CARE-PARTNER [11] is a web-based system that supports healthcare professionals' decision making processes by storing data about diseases, symptoms, etc.

2 Details of Research

In the literature, there is no system that captures the symptoms, diseases, timelines, medical procedures, medications, etc., of various people and applies them to a new person who is in a similar health situation. For example, if a person is feeling tired and 5 months later, he/she is identified as having diabetes and this lead to a disease a year later then this would be useful information for a new person who has the same health pattern. In other words, the issue addressed in this current research is to solve (or apply) the sequence recognition problem [12] to the medical domain. Apart from applying it to the medical domain, a new algorithm is also provided in this research. However, unlike HYREC, there are no user ratings to different items such as the symptoms, diseases, and timelines. A prototype system has been developed and it is named MED-HYREC. Unlike HYREC, in MED-HYREC, a user can correct the mistakes in the inputs he/she has provided to the system. Also, another important difference between HYREC and MED-HYREC is that, in the latter one, each item belongs to a specific class where as in HYREC, all the items belong to the same unspecific class. For example, in MED-HYREC, symptoms is a class, diseases is a class, and so on. Each item in a class is uniquely identified by a sequence number. A range of sequence numbers are assigned to each class and, from these numbers, each item in that class is assigned a number. For example, in MED-HYREC, timelines are fractional numbers between 0 and 120. A timeline represents the time gap, in terms of years, between the previous item and the current item in a plan (plans are explained later on). Once the user enters the number of days, months, and years, this is converted into years. For example 0 represents the fact that there is no time gap between the previous item and current item in that plan. Similarly 0.5 represents six months, and so on. Symptoms are represented as integers and are ranged from 200 to 10,000. That is, MED-HYREC can accommodate 8,800 different symptoms. Diseases are represented as integers and are ranged from 10,001 to 50,000. Medical procedures are represented as integers and are ranged from 50,001 to 100,000. Finally, medications are represented as integers and are ranged from 100,001 to 200,000.

3 System Design

In MED-HYREC, the library of plans is created when the users enter their medical experiences into the system. Symptoms, diseases, timelines, medical procedures, and medications together referred to as items in this work.

The sequence of items (procedures, symptoms, diseases, etc.) of each user is represented as a plan. There is exactly one plan that corresponds to each user of the system. A plan consists of an ordered sequence of states. Each state contains one (and only one) of the items. In addition, each state contains the identity of the user, his/her age, and the date or approximate date on which the item has occurred or performed. Each state in the plan is automatically assigned a sequence number, which is 1 for the 1st state, 2 for the 2nd state, and so on. A state $S1$ precedes another state $S2$ in a plan if the event in $S1$ has occurred prior to that of $S2$. If both the events happened at the same time then the event that contains a symptom precedes the other event. If both the events represent the symptoms, then any one of them can come first and the other one next. The terms "state" and "item" are used interchangeably in the rest of the paper, although in reality an item is part of a state. Each plan is divided and organized as sub-plans and the organization details of these sub-plans are same as that of HYREC [12].

3.1 Identifying the Closest Matches

Let n represents the total number of states in the current user. The case-based reasoning plan recognition (CBRPR) approach [12, 13] works on i steps. The value of i is initially set to n. A conflict set of plans is determined, based on the items entered by the current user. The item, which is not a timeline and immediately follows the sequence and is present in the majority of these sub-plans in the conflict set, is recommended. If more than one such item exists (if there is a tie) then the item that is most recently entered is recommended. The most recently entered one is recommended because new diseases, treatments, etc. are getting discovered and they are more appropriate to recommend than going with old ones. Still if there is a tie then all these different options will be recommended. If there is no item in the conflict set, except time line (if any), then i is reduced by 1 and the latest $(i - 1)$ states of the current user will be considered and the process will be repeated with this new value of i. If $i = 1$ and no item, except time line (if any), follows in the conflict set then nothing will be recommended to the current user. The details of the algorithm are presented. A user has the provision to specify his/her age group. In that case, only the plans (sub-plans), corresponding to the users of that age group will only be considered in the recommendation process.

3.2 Algorithm

MED-HYREC(i)

/* INTIALIZATION */
1. Conflict set $S :=$ Empty.
Pi, ..., P1 are respectively the *i* most recent items entered by the current user. Here, *P1* is the most recent and *Pi* is the least recent among all.
SSP1 := The set of all existing sub-plans that start with *Pi*.
SSP2 := An empty set.

/* DETERMINE THE CONFLICT SET BASED ON THE ITEM SEQUENCE OF THE CURRENT USER*/
2. Repeat the following operations in the specified order until *SSP1* is empty. Randomly select a sub-plan *SP* from *SSP1*. Remove *SP* from *SSP1*. Add *SP* to *SSP2*. Add *SP* to *S* if and only if *SP* contains at least $(i +1)$ states and also the items in the first *i* adjacent states respectively match with *Pi,..., P1*.

/* CHECK THE CONFLICT SET */
3. If *S* is empty and $i = 1$ then do not recommend anything and exit the process. If *S* is empty and $i \neq 1$ then $i = (i -1)$ and go to Step 1. If *S* is not empty then go to Step 4.

/* DETERMINE THE WINNER AND RECOMMEND */
4. Select the item, which is not a timeline and is in the $(i + 1)^{th}$ state and is also presenting in the majority of the sub-plans of *S*. If there is more than one such item then recommend the one that is most recent.

Example: Consider a simple scenario in which a user entered the following:
223, 0.5, 10020, 100023

This input conveys the fact that the user has a symptom (whose identity is 223) and the user waited for six months then the disease, whose identity is 10020, is identified. Then the medication, whose identity is 100,023, is used.
Assume that the conflict is:

223, 0.5, 10020, 100,023, 345
223, 0.5, 10020, 100,023, 276
223, 0.5, 10020, 100,023, 345

In this situation, symptom 345 is a possible symptom that may be noticed by the user and the user is advised to go for a checkup for that symptom. User identities, age, etc. are not shown due to simplicity reasons.
MED-HYREC adjusts to the changes in the medical procedures and medications. For example, if a new medical procedure or a medicine is found to be more useful for some diseases then the system should be able to use that new medical procedure or medication rather than using the old ones. More details are provided in the evaluation studies.

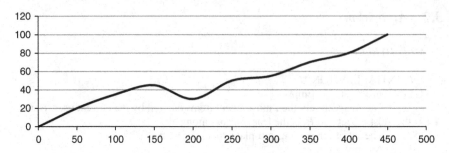

Fig. 2 Number of plans on the X-axis versus acceptance rate on the Y-axis

3.3 Evaluation Studies

MED-HYREC has been evaluated to determine its operational performance in terms of adaptation to change. The relation between the number of plans and the effectiveness of the recommendations is investigated and the results are depicted in Fig. 2.

In this figure, the X-axis represents the number of plans and the Y-axis represents the acceptance rate for the recommendations made by MED-HYREC. After the first 150 plans, the acceptance rate started decreasing monotonically and it eventually fell below the threshold value (55 %) at around the 200th plan. As a result, only the sub-plans that are recorded on or after February 28th 2015 (corresponding to the location (150, 45) in the graph) are considered for the recommendation purposes. Note that the date is not shown in the graph due to simplicity reasons. After the adjustment in the recommendation process (see HYREC [12] for more details), the acceptance rate has quickly increased.

More experimental work is under progress to determine the performance of MED-HYREC when compared to HYREC and a pure case-based reasoning system in which the plans are not divided into pieces.

4 Conclusions

This paper presented a recommendation system in medical domain and it is based on case-based reasoning and sequence recognition problems. The system is very useful for both the patients and for the doctors to compare the current medical conditions of a patient with other patients and to predict the future health conditions, medications, etc. Unlike other case-based systems that are used for specific medical domains, this research considers the entire medical history of a patient to determine the possible health conditions. The algorithm is also much simpler and straight forward when compared to that of HYREC. More experiments need to be conducted in order to further compare MED-HYREC with HYREC and other recommendation systems to determine how fast it can generate recommendations, as the case-base consists of hundreds of thousands of cases.

Acknowledgments The author is very much thankful to Dr. Bobby Granville and Dr. Bhanu Prasad from the Department of Computer and Information Sciences at Florida A&M University (FAMU) for all their help clarifying many questions in the areas of artificial intelligence and case-based reasoning. The author is also thankful to Mr. Samuel Pyne from the School of Business and Industry of FAMU for his help in conducting the experiments. The author is also thankful to Mrs. Latina Banks from the Department of Biology at FAMU for all her support and encouragement.

References

1. Russell, S. and Norvig, P.: Artificial Intelligence: A Modern Approach, 3rd Edition, Pearson, USA (2009).
2. Wikipedia on CBR: https://en.wikipedia.org/wiki/Case-based_reasoning (last visited on July 19, 2015).
3. Shortliffe, E. H.: Computer-Based Medical Consultations: MYCIN. North Holland, New York, NY: Elsevier (1976).
4. Jaulent, M.C., Bozec, C. L., Zapletal, E., and Degoulet, P.: A case-based reasoning method for computer assisted diagnosis in hisopathology. In Artificial Intelligence in Medicine, 239–242. AIME'97 (1997).
5. Perner, P.: An architecture for a cbr image segmentation system. Journal on Engineering Application in Artificial Intelligence 12(6):749–759 (1999).
6. Nilsson, M., Funk, P., and Sollenborn, M.: Complex measurement classification in medical applications using a case-based approach. In Workshop on CBR in the Health Sciences, 63–72. ICCBR'03 (2003).
7. Montani, S., Portinale, L., Leonardi, G., and Bellazi, R.: Applying case-based retrieval to hemodialysis treatment. In Workshop on CBR in the Health Sciences, 53–62. ICCBR'03. (2003).
8. Holt, A., Bichindaritz, I., Schmidt, R., Perner, P.: The Knowledge Engineering Review, Volume 20 Issue 3, Pages 289–292 (2005).
9. Bichindaritz, I. and Marling, C.: Case-Based Reasoning in the Health Sciences: Foundations and Research Directions. Studies in Computational Intelligence, 309, 127–157. doi:10.1007/978-3-642-14464-6_7 (2010).
10. Chattopadhyay, S., Banerjee, S., Rabhi, F.A., Acharya, U.R.: A Case-Based Reasoning system for complex medical diagnosis, Expert Systems, Volume 30, Issue 1, pages 12–20, February (2013).
11. Marling, C., Montani, S., Bichindaritz, I., and Funk, P.: Synergistic Case-Based Reasoning in Medical Domains. Expert Systems with Applications. doi:http://dx.doi.org/10.1016/j.eswa.2013.05.063 (2013).
12. Prasad, B: HYREC: A Hybrid Recommendation System for E-Commerce, Proceedings of the 6th International Conference on Case-Based Reasoning, Springer-Verlag (2005).
13. Kautz, H.: A Formal Theory of Plan Recognition and its Implementation. In Allen, J., Pelavin, R., Tenenberg, J. (eds.): Reasoning About Plans. Morgan Kaufmann, San Mateo, California, USA, 69–125 (1991).
14. Aamodt, A. and Plaza, E.: Case-based reasoning: Foundational issues, methodological variations, and system approaches, AI Communications 7, no. 1 (1994).

Statistical and Linguistic Knowledge Based Speech Recognition System: Language Acquisition Device for Machines

**Challa Sushmita, Challa Nagasai Vijayshri
and Krishnaveer Abhishek Challa**

Abstract Today's speech recognizers use very little knowledge of what language really is. They treat a sentence as if it would be generated by a random process and pay little or no attention to its linguistic structure. If recognizers knew about the rules of grammar, they would potentially make less recognition errors. Highly linguistically motivated grammars that are able to capture the deeper structure of language have evolved from the natural language processing community during the last few years. However, the speech recognition community mainly applies models which disregard that structure or applies very coarse probabilistic grammars. This paper aims at bridging the gap between statistical language models and elaborate linguistic grammars. Firstly an analysis of the need to integrate the conventional Statistical Language Models with the modern Linguistic Knowledge based language models is made, thereby justifying the Statistical and Linguistic Knowledge based Speech Recognition System which is asymptotically error free.

Keywords Statistical language models · Speech recognizer · Linguistic knowledge · Speech recognition

1 Introduction

The aim of automatic speech recognition is to enable a machine to recognize what a human speaker said. A machine that can "hear" can be helpful in many ways. The user can control the machine by voice, which keeps his hands and eyes free for

C. Sushmita (✉) · C.N. Vijayshri
Andhra University, Visakhapatnam, India
e-mail: challa.abstract@gmail.com

C.N. Vijayshri
e-mail: shri.vijayshri@gmail.com

K.A. Challa
Gayatri Vidya Parishad, Visakhapatnam, India
e-mail: com2mass@gmail.com

© Springer India 2016
S.C. Satapathy et al. (eds.), *Information Systems Design and Intelligent Applications*, Advances in Intelligent Systems and Computing 434,
DOI 10.1007/978-81-322-2752-6_60

other tasks, it can save the user from typing vast amounts of text by simply dictating it, the recognized speech can be used to index speech such as broadcast news which allows efficient document retrieval, or the system may even understand what the user intends to do or answer his questions. These examples illustrate that speech recognition is an important aspect of improving human-machine interfaces and thus making machines more usable and user friendly.

It was believed, that as soon as the spectrum of a speech signal could be computed fast enough, the speech recognition problem could be easily solved. Although thousands of researchers around the world worked on the problem for more than half a century, the task must be still considered to be unsolved. In difficult acoustical environments machines perform orders of magnitude worse than humans.

How was such a misinterpretation possible? On one hand the speech recognition problem is often largely underestimated because it is so natural for human beings to listen to others and understand them. We are not aware of the tremendous amount of variability present in a speech signal. We can understand people we never met before, we are able to recognize a huge amount of different words in continuous speech, and we are even able to understand ungrammatical utterances or expressions we have never heard before. We are able to perform so well because we include a wide variety of knowledge sources: we have prior knowledge about the syntax and semantics of a language, we can derive the meaning of new words by analogy, we use situational clues like the course of a dialogue and we have access to all experiences we made in our live and all knowledge about the world we have. Machines cannot keep up with that.

Written language consists of a sequence of discrete symbols, the letters of the alphabet. These symbols are uniquely identifiable and do not interact. The boundaries of a word are well defined as words are separated by spaces. This is still true for the smallest linguistic elements of speech, the phonemes. In written form, these are discrete symbols as well. However, the situation changes dramatically when we are going from written form to spoken form, or more specifically if we look at a speech signal.

A speech signal contains a tremendous amount of variability from several sources. There is no one-to-one relationship between letters or phonemes and their physical realisation in a speech signal:

- The acoustic realisation of a phone largely depends on the individual speaker properties such as sex, vocal tract shape, origin, dialect tone coloration, speaking rate, speaking style (normal, whispering, shouting), mood and health.
- The pronunciation of a particular phone is influenced by its phonetic context (coarticulation). This influence may span several phones and even syllable and word boundaries.
- Allophonic variants and phoneme variations.
- The signal is altered by the room characteristics like reverberation, the microphone characteristics, signal coding and compression, as well as background noise.

In order to convert speech to text, a description of the acoustic events of speech alone is not sufficient. To resolve ambiguities, knowledge about the language at hand is indispensable and plays a very important role in speech recognition.

2   Language Models for Speech Recognition

A language model (LM) is a collection of prior knowledge about a language. This knowledge is independent of an utterance to be recognized. It therefore represents previous knowledge about language and the expectations at utterances. Knowledge about a language can be expressed in terms of which words or word sequences are possible or how frequently they occur.

Language models can be divided into two groups. The criterion is whether the model is data driven or expert-driven:

- **Statistical language models**: If the model is based on counting events in a large text corpus, for example how frequent a certain word or word sequence occurs, the model is called to be a statistical language model. Such a model describes language as if utterances were generated by a random process. It is therefore also known as stochastic language model [6, 7].
- **Knowledge based models**: If the knowledge comes from a human expert the model is called knowledge-based language model. Such linguistic knowledge could for example include syntax, the conjugation of verbs or the declension of adjectives. The basis of such a model does not rely on counting observable events, but rather the understanding of the mechanisms, coherences and regu-larities of a language. If this knowledge is defined by rules, such models are also called rule-based models [2].

Since statistical language models are the most commonly used models, they will be discussed first. Consequently, there is a description of the key idea, the advantages and the limitations of statistical LMs. The limitations will motivate the use of knowledge based models and the approach that was taken in this thesis.

2.1   Statistical Language Models

A statistical LM aims at providing an estimate of the probability distribution P(W) over all word sequences W. It must be able to assign a probability to each possible utterance. The conditional probabilities must be estimated on large amounts of texts related to the recognition task at hand. The number of frequencies that must be counted and stored for this model is prohibitive. The longer the conditioning history gets, more and more strings will never occur in the training data. An obvious solution is to limit the length of the histories by assuming that the probability of each word does not depend on all previous words, but only on the last N − 1 words which leads to the so called N-gram language model.

An N-gram language model assumes that the probability of a word is not influenced by words too far in the past. It considers two histories to be equivalent, if they have their last N − 1 words in common. With decreasing N the approximation gets coarser and the space requirements decrease. The N-gram is currently the most widely used language model in speech recognition [1, 2].

The simplicity of the model, its easy integration into the decoding process and its ability, at least to some extent, to take semantics into account, contribute to its success. It is also attractive because it is completely data driven, which allows engineers to apply it without requiring detailed knowledge about the language at hand.

However, despite of its success, the word N-gram language model has several flaws:

- **False conditional independence assumption**: The N-gram model assumes that a word is only influenced by its N − 1 preceding words and that it is independent from other words farther in the past. It assumes that language is generated by a Markov process of order N − 1, which is obviously not true.
- **Saturation**: The quality of N-gram models increased with larger amounts of data becoming available online. However the improvement is limited due to saturation. Bigram models saturate within several hundred million words, and trigrams are expected to saturate within a few billion words.
- **Lack of extensibility**: Given an N-gram model it is difficult or even impossible to derive a new model which has additional words. The information contained in the model is not helpful to derive N-grams containing new words. Grammars, on the other hand, are able to generalize better because they are based on the underlying linguistic regularities.
- **Lack of generalization across domains**: N-grams are sensitive to differences in style, topic or genre between training and test data. The quality of an N-gram model trained on one text source can degrade considerably when applied to another text source, even if the two sources are very similar.

N-grams fail on constructions like in the following example sentence:

"The dogs chasing the cat bark."

The trigram probability P(bark/the cat) will be very low because on one hand cats seldom bark, and on the other hand because a plural verb (bark) is unexpected after a singular noun (cat). Nevertheless this sentence is completely sound. The verb (bark) must agree in number with the noun (dogs) which is the head of the preceding noun phrase, and not with the noun that linearly precedes it [9].

2.2 Knowledge Based Language Models

Undoubtedly, written language and spoken language follow certain rules such as spelling and grammar. In a knowledge-based approach these rules are collected by experts (linguists) and are represented as a hand-crafted formal system. This system allows deciding if a sentence belongs to the language defined by the rules, and if that is the case, to derive its syntactic structure. The knowledge is explicitly available [4].

In contrast to a statistical LM, no training data is needed for a knowledge-based system. This can be advantageous if no or only a small amount of (annotated) data is available from a specific domain. At the same time this means that the lexicon can be easily extended.

The knowledge based approach faces several problems. One is of course the difficulty to build a formal system which appropriately reflects the phenomena of a natural language. The main problem that a speech recognizer has to deal with is the binary nature of a qualitative language model. If no appropriate measures are taken, the system is only capable of recognizing intra-grammatical utterances. This is quite a strong limitation, since a recognizer should be able to transcribe extra-grammatical utterances as well. The lack of frequencies of a purely rule-based system is disadvantageous if the recognizer has several hypotheses to choose from which are all intra-grammatical. For example, the sentences "How to recognize speech?" and "How to wreck a nice beach?" are both syntactically correct, however the first is a priori more likely and should be preferred.

Thereby there is a need to integrate the conventional Statistical Language Models with the modern Knowledge based language models leading to a Statistical and Linguistic Knowledge based Speech Recognition System which is asymptotically error free [3, 8].

3 Statistical and Linguistic Knowledge Based Speech Recognition System

3.1 N-Grams Derived from a Statistical Grammar

Stochastic grammars have the advantage that they typically have much less parameters than an N-gram model. Stochastic grammars can thus be more reliably estimated from sparse data than N-grams. However, N-grams can be more easily integrated into the decoder without requiring a parser. The idea is therefore to combine the advantage of reliably estimating the parameters of a stochastic grammar with the ease of integration of N-gram models. This is accomplished by estimating N-gram probabilities from a stochastic grammar instead of using the N-gram counts of sparse data. Including natural-language constraints into the decoder can be desirable for two reasons: First, decoding can be more efficient due to the reduced search space, and second, it may improve recognition accuracy. The advantage is that undesired, extra-grammatical sentences can be ruled-out early and that low scored intra-grammatical sentences can be saved from being pruned away. To include a grammar into a Viterbi decoder it must be possible to process the grammar left-to-right as the Viterbi-algorithm runs time-synchronously [1–3, 8].

If the grammar is regular, it can be modelled by a finite state automaton and directly integrated into the recognition network of an HMM recognizer; Some natural language phenomena cannot be described in terms of regular grammars or are more elegantly formulated by a context-free grammar. It is not feasible to compile CFGs into a static, finite state transition network because the number of states could be unmanageably large or infinite [5].

However, due to pruning only a part of the state transition network is active at each point in time, therefore a CFG can be realized as a network by dynamically extending the necessary part of the finite state network.

The system incrementally extends the recognition network of a Viterbi decoder by a NL parser and a unification-based CFG. The recognition network is generated on the fly, by expanding the state transitions of an ending word into all words which can follow according to the grammar. It does so by predicting terminal symbols in a top-down manner; non-terminal symbols on the right-hand-side of context-free rules are expanded until a terminal is found.

The dynamic approach was extended by a probabilistic component. It uses a SCFG to compute a follow set and word transition probabilities for a given prefix string. If the prefix string is parsable the SCFG is used to compute the probability distribution of possible following words. If the string cannot be parsed, the system falls back to bigram probabilities instead.

The idea behind predict and verify is very similar to the dynamic generation of partial grammar networks. The main difference is that in the dynamic generation approach the parser is driven by the HMM decoder, while in the predict and verify approach the emphasis is put on the parser which drives the recognition process. It is based on predicting the next word or the next phone in a top down manner and is also called analysis by synthesis. A word or a phone is assumed to be present if its maximal likelihood over all possible ending points is larger than a threshold [3, 5].

References

1. Brown P.F., Della Pietra V.J, deSouza P.V., Lai J.C., and Mercer R.L.: Class-based n-gram models of natural language. Computational Linguistics, 18(4):467–479 (1992).
2. Brill E., Florian R., Henderson J., and Mangu L.: Beyond N-grams: Can linguistic sophistication improve language modeling? In COLING/ACL 1998, pages 186–190, Montreal, Canada (1998).
3. Beutler R., Kaufmann T., and Pfister B.: Integrating a non-probabilistic grammar into large vocabulary continuous speech recognition. In Proceedings of the IEEE ASRU 2005 Workshop, pages 104–109, San Juan (Puerto Rico), (2005).

4. Beutler R., Kaufmann T., and Pfister B.: Using rulebased knowledge to improve LVCSR. In Proceedings of ICASSP, pages 829–832, Philadelphia (2005).
5. Burshtein D.: Robust parametric modeling of durations in hidden Markov models. In Proc. of ICASSP, volume 1, pages 548–551, Detroit, Michigan U.S.A (1995).
6. Charniak E.: Statistical parsing with a contextfree grammar and word statistics. In Proceedings of AAAI/IAAI, pages 598–603 (1997).
7. Gillick L. and Cox S.: Some statistical issues in the comparison of speech recognition algorithms. In ICASSP, pages 532–535 (1989).
8. Harper M., Jamieson L., Mitchell C., Ying, S. Potisuk G., Srinivasan P., Chen R., Zoltowski C., McPheters L., Pellom B., and Helzerman R.: Integrating language models with speech recognition. In "Proceedings of the AAAI- Workshop on Integration of Natural Language and Speech Processing" (1994).
9. Rosenfeld R.: Two decades of statistical language modeling: Where do we go from here? Proceedings of the IEEE, 88(8):1270–1278 (2000).

Energy Stability in Cloud for Web Page Ranking

Sutirtha Kumar Guha, Anirban Kundu and Rana Dattagupta

Abstract We are going to propose a new approach to implement predictive Web page ranking in cloud environment. User query is transferred to the server side of cloud based search engine. Searching and ranking procedures have been executed on the server side and final output is sent to the client as a response. In this paper, prediction task has been performed based on user behaviour. Available data, resources, and functional support systems have been processed using schedulers. Session, duration, and other user relevant data are combined with the processed information for prediction. Different Web based data repositories are considered as cloud structure for energy stabilization.

Keywords Web page ranking · Predictive ranking · Data cloud · Resource cloud · Support cloud · Information bag controller · Cloud session · Energy stability measurement

S.K. Guha (✉)
Seacom Engineering College, Howrah, West Bengal, India
e-mail: sutirthaguha@gmail.com

A. Kundu
Netaji Subhash Engineering College, Kolkata, West Bengal, India
e-mail: anik76in@gmail.com

R. Dattagupta
Jadavpur University, Kolkata, West Bengal, India
e-mail: ranadattagupta@yahoo.com

S.K. Guha · A. Kundu
Innovation Research Lab, Howrah, West Bengal, India

© Springer India 2016
S.C. Satapathy et al. (eds.), *Information Systems Design and Intelligent Applications*, Advances in Intelligent Systems and Computing 434,
DOI 10.1007/978-81-322-2752-6_61

1 Introduction

1.1 Overview

Web is considered as an important system to acquire and gather important information in our daily life. Web area is congested due to ever increasing dependability over Web. Introduction of search engine is considered as an important tool for indexing and tagging that growing Web database. Different Web pages are connected with the search engine. Matched URLs are displayed based on user query. Matched Web pages are ranked based on different algorithms. More accurate and fast ranking is achieved by predictive ranking approach. Deterministic ranking or searching is achieved based on previous user data.

1.2 Related Work

A method is introduced in [1] to predict the ranking position of a Web page. A set of successive past top-k rankings are presumed and evolution of Web pages is analysed in terms of ranking trend sequences used for Markov Models training [1]. An algorithm is proposed in [2] for conducting web page access prediction. The use of page rank algorithm is extended to predict next page with several navigational attributes [2]. Ranking in cloud environment is proposed in [3]. The ranking operation is performed based on the data as mentioned in [4]. Cloud classification is analysed by accessibility on the specific cloud in [5]. Cloud classifier is analyzed structurally in [6]. SaaS (Software-as-a-Service) is introduced for user specific application using heterogeneous distributed system [7]. K-Erlang distribution method is implemented in cloud computing to enhance the performance [8]. Different self organizational characteristics of intelligent cloud are discussed in [9]. Load balancing in cloud is proposed in [10].

Rest of the paper is organized as follows: Proposed system design is depicted in Sect. 2 along with detailed analytical studies. Experimental results have been depicted in Sect. 3. Section 4 concludes the paper.

2 Proposed Work

Prediction based ranking is introduced in this paper to achieve efficient searching and ranking in cloud environment. Operational activities of proposed system are shown in Fig. 1.

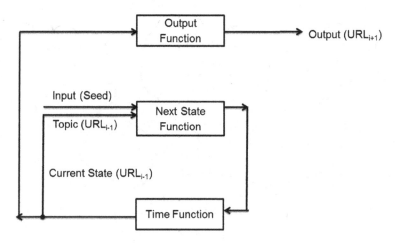

Fig. 1 State diagram of proposed system

Input from user and previous state (URL_{i-1}) are combined and processed through 'Next State Function'. This processed information is sent to 'Time Function'. 'Time Function' sends the information to 'Output Function' to produce predicted output (URL_{i+1}) and 'Next State Function' as URL_i. Operational flow-chart of our proposed system is shown in Fig. 2.

Operational activities of different segments of proposed system are discussed in next sub sections.

Definition 1:   **Data Cloud** Information such as text data, image data, audio data, and video data are stored in a predefined database in server side. This predefined database is addressed here as "Data Cloud".

Definition 2:   **Resource Cloud** Client request is served in server side to provide efficient and accurate performance. Different hardware and software resources are required to execute this operation. These hardware and software resources are resided in "Resource Cloud".

Definition 3:   **Support Cloud** Different support modules are resided in this part of the cloud.

Definition 4:   **Information Bag Controller** Data from "Data Cloud", "Resource Cloud" and "Support Cloud" are sent and merged to perform the job in co-operative manner. Integration and control of data from these three clouds are executed in a controller named as "Information Bag Controller". Raw processed data is sent to the next module as web content that is a combination of required information and unwanted noise.

Definition 5:   **Cloud Mining** Received data from 'Information Bag Controller' is processed to eliminate the noise. Processed information is then searched according to client side request and analyzed.

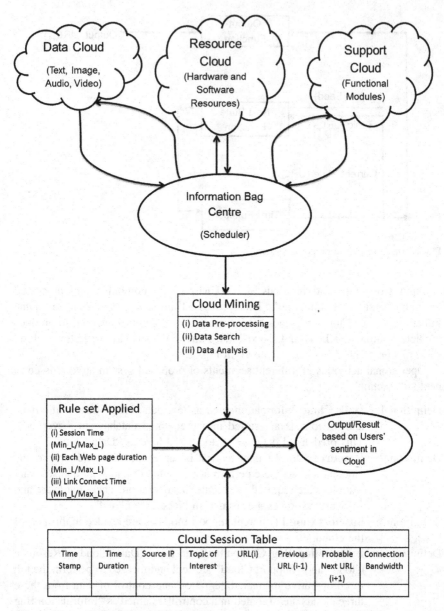

Fig. 2 Operational flowchart of proposed system

2.1 Rules Applied in Cloud

Set of rules is applied into the processed information. Major rules are as follows:

(a) Session Time (Min_L/Max_L) Visitor session on that particular URL is addresses as "Session Time". Upper and lower limit is defined as Max_L and Min_L.

(b) Each Web page duration (Min_L/Max_L) Life time of the Web page is named as "Web page duration". It ranges in between Max_L and Min_L.

(c) Link Connect Time (Min_L/Max_L) It is considered as the time required to connect the Web page. It resided between Max_L and Min_L.

2.2 Cloud Session Table

Different information of cloud is stored in "Cloud Session Table" for predicting user's requirements using time stamp, duration, source IP, topic of interest, URL(i), previous URL(i − 1), probable next URL(i + 1), and connection bandwidth.

2.3 Energy Stability in Cloud

Energy stability measurement (S_M) of the proposed system depends on the following factors:

C_S Cloud session;
S   State;
S_t State transition;
T   Time;
S_e Seed (Starting Point);
E_C Energy of cloud;

∴ Cloud Energy, E_C depends on $S_M(S_e, S, S_t, T, C_S)$
∴ Energy flow in a Cloud is considered as follows:

$$\frac{d}{dt}(E_C) = S_M(S_e, S, S_t, T, C_S) \tag{1}$$

Applying integration on both sides of Eq. (1),

$$\int \frac{d}{dt}(E_C)dt = \int S_M(S_e, S, S_t, T, C_S)dt$$

$$\therefore \int E_C = \int S_M(S_e, S, S_t, T, C_S)dt \qquad (2)$$

On the other hand, cloud session is represented as follows based on "Cloud Session Table" mentioned in Table 1:

$$C_S = f(T_S, T_D, S_{IP}, T_I, URL_i, URL_{i-1}, URL_{i+1}, B_W) \qquad (3)$$

where,

T_S = Time stamp; T_D = Time duration; S_{IP} = Source IP; T_I = Topic of Interest; URL_i = Current URL; URL_{i-1} = Previous URL; URL_{i+1} = Next URL; B_W = Bandwidth;

Here, URL represents a Web page; and, URL_{i+1} is always probabilistic as it depends on the users' choice in real time.

Combining Eqs. (2), (3),

$$\int E_C = \int S_M(S_e, S, S_t, T, f(T_S, T_D, S_{IP}, T_I, URL_i, URL_{i-1}, URL_{i+1}, B_W))dt \qquad (4)$$

$S_M(S_e, S, S_t, T, C_S)$ is realised as follows:

Data flow execution of Fig. 3 is depicted in Algorithm 1.

Algorithm 1: Energy_stability_measurement

Input: Initial State
Output: Next desired State

Step 1: Start session.
Step 2: Initial state is considered as input.
Step 3: Next state is selected based on users' requirement.
Step 4: Session time, Web page duration, Link connect time are applied on selected next state.
Step 5: State transitions are executed.
Step 6: Stop session

Data flow is started from initial state. Next probable state is selected based on users' requirement at next time stamp. Defined protocol sets are then applied on probable next state. State transitions are then executed to get desired state.

Fig. 3 Data flow chart of energy stability measurement (S_M)

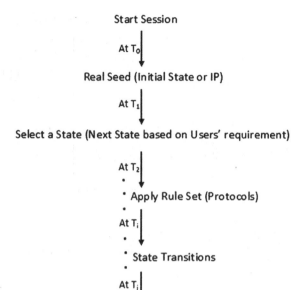

Start Session

At T_0

Real Seed (Initial State or IP)

At T_1

Select a State (Next State based on Users' requirement)

At T_2

• Apply Rule Set (Protocols)

At T_i

• State Transitions

At T_j

• Final (Desired) State

At T_k

Stop Session

Similarly, cloud session (C_S) consists of a cycle of operations initiated by particular users.

$C_S = f(T_S, T_D, S_{IP}, T_I, URL_i, URL_{i-1}, URL_{i+1}, B_W)$ is realized as follows: Data flow execution of Fig. 4 is depicted in Algorithm 2.

Algorithm 2: Cloud_session

Input: User IP
Output: Bandwidth usage

Step 1: Start session.
Step 2: Source IP is considered as input.
Step 3: Topic of interest is searched in present Web link (URL_i)
Step 4: If, topic of interest is found in present Web link (URL_i), then go to Step 5. Otherwise, go to Step 7.
Step 5: Next Web link (URL_{i+1}) is predicted based on present Web link (URL_i) and previous Web link (URL_{i-1}).
Step 6: Total time duration is calculated and go to Step 2.
Step 7: Bandwidth usage for particular case is checked.
Step 8: Stop session.

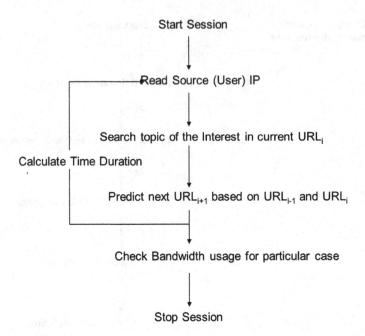

Fig. 4 Data flow chart of cloud session (C_S)

Source IP is considered as input. Matched topic is searched in present Web page (URL_i). Next Web page is predicted based on present (URL_i) and previous Web page (URL_{i-1}). This process is repeated for every matched topic in present Web page. Total bandwidth is measured and considered as cloud session.

3 Experimental Results

Different Web pages are selected at a particular time instance. Web pages from same database based on same query are selected by the same user at different time instances. Prediction success of our proposed method is shown in Fig. 5.

It is evident from the experimental data that accurate and efficient prediction is achieved by our proposed approach.

Number of Users	Number of Age Groups	Number of Time instances
10	02	05

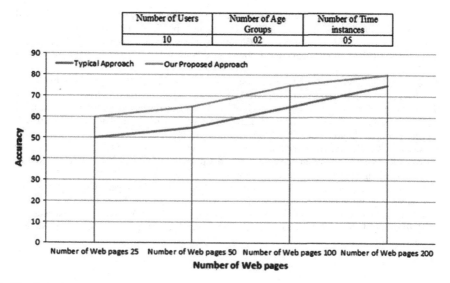

Fig. 5 Prediction success comparison based on different age groups user

4 Conclusion

In this paper, we have proposed a prediction based ranking approach in cloud. Different Web page related information such as data set, resources are considered to predict next Web page selection. Collected data are processed and analysed to produce successful prediction in cloud. Various rules are implemented on data set to achieve better outcome. User query is sent to the server to acquire most feasible result. Our proposed approach is performed on server side based on user sentiment using energy measurement. More accurate prediction is achieved as the procedure is dynamically operated based on previous user data. Hence, accuracy is increased as wider range of data is considered.

References

1. Vazirgiannis, M., Drosos, D., Senellart, P., Vlachou, A.: Web Page Rank Prediction with Markov Models. In: 17th International World Wide Web Conference, Beijing, China, April 21–25, 2015, pp. 1075–1076 (2015).
2. Thwe, P. : Proposed Approach For Web Page Access Prediction Using Popularity And Similarity Based Page Rank Algorithm. In: International Journal of Scientific & Technology Research, Volume 2, Issue 3, March, 2013, pp. 240–246 (2013).
3. Sampath, P., Ramya, D.: Performance Analysis Of Web Page Prediction With Markov Model, Association Rule Mining(Arm) And Association Rule Mining With Statistical Features (Arm-Sf). Journal of Computer Engineering (IOSR-JCE), Volume 8, Issue 5, pp. 70–74 (2013).

4. Kundu, A., Xu, G., Ji, C.: Data Specific Ranking in Cloud. International Journal of Cloud Applications and Computing (IJCAC), Vol. 4, Issue 4, pp. 32–41 (2014).
5. Kundu, A., Xu, G., Ji, C.: Analysis on Cloud Classification using Accessibility. International Journal of Cloud Applications and Computing (IJCAC), Vol.4, Issue 3, 2014, pp. 44–53 (2014).
6. Kundu, A., Xu, G., Ji, C.: Structural Analysis of Cloud Classifier. International Journal of Cloud Applications and Computing (IJCAC), Vol. 4, No. 1, pp. 63–75 (2014).
7. Kundu, A., Ji, C., Liu, R.: Software-as-a-Service using Heterogeneous Distributed System for User specific Applications. International Journal of Cloud Applications and Computing (IJCAC), Vol. 4, No. 1, pp. 15–32 (2014).
8. Banerjee, C., Kundu, A., Agarwal, A., Singh, P., Bhattacharya, S., Dattagupta, R.: Priority based K-Erlang Distribution Method in Cloud Computing. International Journal on Recent Trends in Engineering & Technology, Vol. 10, No. 1, pp. 01–11 (2014).
9. Kundu, A., Xu, G., Ji, C.: Self Organization Behavior of Intelligent Cloud. Journal of Convergence Information Technology (JCIT), Vol. 8, No. 16, pp. 39–47 (2013).
10. Kundu, A., Xu, G., Liu, R.: Efficient Load Balancing in Cloud: A Practical Implementation. International Journal of Advancements in Computing Technology (IJACT), Vol. 5, No. 12, pp. 43–54 (2013).

Web Based System Design Using Session Relevancy

Sutirtha Kumar Guha, Anirban Kundu and Rana Dattagupta

Abstract In this paper, we propose an advanced technique for validating visitors' sessions on Web pages measuring the real-time ranking. The session of a Web page is going to be considered as an important parameter for calculating the Web page ranking. Visitors' session has been computed implementing fuzzy set theory based on Session Quotient (Q_s) and Informative Quotient (Q_i). Session Quotient is the combination of a threshold value (T_HV) and number of visitors. T_HV is considered for individual Web page based on the Web page contents to inspect the session. Informative Quotient is being calculated using number of incoming Web links to the particular Web page using proposed formula. Session validation has been performed implementing fuzzy logic as an application tool.

Keywords Session · Threshold value (T_HV) · Field matching · Pattern matching · Keyword matching index (K_{index}) · Session quotient (Q_s) · Informative quotient (Q_i)

1 Introduction

The ever increasing Web based activities have produced a huge number of Web users for using billions of Web pages. The increasing arrivals of various Web pages and the wide area of available information make the searching lot more time

S.K. Guha (✉)
Seacom Engineering College, Howrah, West Bengal, India
e-mail: sutirthaguha@gmail.com

A. Kundu
Netaji Subhash Engineering College, Kolkata, West Bengal, India
e-mail: anik76in@gmail.com

R. Dattagupta
Jadavpur University, Kolkata, West Bengal, India
e-mail: ranadattagupta@yahoo.com

S.K. Guha · A. Kundu · R. Dattagupta
Innovation Research Lab, Howrah, West Bengal, India

© Springer India 2016
S.C. Satapathy et al. (eds.), *Information Systems Design and Intelligent Applications*, Advances in Intelligent Systems and Computing 434,
DOI 10.1007/978-81-322-2752-6_62

consuming. The concept of Search Engine is a reliable solution for solving delay related problems. Web pages are fetched from the database and displayed with proper ranking in respect of users' queries. Rank of a Web page is determined based on the importance of the page. Informative Web pages are visited by many users. Surfing time of a user on a particular Web page is higher compared to other less important Web pages. Surfing on a particular Web page for a specified duration is considered as visitors' session which is an important indicator of Web page importance. Irrelevant noise creates erroneous visitors' session values. Unwanted access to a Web page increases visitors' session values of the particular Web pages. Erroneous visitors' session values should be filtered confirming the relevance of Web page with respect to users' queries.

Rest of the paper is organized as follows: related works has been discussed in Sect. 2; proposed system framework has been described in Sect. 3; Sect. 4 contains experimental results; Sect. 5 concludes the paper.

2 Related Works

Task of an ideal Search Engine consists of fetching relevant results from predefined databases, and allowing users to finish ongoing tasks [1]. Researches on session analysis and classification have been aided by the introduction of the session track at TREC [2, 3]. The problem of Intrinsic Diversity (ID) in retrieval is characterized and addressed to optimize whole session relevance [4]. Three key problems for ID retrieval are as follows: identifying authentic examples of ID tasks from post hoc analysis of behavioural signals in search logs; learning to identify initiator queries that mark the start of an ID search task; and given an initiator query, predicting which content to pre-fetch and rank are identified and addressed in [4].

Web page session relevancy has been examined by a threshold value in [5]. The threshold value has been a combination of field matching and pattern matching. Unfortunately, these parameters would not be sufficient enough to filter the unexpected noises.

3 Proposed Work

Visitors' session of a Web page for inspection has been introduced in [5] using a threshold value. We have enhanced the procedure of session relevancy in this paper with modified session relevancy inspection method.

The threshold value ($T_H V$) has been measured using Eq. (1) [5].

$$T_H V = \frac{K_{index}}{Data\ Transfer\ Speed} \tag{1}$$

K_{index} is calculated based on Field matching and Pattern matching between the requirement and Web page contents. Field matching value is calculated by implementing hierarchical database concept. Pattern matching value is calculated implementing fuzzy set theory.

K_{index} is calculated based on Eq. (2). Data transfer speed is a dynamic value which is assigned based on data speed between the client and the server in particular time instance. It is measured as bits per second (bps) [5].

$$K_{index} = \frac{(F_M + P_M)}{2} \tag{2}$$

where,

F_M Field Matching Value
P_M Pattern Matching Value

Calculation procedure of field matching value is explained in Algorithm 1.

Algorithm 1: Field_Matching_Value_Measurement

Input: User given string
Output: Field matching value (F_M)

Step 1: Start
Step 2: User given string (U_S) is taken as input from End user
Step 3: Match (U_S, Database)
Step 4: Traverse (Left wing of the Database)
Step 5: If (Match (U_S, Database) == 1)
         then,

$$F_M = \frac{1}{Total\_Path\_Length}$$

Step 6: Else

$$F_M = \frac{1}{(P_L, Distance(N_G, N_{CP}))}$$

where,

P_L Total Path Length
N_G Goal Node
N_{CP} Common Parent Node of successful path and last visited path

Step 7: Stop

Pattern matching between the requirement and the Web page is measured according to Eq. (3) [5].

$$P_M = (F_D \times F_W)$$ (3)

where,
P_M Pattern Matching Value;
F_D Final Decision value according to Fuzzy Set theory;
F_W Pattern Factor ($F_W = n$, $1 <= n < 0$);

Field matching value calculation is represented in Eq. (3).

In this paper, Session Quotient (Q_s) and Informative Quotient (Q_i) have been introduced for inspecting session relevancy for particular Web pages. These quotients are being measured using field matching and pattern matching values, page ranks of inbound and outbound Web pages. Q_s of a Web page is measured dividing the threshold value by average visitors' sessions. Total session time of a Web page has been considered as an important decisive factor for session inspection. Weightage of a Web page is measured by Q_i. Q_i is calculated by the page rank of inbound and outbound Web links. A Web page containing important features has been referred by different Web pages. The value of Q_i is higher for a relevant Web page. These two quotient values are being combined to yield final decision implementing fuzzy.

High K_{index} does not indicate high relevancy of a Web page in proposed approach. K_{index} is measured as a combination of field and pattern matching. High field and pattern matching values could not be able for producing most relevant Web page.

$$Q_S = T_H V \times \frac{\sum_1^n V_S}{V_n}$$ (4)

where,
Q_s Session Quotient of Web page;
$T_H V$ Calculated Threshold value of Web page;
V_s Visitors' session of Web page;
V_n Total Number of Visitors in the Web page;

Visitors' session has been inspected operating Q_s and Q_i. Q_s is measured using Eq. (4). Algorithm 2 shows the step-wise calculations of Q_s.

Algorithm 2: Session_Quotient_Calculation

Variables: $T_H V$, V_s, V_{stotal}, V_n, Result, Temp_var

Step 1: Start
Step 2: Temp_var = $T_H V$

Step 3: Initialize $V_{stotal} = 0$
Step 4: while($V_s \ != 0$)
Step 5: $V_{stotal} = V_{stotal} + V_s$
Step 6: Result = V_{stotal}/V_n
Step 7: $Q_s = T_H V \ X \ Result$
Step 8: Go to step 2

where,
Temp_var   Temporary Variable;
V_{stotal} Summation of all V_S;
Result   A variable storing average session value;

A Web page having informative contents is referred by many other Web pages. Number of inbound Web links is considered as an importance factor.

$$Q_i = \frac{\sum_1^n PR(P_i)}{\sum_1^n PR(P_o)} \tag{5}$$

where,
Q_i Informative Quotient of Web page;
$PR(P_i)$ Page Rank of Inbound Web page;
$PR(P_o)$ Page Rank of Outbound Web page;

Q_i is calculated based on inbound and outbound Web pages as shown in Eq. (5). Algorithm 3 shows the step-wise calculations of Q_i.

Algorithm 3: Informative_Quotient _Calculation

Variables: n_i, PR_{itotal}, PR_{ototal}, n_o, Q_i

Step 1: Start
Step 2: Initialize $PR_{itotal} = 0$ and $PR_{ototal} = 0$
Step 3: while($n_i \ != 0$)
Step 4: $PR_{itotal} = PR_{itotal} + PR(P_i)$
Step 5: while(no != 0)
Step 6: $PR_{ototal} = PR_{ototal} + PR(P_o)$
Step 7: $Q_i = PR_{itotal}/PR_{ototal}$
Step 8: Stop

where,
n_i Number of inbounded Web pages
n_o Number of outbounded Web pages
PR_{itotal} Summation of Page rank of all inbounded Web pages
PR_{ototal} Summation of Page rank of all outbounded Web pages

4 Experimental Results

In our experiment "Pattern Recognition" is considered as the searching string and collected data from a typical search engine is shown in Table 1.

It is depicted from real time data, acquired from a typical Search Engine, that Web page ranking procedure followed in typical Search Engine sometimes are not exhibiting most desired results. It is evident from collected data that "Rank 2" Web page contains information about a novel having same name as searching string. It is expected that technical information regarding the topic of pattern recognition should be queried by searching string "Pattern Recognition". Searching for information regarding any novel having same name would be searched providing the query as "Pattern Recognition Novel". Hence, a Web page having information about the novel seems irrelevant from users' perspective for the searching string "Pattern Recognition".

It is obvious that user is looking for some informative material about the searching string. User session is considered as an important parameter to identify the importance of Web pages in typical search engine environment. User session is calculated based on the session time spend in the particular Web page by users.

Our proposed approach is demonstrated and analyzed through a case study. In this case study 'Pattern Recognition' is considered as the sample query. People from different domain are considered as sample set. A real time case analysis results the following information as shown in Table 2.

Visitor session is calculated based on the real time data. Importance of the Web page is measured based on the calculated session value. Ideal importance of Web page is measured based on the user sentiment.

It is depicted from our case study that relevancy of a Web page with respect to the user query could not be indicated by a longer visitor session. Introduction of

Table 1 Contents of Web pages ranked in typical search engine

Web page	Contents summary
Rank 1 [6]	This page contains tutorial information regarding searching string. This page would be designated useful for user to gather knowledge and information regarding searching string
Rank 2 [7]	This page contains information regarding a novel having same name as searching string
Rank 3 [8]	Advertise of a specific publisher having articles/books of the searching topic is shown in this Web page
Rank 4 [9]	Advertise of a specific publisher having articles/books of the searching topic is shown in this Web page
Rank 5 [10]	Advertise of a specific book selling website articles/books is shown in this Web page
Rank 6 [11]	Advertise of a specific book selling website articles/books is shown in this Web page

Table 2 Comparative study of session for sample Web pages

Web pages	Average visitor session (mins)	Importance of the Web page based on the session	Ideal importance of the Web page based on the relevancy
R1	(10 + 13 + 14 + 7)/4 = 11	Very high	Very high
R2	(11 + 12 + 4)/3 = 9	High	Low
R3	(10 + 8 + 3 + 5 + 9 + 9 + 6 + 6)/8 = 7	High	Moderate
R4	(12 + 3 + 3)/3 = 6	Moderate	Moderate
R5	(7 + 9 + 2 + 4 + 3)/5 = 5	Moderate	Moderate
R6	(6 + 1 + 2)/3 = 3	Low	Moderate

malfunction and spamming Web pages causes deliberate longer session of the particular Web page. Hence, the authenticity of session could not be trustworthy.

Session quotient and informative quotient are introduced for examining the importance of the Web page. Calculation of predicted sessions based on the quotient values is shown in Table 3. Session Quotient and Informative Quotient value is calculated based on the real time data collected from the above stated case study,

It is depicted that Web pages R1, R2, R3, R4, R5 and R6 have different Session Quotient and Information Quotient values. Information Quotient is calculated based on the rank of in bounded and out bounded Web pages. Session Quotient value is calculated based on the user session on the particular Web page. User session data is collected in real time. Hence, different Session Quotient value is obtained for Web page R3, R2, R3, R4, R5 and R6 though all these Web pages have similar type of information. As Session Quotient value depends on user behavior in real time, different value is obtained.

Comparative study of ideal Web page rank and predicted session according to our proposed approach is depicted in Fig. 1.

It is apparent that rank of a Web page depends on the informative value of a Web page. Similarly longer user session is expected in a relevant and informative Web page. Hence, a Web page having high rank is expected to have better session value.

Table 3 Predicted session based on quotient values for sample Web pages

Web pages	Calculated informative quotient value	Calculated session quotient value	Predicted session
R1	0.32	0.22	Maximum session
R2	0.86	0.35	Minimum session
R3	0.35	0.37	Moderate session
R4	0.33	0.37	Moderate session
R5	0.37	0.31	Moderate session
R6	0.31	0.82	Minimum session

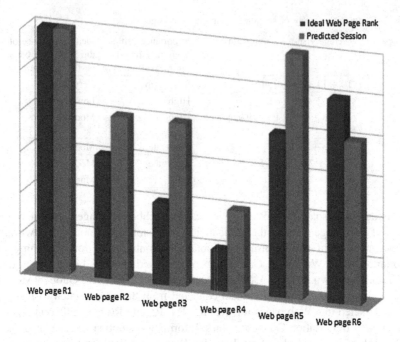

Fig. 1 Comparison of ideal Web page rank and predicted session

It is evident from the experimental data that ideal Web page rank and calculated session are matched as shown in Fig. 1.

The experiments are executed over different dataset and acquired results do not vary too much as shown in Fig. 2.

Fig. 2 Comparison of experimental result acquired from different dataset

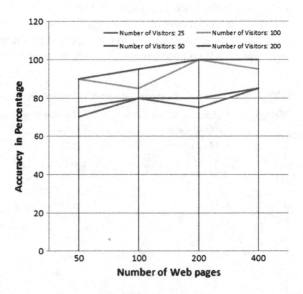

5   Conclusion

An advanced approach with better relevancy and realistic session calculation has been demonstrated in this paper for inspecting the relevancy of different sessions of particular Web pages. Session has been scrutinized based on distinct quotient values. Q_s and Q_i have been calculated for examining based on ideal sessions of a particular Web page. A set of predefined decisions have been performed using calculated quotient values. The session at any time instance is compared with the predicted decisions for checking the importance of session relevancy. Session values have been error less and unambiguous using our proposed technique.

References

1. Agichtein, E., White, R., Dumais, S., Bennett, P.: Search, Interrupted: Understanding and Predicting Search Task Continuation. In: SIGIR '12, Portland, Oregon, USA (2012).
2. Kanoulas, E., Carterette, B., Hall, M., Clough, P., Sanderson, M.: Overview of the TREC 2011 Session Track. In: TREC '11 (2011).
3. Liu, J., Cole, M. J., Liu, C., Bierig, R., Gwizdka, J., Belkin, N. J., Zhang, J., Zhang, X.: Search behaviors in different task types. In: JCDL '10, Brisbane, Australia (2010).
4. Raman, K., Bennett, P. N., Collins-Thompson, K.: Toward Whole-Session Relevance: Exploring Intrinsic Diversity in Web Search. In: SIGIR'13, Dublin, Ireland (2013).
5. Guha, S. K., Kundu, A., Dattagupta, R.: Introducing Session Relevancy Inspection in Web Page. In: Second International Conference on Computer Science, Engineering and Applications (ICCSEA 2012), Delhi, India (2012).
6. http://en.wikipedia.org/wiki/Pattern_recognition.
7. http://en.wikipedia.org/wiki/Pattern_Recognition_(novel).
8. http://www.journals.elsevier.com/pattern-recognition/.
9. http://www.sciencedirect.com/science/journal/00313203.
10. http://www.amazon.com/Pattern-Recognition-William-Gibson/dp/0425198685.
11. http://www.amazon.com/Pattern-Recognition-William-Gibson-ebook/dp/B000OCXGVY.

Design of IOT based Architecture Using Real Time Data

Nivedita Ray De Sarkar, Anirban Kundu, Anupam Bera and Mou De

Abstract We propose an Internet of Things (IOT) based system for handling of real time data fed from the environment to measure distinct activities. We provide unique identifications for each required substance using IOT. In this paper, we have presented and discussed the use of sensors as part of senders and receivers for protecting sensitive areas of specified zones. Furthermore, we have discussed the presence of IOT in preserving the intrusion details of sensitive areas.

Keywords Sensor · Security · Internet of Things (IOT) · IOT Architecture · Intruder detection

1 Introduction

Advent of IOT [1] has provided each substance with unique identification (ID) [2]. We can have smart substances around us, and it can respond based on specific environments. IOT is used for creating intelligent transportation systems [3]. All information would be gathered remotely through different devices such as sensors, actuators, GPS, RFID [4]. Sensors [5] have abilities of continuously monitoring specific area and identifying disparity in environment. In this paper, we provide conceptual usage using set of sensors for monitoring high security area. Receivers

N.R. De Sarkar (✉) · A. Kundu · A. Bera · M. De
Netaji Subhash Engineering College, Kolkata 700152, India
e-mail: nivedita.raydesarkar@gmail.com

A. Kundu
e-mail: anik76in@gmail.com

A. Bera
e-mail: drive.abera@gmail.com

M. De
e-mail: mou.latu@gmail.com

N.R. De Sarkar · A. Kundu · A. Bera · M. De
Innovation Research Lab, Howrah 711103, West Bengal, India

© Springer India 2016
S.C. Satapathy et al. (eds.), *Information Systems Design and Intelligent Applications*, Advances in Intelligent Systems and Computing 434,
DOI 10.1007/978-81-322-2752-6_63

assimilate data for sending over World Wide Web (WWW). IOT paradigm intends to provide a unique ID to everything possible on earth. So, it can provide a virtual view of a huge variety of real life. There has been a noted use of wireless sensor networks (WSN) for long term environmental data acquisition [6]. Substances having unique IDs through IOT should be provided for transmitting information. Microcontrollers typically play a major role in intelligence [7]. Objects could be responded using predefined patterns. IOT is considered to comprise of smart objects having sensor terminals which have capabilities to collect variety of information [8]. The concept of enhancement of security within a community could be entrusted to sensor enabled devices providing information over internet. We already have features of neighbourhood watch [9] which tracks surrounding environment for any abnormal pattern or situation. Sensors could be used to propagate signal or information across a typical city to make it smart [10, 11].

2 System Architecture

2.1 Overview

In this paper, we have designed a system for increasing safety of a high security room in an organization. The room is provided with a set of sensors which are continuously monitoring environment. Our sensor based device typically senses information in real time for transferring data to our network repository. Further data is analyzed with respect to predefined threshold values. Whenever data crosses upper/lower threshold values, specific abnormality is noted in our environment. Initially, sensor communicates with connected circuit board to inform its status. The circuit in turn starts connecting to the neighbourhood controllers through neighbourhood collaboration. Once alarming condition is handled, system gets back to working in its normal mode.

2.2 Architectural Analysis

In our system, we have used two modules such as sender and receiver. The sender side is located in same room whose security is trying to be enhanced in a controlled way. Set of sensors are located within the particular room for checking any abnormal condition.

Data flow through sender part along with all the detailed conditions is depicted in Fig. 1 as an illustration. If particular sensor is active, data is sent to respected 8051 circuit board. If particular 8051 circuit board is not busy, then it should be utilized further. Then, sensor id is added with data and check whether the controller is active. If controller is not busy, then select a particular controller and apply similarity function as required to analyze the situation. Data is further passed through similarity function for pattern checking. If live pattern does not match, flag is set to '1'

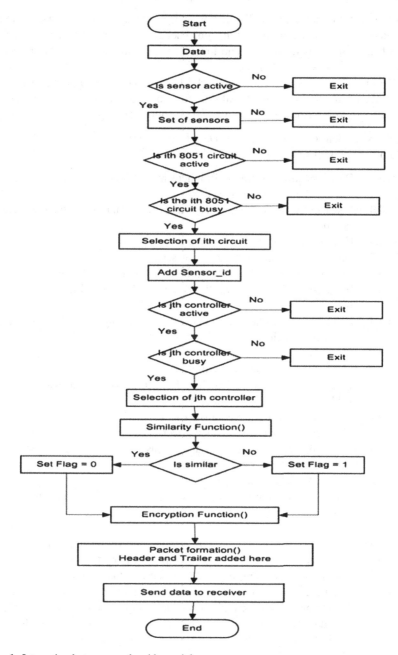

Fig. 1 Interaction between sender-side modules

considering it as abnormal condition. Otherwise, flag is set to '0'. Data is then encrypted using typical encryption function. Then, data is sent for packet formation to add header and trailer. Finally, send function is used to transmit data to receiver module over network (refer Fig. 1).

Algorithm 1 exhibits step-wise approach of our sender-side devices which help to fetch data from environment in real time for monitoring and analyzing purpose, and further send information to receiver-side of our network.

Algorithm 1: Sender_module Input: Data from environment
Output: Data sent to receiver side

Step 1:   Data is received from environment in real time.
Step 2:   Check whether sensor is active
Step 2.1: If sensor (S_x) is active, then sensor receives data from the environment (S_x (data))
Step 2.2: If sensor (S_x) is inactive, then exit()
Step 3:   Data is passed through the set of sensors
Step 4:   Check whether ith 8051 based hardware circuit is active (Ckt_m)
Step 4.1: If the circuit is active, check whether ith 8051 based hardware circuit is busy
Step 4.1.1: If the circuit is idle, then data is passed through the ith 8051 based hardware circuit
Step 4.1.2: If circuit is busy, then exit()
Step 4.2: If the circuit is inactive, then exit()
Step 5:   The sensor id is added by the 8051 based hardware circuit; Add_Sensor_ID($Ckt_m(S_x(data))$)
Step 6:   Check whether the jth controller is active ($Ctrl_n$)
Step 6.1: If the controller is active, then check whether jth controller is busy
Step 6.1.1: If the controller is idle, then data is passed through the jth controller
Step 6.1.2: If the controller is busy, then exit()
Step 6.2: If the controller is inactive, then exit()
Step 7:   Similarity function (Smfn()) checks data for a typical pattern
Step 7.1: If (similar to pattern), then set flag = 0
Step 7.2: If (not similar to pattern), then set flag =1; Chkd_data = Smfn($Ctrl_n$)
Step 8:   The output (Chkd_data) is sent through encryption function (encrfn())
Step 9:   The Chkd_data is encrypted (encrfn(Chkd_data)) using typical encryption procedure, Encr_data = encrfn(Chkd_data)
Step 10:   The Encr_data is passed through packet formation function (pktfn()) for adding the header and trailer, Data_to_send = pktfn(Encr_data)
Step 11:   The data(Data_to_send) is forwarded to the outside world for the receiver
Step 12:   Stop

Data is sent to receiver-side for analyzing purpose. Receiver-side is remotely located as network of server machines. Figure 2 is flowchart of receiver-side. The data_received function receives real time data from sender-side. If kth receiver is

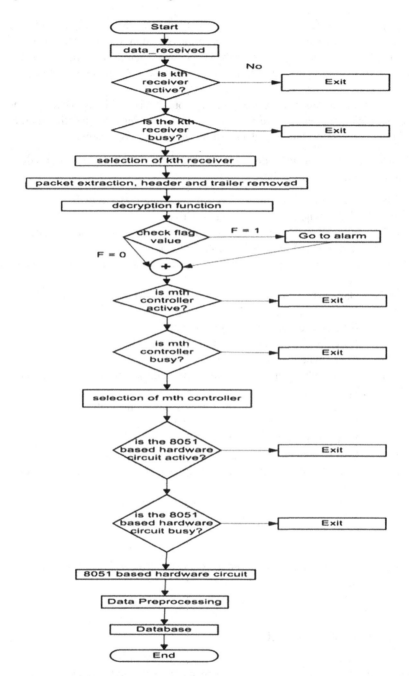

Fig. 2 Interaction between receiver-side modules

active, then check whether it is busy. If the particular receiver is not busy, then data packet is being extracted to apply decryption function further. The check_flag function checks for flag value. If alarm is required, then send alarm information. Overall information is stored within server database through mth controller and corresponding 8051 based hardware circuits.

Algorithm 2 exhibits step-wise approach of our receiver-side devices which help to collect data from sender-side in real time for analyzing and statistical purpose, and further send information to database of our server based network.

Algorithm 2: Receiver_module Input: Data collected from transmitting side
   Output: Send data to database

   Step 1:   Data is received from sender in real time (Data_recvd)
   Step 2:   Check whether kth receiver is active
  Step 2.1: If receiver (R_p) is active, then receiver receives data from sender
  Step 2.2: If receiver (R_p) is inactive, then exit()
   Step 3:   Data_recvd is passed through packet extraction function for removal of header and trailer
   Step 4:   Data_extr is decrypted using typical decryption procedure
   Step 5:   The flag value is checked
  Step 5.1: If flag == 0, then move to next step
  Step 5.2: If flag == 1, then set off alarm function and move to next step
   Step 6:   Check whether the controller is active
  Step 6.1: If the mth controller is active, then check whether mth controller is busy
 Step 6.1.1: If the controller is idle, then data is passed through the mth controller
 Step 6.1.2: If the controller is busy, then exit()
  Step 6.2: If controller is inactive, then exit()
   Step 7:   Check whether 8051 based hardware circuit is active
  Step 7.1: If the circuit is active, then check whether 8051 based hardware circuit is busy
 Step 7.1.1: If the circuit is idle, then send data for pre-processing
 Step 7.1.3: If the circuit is busy, then exit()
  Step 7.2: If circuit is inactive, then exit()
   Step 8:   Data pre-processing()
   Step 9:   Pre-processed data is sent to database using existing methods
  Step 10: Stop

3   Experimental Results

In this section, experimental results have been demonstrated to realize our proposed system in different perspectives. Noise sensor has been used for measuring noise/sound in specific environment within organization using predefined patterns as basis of threshold.

In real time, it has been realized in Fig. 3 that data_to_send is directly proportional to time can be represented as: *data_to_send* ∝ *time*. 'k' is constant introduced for modification of equation as *data_to_send* = *kt*; where t is time. The graph is represented in Fig. 3 with constant values ranging from '1' to '3'. We have achieved *data_to_send* ∝ 1/(*number of* 8051 *circuits*). 'k' is constant which we have introduced to modify equation as *data_to_send* = *k*/(*number of* 8051 *circuits*) as shown in Fig. 4 with constant values ranging from '1' to '3'.

Figures 5 and 6 represent situations based on equation *data_to_send* = *kt*/*number of* 8051 *circuits*. Fig. 5 involves constant values such as "k = 1" and time (t) ranging from '1' to '3'; whereas Fig. 6 involves constant values such as "k = 2" and time (t) ranging from '1' to '3'.

Figure 7 shows relation between time and number of 8051 circuits available in our system. Figure 8 supports the graph for equation *data_to_recv* ∝ *number of queries* for distinct 'k' values ranging from '1' to '3'.

Fig. 3 Data_to_send versus time with 'k' values

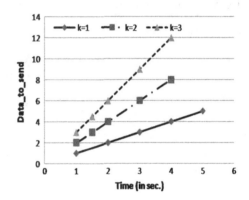

Fig. 4 Data_to_send versus number of 8051 circuits with 'k' values

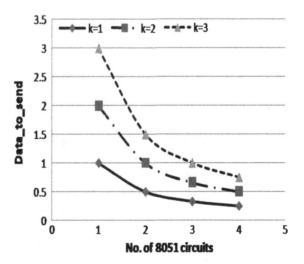

Fig. 5 Data_to_send versus number of 8051 circuits with 'k = 1' and 't' values

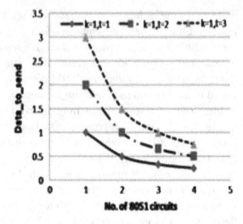

Fig. 6 Data_to_send versus number of 8051 circuits with 'k = 2' and 't' values

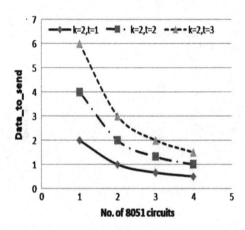

Fig. 7 Time versus number of 8051 circuits

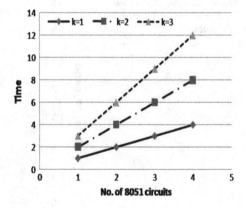

Fig. 8 Data_to_receive versus number of queries for distinct 'k' values

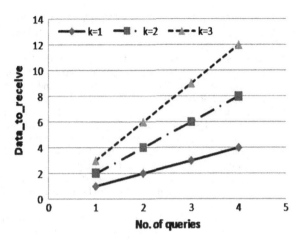

4 Conclusion

We have proposed the design of IOT based architecture using real time data feed with an aim to provide a safe environment for high security zones in organizations without direct human involvements. Our system has incorporated distinct types of sensors for sensing specific kinds of abnormalities in surrounding environment. We have demonstrated the relationships between data, time, and number of 8051 circuits used in our system. The relations between "data received by the receiver-end to be sent to database" and "queries used in processing" in two proposed theorems have been proved and analyzed. Experimental results have helped us to establish the facts described in algorithms and theorems. Overall, our work has utilized the benefits of IOT based sensors for enhancing security features of organizations.

References

1. Gubbi, J., Buyya, R., Marusic, S., Palaniswami, M.: Internet of Things (IoT): A vision, architectural elements, and future directions. In: Future Generation Computer Systems, vol 29, issue 7, pp. 1645–1660, Elsevier (2013).
2. Koshizuka, N., Sakamura, K.: Ubiquitous ID: Standards for Ubiquitous Computing and the Internet of Things. In: IEEE Pervasive Computing, vol. 9, no. 4, pp. 98–101. IEEE CS (2010).
3. Xu, L.D., He, W., Li, S.: Internet of Things in Industries: A Survey. In: IEEE Transactions on Industrial Informatics, vol. 10, no. 4, pp. 2233–2243. (2014).
4. Aragones-Vilella, J., Martínez-Ballesté, A., Solanas, A.: A Brief Survey on RFID Privacy and Security. In: World Congress on Engineering (WCE 2007), vol. II, London, UK (2007).
5. Tozlu, S., Senel, M., Mao, W., Keshavarzian, A., Bosch LLC, R.: Wi-Fi Enabled Sensors for Internet of Things: A Practical Approach. In: IEEE Communications Magazine, pp. 134–143. (2012).

6. Lazarescu, M.T.: Design of a WSN Platform for Long-Term Environmental Monitoring for IoT Applications. In: IEEE Journal on Emerging and Selected Topics in Circuits and Systems, vol. 3, no. 1, pp. 45–54. (2013).
7. Chang, K.L., Chang, J.S., Gwee, B.H., Chong, K.S.: Synchronous-Logic and Asynchronous-Logic 8051Microcontroller Cores for Realizing the Internet of Things: A Comparative Study on Dynamic Voltage Scaling and Variation Effects. In: IEEE Journal on Emerging and Selected Topics in Circuits and Systems, vol. 3, no. 1, pp. 23–34. (2013).
8. Kawamoto, Y., Nishiyama, H., Fadlullah, Z.M., Kato, N.: Effective Data Collection via Satellite-Routed Sensor System (SRSS) to Realize Global-Scaled Internet of Things. In: IEEE Sensors Journal, vol. 13, no. 10, pp. 3645–3654. (2013).
9. Li, X., Lu, R., Liang, X., Shen, X., Chen, J., Lin, X.: Smart Community: An Internet of Things Application. In: IEEE Communications Magazine, pp. 68–75. (2011).
10. Jin, J., Gubbi, J., Marusic, S., Palaniswami, M.: An Information Framework for Creating a Smart City Through Internet of Things. In: IEEE Internet of Things Journal, vol. 1, no. 2, pp. 112–121. (2014).
11. Zanella, A., Bui, N., Castellani, A., Vangelista, L., Zorzi, M.: Internet of Things for Smart Cities. In: IEEE Internet of Things Journal, vol. 1, no. 1, pp. 22–32. (2014).

Performance Evaluation of Classifier Combination Techniques for the Handwritten Devanagari Character Recognition

Pratibha Singh, Ajay Verma and Narendra S. Chaudhari

Abstract In this paper we have applied the classifier combination approach to Devanagari character recognition. Two types of combination models are experimented in this work. The first is based on stacking and the second one is based on parallel combination of classifiers. The chain code and gradient based features are used in this work. Three classifiers namely Linear Discriminent, Quadratic Discriminent and k-nearest neighbor classifier are combined using same features in stacking based approach. In parallel combination three different feature sets namely chain code, gradient based and distance based features are used for the classifiers of similar kind. Various fixed combining rules like sum, max, min, product, median and majority voting are used in both the combination schemes and out of these product rule performs best in most of the cases.

Keywords MLP · Feature extraction · Neural network (NN) · k-Nearest neighbor (k-NN)

1 Introduction

The main objective of pattern recognition is to obtain an improved performance in terms of recognition rate. To accomplish this task various researchers attempted classifier combination strategies. Multiple classifier combination is also known by various other names, such as classifier ensemble, committee of classifiers, hybrid

P. Singh (✉) · A. Verma
IET DAVV, Khandwa Road, Indore 452017, MP, India
e-mail: prat_ibh_a@yahoo.com

A. Verma
e-mail: ajayrt@rediffmail.com

N.S. Chaudhari
VNIT, Nagpur, MH, India
e-mail: nsc0183@gmail.com

© Springer India 2016
S.C. Satapathy et al. (eds.), *Information Systems Design and Intelligent Applications*, Advances in Intelligent Systems and Computing 434,
DOI 10.1007/978-81-322-2752-6_64

methods, cooperative agents, opinion pool etc. These have shown to outperform over single-expert systems for a broad range of applications in a variety of scenarios. While creating an ensemble-based systems, it is generally preferred to combine diverse models (classifiers). There are many reasons for using classifier combination. Firstly it improves the overall accuracy. Secondly it makes the overall classifier more robust. Dietterich [1] given three reasons for justifying classifier combinations namely: statistical reason, computational reason and the representational reason. Combined classifier may not beat a single classifier on a particular data-point, but will surely outperforms most of the individual classifier. Using base classifier performing differently on different datasets and even on different subsets of datasets, creation of an ensemble leads to good generalization performance. Each classifier starts at an arbitrary point to find the solution. So there may be chance of classifier to be stuck in local minima. In classifier combination the local searches starts at various different points and tendency to stuck up in local minima will be reduced. The paper is organized as follows Sect. 1 gives introduction and related work, Sect. 2 describes the model of classifier combiner, Sect. 3 describes the rules for classifier combination, Sect. 4 describes the experimental results and Sect. 5 describes the conclusions.

1.1 Related Work

Combination of classifier can be attempted in various ways. If the input given to the classifier is occupying different input space according to suitability of classifier then the approach is called divide and conquer based classifier combination [2]. In another approach one classifier gives a decision and other classifiers are invoked if the decision obtained from the previous stage is having low confidence, as used by Bhattacharya and Choudhuri [3]. This approach is called the sequential approach. Classifier combination strategies can be divided into two types: 1. Classifier fusion and 2. Dynamic classifier selection. The main difference between the two is that, in the first one the member of ensemble must know the whole feature space while in the later each member is designed to know only a part of whole feature space. In dynamic classifier selection approach selection of the classifier is made for all input patterns. This selection is based on the performance of that classifier which is most likely to give the correct decision for the specific input pattern [4]. The output of classifier gives the decision which is classified as; abstract level, rank level and score/decision level. When each classifier outputs a decision indicating the proximity of the input data to one class, integration can be done at the decision level. Rank level classifier combination is used when the output of each classifier is the subset of possible measures stored in decreasing order of their confidences. A consensus rank is obtained by merging the ranks obtained from individual classifier. Ranking provide more insight into problem of pattern recognition. Highest rank [5], Borda count or Logistic Regression may be used to combine the rank produced by individual classifier. The rank level fusion is simpler to implement compared to score level fusion technique, because normalisation is not required. Unlike classifier scores, the ranking output of several classifiers are comparable; however, the provided ranks reveal less information than classifier scores. Various scenarios and combination strategies are introduced in the literature namely sequential, parallel and hybrid [6]. The sequential combination uses various classification methods such that the classification result of one classifier is used by another. The order in which the classifiers are connected matters a lot in this configuration. In parallel architecture different classifiers are driven by the same set of features or same base classifier driven by different type of features. The order of classifier is not important in this type of configuration. Hybrid method is the combination of serial and parallel architectures. Sequential approaches and selection of classifier has largely being ignored by researchers. Majority of the research in the field of the classifier combination focuses mainly on parallel combination of classifiers with the same/different set of input and fusing their result for the final decision.

Pal et al. [7] applied classifier combination approach for character recognition. Gradient and curvature features are used to drive SVM and MQDF classifier. The improved recognition result of 95.13 % is observed by the combination of classifier for dataset of size 36,172 samples.

Shelke et al. [8] used multistage multi-feature classification approach. In the first stage a pre-classification is done based on the structure of patterns. In the second stage parallel architecture of neural network is applied using three different types of features namely density feature, distance based features and modified wavelet based features. The output of the classifiers are combined using majority voting scheme. The recognition efficiency of the proposed method reported was 97.98 %.

A bench-marking dataset was developed by Bhattacharya et al. [3] of 22,556 images of handwritten Devanagari Numerals. MLP based multi-classifier approach was applied where the pattern rejected by one stage classifier were applied at the input of next stage classifier. The output of such three stages are combined and given as input to another MLP based classifier. The overall recognition rate for the proposed multistage model of the test set was 99.04 %.

Jindal et al. [9] applied Adaboost method for combining multiple MLP classifiers consisting of different number of hidden nodes. Zernike moment based features of different orders were used in this study. The maximum recognition rate for the test set of ISI dataset reported was 98.9 %.

A benchmarking Devanagari dataset of 83,300 isolated characters is developed by Rajiv Kumar et al. [10]. They developed another dataset of 35,000 handwritten numeral samples (15,000 constrained, 5,000 semi-constrained and 15,000 unconstrained). The performance of the numeral dataset was evaluated by them for simple pixel based features and Neural Network based classifier. For this dataset various features based on the density of the zones, the profile based, the Krisch Transform based directional and Wavelet transform based were extracted by them. The various classifiers based on Pattern recognition network (PR), Feed-forward network (FFN), Fitness function network (FFT), Cascade neural network (CCN), and (KNN) k-Nearest neighbor were tested with these features [11]. The classifier ensemble was used for which result of 97.87 % was obtained using weighted majority based classifier combination scheme. For the characters the result [12] of recognition found is 84.03 % for 5.03 % rejection. Das et al. [13] performed experiments on various numeral datasets such as CMATER, ISI, CEDAR using a feature combination approach. A combination of modular principle component analysis (MPCA) based features and Quadtree longest run feature forms the feature vector. SVM based classifier was used. The recognition efficiency for ISI numeral dataset reported was 99.06 %.

2 The Model of Classifier Combination

The classifier combiner is a model which produces a confidence score taking input as posterior probabilities.

Let p_m^n be the posterior score of class n obtained from classifier m for any data instance. Let $p_m = \left[p_m^1, p_m^2 \ldots, p_m^N\right]^T$ then the input to the combiner is $f = \left[p_1^T, p_2^T \ldots, p_M^T\right]^T$ where N is the number of classes and M is the number of classifiers. Outputs of the combiner are N different scores representing the degree of support for each class. Let r^n be the combined score of class n and let $r = [r^1, r^2 \ldots, r^N]^T$ then in pattern recognition, the combiner is defined as a function $g : \mathbb{R}^{MN} \to \mathbb{R}^N$ such that $g = r(f)$. On the test phase, label assignment to a data instance is done as follows:

$$\hat{y} = arg \max_{n \in [N]} r^n \tag{1}$$

where $[N] = \{1, \ldots, N\}$. Block diagram of the classifier combination problem with confidence score outputs is given in Fig. 1.

According to Duin et al. [14] there are various constructs available for combination, which can be classified as:

- Stacked combination
- Parallel combination
- Sequential combination
- Fixed combination
- Trained combination of classifier

There are two types of combination method experimented in this paper: Stacked and parallel combination. *Stacked combining* is method of combining different classifiers for the same feature space. Base classifiers used in stacking are of a different nature. This method is based on meta-learning approach in which the learning takes place by taking input from various learners. Training in this case takes place in two stages; in the first stage the base classifiers are trained on training data and in the second stage the meta classifier training takes place using the output

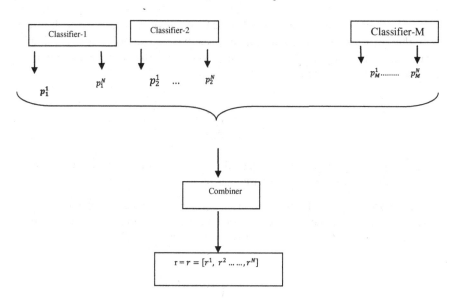

Fig. 1 Block diagram of the classifier combination

of trained base classifiers. The mapping in the second stage of stacking can be of two types (1) Non-trainable (2) Trainable. In the present work, the combination of Linear Discriminent classifier(Bayes Normal-1, ldc), Quadratic Discriminent (Bayes Normal-2, qdc) and K-Nearest Neighbor(k-NN, with k = 1) are used and fixed combination rules as discussed in the next section are used for combination of the decision.

Parallel combining of classifiers is used for combining different feature sets. This may be especially useful if the objects are represented by different feature sets, when they are represented in different physical domains or when they are processed by different types of techniques. Most often base classifiers used in parallel combination are of the same type but it is not necessary.

3 Rules for Combination of Classifier

The non-trainable classifier combiner is used in case of inadequate training data. In this case fixed combination rules can be applied to different type of classifiers using decision combination based approach where the measurement level outputs are combined using combination rule. The various types of combining rules are as follows:

- This first type of combination is called the sum rule or the average rule. It is one of the strongest non-trainable combination types. The basic idea is simply summing, or averaging, the confidence scores of a particular class through base classifiers to obtain the final score of that class,

$$r^n = \frac{1}{M} \sum_{m=1}^{M} p_m^n \tag{2}$$

Sometime trimming is done in calculating the mean for the classifiers that give unusually high or low scores to a particular class.

- The second method of combination is product rule wherein we multiply the scores to calculate the final score. For obtaining the correct score the participating scores need to be calculated accurately using Eq. (3).

$$r^n = \frac{1}{M} \prod_{m=1}^{M} p_m^n \tag{3}$$

This method is very sensitive to the outliers present in the ensemble, because very low or very high score has a huge effect on the resulting score in product.

- The third rule for fixed combination is minimum rule in which final score is calculated as the minimum of participating scores as given by Eq. (4).

$$r^n = \min_m p_m^n \tag{4}$$

- The fourth rule is maximum combiner in which for getting final score we just take the maximum among the classifiers output scores. With this combination, if one base classifier insists on a particular class for a given test instance, final decision assigns the test instance to that class; even if all other base classifiers disagree as given by Eq. (5).

$$r^n = \max_m p_m^n \tag{5}$$

- The fifth rule is median combination rule for which final score is calculated as the median of the candidate scores. This combination is useful for reducing effect of outliers.

$$r^n = \text{median}_m \, p_m^n \tag{6}$$

Majority Voting is another one of the widely used non-trainable combiners other than the rules discussed earlier. To compute the final score for a particular class, here we simply count the number of classifiers selecting that particular class. The rule can also be applied to rank level classifier that outputs class labels or class ranks. In fact, it does not use the scores and just uses the class labels. Therefore if only the class labels are obtained from base classifiers, majority voting is the optimal rule with the following minor assumptions: (1) The number of classifiers are odd and the problem is a binary classification problem, (2) The probability of each classifier for choosing any class is equal for any instance, (3) Base classifiers are independent [15]. Majority Voting based classifier combination is the simplest method in which final decision is that class for which maximum (greater than N/2) participating classifier vote, where N is the number of classifiers.

Fixed combining rules assume that the responses of the base classifiers are comparable. This assumption does not necessarily holds if different classifier architectures are combined, such as Fisher Linear Discriminant (FLD), Support vector machines or neural networks.

4   Experimental Results

We tested the effectiveness of the proposed method for three already available datasets of Devanagari characters, whose details are discussed in next subsections. The training and testing patterns of each dataset are subjected to various preprocessing and feature extraction steps. In preprocessing we have performed size normalization, bounding box representation and conversion of patterns to boundary (contour) representation. Features are calculated by dividing the bounding box into nine partitions according to density based zoning criterion proposed in our previous work [16]. Chain code and gradient based direction features are extracted for each of nine zones.

- **CVPR-ISI Dataset**

This dataset is available for the global research community since 2009 and is developed by CVPR unit of ISI Kolkata. The numeral database includes samples collected from mail pieces and job application forms. The dataset consists of 22,556 images stored in 'tif' format collected from 1049 writers.

- **The CPAR-2012 Dataset**

This dataset is recently developed by Intelligent system group Noida (Centre for Pattern Analysis and Recognition) [10] and one of the largest dataset for Devanagari handwritten characters. It consists of 35,000 images for numerals and 78,400 images of characters. The data is collected from diverse population strata of 2000 writers from various states of India having different religions. The numeral dataset is having 11 classes in this dataset because the pattern corresponding to number '9' can be written in two alternate ways. The character dataset is having 1000 training images and 600 test images of 49 classes each.

For each classifier directional histogram features is obtained by dividing each image pattern into nine zones. We mainly focused on stacked classifier combination and parallel combination approach which are implemented using PRTOOL (Pattern Recognition tool box in MATLAB) [17]. Stacked combining scheme applied to four different classifiers using one feature set at a time. The result of stacked combination scheme is depicted in Tables 4, 5 and 6. All best results for each feature set are printed in bold. The product and median rule frequently scores a best result for stacking based combination. In addition all combined results that are found better than the best individual classifier are underlined. All the combination rules as discussed in Eqs. (2–6) are applied. Features used for the stacked combination are global zone based gradient directional features as it gives best result for almost all datasets. Performance of the individual classifier is given in Tables 1, 2 and 3.

Table 1 Percentage error rate for individual classifier of Devanagari characters CPAR-2012

	Edge			Chain		
	Local	Global	Standard	Local	Global	Standard
qdc	16.74	**15.73**	17.75	21.71	20.27	20.1
ldc	23.6	22.67	25.92	29.36	26.69	29.6
knnc	24.93	23.07	23.9	28.84	26.3	25.8
bpxnc	25.73	27.35	31.04	33.67	29.4	70.7

Table 2 Percentage error rate for individual classifier of Devanagari numerals CPAR-2012

	Edge			Chain		
	Local	Global	Standard	Local	Global	Standard
qdc	4.16	4.09	4.28	6.01	5.4	5.9
ldc	6.2	6.6	7.14	7.6	7.17	8.04
knnc	3.86	**3.27**	3.5	3.82	3.6	3.43
bpxnc	3.93	3.36	3.7	3.9	3.47	3.78

Table 3 Percentage error rate for individual classifier of Devanagari numerals CVPR-ISI

	Edge			Chain		
	Local	Global	Standard	Local	Global	Standard
qdc	3.9	3.33	4.17	4.2	4.1	7.3
ldc	5.3	5.52	7.3	6.6	5.8	4.5
knnc	2.57	**1.89**	2.43	2.8	2.0	2.4
bpxnc	2.84	2.62	2.62	4.03	3.19	9.67

Table 4 Percentage error rate for stacked combination with global edge based gradient features

Dataset	ldc	qdc	knnc	Product	Mean	Median	Max	Min	Vote
CPAR-2012 digit	6.6	4.1	3.3	3.9	3.3	*4.6*	3.3	4.3	3.3
ISI digit	5.5	3.3	1.9	3.1	1.9	*3.9*	1.9	3.3	2.3
CPAR-2012 character	22.7	15.7	23	**14.6**	23	15.4	23	19.3	16.9

Table 5 Percentage error rate for stacked combination with global chain code based features

Dataset	ldc	qdc	knnc	Product	Mean	Median	Max	Min	Vote
CPAR-2012 digit	7.2	5.5	3.7	5.1	3.5	**5.1**	3.7	6	**3.9**
ISI digit	5.8	4.1	2.0	3.9	2	**4.1**	2.0	4.4	2.7
CPAR-2012 character	**26.7**	**20.1**	**26.3**	**18.7**	24.8	19.8	26.3	23.3	20.7

Table 6 Percentage error rate for stacked combination for fuzzy zone based distance features

Dataset	ldc	qdc	knnc	Product	Mean	Median	Max	Min	Vote
CPAR-2012 digit	23.7	11.6	21.1	13	20.7	**12.3**	21	17.7	13
ISI digit	23.2	12.1	25.6	13.1	23.7	13	25.6	17	15.4
CPAR-2012 character	61.6	45.6	66.3	46.9	63.6	46	66.3	57.3	52.9

Table 7 Percentage error rate for Devanagari characters using parallel combination of classifiers

Classifier	Feature global chain	Feature global edge	Feature fuzzy distance	Median	Vote	Max	Min	Product	Mean
ldc	29.59	22.67	61.56	29.59	27.12	61.56	22.64	**19.56**	61.56
qdc	20.09	15.73	45.57	20.10	17.26	45.56	15.52	**12.57**	45.56
knnc	25.8	23.01	66.29	19.43	26.78	21.08	29.74	**17.04**	17.32
bpxnc	33.36	28	90.96	30.30	34.41	**23.1**	69.03	34.74	24.85

The results of parallel combining scheme are given in Tables 7, 8 and 9. Combining rules are applied on the results for a each feature set for each individual classification rule. The best results over the 6 combining rules for each classifier are printed bold. For instance, the Product rule yields the frequently best combining result of all combiners.

For ISI Devanagari numerals performance comparison is made with three previously reported results. The result reported by Das et al. [13] is better but it is tested for small subset of ISI data whereas our method is tested on complete ISI dataset. Result reported by Bhattacharya et al. [3] is tested on complete dataset but the accuracy is obtained with 0.24 % rejection of patterns. Also the classification algorithm used by them is a multistage MLP based classifier and there is a rejection

Table 8 Percentage error rate for ISI Devanagari numerals using parallel combination of classifiers

Classifier	Feature global chain	Feature global edge	Feature fuzzy distance	Median	Vote	Max	Min	Product	Mean
ldc	5.83	5.53	23.17	5.8	5.39	23.17	5.53	**4.58**	23.17
qdc	4.14	3.33	12.09	3.79	**2.87**	12.09	3.28	3.06	12.09
knnc	2.03	1.90	25.61	**1.90**	1.92	25.61	7.89	22.63	22.63
bpxnc	2.52	2.3	12.6	1.90	1.87	2.44	2.93	2.01	**1.92**

Table 9 Percentage error rate for Devanagari numerals-CPAR-2012 using parallel combination of classifiers

Classifier	Feature global chain	Feature global edge	Feature fuzzy distance	Median	Vote	Max	Min	Product	Mean
ldc	7.17	6.60	23.67	7.16	6.55	23.67	6.61	**5.34**	23.67
qdc	5.47	4.09	11.63	5.36	**3.90**	11.63	4.31	3.96	11.63
knnc	3.68	3.27	21.08	**2.74**	3.19	16.46	4.39	3.45	3.46
bpxnc	3.74	3.75	12.38	3.05	3.43	3.10	4.15	2.93	**2.75**

of patterns used by the last stage in multistage MLP. For the single k-NN (K-Nearest Neighbor) the benchmark established by them is accuracy level of 97.26 %, where as in this work our proposed global zone based gradient features gives the recognition accuracy of 98.1 % for k-NN single stage classifier.

Kumar et al. [11] applied the method of classifier ensemble on the character dataset and obtained the recognition rate of 84.03 % for a rejection of 5.3 %. For the same dataset the proposed method yield the recognition rate of 85.4 % with stacking based approach and values and 87.41 % with parallel combiner. For CPAR-2012 character dataset the accuracy improvement of 3.38 % is observed with the proposed feature set along with the parallel combination of classifier.

5 Conclusion

We have experimented various combining rules in this work for recognition of Devanagari handwritten datasets. Two type of combiners can be built for a score level output. The first is trainable and the second is non trainable combiner. The stacked and parallel non-trainable architecture of combiner is build and experimented in this work. The main difference between the stacked combiner and the parallel combiner is that the stacked combiner takes the same input feature for various classifier models. The score outputs from various classifier is used to train the second level output. Whereas in parallel combination scheme the same classifier is used in the different feature domains and the decision combination is done using fixed combining rules. Out of the six combining rules the product combination yields the best result in stacking based approach as well as parallel combination scheme. We have obtained higher accuracy for the CPAR-2012 character dataset. For CPAR-2012 numeral dataset and CVPR-ISI dataset our accuracy is comparable or slightly less but in our method we have not rejected any pattern/sample while the previous reported results are using a rejection of certain percentage of patterns.

Acknowledgments The authors would like to thank Intelligent System Group Noida and CVPR unit of ISI Kolkata for providing the dataset of Devanagari characters and numerals.

References

1. K. Woods, K. Bowyer and W. Kegelmeyer, " Combination of multiple classifiers using local accuracy estimates," *IEEE Transactions on Pattern Analysis and Machine Intelligence,* vol. 19, no. 4, pp. 405–410, 1997.
2. T. K. Ho, J. J. Hull and S. N. Srihari, "Decision in multiple classifier system," *IEEE Transaction in Pattern Analysis and Machine Intelligence,* vol. 16, p. 66, January 1994.
3. S. Chitroub, "Classifier combination and score level fusion: concepts and practical aspects," *International Journal of Image and Data Fusion,* vol. 1, no. 2, pp. 113–135, June 2010.
4. U. Pal, S. Chanda, T. Wakabayashi and F. Kimura, "Accuracy Improvement of Devnagari Character Recognition Combining SVM and MQDF," in *11th International Conference on Frontiers in handwriting recognition,* 2008.
5. S. Shelke and S. Apte, "Multistage Handwritten Marathi Compound Character Recognition Using Neural Networks," *Journal of Pattern Recognition Research,* vol. 2, pp. 253–268, August 2011.
6. U. Bhattacharya and B. B. Choudhary, "Handwritten Numeral Databases ofIndian Scripts and Multistage Recognitionof Mixed Numerals," *IEEE Transaction on Pattern Analysis and Machine Intelligence,* vol. 31, no. 3, pp. 444–457, March 2009.
7. T. Jindal and U. Bhattacharya, "Recognition of Offline Handwritten Numerals Using an Ensemble of MLPs Combined by Adaboost," in *ACM Proc. of the 4th International Workshop on Multilingual OCR (MOCR 2013),* Washington DC, USA, 2013.
8. R. Kumar, A. Kumar and P. Ahmed, "A Benchmark Dataset for Devnagari Document Recognition Research," in *WSEAS Press,* Lemesos, Cyprus, 2013.
9. R. Kumar and K. K. Ravulakollu, "Handwritten Devnagari Digit Recognition: Benchmarking on new dataset," *Journal of Theoretical and Applied Information Technology,* vol. 60, no. 3, Feb 2014.
10. R. Kumar and K. K. Ravulakollu, "On the Performance of Devnagari Handwritten Character Recognition," *World Applied Sciences Journal,* vol. 31, no. 6, pp. 1012–1019, 2014.
11. N. Das, J. M. Reddy, R. Sarkar, S. Basu, M. Kundu, M. Nasipuri and D. K. Basu, "A statistical–topological feature combination for recognition of handwritten Numerals," *Applied Soft Computing,* vol. 2, pp. 2486–2495, April 2012.
12. D. M. T. Robert P.W. Duin, "Experiments with Classifier Combining Rules," in *MCS,* 200.
13. R. Polikar, "Ensemble based systems in decision making," *Ieee Circuits And Systems,* vol. 6, no. 3, pp. 21–45, 2006., 2006.
14. P. Singh, A. Verma and N. S. Chaudhari, *Applied Computational Intelligence and Soft Computing,* 2015.

15. R. Duin, "PRTools 3.0, A Matlab Toolbox for Pattern Recognition, Delft University of Technology," 2000.
16. T. G. Dietterich, "Ensemble methods in machine learning," in *In International Workshop on Multiple Classifier Systems, Springer-Verlag*, 2000.
17. D. Frosyniotis, A. Stafylopatis and A. Likas, "A divide-and-conquer method for multi-net classifiers," *Pattern Analysis and applications*, vol. 6, pp. 32–40, 2003.

An Improved Kuan Algorithm for Despeckling of SAR Images

Aditi Sharma, Vikrant Bhateja and Abhishek Tripathi

Abstract Synthetic Aperture Radar (SAR) is an acquisition tool for coherent imagery used for meteorological and astronomical purposes. The speckle noise diminishes the information and image quality which evokes the necessity of pre-processing of SAR images. Kuan filter is a popular despeckling algorithm among the various local statistics filters. Kuan filter works efficiently within the homogenous regions of the SAR images while penalty is imposed by the edges. This paper presents an improved Kuan filter which combines the concept of gradient and conduction function for despeckling of SAR images. In this, the image is processed by classifying it into three regions i.e., homogenous, non-homogenous and isolated regions respectively; depending upon the estimated value of noise parameters. This approach thereby provides a simple solution to overcome blurring across the edges during despeckling. Simulation results make it evident that the proposed despeckling algorithm yields better results as compared to other primitive filters.

Keywords Local statistics · Anisotropic diffusion · Conduction function · Kuan filter · Despeckling

1 Introduction

SAR image is a high resolution two dimensional image which results from the backscattering of coherent electromagnetic wave. SAR images are used to identify the land cover type, forest mapping, monitoring of land subsidence, wetlands etc.

A. Sharma (✉) · V. Bhateja · A. Tripathi
Department of Electronics and Communication Engineering,
Shri Ramswaroop Memorial Group of Professional Colleges
(SRMGPC), Lucknow 227105, U.P., India
e-mail: aditiii065@gmail.com

V. Bhateja
e-mail: bhateja.vikrant@gmail.com

A. Tripathi
e-mail: abhishek1.srmcem@gmail.com

© Springer India 2016
S.C. Satapathy et al. (eds.), *Information Systems Design and Intelligent Applications*, Advances in Intelligent Systems and Computing 434,
DOI 10.1007/978-81-322-2752-6_65

663

However, the main drawback of these images is the presence of speckle which is caused due to many elemental scatterers with a random distribution within a resolution cell. These images could also contaminated with Gaussian noise; but presently the focus is mainly concentrated toward speckle suppression. The coherent sum of their amplitudes and phase results in a strong fluctuation of a backscattering from one resolution cell to another. Consequently, speckled SAR image is not deterministic but follows an exponential uniform distribution. Speckle is a statistical fluctuation of each pixel in the image of scene which makes the radiometric and textural aspect less efficient for class discrimination [1–6]. For these reasons, pre-processing of SAR images is essential; however, it should not deteriorates the useful information (i.e., point target, texture etc.) during filtering. Adaptive speckle filtering is based on the multiplicative model and consider local statistics. The Local Statistic Mean Variance (LSMV) filter adapts itself as a function of local coefficient of variation and can be enhanced by fixing a minimum value for better speckle smoothing and point target preservation. The coefficient of variation are statistically sensitive to texture and speckle noise strength [5–9]. Kuan filter [10] achieves a balance between averaging and identity filter which depends on the coefficient of variation parameter inside the moving window. The Lee [11] and Kuan filters have the same formulation, as the output image is computed by the linear combination of center pixel and is replaced by the average intensity of the mask. In progression, techniques involving Anisotropic Diffusion (AD) were proposed which employs a variable diffusion coefficient for despeckling SAR images [12–14]. The Perona Malik Anisotropic Diffusion (PMAD) [15] filter uses a variable coefficient of diffusion in standard scale-space paradigm so that it has a larger value in homogenous region. This AD filtering model provides better preservation features but this can be achieved at high computational complexity [16–18]. In this paper, an improved Kuan algorithm has been proposed for despeckling of SAR images. The images are processed based on their classified regions using the calculated values of image and noise variation parameters. This leads to efficient speckle suppression in homogenous area thereby preserving point targets and detailed areas as well as curtail computational complexity. The rest of the paper is structured as follows: Sect. 2 elaborates the proposed despeckling methodology and the Image Quality Assessment (IQA) parameters used for performance evaluation of obtained results. Further, the simulation results and discussions are presented in Sect. 3 whereas the conclusions are drawn in Sect. 4.

2 Proposed Methodology

2.1 Background

Various image restoration and enhancement methods have been designed in both spatial and frequency domains which are based on the criteria like Minimum Mean Square Error (MMSE), Linear Minimum Mean Square Error (LMMSE), Bayesian,

Non-Bayesian, etc. [2, 19]. The adaptive Kuan filter uses a MMSE calculation for estimating the center pixel in the filter window. It is an approximation of Lee filter and uses a simple method for calculating the signal estimate from the local mean (m) and variance (Sd). The Kuan filter operates by averaging the random noise in the flat areas and preserves the edge sharpness using the filter window to measure the spatial activity. The various non-stationary image statistical parameters needed for the Kuan algorithm are effective number of looks, noise variation of coefficient, image variation of coefficient, threshold and weighting function. If μ and σ denotes the global values of computed mean and standard deviation respectively; the aforesaid parameters are determined using the mathematical expressions given under Eqs. (1)–(5) respectively. These parameters depend upon the Non-stationary Mean, Non-stationary Variance (NMNV) image model [10]. The gross structure of an image is described by the non-stationary mean, while the edge and textural information is characterized by the non-stationary variance [4, 10, 15, 16].

$$\text{Effective No. of Looks } L = \frac{\mu^2}{\sigma^2} \tag{1}$$

$$\text{Noise Variation of Coefficient } C_u = \sqrt{\frac{1}{L}} \tag{2}$$

$$\text{Image Variation of Coefficient } C_i = \frac{Sd}{m} \tag{3}$$

$$\text{Threshold } C_{\max} = \sqrt{1 + \frac{2}{L}} \tag{4}$$

$$\text{Weighting Function } w_f = \frac{1 - \left(\frac{C_u}{C_i}\right)^2}{1 + C_u^2} \tag{5}$$

2.2 Proposed Algorithm—Improved Kuan Filter

It is known that various local statistics filters have been used for despeckling SAR images. The adaptive Kuan filter smoothens the homogenous areas but blurs the edges while PMAD filter although provides edge preservation features but this is only achieved at high computational complexity. Thus, the proposed algorithm amalgams the concept of PMAD filter and Kuan filter which yields better despeckling and edge preservation results at low computational complexity. The filtering procedure of the Improved Kuan filter is elaborated further. This involves moving a mask of size w over the speckled SAR image throughout; in this process of spatial filtering. Moreover, the image is characterized into three distinct regions i.e., the homogenous area where the scene feature and pixel intensity is constant, heterogeneous area where the scene and pixel intensity varies and isolated point targets where the pixel intensity abruptly changes. To elaborate further, based on the calculated values of noise and image variation parameters, the Improved Kuan algorithm processes the image as described under; where cp refers to the center pixel and i, j indicates the current pixel. At first, if the value of C_i is less than C_u, it is a homogeneous region and the pixel is processed by Kuan filter [10]. Thus the resulting transformed pixel will be calculated as under within the mask.

$$I_t(i,j) = cp(i,j)w_f + m(1 - w_f) \tag{6}$$

Secondly, if the value of C_i is greater than C_{max}, it is an isolated point and the original pixel values are preserved, i.e.

$$I_t(i,j) = cp(i,j) \tag{7}$$

Lastly, if C_i is between the range of C_u and C_{max}, it is a heterogeneous region and the pixel is processed as per the concept defined by PMAD filter. Thus, the transformed pixel in this region is computed using Eqs. (8)–(10). The PMAD filter uses gradient for the detection of the edges and is computed in 4 directions as follows:

$$\begin{aligned}
\nabla_N I_{i,j} &= I_{i-1,j} - I_{i,j} \\
\nabla_S I_{i,j} &= I_{i+1,j} - I_{i,j} \\
\nabla_E I_{i,j} &= I_{i,j+1} - I_{i,j} \\
\nabla_W I_{i,j} &= I_{i,j-1} - I_{i,j}
\end{aligned} \tag{8}$$

Using the gradients calculated in above 4 directions, the respective conduction coefficients are then estimated using Eq. (9) below.

$$c_N = \left(1 + \left(\frac{\nabla_N I}{Sd\sqrt{2}}\right)^2\right)^{-1} \tag{9a}$$

$$c_S = \left(1 + \left(\frac{\nabla_S I}{Sd\sqrt{2}}\right)^2\right)^{-1} \tag{9b}$$

$$c_E = \left(1 + \left(\frac{\nabla_E I}{Sd\sqrt{2}}\right)^2\right)^{-1} \tag{9c}$$

$$c_W = \left(1 + \left(\frac{\nabla_W I}{Sd\sqrt{2}}\right)^2\right)^{-1} \tag{9d}$$

Now, based on the calculated values of gradient and conduction function, the transformed pixel is modeled as Eq. (10).

$$I_t(i,j) = I(i,j) + \lambda \Delta t (\nabla_E I \cdot c_E + \nabla_W I \cdot c_W + \nabla_N I \cdot c_N + \nabla_S I \cdot c_S) \quad (10)$$

With the above working, it can be assimilated that the proposed Kuan algorithm as explained in Eqs. (6)–(10) is dependent jointly upon the local statistics as well as the parameters of PMAD filter. That is, the coefficient of variation and conduction function are two primary factors which have been incorporated not just to reduce speckle but also to preserve radiometric information content. The procedural working of the proposed Improved Kuan algorithm has been summarized under Algorithm 1.

Algorithm 1: Steps for Despeckling SAR Images using Improved Kuan Filter.

	Begin	
Step 1:	*Input*: Noisy image (I) of size MXN.	
Step 2:	*Input*: Window size (w).	
Step 3:	*Compute*: global parameters : Standard Deviation (σ), Mean (μ)	
	Noise Variation of Coefficient as given in Eq. (1)	
	Effective No. of Looks as given in Eq. (2)	
	Threshold as given in Eq. (4)	
Step 4:	*Process*: Image (I) spatially with window (w).	
Step 5:	*Compute*: local parameters : Standard Deviation (Sd), Mean (m)	
	Image Variation of Coefficient as given in Eq. (3)	
	Weighting Function as given in Eq. (5) within the window.	
Step 6:	If $(C_i \leq C_u)$	Refer Eq. (6)
	If $(C_i \geq C_{max})$	Refer Eq. (7)
	If $(C_u < C_i < C_{max})$	Refer Eq. (8), (9) and (10).
Step 7:	*Process*: the mask sequentially until it reaches the last pixel of the image.	
Step 8:	*Display*: the denoised image (I_t).	
	End	

Lastly, the despeckled SAR image is then processed to carry out IQA in order to assess the performance of the proposed algorithm. The main performance parameters used for the evaluation of despeckled SAR images are Peak Signal to Noise Ratio (*PSNR* in dB) and Structure Similarity Index Map (*SSIM*). *PSNR* [19] defines the quality of the image in terms of the power of the original and denoised images. More the *PSNR*, better is the filtering output. *SSIM* [20] is used to compare luminance, contrast and structure between the original and denoised images. The *SSIM* value should be closer to unity for optimal measure of similarity. Now the proposed algorithm is implemented to obtain its quality parameters so as to check its reliability and validity [9, 21, 22].

3 Results and Discussions

SAR images of a moon crater (named as test image 1 and test image 2) has been taken in the simulation to validate the performance of proposed algorithm. The simulated procedure is initiated by normalizing the input SAR image. The normalized image is then contaminated with simulated speckle noise of different variance's ranging from 0.01 to 0.03. Prior to application of proposed Kuan algorithm on the noisy SAR image; the tuning parameters like mask size ($w = 3 \times 3$), step size ($\Delta t = 0.05$) and $\lambda = 1/n_s$ (where n_s is the number of neighbors from center pixel, generally taken as 4) are optimally determined. The proposed algorithm is efficient to overcome blurring at the edges and preserving subtle details than other conventional filters. The obtained results from test image 1 with the proposed Kuan algorithm are shown in Fig. 1. Subsequently the performance is not degraded by the every 0.01 increment in speckle content with decrement in *PSNR* (in dB) and *SSIM*. Also from Fig. 1, it is evident that along with the smoothing in homogenous areas, the edges are preserved and are visually effective. The performance is determined further using IQA metrics like *PSNR* and *SSIM* which are enlisted in Table 1.

In order to benchmark the performance of the proposed algorithm, a test image 2 of moon crater has been compared with few other filters of the literature commonly being Lee [11], Frost [3], Kuan [10] and PMAD [15] filters as shown in Fig. 2. It can be observed that the Lee filter enhance the visual perception of SAR image but at the cost of over smoothening the edges whereas the Frost filter deteriorates the visual perception by blurring the SAR image. The Kuan filter is an approximation of Lee filter showing similar result as Lee that is blurring at edges but proves to be a better despeckling technique. The aforesaid local statistics filters employ primitive criteria's that averages the pixel intensity of the mask to remove speckle but proves

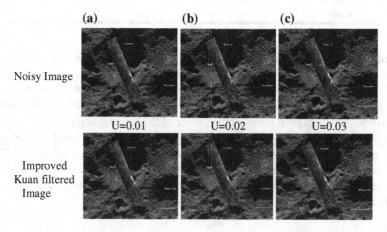

Fig. 1 Despeckled SAR images (test image 1) obtained by the proposed algorithm for different speckle magnitudes (0.01–0.03)

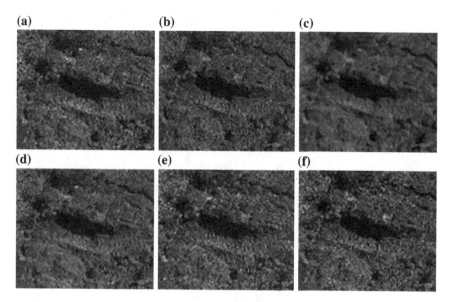

Fig. 2 **a** Speckled Image (test image 2 with speckle variance U = 0.01), Despeckled SAR images using **b** Lee filter **c** Frost filter **d** Kuan filter **e** PMAD filter **f** Improved Kuan filter

Table 1 IQA values at different speckle magnitudes for an improved Kuan algorithm

Speckle variance	PSNR (in dB)	SSIM
0.01	24.5145	0.9708
0.02	23.3691	0.9626
0.03	22.5024	0.9551

inefficient in edge detection and preservation. PMAD is an efficient filtering algorithm for edge detection and uses an edge stopping function for preserving texture and other fine details of the image but at high computational complexity. The proposed algorithm thereby overcomes the above limitations and gives better despeckling results. The Table 2 enlist the IQA parameters at various speckle magnitudes (0.01–0.03) for validating the aforesaid results.

From the above analysis, it is observed that the Improved Kuan gives better values of *PSNR* and *SSIM* as compared to other conventional filters even at higher speckle variance. Therefore, it ensures that the use of gradient and conduction function along with the Kuan filter has proved a useful tool for image smoothing and edge preservation.

Table 2 Performance comparison of the proposed Kuan algorithm with other speckle filters using IQA metrics

Noise variance	PSNR (in dB)					SSIM				
	Lee	Frost	Kuan	PMAD	Improved Kuan	Lee	Frost	Kuan	PMAD	Improved Kuan
0.01	15.7416	17.7405	15.8277	22.0996	24.5145	0.7471	0.8344	0.7824	0.9431	0.9708
0.02	15.1288	16.6990	14.6543	21.1610	23.3691	0.7402	0.8335	0.7776	0.9313	0.9626
0.03	14.4580	16.0478	13.4831	20.4193	22.5024	0.7361	0.8319	0.7721	0.9205	0.9551

4 Conclusion

Speckle is the major undesirable artifact present in SAR images which deteriorates the visual perception of the image. In this paper, the conventional Kuan filter has been improved which is validated by IQA metrics. The proposed algorithm hybridize the concept of Kuan and PMAD filter which comes forth to be a great tool for despeckling of SAR images. Also, the proposed work shows improved results when compared to other conventional filters with better restoration and edge preserving feature. Moreover, the proposed algorithm is quite simple and curtails computational complexity when compared to other advanced Bayesian methods. Therefore, the proposed algorithm defines an interesting approach which turns out to be a remarkable despeckling technique enabling speckle reduction in homogeneous areas, scene feature preservation and absence of artifacts.

References

1. Tupin, F., Maitre, H., Mangin, J.F., Nicolas, J.M., Pechersky, E.: Detection of Linear Features in Sar Images: Application to Road Network Extraction. In: IEEE Transactions On Geoscience and Remote Sensing, 36(2), 434–453, (1998).
2. Moreira, A., Prats-Iraola, P., Younis, M., Kriger, G., Hajnsek, I., Papathanassiou, K.: A Tutorial on Synthetic Aperture Radar. In: IEEE Geoscience and Remote Sensing Magazine, 1 (3), 6–35, (2013).
3. Lopes, A., Touzi, R., Nezry, E.: Adaptive Speckle Filter and Scene Heterogeneity. In: IEEE Transactions on Geoscience and Remote Sensing, 28(6), 992–1000, (1990).
4. A. Jain and V. Bhateja, "A Novel Image Denoising Algorithm for Suppressing Mixture of Speckle and Impulse Noise in Spatial Domain," Proc. of (IEEE) 3rd International Conference on Electronics & Computer Technology (ICECT-2011), Kanyakumari (India), vol. 3, pp. 207–211, April 2011.
5. A. Jain and V. Bhateja, "A Novel Detection and Removal Scheme for Denoising Images Corrupted with Gaussian Outliers," Proc. of IEEE Students Conference on Engineering and Systems (SCES-2012), Allahabad (U.P.), India, pp. 434–438, March, 2012.
6. A. Jain and V. Bhateja, "A Versatile Denoising Method for Images Contaminated with Gaussian Noise", Proc. of (ACM ICPS) CUBE International Information Technology Conference & Exhibition, Pune, India, pp. 65–68, September, 2012.
7. Gupta, A., Tripathi, A., Bhateja, V.: De-speckling of SAR Images via an Improved Anisotropic Diffusion Algorithm. In: Proc. of (Springer) International Conference on Frontiers in Intelligent Computing Theory and Application (FICTA 2012), AISC vol. 199, 747–754, Bhubaneswar, India, (2012).
8. Lopes, A., Nezry, E., Touzi, R., Laur, H.: Structure Detection and Statistical Adaptive Speckle Filtering in SAR Images. In: Int. J. Remote Sensing, 14(9), 1735–1758, (1993).
9. Gupta, A., Tripathi, A., Bhateja, V.: Despeckling of SAR images in Contourlet Domain using a New Adaptive Thresholding. In: Proc. of (IEEE) 3nd International Advance Computing Conference (IACC 2013), Ghaziabad (U.P.), India, 1257–1261, (2013).
10. Kaun, D.T., Sawchuk, A.A., Strand, T.C., Chavel, P.: Adaptive Noise Smoothing for Images with Signal-Dependent Noise. In: IEEE Transactions on Pattern Analysis and Machine Intelligence, PAMI-7(2), 165–177, (1985).

11. Lee, J.S.: Digital Image Enhancement and Noise Filtering by Use of Local Statistics. In: IEEE Transactions on Pattern Analysis and Machine Intelligence, PAMI-2(2), 165–168, (1980).
12. Bhateja, V., Singh, G., Srivastava, A., Singh, J.: Despeckling of Ultrasound Images using Non-Linear Conductance Function. In: Proc. (IEEE) International Conference of Signal Processing and Integrated Networks (SPIN-2014), 722–726, Noida (U.P), India, (2014).
13. Srivastava, A., Bhateja, V., Tiwari, H.: Modified Anisotropic Diffusion Filtering Algorithm for MRI. In: Proc. (IEEE) 2nd International Conference on Computing for Sustainable Global Development (INDIACom-2015), 1885–1890, New Delhi, India, (2015).
14. Bhateja, V., A., Tripathi, A., Gupta, A., Lay-Ekuakille, A.: Speckle Suppression in SAR images Employing Modified Anisotropic Diffusion Filtering in Wavelet Domain for Environment Monitoring . In: Elsevier Measurement Journal, 74, 246–254, (2015).
15. Perona, P., Malik, J.: Scale Space and Edge detection using Anisotropic Diffusion. In: IEEE Transactions on Pattern Analysis and Machine Intelligence, 12(7), 629–639, (1990).
16. Yu, Y., Acton, S. T.: Speckle Reducing Anisotropic Diffusion. In: IEEE Transactions on Image Processing, 11(11), 1260–1270, (2002).
17. Bhateja, V., Singh, G., Srivastava, A., Singh, J.: Speckle Reduction in Ultrasound Images using an Improved Conductance Function based on Anisotropic Diffusion. In: Proc. (IEEE) 2014 International Conference on Computing for Sustainable Global Development, 619–624, (2014).
18. Bhateja, V., Singh, G., Srivastava, A.: A Novel Weighted Diffusion Filtering Approach for Speckle Suppression in Ultrasound Images. In: Proc. of (Springer) International Conference on Frontiers in Intelligent Computing Theory and Application (FICTA 2013), 247, 459–466, Bhubaneswar, India, (2013).
19. A. Jain and V. Bhateja, "A Full-Reference Image Quality Metric for Objective Evaluation in Spatial Domain," Proc. of IEEE International Conference on Communication and Industrial Application (ICCIA-2011), Kolkatta (W.B.), India, no. 22, pp. 91–95, December, 2011.
20. P. Gupta, P. Srivastava, S. Bharadwaj and V. Bhateja,"A New Model for Performance Evaluation of Denoising Algorithms based on Image Quality Assessment," Proc. of (ACM ICPS) CUBE International Information Technology Conference & Exhibition, Pune, India, pp. 5–10, September, 2012.
21. Singh, S., Jain, A., Bhateja, V.: A Comparative Evaluation of Various Despeckling Algorithms for Medical Images. In: Proc. Of (ACMICPS) CUBE International Information Technology Conference & Exhibition, 32–37, Pune, India, (2012).
22. Bhateja, V., Tripathi, A., Gupta, A.: An Improved Local Statistics Filter for Denoising of SAR Images. In: Proc. of (Springer) 2nd International Symposium on Intelligent Informatics (ISI'13), 235, 23–29, Mysore, India, (2013).

GSM Based Automated Detection Model for Improvised Explosive Devices

Rajat Sharma, Vikrant Bhateja and S.C. Satapathy

Abstract The most destructive crimes that are been committed by terrorists, tops the chart in perpetrating the world's worst crimes. Nowadays, explosives are planted at crowded places to create a loss to multitude of people lives. Special bomb detection squad is required to search and diffuse the bomb herein termed as Improvised Explosive Devices (IED's). Yet, there is dearth of independent techniques where detection of IEDs could be carried out in an automated fashion without the usage of man-force. This paper presents an automated approach for simplified detection of IED without the physical presence of bomb diffusion squad. The detection model proposed in this work consists of infrared sensors, AVR programmer, ATMEGA micro-controller and GSM module. The operational principle of the proposed model is based on the idea that an individual/machine carrying IED is detected by the sensors and at the same time this information is forwarded as a text message to the nearest workstations. The demonstration of prototype model reports the effectiveness of the detection approach with minimal complexity.

Keywords Improvised explosive device (IED) · Infrared sensors · GSM module

R. Sharma (✉)
SOPRA STERIA Pvt. Ltd., Noida, U.P., India
e-mail: march10rajat@gmail.com

V. Bhateja
Department of Electronics and Communication Engineering,
Shri Ramswaroop Memorial Group of Professional Colleges
(SRMGPC), Lucknow 227105, U.P., India
e-mail: bhateja.vikrant@gmail.com

S.C. Satapathy
Department of Computer Science and Engineering, ANITS,
Visakhapatnam, A.P., India
e-mail: sureshsatapathy@gmail.com

© Springer India 2016
S.C. Satapathy et al. (eds.), *Information Systems Design and Intelligent Applications*, Advances in Intelligent Systems and Computing 434,
DOI 10.1007/978-81-322-2752-6_66

1 Introduction

The world is growing on the great heights of advancements and in this saga of developments; the explosives have also witnessed enhancements to fulfill the destructive requirements rather than constructive ones. These devices are termed as Improvised Explosive Devices (IED). Thus, IED's are needed to be detected in the fastest possible way, so that "It's not too late". Hence, there exists two types of detection techniques for IEDs so far stated as: bulk detection and trace detection techniques [1]. The major limitation of the existing explosive detection devices lies in the fact that they only serves the purpose of effective detection if the location of the bomb implantation is known via some third party. These processes are cumbersome and time consuming, at times leading to fatal and massive destruction. Remedial solutions are rare, which could confirm pre-detection of the explosive material (being carried by an individual) or planted/installed at some location. There is a need to devise sophisticated explosive detection techniques so that such hazardous objects could be effectively tracked within minimum span of time. Literature in the past is available with diverse techniques developed for the purpose of explosive detection. Roy et al. [2] presented an Associated Particle Technique (APT) system in single-sided geometry comprising of D-T neutron source and bismuth Germanate (BGO) detectors fixed on a portable module. This system was well suited to detect benign samples and explosive simulants under laboratory condition. Silva et al. in their work [3] proposed design and fabrication process for an electrochemical "electronic nose" type sensor. This metal oxide based system focuses on the detection of ammonia as a possible sign of ammonium nitrate based explosives. Bauer et al. [4] have shown that powders themselves scatter the laser radiation used in the excitation of the spectra, making other components more difficult to discern. The preliminary work done by the authors with Laser Induced Breakdown Spectroscopy (LIBS) showed that metal powders are easily detected and identified and that fuel compounds in flash powder mixtures are easily classified with Principal Component Analysis [5–8] into those containing oxygen and chlorine or those containing oxygen and nitrogen. In another work, Fransisco [9] proposed to include dispersing a mixture containing a fluorescent material uniformly over a ground cover, illuminating the ground cover with wavelengths of visible light or Ultraviolet (UV) light causing the fluorescent material to fluoresce in a visible light spectrum, and have detected where the mixture has been disturbed on the ground cover by visually observing inconsistencies in the fluorescent material on the ground cover that is fluorescing to indicate a location of the improvised explosive device. Phelan et al. [10] devised a Stepped-Frequency Radar (SFR), vehicle-mounted forward-looking ground-penetrating radar; designed for high-resolution detection of buried landmines and improvised explosive devices. The work of Kamarul and Shawal [11] is based on Infrared Thermography (IRT) technology. Huri et al. [12] presented a study of four types of pyrotechnics explosives packed in pipe bombs were utilized via sampling exercise. They have determined anionic inorganic constituents of improvised explosives by using ion chromatography. In the same sequence, Peters et al. [13] developed Microfluidic Paper based Analytical Devices (μPADs) for rapid, on-site detection of improvised explosives. Daniel [14] proposed a sensor design for the detection of triacetone triperoxide (TATP). Their design entailed a thermodynamic gas sensor that measures the heat of decomposition between trace TATP vapor and a metal oxide catalyst film. The techniques reviewed and discussed so far are functional only if the location of the implanted explosives is known by the detection. But there are no such automated and sensitive devices which could be placed in overcrowded/suspected to raise an alarm by sensing such hazardous material/object. The work presented in this paper deploys a circuit model consisting of sensors, AVR programmers, multi-controller and GSM module to sense, detect and respond at once via message circulation to the programmed workstations for an earliest possible action to be taken. The remaining part of the paper is structured as follows: Sect. 2 describes the materials and methods covering the proposed methodology and its implementation in presented in Sect. 3. Section 4 concludes the work and outlines the future scope.

2 Materials and Methods

The proposed work demonstrates a prototype working model that could be used to detect and prevent the explosion of IED's with following experimental setup and working methodology generating results through implementation phase. The proposed model is set-up in such a way that it could sense the presence of explosive device by using sensors assembled at its mouth, then the indication message will be forwarded to microcontroller showing that a device has been sensed by the sensor, now the AVR Programmer is also added in the model as it provides the required programming code by which microcontroller will establish its functioning and finally the signal is sent to attached GSM module about the presence of explosive device, thereafter the alarming message is forwarded to included mobile numbers for further actions to be taken.

2.1 Experimental Setup and Hardware Assembly

The proposed model has been assembled by using various electronic devices as mentioned in Table 1. The prototype model has been included with two IR sensors that will act as two nodes at two different sides for the detection purposes.

If signals are sensed by node 1 at one direction, then will inform the microcontroller at once or if the explosive device is been sensed by second sensor then it will also send the immediate signal to the microcontroller. Further, it is added up with one AVR programmer that facilitates the purpose of whole program coding which directs the microcontroller that what and when certain defined operations are to be followed to generate the desired output. Then, main part of the model i.e.

Table 1 Showing the names and quantity of the components used

Name of component	Quantity of the components used
IR sensor	2
AVR programmer	1
ATMega16 microcontroller	1
GSM module	1
12 V power supply battery	1
USB cable	1
Connecting wires	–

ATMega 16 microcontroller with quantity equals to one, is been used to give all the commands and signals wherever and whenever they are required by different components used in the model. Also, one GSM module has been circuited with the remaining set up to initiate the purpose of message sending from the project model to the registered workstations.

2.2 Methodology

The prototype model will come into the active mode when it will be connected with the power supply of 12 V adapter as the GSM module will start its function on getting the power supply. At the same time, the AVR Programmer is to be connected with the USB cable in the provided USB slot so that the AVR Programmer, ATMega16 microcontroller and IR sensor could come into active mode. The circuit diagram shown in the Fig. 1, specifies that two sensors are present each with three terminal nodes as-: OUT–To send the output signal for further processing; GND–Ground terminal and V_{CC}–Positive voltage supply terminal. Now, the above mentioned three terminals of both the sensors are in connection with the terminals of ATMega 16 microcontroller as $-V_{CC}$ terminal of the sensor is connected with +ve volt supply terminal and GND terminal of the sensor is connected with –ve volt supply terminal whereas the OUT terminal of the sensor is connected with Port A of the microcontroller because Port A in the whole microcontroller works as the Analog to digital converter for signal conversion. Now, the AVR programmer is been used, to dump the hex code file so that microcontroller could get its functioning as per the coding is embedded. Now, another component which is been connected with the whole system is GSM Sim 900 module which facilitates the function of sending the message to the mobile numbers registered already in the code. The transmitter and receiver terminals of the GSM module come in connection with the Port D terminals that have Rx (receiver) and transmitter (Tx) terminals where the Rx terminal of the GSM module will be connected to the Tx terminal of the microcontroller and similarly, Rx terminal of the microcontroller will be connected to the Tx terminal of the GSM module. The power supply to the

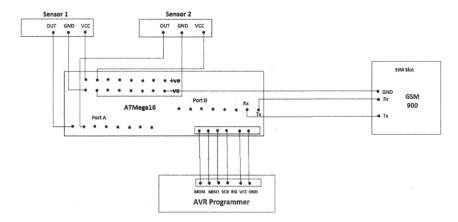

Fig. 1 Circuit diagram of the whole project model

whole project model is given by a 12 V adapter which provides a combination of the potential difference and current as-12 V::1 A. The list of the software tools that have been used in the project are- AVR Studio 4, AVR Dude GUI. The IR sensor after coming in the active mode is constantly sending the infrared signal from its transmitter and if the improvised explosive device is detected then the Infrared rays will be received by the receiver of the IR sensor and thus the signal of sensing an explosive device will be sent to the microcontroller. Figure 1 is showing image of the circuit diagram with connections that have been stated in earlier part.

The IR sensor is having its OUT port connected at the PORT A of the micro-controller so that the analog signal is converted in the digital signal and can be forwarded in the whole circuit as PORT A of the microcontroller is having ADC (analog to digital Converter). Now, the AVR Programmer which is having the dumped hex file consisting of the program which is to be executed on the required signal is activated and the microcontroller executes in the same manner as it's programmed. The program is dumped into the AVR programmer by using the AVR Program loader, once the program is compiled and is executed successfully by the software tool-AVR Studio 4. Now, the ATMega 16 microcontroller sends the signal by its transmitter (Tx) presented on the PORT D to the receiver (Rx) terminal of the GSM module which are connected together via jump in wires. The GSM module on getting the signal from microcontroller will activate its Sendmessage () function as per the programming done to send the message by its antenna to the satellite, indicating that an alarming message which is stored in the string format in the program code should be sent to the registered numbers of the nearest workstations or the control rooms to the detected location of the improvised explosive device,

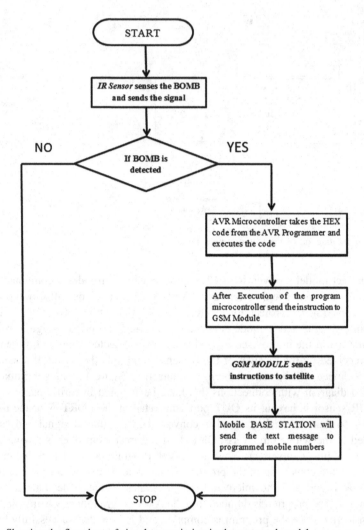

Fig. 2 Showing the flowchart of signal transmission in the proposed model

that are been preinstalled in the program code via base stations. Thus, in the final phase, workstations will at once get the message and will be doing the further precautionary measures with earliest effort to stop the destruction of the thousands of innocent lives. Now, Fig. 2 will show how the flow of signals will travel if an improvised explosive device is detected as well as will show that if no activity is taking place then the system remains idle.

3 Hardware Implementation

The proposed model could be implemented as in the architecture shown in the following figures. The whole circuitry is maintained on a piece of cardboard having sensors fixed at two ends along with AVR programmer, ATMega 16 multi-controller and GSM Module that are connected with each other by usage of jumping wires, having the power supply from 12 V adapter. The photograph of the live working prototype model is shown Fig. 3.

Figure 4, shows the schematic diagram of the whole concept (discussed above in methodology section) which shows the live working of the proposed model when it will be in the working application phase at some location.

The alert messages will be forwarded to the mobile devices of the people as is shown below in Fig. 5 having the screenshot of the mobile showing the message

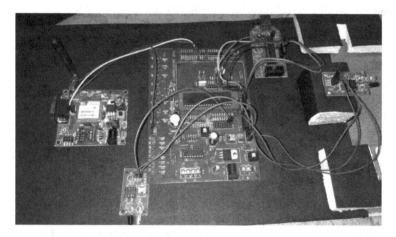

Fig. 3 Snapshot showing the full assembled architecture of the proposed model

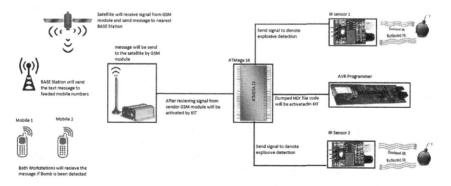

Fig. 4 Operational diagram of the proposed model

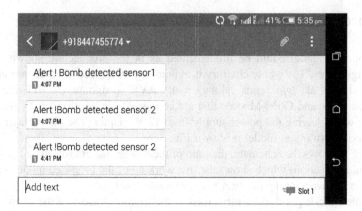

Fig. 5 Screenshot showing the messages that are received from GSM module of project

received on sensing any improvised explosive device passed by that region or vicinity.

The messages forwarded can be seen below:

Alert! Explosive device has been detected by sensor 1 (in case sensor 1 detects the bomb)
Alert! Explosive device has been detected by sensor 2 (in case sensor 2 detects the bomb).

4 Conclusion and Future Scope

The proposed work serves to provide for detection of an improvised explosive device. For this purpose, the device has to be installed in such a manner so as to sense/detect the explosive device; if such explosive content is present in the near vicinity of the device. The process is microcontroller programmed to direct the detection signals to the GSM module. This module is deployed for sending an alarming message to the nearest work stations available (from the detected location). Further for future enhancements, the device could include multiple sensors to detect different hazardous elements like metal, explosive chemicals (TNT, liquid explosives) etc.

References

1. Details about the present scenario of explosive device detection http://www.nap.edu/read/10998/chapter/7.
2. Roy, T., Yogesh K., Mayank S., Agrawal A., Bajpai S., Patel T., and Sinha A.: Associated particle technique in single sided geometry for detection of explosives. In: Applied Physics Letters 106, no. 12: 124103 (2015).

3. Silva, S. M., Jorge D. G., Cesar A. Hernandez, and Johann F. Osma.: Design and fabrication of a sensor for explosives as a first step to an IED detection device. In: 9th Ibero- American Congress on Sensors (IBERSENSOR), 2014 IEEE, pp. 1–4., (2014).
4. Bauer, A. JR, Farrington P. M., Kellen S., and Miziolek A. W.: Laser-induced breakdown spectroscopy and spectral analysis of improvised explosive materials. In: SPIE Sensing Technology Applications, International Society for Optics and Photonics pp. 91010 M– 91010 M, (2014).
5. Krishn A., Bhateja V., Himanshi and Sahu A.: Medical Image Fusion using Combination of PCA and Wavelet Analysis. In: Proc. 3rd (IEEE) International Conference on Advances in Computing, Communications and Informatics (ICACCI-2014), Gr. Noida (U.P.), India, pp. 986–991, (2014).
6. Krishn A., Bhateja V., Himanshi and Sahu A.: PCA based Medical Image Fusion in Ridgelet Domain, In: Proc. (Springer) 3rd International Conference on Frontiers in Intelligent Computing Theory and Applications (FICTA-2014), Bhubaneswar, India, vol. 328, pp. 475–482, (2014).
7. Himanshi, Bhateja V., Krishn A., and Sahu A.: An Improved Medical Image Fusion Approach Using PCA and Complex Wavelets. In: Proc. (IEEE) International Conference on Medical Imaging, m-Health & Emerging Communication Systems (MEDCom-2014), Gr. Noida (U.P.), pp. 442–447, (2014).
8. Himanshi, Bhateja V., Krishn A., and Sahu A, : Medical Image Fusion in Curvelet Domain Employing PCA and Maximum Selection Rule, In: Proc. (Springer) 2nd International Conference on Computers and Communication Technologies (IC3T-2015), Hyderabad, India, Vol. 1, pp. 1–9, (2015).
9. Velasco, F.: Method and system to detect improvised explosive devices. In: U.S. Patent 8,813,627, (August 26, 2014).
10. Phelan, B. R., Gallagher K. A., Sherbondy K. D., Ranney K. I., and Narayanan R. M..: Development and performance of an ultrawideband stepped-frequency radar for landmine and improvised explosive device (IED) detection. In: Sensing and Imaging 15, no. 1 pp. 1–12 (2014).
11. Ghazali, Hawari K., and Jadin Mohd S.: Detection Improvised Explosive Device (IED) Emplacement Using Infrared Image. In: Proceedings of the 2014 UKSim-AMSS, 16th International Conference on Computer Modelling and Simulation, pp. 307–310 (2014).
12. Huri, M. Afiq M., Ahmad Umi K., and Mustafa O..: Determination of anionic profiling of improvised explosive devices from different types of low explosive material. In: Malaysian Journal of Fundamental and Applied Sciences 10, no. 3 (2014).
13. Peters, Kelley L., Corbin I., Lindsay M. K., Kyle Z, Blanes L., and Bruce R. M.: Simultaneous colorimetric detection of improvised explosive compounds using microfluidic paper-based analytical devices (μPADs). In: Analytical Methods 7, pp. 63–70 (2015).
14. Mallin, D.: Increasing the selectivity and sensitivity of gas sensors for the detection of explosives. In: PhD diss., University Of Rhode Island, (2015).

Review on Video Watermarking Techniques in Spatial and Transform Domain

Garima Gupta, V.K. Gupta and Mahesh Chandra

Abstract This paper presents a technical review on various watermarking techniques applied to different videos. In this paper watermarking techniques are categorized according to two domains: spatial domain and transform domain. These techniques are studied and research progress in both the techniques in the field of watermarking has been shown in this paper. A summary of various watermarking technique in both domain are also presented in tabular form which provides an overview about the recent research work going on about video watermarking.

Keywords H.264/AVC video standard · Spatial domain watermarking · DCT · DWT

1 Introduction

Now a day's most of the data can be easily copy-pasted. So it is a challenge to maintain the ownership of that data because of easy editing of the data. Data can be text, audio, image or video. Digital watermarking may be a solution for such kind of data piracy. It can work as a copyright protection tool. Process of hiding a watermark is defined as watermarking. Here the watermark message is related to owner of that data in a digital form such as an image, a song or a video to maintain the ownership of that data. Hidden signal can either be audio or image. Watermarking method has its own importance as it is a method for copyright protection. Importance of video watermarking can be understood by the example

G. Gupta (✉) · V.K. Gupta
ECE Department, IPEC, Ghaziabad, India
e-mail: garima1204@gmail.com

V.K. Gupta
e-mail: guptavk76@gmail.com

M. Chandra
ECE Department, BIT, Mesra, India
e-mail: shrotriya69@rediffmail.com

© Springer India 2016
S.C. Satapathy et al. (eds.), *Information Systems Design and Intelligent Applications*, Advances in Intelligent Systems and Computing 434,
DOI 10.1007/978-81-322-2752-6_67

683

that owner 'A' of some video uploads his video on YouTube with advertisement. After some time video of A becomes popular. Then 'A' inserts his watermark in the video. He registers his watermark with the legal authority and stores key and secret information at its own to a safe place. However 'B' obtains illegal copy of video and uploads it with his name and earns revenue from it. When 'A' came to know about the fact then he goes to the legal authority shows his embedded secret information that is the proof of authentication and proves that he is the actual owner of the video. Then legal authority communicates with YouTube and tells the truth and YouTube redirects the revenue to person 'A'.

Another application of watermarking may be in mobile phone for providing security as suggested by Honggang Wang et al. between users and mobile media cloud [1]. Watermarking system will have an embedder for hiding data and a decoder to extract it back. Extraction can be either blind or non-blind. A good amount of research work has already been done in digital watermarking like audio and image. There is a lot of scope left for video watermarking as it has both audio and visual features and to decide which feature will be best for hiding data requires a good amount of review and research. Basic concept of video watermarking has been shown in Fig. 1.

Various watermarking techniques have been proposed by a number of researchers in the last few years. Anu Pramila et al. [2] have presented a method for using watermarking technique in print media by embedding a watermark where embedded watermark can be detected blindly with camera phone. Tiziano Bianchi et al. [3] discussed issues for secure watermarking in multimedia like secure server side embedding, secure and safe client side embedding. They proposed that system scalability problem can also be worked out to make the watermarking system secure and robust. YiqiTew et al. [4] has given an overview about various hiding methods in H.264/AVC videos, i.e. the video in compressed form. Lot of work has been done on audio and image watermarking separately. However, few analyses and comparison has been done on the methods of video watermarking based on its domain. This paper presents the conventional watermark to understand the previous research work and to find out the possibility of future work. Now a day's the transfer or

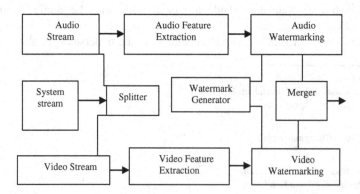

Fig. 1 Basic concept of video watermarking

frequency domain watermarking methods are more popular as compared to that of the spatial domain as transfer domain methods provide more robustness for watermarking. Most popular method in transform domain is DCT-based method.

2 Techniques of Watermarking

Explanation of different watermarking techniques has been given in this section.

2.1 Spatial Domain Techniques

Here, watermark signal is embedded to the message video directly. It is done by changing the pixel values of original video. A number of spatial domain watermarking methods are explained by researchers. Some of these methods are least significant bit coding technique, predictive coding technique, spread spectrum coding technique and patchwork watermarking technique.

LSB coding technique is first technique applied to image and audio watermarking. Here the sequence is watermarked by embedding watermarking bits. These bits are embedded in accordance with a key known as secret key. Also, it acts like a safe and secure password for the owner. The disadvantage with LSB coding is that if only LSB coding is applied then the retrieval of the watermark cannot be done properly. It also contains a noise component.

Therefore, predictive coding scheme is used in spatial domain for further improvement. In this technique correlation between adjacent pixels are used. Watermark is embedded in a set of pixels. Alternate pixels are replaced by the difference between the adjacent pixels. Predictive coding scheme shows better robustness as compared to LSB coding.

Another method that is widely used is spread spectrum coding technique. In this method, messages are encoded with sequences of symbols. Symbols are represented by a signal referred to as chip. Typically chips are pseudo-random sequence of 0's and 1's. For inserting watermark, perceptually significant coefficients are more reliable as observed by Cox et al. [5].

Last one that is generally used in spatial domain is patchwork based method. In patchwork watermarking method, the image is divided to find the two subsets of the image. Here brightness of one subset is incremented by some factor. Also brightness of another subset is decremented by the same factor. This type of technique is mostly used for audio watermarking and can also be combined with DCT resulting in transform domain.

Metaliya Viral et al. [6] embeds text in the frames of video. Venugopala et al. [7] embeds image in video by dividing image into bit plane and to embed that according to scene change. A tabular review on different video watermarking techniques in spatial domain are shown in Table 1.

Table 1 Review on different video watermarking techniques in spatial domain

S. no	Spatial domain techniques			
	Name of paper	Name of author	Year of publication	Description
1.	Robust digital video watermarking using reversible data hiding and visual cryptography [8]	Manoj Kumar, Arnold Hensman	2013	Both host video and information to be embedded is divided into different frames in such a way that approximately whole video has uniform distribution of watermark to be embedded. This approach provides high robustness against common signal processing attacks like frame dropping, frame averaging, contrast and color enhancement, cropping, filtering attacks, adding some noise in the frames, change of resolution etc.
2.	Reversible data hiding in videos using low distortion transform [9]	M. Jeni, S. Srinivasan	2013	In this method authors extracted feature values first from the video. As feature value is made of pixels so author use spatial domain watermarking and embed secret message in feature value
3.	2D—Spread spectrum watermark framework for multimedia copyright protection [10]	Vladimir BÁNOCI, Martin BRODA, Gabriel BUGAR, Dusan LEVICKÝ	2014	This is a 2D modulation method in time domain in which orthogonal properties of PN sequences are used. Author embeds watermark in 2 levels. First watermark is embedded in frames by PN sequence spreading and second modulation is performed with in the frames

(continued)

Table 1 (continued)

S. no	Spatial domain techniques			
	Name of paper	Name of author	Year of publication	Description
4.	Video watermarking by adjusting the pixel values and using scene change detection [7]	Venugopala P S, H. Sarojadevi, Niranjan N. Chiplunkar	2014	Authors embed 8 bit-plane images, obtained from single gray scale watermark image, into different scenes of a video sequence. Method is based on blind watermarking for uncompressed video. This technique is robust against attacks such as frame dropping, temporal shifts and addition of noise

2.2 Transform Domain Techniques

Watermark embedding after conversion of video from time domain to frequency domain results in transform domain techniques. There are a number of techniques in transform domain while DFT and DCT are being explained here. DFT is applied to complex numbers whereas DCT uses just real numbers. DCT and DFT are equivalent for real input data with even symmetry.

2.2.1 Discrete Cosine Transform (DCT) Based Technique

Energy compaction property of DCT makes it a useful technique for watermarking. Thus the complete image is divided into several distinct frequency bands. Therefore, embedding the watermark in the desired area of the image is now very easy. In DCT domain, a large part of energy is concentrated in low frequencies band. Also, the human eye perception for low frequencies are better as compared to high frequencies and the chances of the watermark being perceptible is high in low frequencies. Where as high frequencies are prone to attacks such as compression and scaling. So, a tradeoff is required. Sonjoy Deb Roy et al. [11] proposed a watermarking method for videos in DCT domain in real time implementation on FPGA. As DCT is also used during image compression and now a day's most widely used compression standard is H.264/AVC. Now a days, lot of research is going on in this field as it is an attractive way of hiding information because it provides confidentiality of the content. DawenXu et al. [12] proposed a method in which data hiding is done in encrypted H.264/AVC by substituting code words. YiqiTew et al. [4] also proposed the method for embedding data in compressed domain.

2.2.2 Discrete Wavelet Transform (DWT) Based Technique

Initially all signals are generally represented in time domain. But time domain representation cannot deliver complete information because it does not show different frequencies present in the signal. Frequency domain provides the details of different frequency components in the signal which are very useful in some medical applications. The frequency spectrum of a signal is basically the spectral components of that signal. But, frequency domain does not provide information about the time when these frequencies exist. The solution of both the problems is wavelet transform. Time-frequency representation of the signal can be achieved by wavelet transform at the same time. Wavelet transform can provide better information as compared to Fourier transform and short-time-Fourier-transform. In STFT, a longer time signal is divided into shorter segments of equal length. Then Fourier transform is separately applied on each shorter segment. This reveals the Fourier spectrum on each shorter segment. One then usually plots the changing spectra as a function of time. Tamanna et al. [13] proposed a method for video watermarking based on 3-level DWT. Table 2 shows summary of various watermarking technique papers in the transform domain as well as details of some recent work in video watermarking.

There are three methods for watermark embedding. One is directly embedding watermark in the original video without any modification. This method will simply give rise to spatial domain watermarking. Other method can be embedding watermark at the time of compression. This method of embedding is much more robust as compared to previous method and will be considered in transform domain. As during compression we use DCT in case of image, so by changing the coefficient values at that time of compression embedding will be done. In this method watermark detection is done after decompression. So every user can use it at receiving end, if there is no tempering involved during transmission and identity of owner will be maintained. Third method of embedding watermark is to embed the watermark after compression. This type of embedding is not easily detectable as well as various attacks will not degrade the video which is the main advantage of this method. But one drawback of this method is that watermark extraction is done before de-compression, so it is difficult to be used in day to day life for transferring video on internet. Due to this, at the receiving end it will be more time consuming and every receiver should have a de-watermarking system. If we have a watermarking system in which de-compression is done after extraction of video, then such method will be more suitable in terms of robustness and may perform better against various intentional and unintentional attacks.

Frame dropping, averaging, cropping, median filtering and sometimes attack by unauthorized user are intentional attacks while addition of noise and compression are unintentional attacks. Unintentional attacks will occur generally at the time of transmission of video.

Table 2 Review on different video watermarking techniques in transform domain

S. no	Transform domain techniques			
	Name of paper	Name of author	Year of publication	Description
1.	Hardware implementation of digital watermarking system for video authentication [14]	Sonjoy Deb Roy, Xin Li, Yonatan Shoshan, Alexander Fish and OrlyYadid-Pecht	2013	Authors presented a hardware implementation of digital watermarking system that can insert invisible, semi-fragile watermark information into compressed video streams in real time. They used FPGA for implementation
2.	An overview of information hiding in H.264/AVC compressed video [4]	YiqiTew, KokSheik Wong	2014	Author has provided an overview on H.264 video compression standard. Also various hiding venues in compressed domain are summarized
3.	Data Hiding in Encrypted H.264/AVC Video Streams by Codeword Substitution [12]	DawenXu, Rangding Wang, and Yun Q. Shi	2014	In this paper authors embed watermark after compressing the video by bit substitution method, provided the video file size is strictly preserved. Since data hiding is completed entirely in the encrypted domain, this method preserves the confidentiality of the content completely
4.	Non-blind structure-preserving substitution watermarking [15]	Thomas Stütz, Florent Autrusseau, and Andreas Uhl	2014	In this method embedding is done after H.264 compression by simple bit substitution. It offers significantly increased marking space and high robustness to re-compression

(continued)

Table 2 (continued)

S. no	Transform domain techniques			
	Name of paper	Name of author	Year of publication	Description
5.	Digital watermarking in video for copy right protection	Anita Jadhav, Megha Kolhekar	2014	In this method original video is divided into RGB frames and then into YCbCr frames. Then watermark is embedded in Y component. Author implemented dynamic 3D-DCT, compared the results with static DCT. It was observed that, although, the MSE and the frame error rate, values do not change much for both but there is observable change in visual quality

3 Conclusion

In this paper a number of video watermarking techniques have been studied. In some techniques watermark is inserted in the raw video before compression either in spatial or transform domain. In some methods watermarks has been inserted after or during compression. H.264 is the standard used for video now a day. After compression if data is hidden then it can be extracted by two methods- either before decompression or after decompression. So it is a good choice to hide the data in the video after compression as it may provide better security for mobile. At the same time, method should work well for both intentional and unintentional attacks. There is a lot of research scope in this field which may result into a robust video watermarking technique.

References

1. Honggang Wang, Shaoen Wu, Min Chen, Wei Wang, "Security protection between users and mobile media cloud," IEEE communication magazine, Volume: 52, no. 3, pp. 73–79 March 2014.
2. Anu Pramila, Anja Keskinarkaus, Tapio Seppanen, "Toward an interactive poster using digital watermarking and a mobile phone camera," *Signal, Image and Video Processing,* Volume 6, Issue 2, pp. 211–222, 2012.
3. Tiziano Bianchi, Alessandro Piva, "Secure watermarking for multimedia content protection," IEEE signal processing magazine, Vol. 30, no. 2, pp. 87–96, March 2013.

4. Yiqi Tew, KokSheik Wong, "An Overview of Information Hiding in H.264/AVC Compressed Video," IEEE transactions on circuits and systems for video technology, vol. 24, no. 2, pp. 305–319, February 2014.
5. I. Cox, M. Miller, J. Bloom, J. Fridrich, and T. Kalker, Digital Watermarking and Steganography, 2nd ed. San Francisco, CA, USA: Morgan Kaufmann Publishers Inc., 2008.
6. Metaliya Viral G, Deepak Kumar Jain, Sardhara Ravin, "A Real Time Approach for Secure Text Transmission Using Video Cryptography," Proceeding of IEEE Fourth International Conference on Communication Systems and Network Technologies, pp. 635–638, 2014.
7. Venugopala P S, H. Sarojadevi, Niranjan N. Chiplunkar, Vani Bhat, "Video Watermarking by Adjusting the Pixel Values and Using Scene Change Detection," Proceeding of IEEE fifth International Conference on Signals and Image Processing, pp. 259–264, 2014.
8. Manoj Kumar, Arnold Hensman, "Robust Digital video watermarking using reversible data hiding and visual cryptography," Proceedings of 24th IET Irish Conference on Signals and Systems (ISSC 2013), pp: 1–6, 2013.
9. M. Jeni, S. Srinivasan, "Reversible Data Hiding in Videos Using Low Distortion Transform" Proceeding of IEEE conference on Information Communication and Embedded Systems (ICICES), pp. 121–124, 2013.
10. Vladimír BÁNOCI, Martin BRODA, Gabriel BUGÁR, Dušan LEVICKÝ, "2D - Spread Spectrum Watermark Framework for Multimedia Copyright Protection," Proceeding of 24th IEEE Conference on Radioelektronika (RADIOELEKTRONIKA), pp. 1–4, 2014.
11. Sonjoy Deb Roy, Xin Li, Yonatan Shoshan, Alexander Fish, "Hardware Implementation of digital watermarking system for video authentication," IEEE transactions on circuits and systems for video technology, vol 23, no. 2, pp. 289–301, Feb. 2013.
12. Dawen Xu, Rangding Wang, and Yun Q. Shi, "Data Hiding in Encrypted H.264/AVC Video Streams by Codeword Substitution," IEEE transactions on information forensics and security, vol. 9, no. 4, pp. 596–606, April 2014.
13. Tamanna Tabassum, S.M. Mohidul Islam, "A Digital Video Watermarking Technique Based onIdentical Frame Extraction in 3-Level DWT," Proceedings of 15th International Conference on Computer and Information Technology (ICCIT), pp 101–106, 2012.
14. A. K. Verma, Mayank Singhal, C. Patvardhan, "Robust Temporal Video Watermarking Using YCbCr Color Space in Wavelet Domain," Proceedings of IEEE 3rd International conference on Advance Computing Conference (IACC), pp. 1195–1200, 2013.
15. Thomos Stutz, Florent Autrusseau, Andreas Uhl, "Non blind structure preserving substitution watermarking of H.264/CAVLC inter frames," Proceedings of IEEE on multimedia, vol. 16, no. 5, pp. 1337–1349, August 2014.

Musical Noise Reduction Capability of Various Speech Enhancement Algorithms

Prateek Saxena, V.K. Gupta and Mahesh Chandra

Abstract This paper presents a comparative analysis of spectral subtraction and Weiner denoising techniques for musical noise reduction. The iterative spectral subtraction method provides least musical noise generation applied in different noisy environments. The method of musical noise production is traced by observing the change in the kurtosis ratio of noise spectrum using different denoising techniques for different noisy signal. A MATLAB simulation is performed for four different noisy environments car noise, babble noise, operation room noise and machine gun noise at −10, −5, 0, 5 and 10 dB input SNR levels. It is observed that wiener based methods provide more improvement in SNR as compared to spectral subtraction based methods. But at the same time musical noise generation is more in wiener based methods. The wiener based method HRNR gives a maximum 35.77 dB improvement in SNR for car noise at −10 dB input SNR level. Iterative spectral subtraction gives the minimum value of kurtosis ratio for all noises at all input SNR level.

Keywords Speech enhancement · Musical noise · Spectral subtraction · Iterative spectral subtraction · Geometrical approach · DD approach · TSNR · HRNR

1 Introduction

The speech quality is deteriorated under adverse noise conditions in hearing aids and mobile phones. Therefore noise reduction requires more attention by various researchers. A commonly and efficient method used for noise reduction is spectral

P. Saxena (✉)
Department of ECE, RVIT Engineering College, Bijnor, India
e-mail: prateeksaxenas@rediffmail.com

V.K. Gupta
Department of ECE, IEC, Ghaziabad, India
e-mail: guptavk76@gmail.com

M. Chandra
Department of ECE, BIT, Mesra, Ranchi, India
e-mail: shrotriya69@rediffmail.com

© Springer India 2016
S.C. Satapathy et al. (eds.), *Information Systems Design and Intelligent Applications*, Advances in Intelligent Systems and Computing 434,
DOI 10.1007/978-81-322-2752-6_68

subtraction method. This method gives reduction in both noise performance as well as computational complexity [1–5]. Whereas Weiner based denoising techniques like TSNR (two step noise reductions) and HRNR (harmonic regeneration noise reduction) effectively removes the reverberation effect. Classic short-time noise reduction techniques removes noise as well as introduces harmonic distortion. For example, TSNR enhances speech but introduces harmonic distortion because of the unreliability of estimators for low signal-to-noise ratios [6]. A significant improvement is brought by HRNR compared to TSNR. However, the major disadvantage of these methods is Musical noise generation due to non-linear signal processing. This provides a significant distortion in the speech quality and intelligibility. This paper is an extension of the paper titled "Comparative analysis of speech enhancement methods" [7] with a comparison of musical noise reduction capability of various speech enhancement methods.

Spectral subtraction method [7] provides an efficient noise reduction for low musical noise. The amount of musical noise generation and the difference between higher-order statistics of the power spectra before and after nonlinear signal processing shows a higher correlation [8, 9]. Here, amount of musical noise generated by the spectral subtraction [10], iterative spectral subtraction method [11–14] geometrical approach for spectral subtraction [15], Weiner based denoising techniques like TSNR and HRNR [16, 17] are compared for musical noise reduction capability.

2 Mathematical Analysis of Musical Noise Generation via Higher-Order Statistics

The amount of musical noise generation is strongly correlated with different isolated power spectral components and the isolation level of these components [18]. A Higher order statistics called Kurtosis is adopted to measure these isolated components among all components. A higher value of kurtosis signifies a signal with many isolated components. However, the calculation of kurtosis is not sufficient to measure the amount of musical noise generation. Therefore, the change in kurtosis between signals before and after signal processing is used to identify only the musical-noise components. Therefore the *kurtosis ratio* is used as a measure to estimate musical noise defined as

$$kurtosis\ ratio = \frac{kurt_{proc}}{kurt_{org}} \tag{1}$$

where $kurt_{proc}$ = kurtosis of processed signal, $kurt_{org}$ = kurtosis of observed signal
The kurtosis *ratio* increases with increment in amount of musical noise.

3 Speech Enhancement Algorithms

Two different class of enhancement algorithms are presented, out of which three are spectral subtraction based methods and other two are Weiner based methods. Noisy speech signal is given by Eq. (2).

$$y(n) = s(n) + d(n) \tag{2}$$

where $s(n)$, $d(n)$ and y(n) represent the pure speech signal, uncorrelated additive noise and the degraded speech signal respectively [10].

3.1 Spectral Subtraction

The principal of spectral subtraction method [10] is to achieve the estimated clean signal spectrum by the subtraction of an estimated noise spectrum from the corrupted speech signal spectrum. The estimation of noise spectrum taken and updated during the silence periods when the signal is not present i.e. in presence of noise only. The noise is assumed to be additive, stationary or near stationary. The Eq. (2) can be converted to Eq. (3) after Fourier transform.

$$Y[w] = X[w] + D[w] \tag{3}$$

Magnitude and phase of Y[w] can be expressed as follows

$$Y[w] = \left| y[w].e^{j\phi_y} \right| \tag{4}$$

where $|Y(w)|$ = magnitude spectrum, ϕ_y = phase spectra of the corrupted noisy speech signal. Noise signal can be expressed in transformed domain as follows

$$D[w] = \left| D[w].e^{j\phi_y} \right| \tag{5}$$

Here clean speech signal is estimated by subtraction of noise spectrum from the noisy speech spectrum as given in Eq. (6).

$$X(w) = [|Y(w)| - |D(w)|].e^{j\phi_y} \tag{6}$$

The unknown noise spectrum $|D(w)|$ is calculated by the average value in absence of speech signal. Spectral subtraction method is represented in Fig. 1.

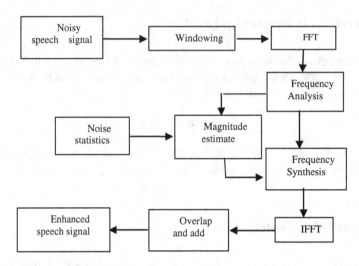

Fig. 1 Magnitude spectral subtraction block diagram

3.2 Iterative Spectral Subtraction

The only drawback of spectral subtraction is that a clear narrowband of noise still remains in the spectrum, even if our estimate of noise is correct. To overcome the drawback of spectral subtraction of weak signals another approach in which spectral analysis is iteratively applied on the signals, commonly known as Iterative Spectral Subtraction methods [11–14].

3.3 Spectral Subtraction Using Geometrical Approach

Geometric approach [15] is used to overcome the problem of spectral subtraction algorithm. This method involves the estimation of phase differences between the noisy signals and noise.

3.4 Decision-Directed (DD) Approach

The characteristics of this estimator has been tested by decision-directed (DD) approach proposed by Ephraim and Malah [19]. The main disadvantage of DD approach is to introduce the reverberation effect. Reverberation effect is minimized by Two-Step-Noise-Reduction (TSNR) technique as well it keeps the benefit of DD method. But TSNR, introduce harmonic distortion in enhanced speech because of the unreliability of the estimator for small SNR. To remove these harmonic distortions Harmonic Regeneration Noise Reduction (HRNR) is implemented [16, 17, 20].

4 Simulation Results and Discussion

All algorithms are implemented and simulated for speech enhancement in MATLAB. Then these algorithms are compared for four different noises- car noise, F16 noise, operation room noise and machine gun noise. Sound quality is evaluated and compared on the basis of their improved output SNR and the higher order statistics by finding Kurtosis ratio before and after the signal processing. One sample "YAHA SAI LAGHBAG PANCH MEAL DAKSHIN PASCHIM MAI KATGHAR GAON HAI"] has been used to check performance from our database [21].

The noisy version of this sentence was prepared by adding car noise and F16 noise from NOISEX-92 database [22] to this clean sentence at -10, -5, 0, 5 and 10 SNR levels. In spectral subtraction methods, $\beta = 1.1$ and $\eta = 0.8$ are taken for implementation where β and η are over subtraction factor and spectral floor parameter respectively. Residual noise and the perceived Musical noise are controlled by parameter β. A small value of β means the audible musical noise but the reduced residual noise. Also for a large β, residual noise will be audible but the musical issues will be reduced due to spectral subtraction. Also the amount of speech spectral distortion is greatly affected by the parameter α. The resulting signal will be highly distorted for large α. As well as the signal is suffered with poor intelligibility. Noise remains in enhanced speech signal for small value of α. In Geometrical based analysis parameter α is taken as 0.98 and parameter β is taken as 0.6. Similarly in Weiner based algorithms the value of parameter α at 0.98 gives the optimum result for enhanced speech. Simulation results are shown in the Tables 1, 2, 3 and 4 for the car noise, babble noise, operation room noise and machine gun noise respectively. Figure 2 shows average improvement in SNR at all input SNR level for all noises for each enhancement method.

Figure 3 shows average kurtosis ratio at all input SNR level for all noises for each enhancement method. It is observed from these figures that wiener based methods gives better results than basic spectral subtraction based methods in terms of increase in output SNR. It is observed from the Fig. 2 that output SNR level of HRNR algorithm gives the best result among all the algorithms at all input SNR level for all noises. Iterative spectral subtraction gives improvement in SNR with lesser musical noise generation as compared to other methods due to lowest kurtosis ratio than others.

Table 1 Output SNR and Kurtosis ratio for car noise

Output SNR in dB						
Methods	Parameters	−10 dB	−5 dB	0 dB	5 dB	10 dB
SS	Output SNR	−4.9	−0.3	5.4	11.42	17.5
	Kurtosis ratio	49.4	49.28	48.97	48.49	47.69
IS	Output SNR	−3.62	0.44	2.65	5.24	13.57
	Kurtosis ratio	10.14	14.16	15.31	10.8	14.32
GA	Output SNR	−2.99	6.5	12.62	17.86	22.81
	Kurtosis ratio	22.74	22.71	22.67	22.79	22.99
TSNR	Output SNR	6.61	12.45	17.82	21.9	23.83
	Kurtosis ratio	43.54	43.71	40.76	17.18	4.79
HRNR	Output SNR	25.77	30.43	33.35	27.53	25.35
	Kurtosis ratio	17.23	20.38	23.19	24.61	25.44

Table 2 Output SNR and Kurtosis ratio for babble noise

Output SNR in dB						
Methods	Parameters	−10 dB	−5 dB	0 dB	5 dB	10 dB
SS	Output SNR	2.24	3.2	5.86	10.65	16.6
	Kurtosis ratio	3.15	3.21	3.23	3.2	3.18
IS	Output SNR	10.96	6.74	9.99	14.48	21.08
	Kurtosis ratio	0.51	1.31	1.47	1.35	1.53
GA	Output SNR	−0.028	1.85	6.52	13.41	20.28
	Kurtosis ratio	1.98	2.01	2.05	2.09	2.12
TSNR	Output SNR	−1.6	0.61	5.48	12.07	17.97
	Kurtosis ratio	2.25	2.21	2.24	2.19	2.1
HRNR	Output SNR	−2.85	0.01	5.27	12.04	18.13
	Kurtosis ratio	2.47	3.01	3.09	3.44	4.01

Table 3 Output SNR and Kurtosis ration of operation room noise

Output SNR in dB						
Methods	Parameters	−10 dB	−5 dB	0 dB	5 dB	10 dB
SS	Output SNR	−6.58	−2.21	5.45	12.39	19.03
	Kurtosis ratio	5.73	5.71	5.33	5.21	5.09
IS	Output SNR	−0.77	−1.47	3.48	12.79	22.7
	Kurtosis ratio	1.33	1.32	1.35	1.38	1.41
GA	Output SNR	−1.41	0.3	1.89	11.55	18.58
	Kurtosis ratio	8.51	8.4	8.47	8.38	8.21
TSNR	Output SNR	0.87	7.89	15.53	22.47	28.11
	Kurtosis ratio	17.77	17.36	17.66	17.82	17.28
HRNR	Output SNR	14.97	21.98	29.62	36.52	39.09
	Kurtosis ratio	2.81	3.3	4.9	6.8	8.11

Table 4 Output SNR and Kurtosis ration for machine gun noise

Output SNR in dB						
Methods	Parameters	−10 dB	−5 dB	0 dB	5 dB	10 dB
SS	Output SNR	−7.67	−0.13	1.89	5.71	10.15
	Kurtosis ratio	2.87	2.88	2.89	2.94	2.97
IS	Output SNR	−8.37	−1.2	0.56	5.76	12.86
	Kurtosis ratio	0.52	0.89	1.01	0.82	0.99
GA	Output SNR	−4.6	−2.01	4.94	10.92	16.77
	Kurtosis ratio	2.84	2.85	2.89	2.95	3.01
TSNR	Output SNR	−4.69	4.89	11.66	17.57	23.57
	Kurtosis ratio	4.76	4.75	4.88	4.92	4.99
HRNR	Output SNR	20.98	27.4	33.75	39.43	41.4
	Kurtosis ratio	4.76	4.75	4.88	4.92	4.99

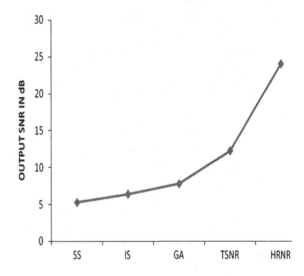

Fig. 2 Improvement in output SNR for various speech enhancement algorithms

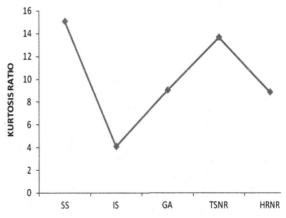

Fig. 3 Kurtosis ration variations in speech enhancement algorithms

5 Conclusion

Higher-order statistics has been used for implementation of musical-noise-generation analysis for nonlinear noise reduction. The HRNR Weiner based algorithm provided the best output SNR among all algorithms at all input SNR levels. It is observed from the higher order statistics that iterative spectral subtraction has less kurtosis ratio such that it enhanced the signal with least musical noise generation.

References

1. P. C. Loizou, Speech Enhancement Theory and Practice, Boca Raton, FL: CRC, Taylor & Francis Group, 2007.
2. S. F. Boll, "Suppression of acoustic noise in speech using spectral subtraction", IEEE Trans. Acoust., Speech, Signal Process., vol. 27, no. 2, 1979, pp. 113–120.
3. M. Berouti, R. Schwartz, and J. Makhoul, "Enhancement of speech corrupted by acoustic noise", Proc. ICASSP, 1979, pp. 208–211.
4. R. McAulay and M. Malpass, "Speech enhancement using a soft-decisionnoise suppression filter",IEEE Trans. Acoust., Speech, Signal Process., vol. 28, no. 2, 1980, pp. 137–145.
5. R. Martin, "Spectral subtraction based on minimum statistics", Proc. EUSIPCO, 1994, pp. 1182–1185.
6. Cyril Plapous, Claude Marro, and Pascal Scalart, "Improved Signal-to-Noise Ratio Estimation for Speech Enhancement" IEEE Transactions on Audio, Speech, and Language Processing, Vol.14, Issue 6, 2006, pp. 2098–2108.
7. Pankaj Goel, Prateek Saxena, V.K. Gupta, Mahesh Chandra, "Comparative analysis of speech enhancement methods", proc.10th IEEE Int.Confrence on Wireless and Optical networks, 2013, pp.1–5.
8. Y. Uemura, Y. Takahashi, H. Saruwatari, K. Shikano, and K. Kondo, "Automatic optimization scheme of spectral subtraction based on musical noise assessment via higher-order statistics", Proc. Of Int. Workshop. Acoust. Echo and Noise Control, 2008.
9. Y. Uemura, Y. Takahashi, H. Saruwatari, K. Shikano, and K. Kondo, "Musical noise generation analysis noise reduction methods based on spectral subtraction and MMSE STSA estimation", Proc. Of ICASSP, 2009, pp. 4433–4436.
10. Purav Goel, Anil Garg, "Developments in spectral subtraction for speech enhancement," International Journal of Engineering Research and Applications, Vol. 2, Issue 1, 2012, pp. 055–063.
11. K. Yamashita, S. Ogata, and T. Shimamura, "Spectral subtraction iterated with weighting factors," Proc. IEEE Speech Coding Workshop, 2002, pp. 138–140.
12. Kiyohiro Shikano, and Kazunobu KondoK. Yamashita, S. Ogata, and T. Shimamura, "Improved spectral subtraction utilizing iterative processing", IEICE Trans. A, vol. 88, no. 11, 2005, pp. 1246–1257.
13. M. R. Khan and T. Hassan, "Iterative noise power subtraction technique for improved speech quality," Proc. of Int. Conf. Elect. Comput. Eng, 2008, pp. 391–394.
14. X. Li, G. Li, and X. Li, "Improved voice activity detection based on iterative spectral subtraction and double thresholds for CVR," Proc. of Workshop Power Electron. Intell. Transport. Syst., 2008, pp.153–156.
15. Yang Lu, Philipos C. Loizou, "A geometric approach to spectral subtraction," Speech Communication, Vol. 50,2008, pp. 453–466.

16. C. Plapous, C. Marro, P. Scalart, and L. Mauuary, "A Two-Step Noise Reduction Technique," *IEEE Intl. Conf. Acoust., Speech, Signal Processing, Canada*, Vol. 1, 2004, pp. 289–292,
17. C. Plapous, C. Marro, and P. Scalart, "Speech Enhancement Using Harmonic Regeneration,"*IEEE Intl. Conf. Acoust., Speech, Signal Processing, USA*, Vol. 1, 2005, pp. 157–160.
18. Ryoichi Miyazaki, Hiroshi Saruwatari, Takayuki Inoue, Yu Takahashi, "Musical-Noise-Free Speech Enhancement Based on Optimized Iterative Spectral Subtraction", IEEE Transactions on Audio, Speech, and Language Processing, VOL. 20, NO. 7, 2012, pp. 2080–2094.
19. Y. Ephraim, and D. Malah, "Speech Enhancement Using a Minimum Mean-Square Error Short-Time Spectral Amplitude Estimator," IEEE Trans. Acoust., Speech, Signal Processing, Vol. 32, No. 6, 1984, pp. 1109–1121.
20. O. Capp´e, "Elimination of the Musical Noise Phenomenon with the Ephra¨ım and Malah Noise Suppressor," *IEEE Trans. Speech and audio Processing*, Vol. 2, No. 2, 1994, pp. 345–349.
21. Samudravijaya K *et. al.*, Hindi Speech Database, *Proc. ICSLP00*, Beijing, China, CDROM 00192.pdf.
22. A. Varga, H. J. M. Steeneken, D. Jones, "The noisex-92 study on the effect of additive noise on automatic speech recognition system," Reports of NATO Research Study Group (RSG.10), June 1992.

Model with Cause and Effect
for MANET (M-CEM)

Sharma Pankaj, Kohli Shruti and Sinha K. Ashok

Abstract Mobile ad-hoc network has gained popularity in recent years. There are several challenges faced by the researchers in this area. The most common research issue is the performance evaluation and enhancement. As we know that the performance of the MANET cannot be evaluated only with the selection of the protocol because the routing performance needs environment to operate. There are many network environmental conditions i.e., causes (e.g. Node mobility, number of nodes, pause time, network size etc.), which influence the performance of MANET. So it needs to emphasize to detect the possible causes (environmental causes) and their effect on performance. In this paper we have focused on this issue by developing a **Model with Cause and Effect for MANET (M-CEM)**. The model has been tested on multiple scenarios and some serious causes and their effects were detected in MANET performance. The model is based on Fuzzy-AHP (Analytical Hierarchy Process) for decision making. Fuzzy Decision Map (FDM) has been used by evaluating weightage of possible causes. The model is simulated under NS2.34 and tested for various imaginary network environmental conditions. The model was found satisfactory to handle uncertainty, vagueness and imprecise information of various possible network causes.

Keywords Fuzzy · M-CEM · COA · FDM etc.

S. Pankaj (✉) · S.K. Ashok
Department of Information Technology, ABES Engineering College,
Ghaziabad, UP, India
e-mail: pankaj.sharma@abes.ac.in

S.K. Ashok
e-mail: aksinha_1@yahoo.com

K. Shruti
Department of Computer Science, Birla Institute of Technology, Noida, UP, India
e-mail: shruti@bitmesra.ac.in

© Springer India 2016
S.C. Satapathy et al. (eds.), *Information Systems Design and Intelligent Applications*, Advances in Intelligent Systems and Computing 434,
DOI 10.1007/978-81-322-2752-6_69

1 Introduction

A number of protocols have been proposed with lots of advancement to improve the performance of network doing communication using route establishment and exchange of message in network. A Mobile Ad hoc Network (MANET) is self-configured infrastructure less network of mobile nodes. Each node has the freedom to join, leave and move creating a highly dynamic environment [1]. As optimization of protocol depends on multiple cause involved in formation and communication of network. So performance evaluation of any routing protocol is based on multi criteria decision making concept. An approach MCDM [2] of solving decision problems involving many criteria, factors or objectives. The fundamental characteristics of the goals are that they are often conflicting to one another. For instance, we need a bigger house but we can only afford at a cheaper price. DSR [3] is based on "on-demand behaviour" means routing activities initiated only when any node wants to send a data packet. Main feature of DSR is source routing that is routes are stored in cache and every data packet contain complete route to destination in its header. **Ad-hoc-on Demand Distance vector (AODV)** [3, 4] is also based on "on-demand behaviour". Whenever any node wants to send data packet, it initiate the route discovery process by flooding the RREQ in network. When a node is itself a destination or knows the route to destination, it generates the RREP and forward it using reverse route entries towards originating node. Originating node updates its routing table with routing information having highest Sequence Number. **Optimized link state routing (OLSR)** [5] uses the hop by hop routing means every node uses its most updated information to decide the route for routing of a data packet. It has the routes available whenever needed. It does not broadcast the request control message like DSR and AODV, selects some nodes in neighbourhood for propagation of control message; results into reduction in no. of retransmission. Saeed [6] suggested an intelligent routing system using neural network for modelling and genetic algorithm as an optimisation. Intelligent model is trained by considering the causes that affect the performance as input to neural network. Then neural network (ANN) works as input to the genetic algorithm which gives the decision about the optimum protocol based on the priority sets for performance parameters. If neural network incorporated with some fuzzy technique, it may leads to better result as fuzzy logic deals with uncertainty and imprecision [6–10].

2 Model with Cause and Effect for Manet (M-CEM)

To evaluate the possible causes on performance of MANET under DSR and AODV routing protocol, an intelligent model is proposed which is based on multi criteria decision making algorithm. **M-CEM** is developed using various AHP algorithm and Fuzzy Decision Map (FDM). Model is shown in Fig. 1.

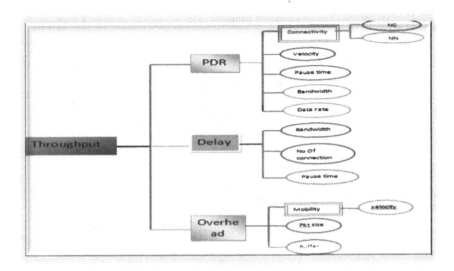

Fig. 1 M-CEM architecture

A. **Network environmental causes**

There are several network environmental conditions or causes. For development of the proposed model we have used some of those conditions.

We have considered four environmental causes:

  i. No. of nodes (NN): variation in node density (70–155) to represent network size.
  ii. No. of connection (NC): variation in no. of connections (10–100) to represent the active nodes.
  iii. Pause time (PT): variation in pause time (3–60).
  iv. Node Mobility (NM): variation in speed for mobility (3–60).

B. **Performance causes affected by causes**:

There may be several causes used in many research works earlier. The model proposed in this paper consists of three performance evaluation metrics are considered.

  i. Throughput (bits/s)—number of bits transferred per unit time.
  ii. Packet delivery ratio (PDR)—success rate of information delivered.
  iii. Control Overhead—Overhead of packets for achieving throughput.

C. **Methodology**:

Considering various influencing causes in network, a model using Analytical hierarchy (AHP) technique [4, 5, 11] is developed. The architecture of the M-CEM is shown in Fig. 1 which is designed using IDS (intelligent decision system) software. Cause and effect relationship is analyzed among multiple influencing causes.

Table 1 Fuzzy scaling of linguistic items

Linguistic terms	Fuzzy scale
Equal throughput	(1, 1, 1)
Equally moderate throughput	(1, 2, 3)
Moderate throughput	(1, 3, 5)
Moderate strong throughput	(2, 4, 6)
Strong throughput	(3, 5, 7)
Strongly very strong throughput	(4, 6, 8)
Very strong throughput	(5, 7, 9)
Extreme throughput	(6, 8, 9)

Table 2 Linguistic values for cause and effect

Linguistic terms	Values using TFN
Weak impact	(0, 0, 0.25)
Medium weak impact	(0, 0.25, 0.5)
Medium impact	(0.25, 0.5, 0.75)
Medium strong impact	(0.5, 0.75, 1)
Strong impact	(0.75, 1, 1)

Table 3 Final global matrix for PDR

Parameters	Fuzzy weight values	COA	Rank
NN	(0.776, 0.828, 0.855)	0.8199	2
NC	(1, 1, 1)	1	4
PT	(0.638, 0.635, 0.660)	0.6488	1
M	(0.670, 0.638, 0.7)	0.6849	3

For analyzing the cause and effect among the network environmental conditions and performance evaluation causes Weight matrix is computed. The computation is performed using the Eqs. (1–8). The process of computation follows a sequence of fuzzy scaling, local weight matrix of causes are evaluated are generated which is shown in Table 1. The linguistic computation is shown in Table 2. Final global matrix according to the pairwise computation matrix represents the pairing among the network environmental causes shown in Table 3. The computation of weight is an iterative process and the final global matrix is obtained using equations.

$$W(l1)_i = \sum p_{ij} = \left(\sum a_{ij}, \sum b_{ij}, \sum c_{ij}\right), i = 1\ldots n \quad (1)$$

$$W(l1) = \left[\left(w(l1)_i^a, w(l1)_i^b, w(l1)_i^c\right)\right]^T, i = 1\ldots n \quad (2)$$

Get the highest fuzzy number α of level-1 fuzzy weights and modify it as in Eq. (4).

$$\alpha = (\alpha^a, \alpha^b, \alpha^c) \tag{3}$$

where $\alpha^a = max\{w(L1)^a\}$, $\alpha^b = max\{w(L1)^b\}$

$$\alpha^c = max\{w(L1)^c\}$$

$$w(L1)_n = w(L1)/\alpha \tag{4}$$

Fuzzy cognitive map is required to stipulate the impact among various environmental parameters. It is obtained using some linguistic values. The weight matrix is calibrated using the following equation:

$$C^{(t+1)} = f(C^{(t)} \cdot E) \tag{5}$$

where $c^{(t)} = I_{n \times n}$ and E is the fuzzy weight matrix obtained using linguistic values.

The Eq. (5) is the calibration of weight matrix and it is iterated until it reaches to some constant values, the modified weight matrix as per below:

$$\mu = (\mu^a, \mu^b, \mu^c) \tag{6}$$

where $\mu^a = max\{C^a\}$, $\mu^b = max\{C^b\}$ $\mu^c = max\{C^c\}$

$$C_n = C/\mu \tag{7}$$

weight obtained in steps 1 and 2 is merged using the following equation:

$$W_{final} = W(L1) + C_n W(L1)_n \tag{8}$$

Finally the rank of each environmental cause based on the final weights are assigned to them corresponding to rank ratio. The demand protocols (DSR, AODV) are simulated on ns2.34 simulator. For assigning weights to each environmental parameter in the architecture model shown in Fig. 1, Fuzzy Decision Map (FDM) is used. Final global matrix is obtained by using Eq. (8), the parameters are ranked according to COA (Centre of Area) values which are shown in Tables 3, 4 and 5 for PDR, DELAY and Throughput respectively.

The model observation is simulated and tested under DSR protocol for evaluating the effect on PDR, Delay and Overhead. The effect of PDR is shown in Fig. 2, which shows the Standard Error (SE) computed with the variation of various causes and Pause Time is having the major influence which shows the higher error rate. Similarly, we can perform the model observation for all the performance evaluation parameters.

Table 4 Final global matrix for delay

Parameters	Fuzzy weight values	COA	Rank
NN	(0.867, 0.666, 0.697)	0.6438	4
NC	(0.688, 0.774, 0.790)	0.7508	1
PT	(0.647, 0.765, 0.801)	0.7381	2
M	(1, 1, 1)	1	1

Table 5 Final global matrix for overhead

Parameters	Fuzzy weight values	COA	Rank
NN	(0.664, 0.711, 0.739)	0.7051	4
NC	(0.971, 0.981, 0.977)	0.9767	1
PT	(1, 1, 1)	1	2
M	(0.8437, 0.8642, 0.8892)	0.8654	3

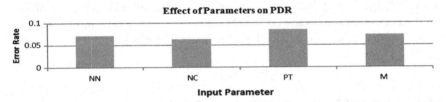

Fig. 2 Influence of pause time on PDR

3 Conclusion

The proposed model for MANET has been tested for various imaginary scenarios similar to real environment based on the outcome of Table 3 (PDR), Table 4 (Delay) and Table 5 (Overhead). Table 3 shows that the Pause Time (PT) is one of the cause which is having the highest rank means it is the major cause that will affect the PDR. So it needs to control to achieve the optimum PDR. Table 4 shows another cause i.e. node mobility as the highest rank holding, and it is cause for delay in the timely delivery of the information. Finally, the Table 5 shows the cause maximum number of active connections which will effect on overhead. The nodes may join or leave the network frequently, consequently the active connection may vary. So all the three causes discussed so far i.e., variation in active nodes, variation in node mobility and variation in pause time are the possible causes which may affect the throughput. The standard error has been shown only for PDR, which is shown in Fig. 2. Similarly the standard error can be calculated for remaining two causes. Thus the throughput of MANET can be defined as directly proportional to PDR and inversely proportional to Delay and Overhead. In this paper, the proposed model is simulated and tested for all the performance evaluation parameters and found satisfactory. So lots of uncertain and imprecise information existed in the dynamic network, due to varying nature of causes. The model **M-CEM** is found good to address these causes and their effects on throughput in MANET regardless of protocols.

References

1. Broach et al., "A performance comparison of Multi hop wireless Ad Hoc Network Routing Protocols," PROC. Mobicom'98.
2. Martin Aruldoss et al. "A Survey on Multi Criteria Decision Making Methods and Its Applications," AJIS 2013 Science and education Publishing.
3. Charles et al., "Performance Comparison Of Two On Demand Routing Protocols For Ad Hoc Networks," IEEE 2001.
4. Charles E. Perkins, Elizabeth M. Royer, "Ad-hoc On-Demand Distance Vector Routing" IEEE 2008.
5. T. L. Satty, Multicriteria Decision Making: The Analytical Hierarchy Process: planning, priority Setting, resource allocation. Pittsburgh, PA, USA: RWS,1990.
6. N. H. Saeed, M. F. Abbod, H. S. AI Raweshidy, "Intelligent Manet Routing System", 978-0-7695-3096-3/08 2008 **IEEE**.
7. Siddesh Gundagatti Karibasappa, K.N. Muralidhara, "Neuro Fuzzy Based routing International Conference on Industrial and Information Systems, ICIIS 2011, Aug. 16 19, 2011 IEEE.
8. M. N. Doja, Bashir Alam, Vivek Sharma, "Analysis of reactive Routing Protocol Using Fuzzy Inference System" 164-169, **Elsevier** 2013 AASRI Procedia.
9. Tarun Varshney, Aishwarya Katiyar, pankaj Sharma, "Performance improvement of MANET under DSR protocol using Swarm Optimization", 978-1-4799-2900-9/14 2014 **IEEE**.
10. Arun Biradar, Dr Ravindra C. Thool, "Reliable Genetic Algorithm Based Intelligent Routing For MANET", 978-1-4799-3351-8/14 2014 **IEEE**.
11. Basem Mohamed Elomda, Hesham Ahmed Hefny Hesham Ahmed Hassan, "MCDM method based on improved fuzzy decision map", 978-1-4799- 2452-3/13/$31.00 ©2013 IEEE.

Parameters Quantification of Genetic Algorithm

Hari Mohan Pandey

Abstract This paper presents the importance of parameters tuning in global optimization algorithms. The primary objective of an experiment is to recognize the process. The experiments are carried out to learn the effect of various factors at different levels. Hence, identifying the optimal parameters setting is important for robust design. One of the most popular global optimization algorithms: genetic algorithm is considered in this study. The domain of inquiry is travelling salesman problem. The present study employs the Taguchi method that involves the use of an orthogonal array in the estimation of the factors. Taguchi approach has been widely applied in experimental design for problems with multiple factors. The use of Taguchi design is a novel idea—leads to efficient algorithms—can find a satisfactory solution in a few iterations, which improves the convergence speed and reduces the cost. Experimental results show that the Taguchi design is less sensitive to initial value of parameters. Two versions of genetic algorithms (with tuning and without tuning) are implemented. The analysis shows the superiority of genetic algorithm with tuning over genetic algorithm without tuning.

Keywords Genetic algorithm · Travelling salesman problem · Taguchi method · Robust design

1 Introduction

Experimental design (ED) approaches are applicable in initiating statistical control of a process. For instance, let us consider a control chart shows the process is out of control and process has numerous manageable input variables. It might be very strenuous to control the process until we know the important input variables. ED

H.M. Pandey (✉)
Department of Computer Science & Engineering, Amity University,
Sector-125, Noida, Uttar Pradesh, India
e-mail: profharimohanpandey@gmail.com

© Springer India 2016
S.C. Satapathy et al. (eds.), *Information Systems Design and Intelligent Applications*, Advances in Intelligent Systems and Computing 434,
DOI 10.1007/978-81-322-2752-6_70

711

methods can be utilized to find the influential process variance. ED is certainly an important tool in engineering for enhancing a production process. In the development of new processes—has vast applications prior in development of process that can improve production and yield: reduce development time, less change and closer conformance to optimal, and minimize cost. ED methods have an important role in engineering design, where existing ones can be improved and new products are developed. Design of experiments (DOE) is one such method used in statistical analysis [1]. EDs are used during development and starting levels of production—helps the experimenter to know output when the input settings are modified. DOE helps in finding the interaction between the factors. DOE is used to consider the factors that are responsible for the changes in the output and represent them in mathematical form. It is also helpful in studying a large number of factors by running the less number of experiments. Statistical experimental methods are utilized in DOE to find the best factors and its appropriate level for optimization or process optimization. DOE constitutes various approaches such as factorial design, fractional factorial design, response surface design, etc. The Taguchi design has been widely applied in industry for the purpose of finding the appropriate parameter value in achieving the goals [2, 3]. Different factors that are related to the results are under the user's control are selected and varied over two or more levels. Experiments are designed considering an orthogonal array produces the effect of each primary factor. The Taguchi approach involves an analysis—reveals the significance of the factors in achieving the goals and directs towards adjusting the factors to improve the quality of the results. The control over finding the results will be best obtained by changes in the primary factors in the direction shown by the analysis. One of the most popular global optimization algorithms, namely, genetic algorithm (GA) is chosen for the simulation purpose over travelling salesman problem (TSP). GA is search and optimization technique proposed by Holland. GA's performance is largely depends on its operators: crossover and mutation—participate actively during the reproduction process [4]. Selection of population performs at two places (in the beginning and after each iteration) also plays a significant role. The selection of the population is known as selective pressure and if not selected intelligently—difficult to achieve global optima. Therefore, crossover rate, mutation rate, population size and problem specific details are an important factor in controlling the power of GA. This paper presents the Taguchi method in finding the best combination of parameter value in finding the global optima. The best combinations were obtained to perform the final experiment. Two variations of GA: GA with tuning and GA without tuning are implemented—results are collected and compared that shows the superiority of GA with tuning over GA without tuning.

The paper comprises of five sections. Initially starting with the overview of the robust design methods and its important in experimental design for achieving the best results. Section 2 represents the basic concept involved in TSP. The purpose of Sect. 3 is to present the summary of global optimization problem and GA. This section also explains the basic configuration of GA. Section 4 presents the simulation model where you can get the working of the Taguchi method and how it is used in identifying the best parameter combination in robust experiment design. It also presents the comparative analysis of the two versions of GA considered to conduct the experiments. Lastly, the conclusion is drawn in Sect. 5, whereas reference section shows the important literatures—can be utilized in robust design.

2 Travelling Salesman Problem

Travelling salesman problem (TSP) is NP-hard combinatorial optimization problem —finds the shortest path, which covers all the cities not more than once and ends with the initial city when the distances between every two cities are given [5]. The time complexity of this problem is less than $O(n!)$ [6]. From language of graph theory, it can be defined as follows: when an undirected weighted graph $G(V, E)$, let $V = \{1, 2, \ldots, n\}$, E and $d_{ij} = (i, j = 1, 2, \ldots, n)$ be the vertex set, the edge set comprises of connected vertices and the weight of edges (i, j) then amount of a loop is the total of weights of all edges [7]. TSP is intended to find minimum cost that represent shortest path moves with all the vertices and the same initial and final vertex.

3 Global Optimization Algorithm: Genetic Algorithm

The main aim of optimization problem is to minimize or maximize a real function by appropriately selecting input values from an allowed set and function value is computed [8]. The optimization theory generalization and methods to various establishments constitutes a major field in applied mathematics. Generally, predicting apt accessible values of an objective function mentioned a domain which is predefined, also a diverse class of various types of objective functions and various domains [9]. A representation of an optimization problem is given below: Given: a function from some set A to the real numbers $(f : A \rightarrow R)$ Sought: An element x_0 which belongs to set A such that $f(x_0)$ is greater than or equal to $f(x)$ for every x in set A called maximization or $f(x_0)$ is less than or equal to $f(x)$ for every x in set A called minimization (Fig. 1).

GA is heuristic adaptive search algorithm based on the evolution and Darwin's theory of "Survival of the fittest". It imitates few processes of nature. The main idea of GA is to solve optimization problems [10]. GA is more robust than traditional AI approaches. There exists three steps need to be implemented in GA after the arbitrary generation of starting population, which includes: selection, crossover and mutation. Selection aims to fill the population with duplicates of fittest individuals from the given population. For the algorithms to converge on a good or a sub optimal solution selection and crossover operators will be used. Mutation induces a random walk through the search space [11]. Selection and mutation creates a parallel, noise tolerant algorithm. Selection is to find the fittest; crossover is used for

Fig. 1 Flowchart for simple
genetic algorithm

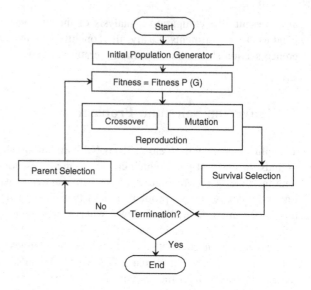

mating between individuals and mutation for arbitrary changes. Although, GA is a
popular method in solving global optimization (GO) problems, but suffers due to
premature convergence. Premature convergence is the situation when the diversity
of the population decreases as it converges to local optima. Hence, to find the global
optimum solution and to explore the search space diversity management is must
[12, 13]. Pandey et al. [14] presented a comprehensive survey of the approaches
proposed to address this issue. It was discussed in [14] that GA's performance
largely depends on problem difficulty, population diversity, search space, initial
population, selection pressure, fitness function and number of individuals.
Typically, selection in GA done in two places: in the beginning known as the initial
population selection, and selection of population for the next generation. Crossover
and mutation probabilities play a significant role in maintaining the diversity the
guides the search process. As discussed, population choice, GA operators (crossover and mutation) and other probabilities significantly affects the GA's performance as they interact in a complex manner. A better parameter setting can be
applied to find the best solution for GO problems [5].

4 Simulation Model

The TSP has been solved using GA. The results are used to analyze the performance of a GA in finding the solution for TSP and find the combination of factors
and its levels which results in optimal output. GA's performance largely depends on
its operators—crossover rate (CR), mutation rate (MR). Population size (PSIZE),
number of generations (GEN) also plays a significant role in GA's performance.

Table 1 Orthogonal array and Taguchi signal-to-ratio for robust experiment design

Exp. no.	MR	CR	PSIZE	CITIES	GENS	Mean	Std. dev	SNR
1	0.015	0.5	100	20	100	51.25	0.777817	−34.1944
2	0.015	0.5	100	40	200	122.8	1.979899	−41.7845
3	0.015	0.8	200	20	100	40.2	2.262742	−32.0914
4	0.015	0.8	200	40	200	125	1.697056	−41.9386
5	0.3	0.5	200	20	200	57.3	1.272792	−35.1642
6	0.3	0.5	200	40	100	117.15	1.484924	−41.3752
7	0.3	0.8	100	20	200	47.05	1.767767	−33.4543
8	0.3	0.8	100	40	100	144.8	2.262742	−43.2159
Response (SNR (smaller is better))								
Low	−37.50	−38.13	−38.16	−33.73	−37.72	MR:CR:PSIZE:CITIES: GENS = 0.015:0.5:100:20:100		
High	−38.30	−37.68	−37.64	−42.08	−38.09	MR:CR:PSIZE:CITIES: GENS = 0.3:0.8:200:40:200		
Delta	2	4	3	1	5	Rank: based on delta (1: highest and 5: lowest)		
Solution combination	0.3	0.5	100	40	100	SNR (smaller is better)		−43.2159

In TSP, numbers of cities (CITIES) are also considered an important factor in overall system performance. For robust design, five factors with two levels are considered outlined: MR = [0.015, 0.3], CR = [0.5, 0.8], PSIZE = [100, 200], GEN = [100, 200], and CITIES = [20, 40]. Table 1 represents the Taguchi design, each row specifies possible combinations of factor levels to conduct the experiment. The Taguchi method uses an orthogonal array referred as Taguchi orthogonal arrays —require only a fraction of full factorial design, so that time can be saved in identifying the appropriate factor level in less time. In the present scenario, the orthogonal array design consists of 5 factors, 2 levels and 8 runs—can be represented as L8 (2 ** 5). To analyze the Taguchi design one need to understand the effect of control and noise factors on the response, then select the best combination of factor setting for conducting the experiment/process. Equation (1) is used to evaluate signal-to-noise ratio (SNR).

$$SNR_i = -10 \log \left(\sum_{u=1}^{N_u} \frac{y_u^2}{N_i} \right) \tag{1}$$

where, i = experiment number, u = trial number, N_i = number of trials for the experiment, and y_u = number generations taken in each trial to reach to the solution. TSP is combinatorial problem and GA's nature is deterministic, hence executions were performed ten times and average is taken to present the final response. Figure 2a–f graphically represents the effects of various factors considered. It is very much clear from Fig. 2b that each factor has some effect on the final response, but 'CITIES' (rank: 1, delta = 1 from Table 1) has larger effect, whereas 'GEN' (rank: 5, delta = 5 from Table 1) has smallest effect.

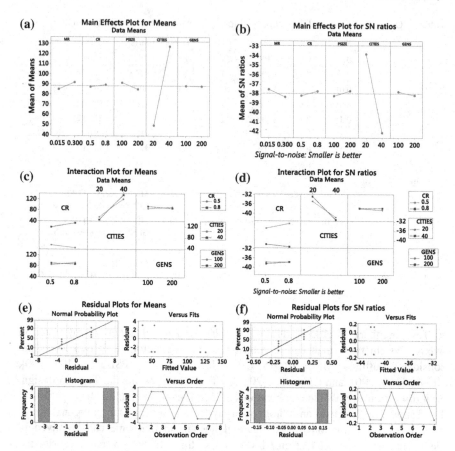

Fig. 2 **a** Main effect plot for means. **b** Main effect plot for SNR. **c** Interaction plot for means. **d** Interaction plot for SNR. **e** Residual plot for means. **f** Residual plot for SNR

The term 'delta' ranks the effect of various factors based on the SNR (smaller is better) value. A factor is said to show a significant effect when its graph deviates from the baseline—indicates that the difference between the means at two different levels is high. It can be seen, factors 'CITIES, MR, CR and PSIZE shows remarkable effects on the result whereas GEN show negligible effect. Sometime it is also possible that two or more factors might be dependent on each other affects the results. Interdependency effect was also analyzed leads to a conclusion that CITIES and CR have strong interaction effect while CR alone shows the very less effect. The interaction effect for means and SNR are represented in Fig. 2c, d respectively. Figure 2e, f shows the residual plots for both means and SNR respectively—presents the impact on the outcome. There exist four plots included: normal probability plot, residual plot, residual versus fitted value plot, histogram plot and residual versus observation order plot. The normal probability plot should roughly form a straight line to explain the residuals are normally distributed—used for easy classification,

Table 2 Analysis of variance for SN ratios

Source	DF	Seq SS	Adj SS	Adj MS	F	P
MR	1	1.280	1.280	1.280	6.12	0.245
CR	1	0.413	0.413	0.413	1.98	0.394
PSIZE	1	0.541	0.541	0.541	2.58	0.354
CITIES	1	139.529	139.529	139.529	667.08	0.025
GENS	1	0.268	0.268	0.268	1.28	0.461
CR*CITIES	1	4.216	4.216	4.216	20.16	0.140
Residual error	1	0.209	0.209	0.209		
Total	7	146.457				

Table 3 Analysis of variance for means

Source	DF	Seq SS	Adj SS	Adj MS	F	P
MR	1	91.5	91.5	91.5	1.20	0.470
CR	1	9.1	9.1	9.1	0.12	0.787
PSIZE	1	86.1	86.1	86.1	1.13	0.480
CITIES	1	12,320	12,320	12,320	162.21	0.050
GENS	1	0.2	0.2	0.2	0.00	0.968
CR*CITIES	1	327.0	327.0	327.0	4.31	0.286
Residual error	1	76.0	76.0	76.0		
Total	7	12910.5				

but unfortunately it is not—since the size of the data is very small. The residual versus fitted plots examine whether the assumption of constant variance is violated or not. As the residuals are randomly distributed on both sides of the baseline—assumption of constant variance does not hold. The general characteristics of residual such as typical values and shape can be analyzed using histogram plot clearly indicate that there are outliers present in the data. The residual versus order of data is collected and can be used to find time related effects and to find non-random errors.

The detailed analysis report of variance for both SNR and means are represented in Tables 2 and 3 respectively. The corresponding F-value and P-value are presented for each factor. For a factor to be significant—its corresponding p-value should be less than 0.05 at 95 % confidence level. From Table 2, p-value for CITIES is less than 0.05 ($0.025 < 0.05$) has a significant effect on the experiment, whereas if obtained p-value is close to 0.05 then the factor is related and shows some effect on the experiment. If the p-value turns out to be greater than 0.05 then there is no significant effect of that factor on the outcome. The above analysis produces the best possible combination values of the factors considered and the one have a more significant effect on the final outcome. The best possible combination was found based on the value of the means and SNR presented. A comparison is done to find the best experimental design gives optimal results in less time. Results differ in each run due to the stochastic nature of GA and TSP is a combinatorial problem. The best

Fig. 3 Distance versus
experiment number chart for
GA with tuning and GA
without tuning

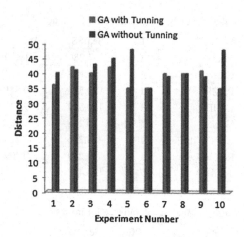

tuned combination for solving TSP via GA is given as: [MR:CR:PS:CITIES:
GEN] = [0.3:0.5:100:40:100] since SNR = −43.2159 (smaller is better). Figure 3
presents the distance versus experiment number chart for TSP applying GA with
tuning and GA without tuning. The average distance values obtained are 38.6 and
41.8 for GA with tuning and GA without tuning respectively—indicates that the
average distance travelled is less (38.6 < 41.8) in case of tunes GA.

5 Conclusions

This paper analyzes the experimental approaches—are used in finding the appropriate parameter values affects the performance of an algorithm. An experimental
design method known as Taguchi method was applied. The analysis reports are
presented thoroughly and discussed in finding the factors—have significant effect
on the outcome and the values at which the algorithm produces the optimum results.
Two version of GAs were implemented—shows the superiority of GA with tuning
over GA without tuning. Algorithm after tuning takes comparatively less time than
unturned algorithm. An obvious outcome of this study is—similar approach can be
applied on various complex combinatorial problems—could not be solve in polynomial time, computational cost can be decreased with better result quality.

References

1. Akbaripour, H., and E. Masehian. "Efficient and robust parameter tuning for heuristic
   algorithms." Int. J. Ind. Eng 24.2 (2013): 143–150.
2. W. Y. Fowlkes and C. M. Creveling, Engineering Methods for Robust Product Design: Using
   Taguchi Methods in Technology and Product Development. Reading, MA: Addison-Wesley,
   1995.

3. A. S. Hedayat, N. J. A. Sloane, and J. Stufken, Orthogonal Arrays: Theory and Applications. New York: Springer-Verlag, 1999.
4. Pandey, Hari Mohan, Anurag Dixit, and Deepti Mehrotra. "Genetic algorithms: concepts, issues and a case study of grammar induction." *Proceedings of the CUBE International Information Technology Conference.* ACM, 2012.
5. Ye, Gao, and Xue Rui. "An improved simulated annealing and genetic algorithm for TSP." *Broadband Network & Multimedia Technology (IC-BNMT), 2013 5th IEEE International Conference on.* IEEE, 2013.
6. Zhang, Liping, Min Yao, and Nenggan Zheng. "Optimization and Improvement of Genetic Algorithms Solving Traveling Salesman Problem." *Image Analysis and Signal Processing, 2009. IASP 2009. International Conference on.* IEEE, 2009.
7. Wei, Gaoqi, and Xiaoyao Xie. "Research of using genetic algorithm of improvement to compute the shortest path." *Anti-counterfeiting, Security, and Identification in Communication, 2009. ASID 2009. 3rd International Conference on.* IEEE, 2009.
8. Johnson, David S., and Lyle A. McGeoch. "The traveling salesman problem: A case study in local optimization." *Local search in combinatorial optimization* 1 (1997): 215–310.
9. Louis, Sushil J., and Gregory JE Rawlins. "Predicting convergence time for genetic algorithms." *Foundations of Genetic Algorithms* 2 (1993): 141–161.
10. Yuan, Lihua, Yuming Lu, and Ming Li. "Genetic Algorithm Based on Good Character Breed for Traveling Salesman Problem." *Information Science and Engineering (ICISE), 2009 1st International Conference on.* IEEE, 2009.
11. Tsai, Jinn-Tsong, Jyh-Horng Chou, and Tung-Kuan Liu. "Tuning the structure and parameters of a neural network by using hybrid Taguchi-genetic algorithm."*Neural Networks, IEEE Transactions on* 17.1 (2006): 69–80.
12. Friedrich, Tobias, et al. "Analysis of diversity-preserving mechanisms for global exploration*." *Evolutionary Computation* 17.4 (2009): 455–476.
13. Whitley, Darrell. "An overview of evolutionary algorithms: practical issues and common pitfalls."*Information and software technology* 43.14 (2001): 817–831.
14. Pandey, Hari Mohan, Ankit Chaudhary, and Deepti Mehrotra. "A comparative review of approaches to prevent premature convergence in GA." *Applied Soft Computing* 24 (2014): 1047–1077.

Emotion Recognition: A Step Ahead of Traditional Approaches

Surbhi Agarwal, Madhulika Bhatia and Madhurima Hooda

Abstract Emotion recognition is an intriguing issue these days. It affects essential applications in numerous regions for example surveillance, defense, financial services etc. Determining a particular expression from face images effectively is a crucial venture. In this paper, we have demonstrated a novel approach to recognize emotions displayed in video sequences. The authors have considered seven basic emotions measuring factors: anger, fear, disgust, happiness, sadness, surprise and neutral. These factors are constantly encountered in our day to day life. The focus of this paper is towards contemplates a combination of extended biogeography based optimization algorithm, support vector machines and local binary patterns to obtain the best possible results.

Keywords Support vector machine · Biometrics · Feature extraction
Viola jones · Linear binary pattern · Extended

1 Introduction

Facial feature recognition has developed as a vital biometric procedure in issues like validation, access to assets, reconnaissance and so forth. The variability that exists regarding light, arrangement, emotion etc. makes this process a little cumbersome [1, 2]. Due to many similarities present on a human face, a good discrimination becomes tedious [3, 4]. Numerous strategies have been proposed and utilized as such. Contours are extracted by using various methodologies such as local feature matching, appearance based, knowledge based, feature based, template matching and holistic matching methods. Template matching methods center their focus on

---

The original version of this chapter was revised: Incorrectly published author name has been corrected. The erratum to this chapter is available at https://doi.org/10.1007/978-81-322-2752-6_77

S. Agarwal (✉) · M. Bhatia · M. Hooda
Azamgarh, India
e-mail: agarwal.surbhi02@gmail.com

© Springer India 2016
S.C. Satapathy et al. (eds.), *Information Systems Design and Intelligent Applications*, Advances in Intelligent Systems and Computing 434,
DOI 10.1007/978-81-322-2752-6_71

intensity of the pixels. In this paper a novel approach by combining many methodologies has been proposed. This aims at utilizing the best traits of every mono approach that has been used in the combination. The prevalent expressive capacity present with human beings is natural. This gives them a standout amongst even the most capable, dynamic and adaptable methods of expressing ourselves. The framework introduced in this paper identifies frontal faces in the video sequences based on the seven basic emotions mentioned above. These emotions are used to express our feelings along with a continuous comprehension of the data that is passed on [5]. We have used support vector machines to classify the expressions in the video sequence belonging to Cohn kannade database [6]. With the use of optimization algorithm the person whose emotions are being judged is also detected.

1.1 Applications of Face Recognition

1.1.1 Law Enforcement and Justice Solutions

To beat world's ever-advancing criminals, today's law enforcement agencies are seeking for innovative technologies, so as to stay a step ahead of them. FRS holds the charge of developing technologies that can make the job of a law enforcement officer easier.

2 Techniques Used in the Framework

2.1 Extended Biogeography Based Optimization Algorithm (Extended BBO)

Extended BBO is an expansion to BBO calculation that was created by Simon [7, 8]. In BBO the immigration rate i.e. the rate at which the island is betrayed and resettlement rate i.e. the rate at which island gets swarmed are considered [7]. In extended BBO as proposed by [8] extra parameters have been presented, for example, suitability index variable (SIV), habitat suitability index and fitness value, distance calculation, extinction rate, dependency factor and relevance factor [8, 9]. The features are attained as per the SIV. Distance calculation compares to the measure of Euclidean separation. Relevance factor kills the hesitant and undesirable features from the face acknowledgment process [8, 9].

2.2 Support Vector Machines (SVM)

Support vector machines are utilized for universally useful pattern acknowledgment [10]. It characterizes the information by utilizing an induced function called hyperplane which meets expectations by making an edge between the distinctive groups. The edge is a set of those pixels which exists between object and background, object and object, region and region [11]. The information to be gathered can be paired or multiclass. The hyperplane not just guides in conveying the danger of misclassification to a base, additionally functions admirably for untrained specimens fitting in with the test set [10]. In this paper SVM has been utilized for binary arrangement.

2.3 Linear Binary Pattern (LBP)

LBP is based on template matching method [12]. Human face dependably comprises of a few features that are more expressive than the others e.g. eyes, lips and so on [13]. Consequently on the off chance that we see the face picture as an arrangement of pixels, then feeling acknowledgment will be more reliant on a few pixels and less subject to others. LBP meets expectations by characterizing a picture regarding pixels. Each time a 3 × 3 area is considered. The center pixel is thought to be the threshold. All the neighboring pixels having a worth more prominent than this limit are numbered as 1 and the pixels having esteem not exactly the edge are numbered as 0. Consequently a string of paired digits gets produced. Utilizing this twofold string a shifting recurrence of 1's and 0's is acquired. Eventually either the centralization of 1's is more than 0's or the other way around. Subsequently a histogram is produced as need be. These histograms help in the partition of diverse features from the face [14, 15]. Further, by normalizing the histograms highlights are removed to fill the need of classification (Fig. 1).

2.4 Viola Jones

This algorithm is utilized to recognize confronts (faces) continuously at a rapid [16]. It incorporates inside itself the advantages of Adaboost and cascade classifiers. The cascade classifier property empowers the segregation between highlights, for

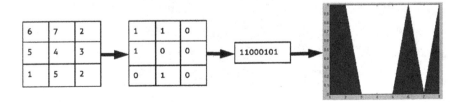

Fig. 1 Example of LBP

example, nose, eyes, lips and so forth. Adaboost scales the execution by having the capacity to characterize the information that has been misclassified by other calculations by investigating the frail base classifiers. E.g. given a chance, say one needs to distinguish a face initially embedded into a background. Viola jones calculation helps in identifying the face by taking out the undesirable data i.e. background. Progressively it comes down to finding the sharp features, for example.

3   Experimental Setup

We isolated the whole process into the accompanying stages that run parallel. Cohn kannade dataset was used for training and testing. Some sample images of the video sequences belonging to cohn kannade dataset are as follows (Fig. 2).

3.1   Phase–I

The pictures of Cohn kannade dataset are passed through Gabor kernels which redress the alignment and enlightenment of the pictures subsequently issuing them a proper orientation. Further, it is passed through principle component analysis for ideal extraction of highlights (features). For a considerably more precise highlight extraction the pictures are passed through extended biogeography based optimization algorithm. At this stride the pictures get prepared and store the outcomes.

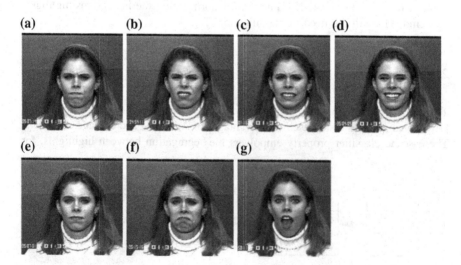

Fig. 2 **a** Anger. **b** Disgust. **c** Fear. **d** Happy. **e** Neutral. **f** Sad. **g** Surprise

We can later on pass an untrained picture and the distinguishing proof of its right subject will happen [9, 16, 17].

3.2 Phase–II

The second stage lives up to expectations in parallel with the first stage by applying viola jones calculation on the dataset. Viola jones crops the coveted piece of the picture by taking out the pointless foundation. Linear binary pattern creates the histogram and permits the highlights to be prepared by classifier. Ultimately support vector machine trains the chosen pictures with the edited part by performing a binary classification utilizing its hyper plane. The untrained pictures are tried later on [13].

3.3 Phase–III

Stage III goes about as a connector between the beforehand portrayed countenances. In this stage we can utilize the untrained pictures to focus the exactness of the structure. When the subject under test is dictated by extended BBO, the face is sent to the system trained by SVM. At long last with the assistance of stage II the expression of the face is distinguished. Subsequently stage I decides the subject (out of the 40 subjects) that is being tried by lessening the time. Stage 2 prepares the system for deciding the articulation of that subject. Stage 3 is utilized to check the accuracy (Fig. 3).

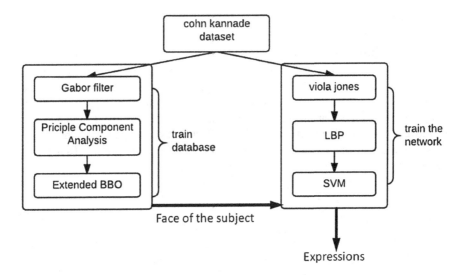

Fig. 3 Block diagram for experimental setup

Fig. 4 ROC curve

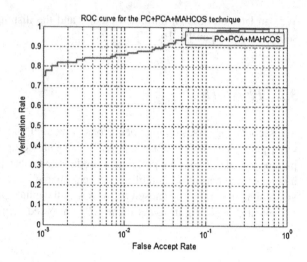

4 Results and Discussion

The Cohn Kannade database utilized comprises of 40 subjects. Every subject contains numerous pictures displaying different emotions. We have restricted our database to 10 pictures in every subject. Training has been performed on 9 pictures of every subject leaving one untrained picture in every subject. On testing the tenth picture for its declaration the outcomes received are as indicated in the subsequent tables and curves.

4.1 ROC Curve

To show effectiveness of face acknowledgment calculation in view of extended species abundance model of biogeography an ROC bend has been plot in Fig. 4, with confirmation rate at Y pivot and False Rejection Rate at X pivot. Productivity can without much of a stretch be seen at the beginning stage in results, it can be plainly demonstrated that preparation database with Extended BBO outflanks the approaches used without extended BBO. Result acquired from methodology demonstrates high check rate than by only Gabor and PCA calculation (Tables 1, 2 and 3).

Table 1 Techniques used in the framework

Comparison	Support vector machine	Linear binary pattern	Viola jones
Proposed by	Vladimir N. Vapnik, Corrina Cortea in 1993	T. Ojala, M. Pietikäinen, and D. Harwood in 1994	Paul Viola and Michael Jones (2001)
Approach	Supervised learning approach	Texture spectrum model	Integrates Adaboost and cascade classification
Application	Text categorization, bioinformatics, hand written character recognition	Image retrieval, visual inspection, motion analysis, environment modelling	Face detection, object detection
Limitations	Inseparable data cannot be classified by SVM	It does not specify any methodology to improve accuracy by some more percentage under high localization errors	It does not work very accurately for side face detection. Moreover the results get altered in the presence of very high or very low lighting effects
Usefulness	It eradicates the problem of overfitting. It is a non-probabilistic binary classifier that works well for extracting emotions in a face image or text	LBP exhibits a robust nature. It is independent of gray level variations in the image. Results generate are appropriate even in the presence of localization errors but with a margin of improvement	It scales the features by keeping the image size constant as it only crops and enlarges the wanted segment. This algorithm gives a better efficiency in less time for extracting features

Table 2 Confusion matrix

Emotions	Anger	Disgust	Fear	Happiness	Sad	Surprise	Neutral
Anger	60.2	4.2	23.4	0	0	5.2	7
Disgust	5.2	80.19	2.7	0	0	12	0
Fear	1.2	20.4	58.3	0	15.1	5	0
Happiness	0	0	0	95	0	5	0
Sad	0	16.2	1.2	0	68	14.6	0
Surprise	1.2	2.4	0	2.2	0	94.2	0
Neutral	8.2	0	0	0	0	0	91.8

Table 3 Average confusion matrix

Emotions	Neutral	Positive	Negative	Surprise
Neutral	91.8	0	8.2	0
Positive	0	95	0	5
Negative	1.75	0	89.05	9.2
Surprise	0	2.2	3.6	94.2

5 Conclusion

Separating emotive data by advanced instruments gives the security framework a help. Henceforth we have shown expression recognition using optimization and classification algorithms. The mix of procedures that have been utilized in the paper live up to the expectations precisely on the database used. To affirm its precision on continuous feature successions multiclass support vector machine must be connected on the image set. Stage 1 and stage 2 can work autonomously and produce results for their particular capacities. Subsequent to consolidating the two stages and building a scaffold between them as specified in the stage 3 a more enhanced working with better results is created. In the future work the centre will be fixated on enhancing the computational unpredictability. The accentuation will likewise be laid on a decent joining of voice and motion acknowledgment alongside expression location.

References

1. Struc, Vitomir, Rok Gajsek, and Nikola Pavesic. "Principal Gabor filters for face recognition. "*Biometrics: Theory, Applications, and Systems, 2009. BTAS'09. IEEE 3rd International Conference on.* IEEE, 2009.
2. Surbhi Agarwal, Madhulika. "A collation: Implementing mono and hybrid approaches for feature extraction" ISSN 0973-4562 Volume 10, Number 6 (2015) pp. 15671–15678 *International Journal of Applied Engineering Research* (IJAER),10, 15671–15678 (2015).
3. Sebe, Nicu, et al. "Emotion recognition using a cauchy naive bayes classifier. "*Pattern Recognition, 2002. Proceedings. 16th International Conference on.* Vol. 1. IEEE, 2002.
4. Madhurima, Madhulika. "Object tracking in a video sequence using Mean-Shift Based Approach: An Implementation using MATLAB7. "*IJCEM International Journal of Computational Engineering & Management* 11 (2011).
5. Michel, Philipp, and Rana El Kaliouby. "Real time facial expression recognition in video using support vector machines." *Proceedings of the 5th international conference on Multimodal interfaces.* ACM.
6. Kanade, T., Cohn, J. F., & Tian, Y. (2000). Comprehensive database for facial expression analysis. Proceedings of the Fourth IEEE International Conference on Automatic Face and Gesture Recognition (FG'00), Grenoble, France, 46–53.
7. Dan Simon, "Biogeography based optimization", IEEE Transactions on Evolutionary Computation, Vol. 12, No. 6, 2008.
8. Goel, Lavika, Daya Gupta, and Vinod Panchal. "Extended Species Abundance Models of Biogeography Based Optimization." *Computational Intelligence, Modelling and Simulation (CIMSiM), 2012 Fourth International Conference on.* IEEE, 2012.
9. Daya Gupta, Kanishka Bansal "A Novel approach fro face recognition based on extended species of abundance model of biogeography". *second international conference of emerging research in computing, information, communication and applications (ERCICA-2014),* ELSEVEIR, vol 2, 2014.
10. Guo, Guodong, Stan Z. Li, and Kap Luk Chan. "Face recognition by support vector machines." *Automatic Face and Gesture Recognition, 2000. Proceedings. Fourth IEEE International Conference on.* IEEE, 2000.

11. Bhatia, M., et al. "Implementing Edge Detection for Medical Diagnosis of a Bone in Matlab." Computational Intelligence and Communication Networks (CICN), 2013 5th International Conference on. IEEE, 2013.

12. Ahonen, Timo, Abdenour Hadid, and Matti Pietikäinen. "Face recognition with local binary patterns." *Computer vision-eccv 2004*. Springer Berlin Heidelberg, 2004. 469–481.

13. Shan, Caifeng, Shaogang Gong, and Peter W. McOwan. "Facial expression recognition based on local binary patterns: A comprehensive study." *Image and Vision Computing* 27.6 (2009): 803–816.

14. Maturana, Daniel, Domingo Mery, and Alvaro Soto. "Face recognition with local binary patterns, spatial pyramid histograms and naive Bayes nearest neighbor classification." *Chilean Computer Science Society (SCCC), 2009 International Conference of the*. IEEE, 2009.

15. Hadid, Abdenour. "The local binary pattern approach and its applications to face analysis." *Image Processing Theory, Tools and Applications, 2008. IPTA 2008. First Workshops on*. IEEE, 2008.

16. Štruc, Vitomir, and Nikola Pavešić. "The complete gabor-fisher classifier for robust face recognition." *EURASIP Journal on Advances in Signal Processing* 2010 (2010): 31.

17. Štruc, Vitomir, and Nikola Pavešić. "Gabor-based kernel partial-least-squares discrimination features for face recognition." *Informatica* 20.1 (2009): 115–138.

Evaluation of Genetic Algorithm's Selection Methods

Hari Mohan Pandey, Anupriya Shukla, Ankit Chaudhary and Deepti Mehrotra

Abstract The focus of this paper is towards analyzing the performance of various selection methods in genetic algorithm. Genetic algorithm, a novel search and optimization algorithm produces optimum response. There exist different selections method available—plays a significant role in genetic algorithm performance. Three selection methods are taken into consideration in this study on travelling salesman problem. Experiments are performed for each selection methods and compared. Various statistical tests (F-test, Posthoc test) are conducted to explain the performance significance of each method.

Keywords Genetic algorithm · Ranking selection · Tournament selection · Roulette wheel selection · Travelling salesman problem

1 Introduction

Genetic algorithm (GA) is population based search and optimization algorithm proposed by Holland [1]. Reproduction operators such as crossover and mutation play an important role in GA's performance and maintain diversity in the population

H.M. Pandey (✉) · A. Shukla · D. Mehrotra
Department of Computer Science & Engineering, Amity University,
Sector 125, Noida, Uttar Pradesh, India
e-mail: profharimohanpandey@gmail.com

A. Shukla
e-mail: anupriyashukla2603@gmail.com

D. Mehrotra
e-mail: dmehrotra@amity.edu

A. Chaudhary
Department of Computer Science and Electirical Engineering,
Truman State University, Kirksville, USA
e-mail: dr.ankit@ieee.org

© Springer India 2016
S.C. Satapathy et al. (eds.), *Information Systems Design and Intelligent Applications*, Advances in Intelligent Systems and Computing 434,
DOI 10.1007/978-81-322-2752-6_72

731

—helps in achieving the global optima. There exists various selection techniques proposed includes roulette wheel, rank based, tournament, steady state, Boltzmann and elitism. Each of these selection techniques has their own ways of selection of populations at the initial stage and after each iteration (selection of population for next generation). The selection of population is known as selection pressure—if not selected intelligently leads to slow convergence rate and premature convergence [2].

Researchers studied the performance of GA via different selection techniques. Typically, GA's performance is evaluated using two factors: convergence rate and the number of generations required to reach to optimal solution. In [3] results of GA was compared for proportional roulette wheel and rank-based roulette wheel selection techniques conclude that rank-based method outperformed proportional roulette wheel considering total number of generations in reaching to optima with some observations made outlined as: rank-based selection is faster, more robust and certainty towards the optima. The comparison of proportional roulette wheel with tournament selection was presented in [4] explains the superiority of tournament selection over proportional roulette wheel. Two versions of rank-based selection probabilities, namely, linear ranking and exponential ranking examined against tournament selection considering the convergence time of GA—concluded that tournament selection is a better choice over rank-based selection, since repeated tournament selection is much faster than sorting the population and assigning rank-based probabilities [4]. Three different GA's selection methods: deterministic, tournament and roulette wheel were applied to inspect the performance of PCB inspection system—discovered that deterministic method needs lowest number of generation in reaching to the highest fitness [5]. A new selection approach known as sexual selection was proposed in [6]—was compared with most commonly used selection strategies—results that sexual selection either outperformed or performed better than roulette wheel selection and tournament selection when no fitness scaling was applied on the more difficult test cases. Goldberg and Deb [7] proposed a comprehensive studies four (proportional, ranking, tournament steady state) selection methods considering the solutions of differential equations. The focus of [7] was towards expected fitness ratio and convergence time—concluded that ranking and tournament selection outperformed proportional selection of maintaining steady pressure towards convergence. The study of [7] was extended considering linear ranking selection and stochastic binary selection leads to a conclusion that both have identical expectations—but they recommendation was drawn for binary tournament selection due to the efficiency in time complexity. Although significant works are conducted showing the importance of selection methods and their comparison but none of the research shows the statistical comparisons to represent the performance significance of various selection techniques. This paper presents the comparison of three selection techniques: tournament selection, roulette wheel selection and ranking selection. The domain of inquiry is travelling salesman problem (TSP)—a combinatorial problem implemented using GA considering all three selection methods. Experiments are conducted and results are collected for the analysis purpose. Statistical tests such as f-test and posthoc tests are conducted to examine the performance significance of selection methods considered.

The rest of the paper is organized as: Sect. 2 represent the GA applied for TSP. The importance of selection techniques with the discussion of three selection methods are reported in Sect. 3. The discussion of experiments, parameter tuning, results, statistical tests and discussion are drawn in Sect. 4. Lastly, conclusion is of the study in given in Sect. 5.

2 Genetic Algorithm and Travelling Salesman Problem

This section provides the general insight of GA for solving TSP. GA is an optimization method based on Darwinian principle of "survival of the fittest" proposed by Holland [1]. It employs stochastic approach to randomly search for optimum solution for the specified problem. Each individual also referred as chromosome—a member of population represents a potential solution. GA supports a number of possible chromosome representations. Fitness is the measure's the quality of solution in GA. GA supports two most important operators: crossover and mutation—helpful in maintaining the diversity of the population and guide the searching towards the global optima. TSP is finding a Hamiltonian cycle with minimum cost [8]. There exist a number of cities present in TSP, where each pair of cities has a corresponding distance. The goal is to visit all the cities with total distance travelled should be minimized. The procedure for solving TSP using GA is represented in Fig. 1.

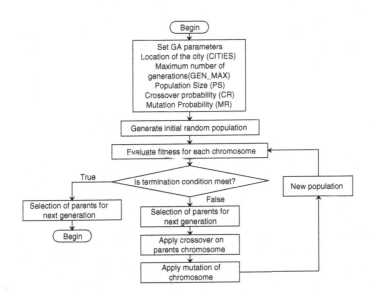

Fig. 1 Genetic algorithm for solving TSP

The GA process begins by initializing important parameters such as maximum number of generations (GEN_MAX), population size (PS), crossover probability (CP) and mutation probability (MP). In TSP location of cities plays an important role, therefore considered. An initial random population is created and evaluation of each chromosome is performed. The population is then checked for termination (maximum number of generation or optimum results) condition—if achieved then displays the results and stop otherwise transformed into a new population or next generation applying three GA operators: selection, crossover and mutation. Selection operation is performed to select couple of parents in order to procreate offspring by crossover and mutation. The new offspring contains a higher proportion of characteristics produced by the 'good' chromosome of previous generation— helps in spreading the good characteristics over the population—which then mixed with other good characteristics in iterative manner. This process continues until GA reach to best solution present in the solution space.

3 Selection Strategies

The primary objective of selection technique is to pick the chromosomes in the current generation—will participate to reproduce the offspring with hopes that next generation chromosome will have higher fitness. Hence, selection plays a signifi- cant role in GA performance—it is important to formulate selection operator that should ensures better member with higher fitness—have a greater probability to be selected for mating, but in some situation worse member still have a small prob- ability to be selected—results in premature convergence or converge at local optima. There exists various selection techniques each has different way of calcu- lating selection probability. The detailing of various selection methods can be seen in [7–11]. In this section, we bring to light an overview of three important selection methods: tournament selection, roulette wheel selection and ranking selection.

Tournament selection is the most popular method in GA due to its efficiency and simple nature [7]. A total n-individuals are picked randomly from the whole pop- ulation—compete against each other in a tournament referred as tournament size commonly set to 2. The highest fitness individual wins and will be selected as one of the next generation population. It gives a fair chance to each individual to be selected, hence maintain diversity.

In roulette wheel selection method individual get selected based on the certain probability directly proportional to the fitness value. Obviously, the individual have higher fitness has more probability of being selected, where the fittest individual occupies the largest segment within the roulette wheel.

In rank based selection the probability of an individual being selected is based on the fitness rank relative to the whole population. It first sort the individual chro- mosome in the population based on their fitness and then computes selection probabilities as per the rank rather than individual fitness value. It utilizes a function that maps the indices of individuals in the sorted list of the selection probabilities.

4 Simulation Model

Experiments are conducted to test the performance significance of selection strategies. Net Beans IDE 7.0.1, Intel Core TM2 processor 2.8 GHz with 2 GB RAM is used. GA's performance is critical to the parameters guides the overall search process, therefore extensive tuning is performed employing orthogonal array and Taguchi signal to noise ratio. For experiment design five factor at two levels were performed give as: crossover rate (CR) = [0.5, 0.8], mutation rate (MR) = [0.3, 0.6], population size (PS) = [120, 180], number of cities (CITIES) = [40, 60] and maximum generations (MGEN) = [100, 200]—the following setting produced the best results [CR: MR: PS: CITIES: MGEN] = [0.8: 0.6: 60: 100: 100].

Figure 2a represents the comparison chart for each selection method implemented. Having seen the Fig. 2a, it is very much clear that average distance for

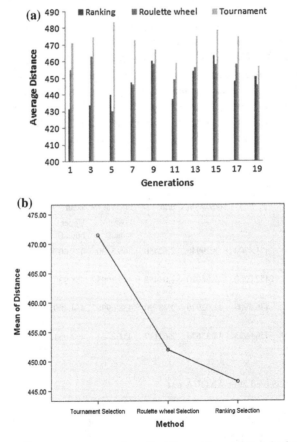

Fig. 2 a Average distance versus generations chart for ranking, roulette wheel and tournament selection methods. **b** Profile plot for estimated marginal mean for each selection methods implemented

tournament selection is higher indicates the worst performance whereas ranking selection produces least average distance—but this much information is not sufficient to reach to some conclusion. Hence, statistical tests are conducted.

F-test is conducted considering the hypotheses *"there is no significant difference in the mean of samples at the 5 % level of confidence"*. Table 1 shows the descriptive analysis whereas Fig. 2b graphically represents the means of three selection methods—X-axis shows the selection techniques and Y-axis presents the estimated marginal means. The result of F-test is reported in form of ANOVA table (see Table 2) indicates that significance value ($p = 0.000$) comparing the group level is less than 0.05—leads to rejection of the hypothesis. The results received from F-test do not indicate which selection method is responsible for the difference.

Therefore, posthoc tests (compare individual approach results with each other) are conducted. LSD and TukeyHSD tests are conducted leads to a conclusion that tournament selection is mainly responsible for the difference because the average distance for TSP received by employing tournament selection is higher than other two selection methods—is verified by applying homogeneity test and results of homogeneous subset test is reported in Table 4. It can be seen that ranking selection and roulette wheel selection belongs to the same set at 0.436 level of significance whereas tournament selection is in separate group at 1.000 level of confidence. From Table 3 one can see that the asterisk (*) mark given next the mean difference indicates that difference is significant. LSD test produces results accurately—is very sensitive to violation to the assumptions of ANOVA therefore most likely to lead to Type-I error i.e. rejecting null hypothesis when it is true. To alleviate this problem, TukeyHSD test is conducted.

Table 1 Descriptive analysis

	N	Mean	Std. deviation	Std. error	95 % confidence interval for mean		Minimum	Maximum
					Lower bound	Upper bound		
Tournament selection	10	471.4700	8.29940	2.62450	465.5330	477.4070	457.00	483.80
Roulette wheel selection	10	452.0200	9.66020	3.05482	445.1095	458.9305	430.10	463.10
Ranking selection	10	446.5900	11.02970	3.48790	438.6998	454.4802	431.40	463.70
Total	30	456.6933	14.35673	2.62117	451.3324	462.0542	430.10	483.80

Table 2 Results received after ANOVA test

	Sum of squares	df	Mean square	F	Sig.
Between groups	3422.673	2	1711.336	18.087	0.000
Within groups	2554.686	27	94.618		
Total	5977.359	29			

Table 3 Multiple comparison test: Posthoc-LSD test Tukey HSD test

Test type	(I) Method	(J) Method	Mean difference (I-J)	Std. error	Sig.	95 % confidence interval	
						Lower bound	Upper bound
Tukey HSD	Tournament selection	Roulette wheel selection	19.45000[a]	4.35013	0.000	8.6642	30.2358
		Ranking selection	24.88000[a]	4.35013	0.000	14.0942	35.6658
	Roulette wheel selection	Tournament selection	−19.45000[a]	4.35013	0.000	−30.2358	−8.6642
		Ranking selection	5.43000	4.35013	0.436	−5.3558	16.2158
	Ranking selection	Tournament selection	−24.88000[a]	4.35013	0.000	−35.6658	−14.0942
		Roulette wheel selection	−5.43000	4.35013	0.436	−16.2158	5.3558
LSD	Tournament selection	Roulette wheel selection	19.45000[a]	4.35013	0.000	10.5243	28.3757
		Ranking selection	24.88000[a]	4.35013	0.000	15.9543	33.8057
	Roulette wheel selection	Tournament selection	−19.45000[a]	4.35013	0.000	−28.3757	−10.5243
		Ranking selection	5.43000	4.35013	0.223	−3.4957	14.3557
	Ranking selection	Tournament selection	−24.88000[a]	4.35013	0.000	−33.8057	−15.9543
		Roulette wheel selection	−5.43000	4.35013	0.223	−14.3557	3.4957

[a]The mean difference is significant at the 0.05 level

Table 4 Homogenous subset test

	Method	N	Subset for alpha = 0.05	
			1	2
Tukey HSD[a]	Ranking selection	10	446.5900	
	Roulette wheel selection	10	452.0200	
	Tournament selection	10		471.4700
	Sig.		0.436	1.000

Means for groups in homogeneous subsets are displayed
[a]Uses Harmonic Mean Sample Size = 10.000

5 Conclusions

This paper analyzes the performance of three selection strategies of GA. A case of TSP was considered and implemented using GA. Extensive parameters tuning were done before performing the final experiments. This paper not only report the comparison of results obtained experimentally but also report the results of various statistical tests such as F-test and posthoc tests. A comprehensive discussion is presented on the statistical test conclude that rank based selection outperformed other two selection methods in terms of quality of results and convergence time. The performance of tournament selection is found worst whereas roulette wheel selection shows average performance. We believe that results and discussion reported in this paper will be helpful in selecting the appropriate selection method in conducting the experiments.

References

1. Holland, John H. *Adaptation in natural and artificial systems: an introductory analysis with applications to biology, control, and artificial intelligence.* U Michigan Press, 1975.
2. Pandey, Hari Mohan, Ankit Chaudhary, and Deepti Mehrotra. "A comparative review of approaches to prevent premature convergence in GA." *Applied Soft Computing* 24 (2014): 1047–1077.
3. J. Zhong, X. Hu, M. Gu, J. Zhang, "Comparison of Performance between Different Selection Strategies on Simple Genetic Algorithms," *Proceeding of the International Conference on Computational Intelligence for Modeling, Control and automation, and International Conference of Intelligent Agents, Web Technologies and Internet Commerce,* 2005.
4. B. A. Julstrom, It's All the Same to Me: Revisiting Rank-Based Probabilities and Tournaments, Department of Computer Science, St. Cloud State University, 1999.
5. S. Mashohor, J. R. Evans, T. Arslan, Elitist Selection Schemes for Genetic Algorithm based Printed Circuit Board Inspection System, Department of Electronics and Electrical Engineering, University of Edinburgh, 974–978, 2005.
6. K. S. Goh, A. Lim, B. Rodrigues, Sexual Selection for Genetic Algorithms, *Artifial Intelligence Review* 19: 123–152, Kluwer Academic Publishers, 2003.
7. D.E. Goldberg and K. Deb, A comparative analysis of selection schemes used in genetic algorithms, in: G.J.E. Rawlins (Ed.), Foundations of Genetic Algorithms, Morgan Kaufmann, Los Altos, 1991, pp. 69–93.
8. Horowitz E., Sahani S, and Rajasekaran S, 2007. Fundamentals of Computer Algorithm, University Press, 2007.
9. Handbook of Evolutionary Computation, IOP Publishing Ltd. and Oxford University Press, 1997.
10. T. Blickle, L. Thiele, A Comparison of Selection Schemes used in Genetic Algorithms. TIK-Report, Zurich, 1995.
11. J. E. Baker, "Adaptive selection methods for genetic algorithm," *Proceeding of an International Conference on Genetic Algorithms and Their Applications,* 100–111, 1985.

An Amalgamated Strategy for Iris Recognition Employing Neural Network and Hamming Distance

Madhulika Pandey

Abstract Biometric comprises of strategies for particularly perceiving people based upon one or more inherent physical or behavioral characteristics. Iris recognition system is one of the fundamental techniques that are used in biometrics for access control, identification system. It is essentially a pattern distinguishment technique that utilizes iris structures and patterns that are measurably novel, with the goal of user identification. It is relentless for the term of the life and serves as a living visa or a code word that one need not remember and recall however is present always. This study concentrates on the novel approach that emphasizes on the characterization methodology of the iris designs by utilizing a collaborative methodology of neural networks and hamming distance. The proposed system additionally uses the support vector machine with the end goal of grouping of the iris as the left iris design or as the right iris of a person.

Keywords Iris recognition · Neural network · Hamming distance · Support vector machine · Segmentation · Normalization · Feature extraction · Classification

1 Introduction

In today's scenario, possibility of unauthenticated intrusion into the system is one of the greatest threats encountered by the IT systems. User authentication systems based on passwords, identification cards can easily be forged and misused hence raise the need of more robust and secured systems. Technologies focusing on

---

An erratum of this chapter can be found under
DOI 10.1007/978-81-322-2752-6_76

M. Pandey (✉)
Pune, India
e-mail: madhulika.pandey02@yahoo.in

© Springer India 2016
S.C. Satapathy et al. (eds.), *Information Systems Design and Intelligent Applications*, Advances in Intelligent Systems and Computing 434,
DOI 10.1007/978-81-322-2752-6_73

biometrics have the potential utilization in distinguishing people and hence managing the access to authenticated areas [1]. Biometric traits attempts to address and prevent the computer based frauds and thefts. They attempt to identify an individual on the basis of his behavioral or physiological trait thereby differentiating in between an authorized person and defrauder [2]. Accuracy rate and search space contributes as an vital parameters for the achievement of the reliable recognition framework. As the size of the databases, identifying an imposter becomes a difficult task hence it becomes important to recognize a procedure which can seek the right individual within the specified duration [3].

The biometric recognition process operates in two phases namely enrollment and verification. In enrollment phase, the users biometric is captured, its unique reference template is extracted and stored into the database. During verification phase, users biometric is captured and converted into the test template and is compared with the reference template stored into the database. If a match occurs, the user is considered to be authentic else remains he unidentified as shown in Fig. 1.

Fig. 1 Biometric recognition system

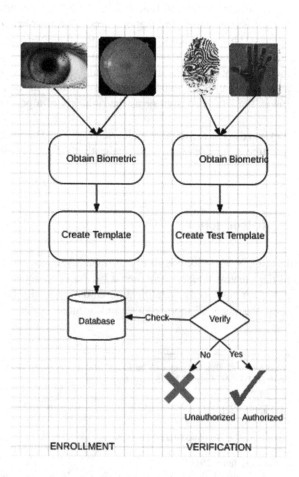

In this study, we emphasize on iris biometrics because of its stability, distinctiveness, high precision and statistically uniqueness in the features among the different biometric applications. The estimated likelihood for the presence of two comparative irises is 1 in 10^{72} [4]. It has been progressively promoted by both private and governmental bodies due to its high precision and stability and is regarded as biometric signature [5].

Iris is a ring shaped shaded architecture near the pupil having phenomenal structure providing many minute traits such as stripes, freckles, coronas etc. [1]. Iris is a protected organ that is externally visible. Its epigenetic structures stay durable for the entire duration of the life. This distinctive feature of iris makes it a potential candidate to be utilized as a biometric for personal identification. Various digital image processing approaches are utilized for extracting iris from the eye picture, encode it and store it as a biometric template [6]. This biometric template represents the novel features that can be compared with other templates that are already enrolled in the database by using a suitable classifier. When classification has to be done using a iris recognition system, a photo of the user's eye is taken and after that an iris template is made from the image. This iris template is contrasted with other templates until a coordinating layout is discovered and the individual is distinguished; or if no match is discovered then the individual stays unidentified [7].

Iris recognition additionally has detriments as well. Eyelid and eyelashes tends to obstruct few sections of the iris. Iris and pupil boundaries sometimes fail to form complete circular geometry thereby failing to form concentric centers. In the event when boundary of the iris is approximated as of circular geometry, a few sections of sclera and pupil may be displayed in the iris region. All these attributes impact the conduct of iris recognition framework. To tackle these issues, a powerful strategy is expected to uproot the impact of all noises to the greater extent [3].

In this paper, aggregation of two classification techniques has been used. For image preprocessing, morphological approach has been used. Feature extraction phase uses HAAR wavelet up to 3 level decomposition and 1D log gabor filter. For classification phase, neural network is used as a primary classifier and for secondary classifier Hamming Distance is chosen. Errors obtained from the primary classifier are compared with the pre set threshold value. If the error lies within the range then results from primary classifier are indicated else results from the secondary classifier are indicated. Figure 2 shows the proposed idea.

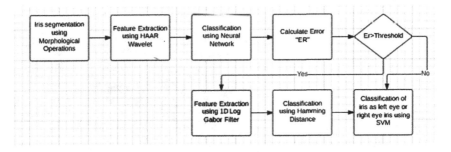

Fig. 2 Flow mechanism of the proposed hybrid approach

2   Iris Recognition Mechanism

Iris recognition framework constitutes image acquisition, image preprocessing, image segmentation, normalization, feature extraction and classification as its stages [8].

2.1   *Image Acquisition and Preprocessing*

Image acquisition is a standout amongst the most essential step and integral component for getting valuable result. A clear and good picture disposes of the methodology of noise removal furthermore helps in removing errors in computation. This project utilizes the pictures given by CASIA and are utilized singularly with the end goal of iris recognition programming examination and usage. For analysis of medical images various image processing techniques are available and to statistical analysis can also be done [9].

2.2   *Iris Segmentation*

In the proposed work, morphological approach has been considered for the segmentation stage. Under this process, an iris image is transformed into black and white image and then is complimented. Then the operations are applied to remove the artifacts present in the image as well as at the border. Figure 3 shows the segmented iris image.

2.3   *Normalization*

Normalization helps in making the representation universal with respect to iris optical size, pupil position to all templates with the identical and comparable

Fig. 3 Segmented iris image

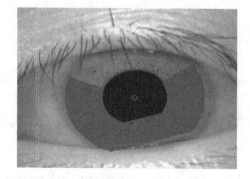

Fig. 4 Normalized iris image

dimensions [10]. After the segmentation stage, the iris region is transformed to have fixed settled dimensions [11]. The Daugman rubber sheet model is used for the normalization stage which maps the pixels of the iris from Cartesian coordinate to polar coordinate (r, Θ) by considering reference point as the center of the pupil as shown in Eqs. (1), (2) and (3) [12].

$$I(x(r, \Theta), y(r, \Theta)) \rightarrow I(r, \Theta) \tag{1}$$

$$x(r, \Theta) = (1 - r)x_p(\Theta) + rx_i(\Theta) \tag{2}$$

$$y(r, \Theta) = (1 - r)y_p(\Theta) + ry_i(\Theta) \tag{3}$$

I(x, y) denotes the image of the iris where (x, y) and (r, Θ) denotes the Cartesian coordinates and polar coordinates respectively; (x_p, y_p) denotes the coordinates of the pupil and (x_i, y_i) denotes the coordinates of the iris with the direction of Θ [6]. Figure 4 shows the normalized view of the segmented iris image.

2.4 Feature Extraction and Classification

Once the segmentation takes place, the significant texture data needs to be extricated. In this proposed work, HAAR wavelet and 1D Log Gabor Filter has been used for feature extraction. For classification, neural network is treated as a primary classifier whereas hamming distance is adopted as secondary classifier. HAAR wavelet formulates a feature vector which services neural network and 1D Log Gabor filter's feature vector services hamming distance as they are well suited for the respective classifiers. On the basis of comprehensive testing, united approach of neural network and hamming separation has preferable distinguishment precision over using a solitary strategy [11].

2.4.1 Feature Extraction

HAAR Wavelet

Wavelets can also be utilized to interpret the information present across the iris. Wavelets possesses the preference over conventional Fourier transforms as their frequency data is confined in wavelet, therefore permitting features which happens at same location as well as the resolution to be coordinated up [11]. For feature extraction HAAR wavelet is enforced to the 64 * 512 sized normalized image

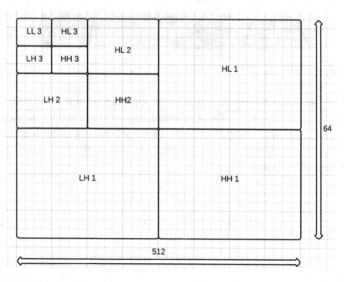

Fig. 5 Haar wavelet decomposition

successively. The image is partitioned into four sub areas namely Approximation coefficients (LL), Vertical coefficients (LH), Horizontal Coefficients (HL) and Diagonal coefficients (HH) where LL incorporates most of the energy as shown in the Fig. 5. The low frequency (LL) region serves a image for the next iteration where wavelet transform can be applied to the significant areas. Here, LL3 is obtained as a result of successive wavelet transformation for 3 times. The information contained in LL3 has 8 * 64 = 512 features. Haar wavelet function $\psi(t)$, its scaling function $\varphi(t)$ are represented as [12]:

$$\psi(t) = \begin{array}{ll} 1 & (0 < t < 1/2) \\ -1 & (1/2 \le t < 1) \\ 0 & \text{otherwise} \end{array} \tag{4}$$

$$\varphi(t) = \begin{array}{ll} 1 & (0 \le t \le 1) \\ 0 & \text{otherwise} \end{array} \tag{5}$$

1D Log Gabor Filter
The process of convolution of the normalized iris area along with 1D Log-Gabor filter leads to feature encoding. The obtained 2D pattern of the normalized region is fragmented into the series of 1D signal that are further convolved with the 1D Gabor filters. Filtering of Image is performed is used to remove noises from the image is collected [13]. After the process of filtering, its output is phase quantized up to four levels. Once the encoding process is completed, bitwise template and noise mask is produced which are used for matching.

2.4.2 Classification

Neural network consists of interconnected neurons that accepts the inputs, processes it and produces the results accordingly. Fault tolerance, self organizing and adaptive nature characteristics of neural network make it a suitable candidate for classification in the era of artificial intelligence. It offers many models that help in the formation of intelligent systems. Feed forward neural network is a standout amongst the most prominent models of neural networks with input layer, intermediate hidden layer and output layer. Values are presented to the input layer which distributes them further to hidden layer to perform necessary computations. Hidden layer further forwards the process by passing the values to the output layer from which the final result is obtained.

Hamming distance quantifies the number of bits that are similar in between the patterns which identifies patterns are of the similar iris or of the different ones [14]. Hamming distance for the bit patterns A and B can be expressed as the total of the differing bits over the total bits in bit pattern N as represented in Eq. (6). Differing bits can be characterized as XOR operation between A and B [15, 16].

$$HD = \frac{1}{N} \sum_{i=1}^{N} A_j(XOR)B_j \qquad (6)$$

In this work, the iris picture is classified using consolidated Neural system and Hamming Distance based methodology. At the point when an iris picture is obtained, its 512 most significant features are separated by HAAR wavelet. These extracted features are utilized for the training and testing purpose of the neural system. In the event that neural system neglects to perform right classification the further classification is carried out by utilizing Hamming Distance.

3 Experiment and Results

The proposed hybrid approach was implemented using MATLAB 7.12.0(R2011a) on Windows7, Intel Core i3 Processor [17]. The experiment was performed by taking 11 users into consideration. 3 images of left eye as well as of right eye of each user were considered for training purpose. The approach was tested against 11 test images.

The accuracy was determined in terms of false rejection ratio. False rejection ratio signifies the probability of not identifying an enrolled user and rejecting him. Further, neural network and hamming distance in combination informs about the user to which the test image belongs.

When feed forward neural network was solely utilized for classification along with SVM to distinguish the iris as of left eye pattern or right eye pattern, rate of precision accomplished was 63.6363 % and false rejection rate was 0.2727. When

Table 1 Comparative results

Approach	Accuracy on trained samples (%)	Accuracy on untrained samples (%)
Neural network + SVM	65.15	63.63
Proposed hybrid approach	97.81	81.8181

feed forward neural network was utilized with hamming distance as a compound strategy in addition with SVM, estimate of accuracy obtained was 81.8181 % and false rejection rate was 0.1818. When 66 training samples, 6 for each user were presented as input to the neural network for classification along with SVM for intra eye classification then the accuracy of 65.15 % was achieved where as in case of proposed hybrid approach precision up to 97.81 % was achieved (Table 1).

4 Conclusion

This paper exhibits a novel productive methodology for feature extraction and recognition of an iris image. Here, feature extraction mechanism uses HAAR wavelet and 1D Log Gabor channel have been utilized for highlight extraction. These extricated features were used for identification of the iris using collaborative approach of Neural Network and Hamming Distance along with SVM for intra-iris classification and the results were compared with neural network approach. The proposed technique has preferable distinguishment ability over employing Neural Network or Hamming separation alone. Also, SVM correctly classifies the iris as left eye or right eye to a great extent. It also reduces the false rejection ratio in contrast with using neural network solely. It is additionally observed that the productivity has been expanded when we utilized separate feature extraction systems for Neural Network and Hamming Distance.

References

1. de Martin-Roche, D., Carmen Sanchez-Avila, and Raul Sanchez-Reillo. "Iris recognition for biometric identification using dyadic wavelet transform zero-crossing." *Security Technology, 2001 IEEE 35th International Carnahan Conference on.* IEEE, 2001.
2. Chirchi, Vanaja Roselin E., Dr LM Waghmare, and E. R. Chirchi. "Iris Biometric Recognition for Person Identification in Security Systems. "*International Journal of Computer Applications (0975–8887)* 24.9 (2011): 1–6.
3. Meetei, Thiyam Churjit, and Shahin Ara Begum. "A Comparative Study of Feature Extraction and Classification Methods for Iris Recognition. "*International Journal of Computer Applications* 89 (2014).
4. Flom, L., and Safir, A., "Iris Recognition System," US Patent 4 641 394, 1987.

5. Daftry, Shreyansh. "Artificial Neural Networks based Classification Technique for Iris Recognition." *International Journal of Computer Applications* 57.4 (2012).
6. Rampally, Deepthi. *Iris recognition based on feature extraction*. Diss. Kansas State University, 2010.
7. Boyd, Michael, et al. "Iris recognition. "*Imperial College London, Inglaterra* (2010).
8. Tania, U. T., S. M. A. Motakabber, and M. I. Ibrahimy. "Edge detection techniques for iris recognition system. "*IOP Conference Series: Materials Science and Engineering*. Vol. 53. No. 1. IOP Publishing, 2013.
9. Madhulika, Bansal Abhay, Yadav, Divakar, Madhurima. Survey and Comparative Study on Statistical Tools for Medical Images. Advanced Science Letters. 2015 Jan; 21(1): 74–77.
10. Patil, Shankargouda M., and B. K. Sarojini. "An Efficient Iris Recognition System using Phase-based Matching".
11. Rai, Himanshu, and Anamika Yadav. "Iris recognition using combined support vector machine and Hamming distance approach. "*Expert systems with applications* 41.2 (2014): 588–593.
12. Ramkumar, R. P., and S. Arumugam. "A novel iris recognition algorithm." *Computing Communication & Networking Technologies (ICCCNT), 2012 Third International Conference on*. IEEE, 2012.
13. Bhatia, Madhulika, Yadav, D, Madhurima, Gupta, P, Kaur, G, Singh, J, Gandhu, M, Singh, A "Implementing Edge Detection for Medical Diagnosis of a Bone in Matlab." *Computational Intelligence and Communication Networks (CICN), 2013 5th International Conference on*. IEEE, 2013.
14. Gupta, Sudha, et al. "Iris Recognition System using Biometric Template Matching Technology. "*International Journal of Computer Applications* 1.2 (2010): 1–4.
15. Daugman, John G. "Biometric personal identification system based on iris analysis." U.S. Patent No. 5,291,560. 1 Mar. 1994.
16. Jain, Yogendra Kumar, and Manoj Kumar Verma. "Comparison of Phase Only Correlation and Neural Network for Iris Recognition. "*International Journal of Computer Science Issues (IJCSI)* 9.1 (2012).
17. MATLAB 7.12.0 (2011a) of MathWorks, Inc, USA, "MATLAB" software's.

A Comparative Investigation of Sample Versus Normal Map for Effective BigData Processing

Shyavappa Yalawar, V. Suma and Jawahar Rao

Abstract MapReduce is an effective tool for the parallel-processing of data. A major problem in practice, MapReduce Skew of the data: imbalance amount of data for each task consigned. Because some of the tasks to last much longer than other, and can greatly affect performance. A scale that is lightweight strategy for data skew problem solving Applications, to the reducer side in MapReduce. In contrast to previous work scale is no need to scan in front of a series of input data or to prevent the overlap between the maps and reduce phases. System uses innovative idea take samples, which can achieve a high level of calculation and produce accurate approximation for the distribution of intermediate data by scanning only a small portion of the data on intermediate Map of the normal processing. It allows the reduction of tasks, to start the copying once the sample map functions selected (only a small part Map of tasks that have been fully spent for the first time). It supports split large clusters make connotations when applied and the total Output data is set up. System is implemented in Hadoop and Our experiments show that the implementation of some popular applications to speed up on negligible and can speed up Factor 4.

Keywords Bigdata · Mapreduce · Dataskew · Straggler

S. Yalawar (✉)
Post Graduate Programme, Computer Science and Engineering, Department of Information Science and Engineering, Dayananda Sagar College of Engineering, Bangalore, India
e-mail: shyavappayalawar@rediffmail.com

V. Suma
Department of Information Science and Engineering, Research and Industry Incubation Centre (RIIC), Dayananda Sagar College of Engineering, Bangalore, India
e-mail: sumavdsce@gmail.com

J. Rao
Department of Industrial Engineering and Management,
Dayananda Sagar College of Engineering, Bangalore, India
e-mail: jawahar_rao@yahoo.com

© Springer India 2016
S.C. Satapathy et al. (eds.), *Information Systems Design and Intelligent Applications*, Advances in Intelligent Systems and Computing 434,
DOI 10.1007/978-81-322-2752-6_74

1 Introduction

For every two years data is doubling up because of Unstructured data are simply data that does not suitable into traditional relational database systems; the term Unstructured Data includes log files, emails, word documents, multimedia, video, PDF files, excel sheets, messaging text, images and graphics, graphs, charts, GPS records, and social media data which contains all the other elements on a huge scale. This huge collection of the data is nothing but a big data [1–13].

MapReduce has verified that one as an effective tool in order to process such huge data sets. There are various computing frameworks to process this data. Ex: Google Map Reduce, Microsoft Dryad and Apache hadoop. MapReduce has proven to be an effective tool to process this data. Apache hadoop is an open source and is broadly used. In a MapReduce, the job finishing point is decided by the slowest running task known as straggler. It has ability to stoppage the progress of the entire job. Data skew is the important task in MapReduce, to avoid the straggler problem data skew should perform well. Data skew means dividing the complete task or job equally to all the nodes.

MapReduce job completion time is still dependent on the work of the slowest running tasks. When a task consumes more time when compared to other (so-called backward) ingesting, it has the ability to sluggish down the progress of the entire job. Later in the image can for various reasons, including data skew is important soon.

The main reason relates to skew the data recorded in the real time are often distorted; User don't know the circulation of the expected data. It should be noted that this problem will not be speculative in MapReduce implementation of strategies for solving. Data migration is not a new problem related MapReduce Some literature studies at the beginning of parallel databases, but only to join and aggregation operations. While some of them after use of MapReduce, Developer have to develop their own data offset remission in most cases, more specific application. Hadoop map Reduce implementation of the standard hash function using static branch out of the intermediate data.

We create a procedure for taking samples of the new data sets of bigdata general knowledge. This method with a high degree of parallelism and little effort, cannot be attained a good approximation for the circulation of the data set temporarily.

We use an innovative approach for move between tasks, and the reduction of Split of large badges for compensation if they support the approval of the application semantics. In addition, the use of our method LIBRA each reducer process about the same amount of data.

Although the predictable of the simple heterogeneous computing platform that can be expand the distribution of the workload on the basis of the enhanced data accordingly, provide the best performance in the absence of Data skew.

LIBRA in Hadoop and assess its performance to some common applications. The experimental results have a clear picture; Libra can improve by up to a factor of 4 time carry out the task.

2   Proposed System

In this part of the proposed method that has the ability to approach system, and to achieve a solution Data skew for general applications. The map Reduce framework by us to adopt to implemented system is in Hadoop—1.0.0. The design goals listed below:

- *Total order*
- *The separation of a enormous data*
- *Transparency*
- *Parallelism*
- *Accuracy.*

2.1   System Overview

System architecture consists of HDFS, Map phase, Master node and Reduce phase.

- *Loading the input into the HDFS*: The HDFS is the distributed file system for hadoop which acts as a storage device. It contains Namenode, Datanode and secondary namenode before beginning to the data process it divides the large input data block into the small blocks so that the namenode assigns them to process the data and it saves the replica of processing data, in case the namenode goes down or offline the secondary namenode will start processing so that this is main advantage of the hadoop, that is the data will not miss in any point.
- *Mapper phase*: The mapper phase consists of two stages one is sample map and other is normal map, the system considering the 20 % of sampling data, according to the requirement we can change the percentage of the sample. Map phase is used to filtering and sorting of the data than it gives its output to the master node.
- *Master node*: The master node takes the output of the map phase then it decides and assigns the job to the reducer which block is to going to which reducer all these things can be handled by master node. In this system the master node only considering the 20 % of the data as a sample and according its output it is estimating data distribution for all other remaining data to process then it elect the partition and notifies to the worker node to process information. By this the map phase completes all the data processing without any overhead.
- *Reducer phase*: Now the output of mapper is goes to reducer it does the summary operation for sorting and shuffling the generated output by both phase and finally sends the output to the HDFS (Fig. 1).

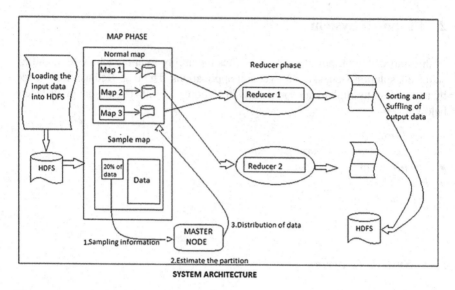

Fig. 1 System architecture represents the complete work flow

2.2 Dividing the Large Cluster

In a MapReduce framework the whole cluster appears in a single reducer to process the data due to this the data skew problem may occur. To avoid the particular data skew on the reducer side, the system splitting the large cluster. As shown in the above figure: example of large cluster splitting, the intermediate data taken from the network traffic analysis data set the name of the country and the count of the network devices is pair in the above figure. There are three countries Finland, Argentina and Singapore and the counts are 10,000, 4000 and 2000 respectively. Network traffic analyser is a data set in my project I have taken only 2 GB of the data (Fig. 2).

To solve the data skew on the reducer side we are trying split the large cluster into two so that both the reducers use same amount of workload. By dividing the count equally all the tasks can process at the same time both the reducer process the complete count by parallel. Another method to process the data is broadcast join but compare to this joining technique the cluster splitting is gives better result. By using the large cluster splitting in the above example the 60 % of the large data is optimized to the reducer 1 and the rest of reducer 2 data. By using this strategy lager cluster splitting it provides more flexibility in data skew mitigation.

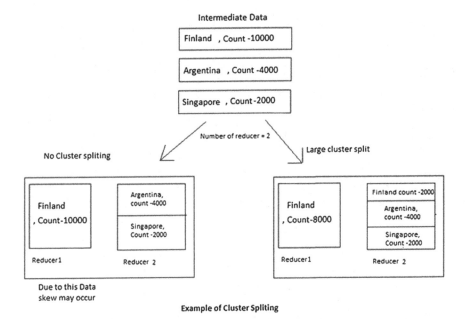

Fig. 2 Example of cluster splitting by using the network traffic analyzer data

3   Result

There are types of partitions in hadoop for data processing we are considering some of them—Key, Hash and Binary. The hash partition is default one to process the data. So we are processing our data set with all the three types of partitions. The above graph represents the job execution time with respect to the partitions. Key partition takes more time than the other strategy. By observing the above graph we can say that hashing technique better technique to process the data (Figs. 3 and 4).

Fig. 3 Result of the different partitioner job execution time

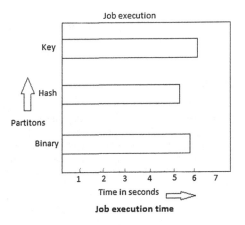

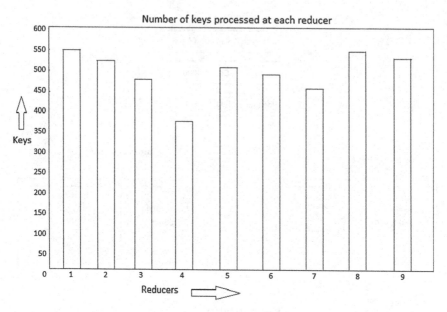

Fig. 4 Represents the 9 reducers taken as boundary versus number keys processed by the each reducer using three partition. Each *block* above denotes the average of the three partitions

4 Conclusion

Data skew mitigation is important in improving MapReduce performance. The technique LIBRA has proposed which implements a set of innovative skew mitigation strategies in an existing MapReduce system. This paper was presented a system; Where the input data is loaded to the HDFS and then map phase will execute the master node collects the information of the sample map through that information the master node estimate the data distribution for overall data by this strategy overhead is reduced and data skew on the reducer side is also solved. A unique feature of this paper is it supports for a large block and cluster split—which is adjusted to the Heterogeneous environments.

References

1. J. Dean and S. Ghema "Mapreduce: simplified data processing on large clusters" Communication acm volume 51, Jan 2008.
2. "Apache Hadoop, http://lucene.apache.org/hadoop/".
3. M. Isard, Y. Yu, and D. Fetterly "Dryad" European conference on Compter Systems 2007.
4. Y. Kwon, M. Balazinka and Howe "A study of skew in MR Applications" Cirrus 2011.

5. C. B. Walton, A.G. Dale, and R. M. Jenevein, "A Taxonomy and Performance model of data skew effects in parallel joins". International Conference on very large databases (VLDB), 1991.
6. D. DeWitt, J. Naughton, A. Schneider, and S. Seshadri, "Practical Skew Handling in Parallel Joins", VLDB, 1992.
7. J. Stamoas and C. Young "Symmetric Fragment and Replicate algorithm for Distributed Joins", IEEE TPDS Vol. 4 1993.
8. V. Poosala and Y. Ioannnidis, "Estimation of query-result distribution and its application in parallel-join load", VLDB, 1996.
9. Y. Xu and P. Kostamaa. "Efficient outer join data skew handling in parallel DBMS", VLDB vol. 2.2 2009.
10. S. Acharya, P. B. Gibbons and V. Poosala, "Congressional Samples for Approximate Answering of Group-by Queries", International Conference on Mgmt of Data 2000.
11. A. Shatdal and J. Naughton, "Adaptive Parallel Aggregation Algorithms", ACM Sigmod International conference on Mgmt of Data 1995.
12. J. Rolia and B. Howe, "Skew–Resistant Parallel Processing of Feature–Exatracting Scientific User-Defined Functions", ACM Symposium on Cloud Computing, 2010.
13. Qi Chen Yao and Zhen Xiao "LIBRA: Lightweight data skew mitigation in mapreduce", IEEE Transactions on parallel and distributed systems.

Big Data Management System
for Personal Privacy Using SW and SDF

Kashinath Sarjapur, V. Suma, Sharon Christa and Jawahar Rao

Abstract In this world of internet and social network, privacy is the one word that concerns everyone. All the data concerned to a person will get updated in the web and is available at ease. Hospitals and health centers when computerizing their center will knowingly or unknowingly be a part of this. Health related data is a very sensitive data that people are reluctant to disclose. Hospitals should go an extra mile to preserve the privacy of their clients. The techniques that are available in preserving privacy don't serve its purpose. Thus a novel method completely new in the field of big data is introduced in privacy preservation namely personalized anonymity. The central idea of this technique can be distributed into two main components. The component one of the workdeal with attributes in the patient data which is used as a flag and can be used to differentiate sensitive attribute. The attributes include sensitive disclosure flag (SDF) as well as sensitive weigh (SW). The second component deals with a new demonstration called Frequency Distribution Block (FDB) and quasi-identifier Distribution Block (QIDB), which uses the SW and SDF for anonymity. The paper provides an overview of personalized anonymity technique in medical big data which in turn enhances the privacy of users.

Keywords Big data management · Privacy preservation · Sensitive disclosure flag · Sensitive weigh · Frequency distribution block

K. Sarjapur (✉) · V. Suma · S. Christa
Department of Information Science and Engineering,
Dayananda Sagar College of Engineering, Bangalore, Karnataka, India
e-mail: kashi.sarjapur@gmail.com

V. Suma
e-mail: sumavdsce@gmail.com

S. Christa
e-mail: sharonchrista@gmail.com

J. Rao
Department of Industrial Engineering and Management,
Dayananda Sagar College of Engineering, Bangalore, Karnataka, India
e-mail: jawahar_rao@yahoo.com

© Springer India 2016
S.C. Satapathy et al. (eds.), *Information Systems Design and Intelligent Applications*, Advances in Intelligent Systems and Computing 434,
DOI 10.1007/978-81-322-2752-6_75

757

1   Introduction

The concept of Big Data started evolving from the first decade of twentyfirst century. Big data can be anything that covers large or complex datasets. Big data analytics came to picture since conventional data analysis become impossible to work on large complex data. It was well received and embraced by computing giants like Google, eBay; LinkedIn and Facebook as well as startup firms. By the introduction of the big data concept the large amounts of data stagnating in the data repositories were put to good use. The impact of analytics in big data can be understood by considering the following real world scenario; the deciphering of human genome. It took originally 10 years to fully process and decipher the human genome. With the use of big data analytics, it can be achieved in a week.

The characteristics that make big data stand out is its volume, velocity and variety. With the exponential revolutionization of data storage technology, huge amount of data can be stored and processed. Data itself varies in different types from video data to image data to text data. Every look and corner of internet provides data in one or the other form. An enterprise system, the size of data can range from terabytes to petabytes. This large volume of data can itself can be termed big data. From batch data to periodic data the swiftness in which data is available to user has changed drastically. In this social media era, real time data is available at our fingertip. This is what is meant by velocity of data. It is not just the amount of data or the speed in which data is made available that is reformed. The wide variety of data that is available to the user and the scope it provides is enormous. From traffic pattern to the climate to medical history to music download, any range of data is being made available. The data varies in its type and format. This is what is meant by variety attribute of big data.

The scope presented by medical data in research is vast. Any kind of data related to medical field is readily available in different health organizations and centers. Since a wide variety of research is going on in the promotion of community health, data is in turn provided to the researchers voluntarily. Personal data of patients can be used by researchers in their research to improve the diagnosis of various disease patterns. All range of information about the patient will be available at the hospitals like name, disease, address, birth date, zip code, providing it to the researcher will be very concerning to a large section of patients. From the medical data the identity of a patient can be easily found by cross checking it with public data records. Thus sensitive data of patients can be made easily available.

Privacy preservation is one of the important research area, since it has a lot of loop holes. Releasing sensitive data is a very delicate issue especially in case of medical data. In most of the cases people are not ready to disclose information. In that case hospitals should go an extra mile to preserve the privacy of their clients. The techniques that are available in preserving privacy lacks some core functionalities. To overcome the lacuna, a unique method completely new in the field of big data is introduced in privacy preservation namely personalized anonymity (Fig. 1).

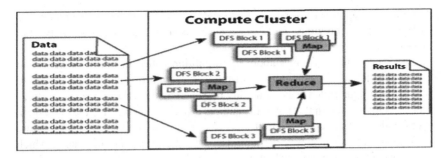

Fig. 1 MapReduce architecture

Fig. 2 HDFS architecture

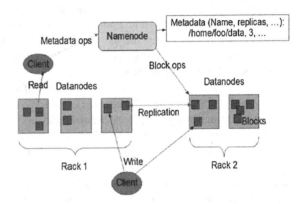

The central idea of personalized anonymity technique introduced for medical related big data can be distributed into two important methods. The first section of the work deals with attributes in the patient data which is used as a flag and can be used to differentiate sensitive attribute. The attributes include sensitive disclosure flag (SDF) as well as sensitive weigh (SW). The second component deals with a new demonstration called Frequency Distribution Block (FDB) and quasi-identifier Distribution Block (QIDB), which uses the SW and SDF for anonymity (Fig. 2).

This research aims to provide an overview of the privacy preservation techniques in medical big data. The organization of this paper is as follows: Sect. 2 is literature survey that briefing about the related work carried out in this domain by various researchers. Section 3 explains the design model for the personalized anonymity system. Section 4 provides implementation and results while Sect. 5 summarizes of the entire work.

2 Literature Survey

Kiran et al., discuss a method called personalized anonymization in which a guard node is made use of to specify whether the data owner is ready to disclose its sensitivity depending upon which anonymization will be done. Here the authors present a novel approach for personalized privacy preservation that overcomes the disadvantages of previous techniques such as k-anonymity, l-diversity and t-closeness [1].

Kristen LeFevre et al., talk about the algorithm called incognito which is an efficient full domain k-anonymity algorithm. Prevention of join attack is done by using the technique called as k-anonymity by generalizing and suppressing portions of released microdata. In this paper the authors discuss method called as full domain generalization [2]. Benjamin Fung et al., here the authors aim at finding k-anonymization not in a way of minimising the data distortion optimally, which keeps classification structure. Their experiments in real-life data shows that classification quality can be retained even for extremely restraining anonymity requirements. So small program with the classification; although a masking deletes some beneficial classification structures, substitute structures help to emerge. Therefore all data items are not as useful for the classification and least useful data elements provides space for anonymizing the data without compromising the utility [3].

In this paper authors discuss about two simple attacks that a k-anonymized dataset has some fine, but brutal privacy problems. If there is any slight diversity in those sensitive attributes then the attacker can discover its sensitive attributes. They provide a thorough analysis of the two attacks and propose a powerful privacy policy called l-diversity. And they show that l-diversity can be efficiently implemented. At last author concludes that k-anonymity datasets allow strong attacks due to lack of diversity in sensitive attributes [4]. Ninghui Li et al., here the authors tell that each equivalence class contains at least k records which are required for the k-anonymity for publishing microdata. In recent times some authors discovered that attribute disclosure is not prevented by k-anonymity. l-diversity has been considered to resolve this, at least l well represented values for each sensitive attribute should be contained in each equivalence class. In this paper authors discuss limitation of the l-diversity. They say that attribute disclosure is not necessarily prevented by it. They introduce a new notion called t-closeness, which requires the threshold t should be as minimum as the distance between two distributions [5].

3 Design Model

k-anonymity l-diversity, t-closeness, are some of the techniques currently in use for medical records for anonymity, and respectively each of it has certain drawbacks. It contains the sensitivity disclosure and useful data loss. To overcome the above mentioned lacuna, personalized privacy preservation is introduced. The record owner's consent is considered for data disclosure. In the record holder can indicate his privacy by specifying with respect to a guard node. The standards of it is built on a hierarchy based Sensitive Data outlined by Publisher. The main disadvantage is that the sensitive attribute distribution is not considered, while anonymizing. To overcome this new privacy preserving technique is proposed.

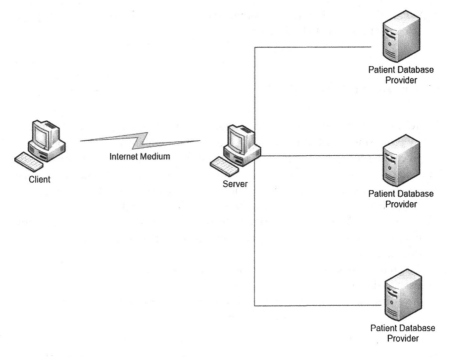

Fig. 3 Block diagram

MapReduce is distributed and parallel programming model used to process and generate huge data sets. Here MapReduce programming model is used for processing the collected patient data (Fig. 3).

The above block diagram shows how system works. Initially the data will collect from the patients and uploaded to the server. From the server any researcher can access the data and use it for his research.

The working of the system involves a centralized server, and patients, their details are collected from various hospital servers. The main aim is to protect the sensitive data from attackers. To achieve it T* is constructed which is a generalized table. In that table, the diversity of overall distribution in FDB should maintain such that it is equal to the distribution of each QIDB. It is used in minimalisation. The assumptions are, attribute S, and s is its value. There are number of tuples present in relation T.

The variables used are THρn-minimum number of records in T, THρiter-maximum number of iterations that must be performed, THρsuppr-minimum number of sensitive values for suppression, THρdisc-minimum number of sensitive values for disclosure, THρacc-minimum threshold that can be added or subtracted. The execution of the process starts by collecting THρn, THρiter, THρsub, THρdis, THρacc. Then total sensitive values present in the data are collected. If sw(n) < THρn then exit otherwise calculate the difference between QIDB and probability using QIDB(±)

THρacc. If the value is less than probability then disclose the data otherwise suppress and fetch the next record of the patient.

4 Results and Discussion

The results were evaluated based on a number of simulations. The output obtained is very promising. A set of records with all possible combinations were considered for processing the patient data. The results are provided in terms of graphs. The following snapshots of the project shows the results obtained by our work. First screenshot shows delay in milliseconds to process the data using k-anonymity technique. Second screenshot shows the delay in milliseconds to process the data using the personalized anonymization technique using sw-sdf. Here we can clearly tell that time taken by our method is more efficient than the previously available methods (Fig. 4).

The graph shows the time delay in the k-anonymity method. It takes 800 ms to process the data (Fig. 5).

The graph shows time delay for sw-sdf method. It is the efficient way to process the data as it takes only 200 ms to process it.

Fig. 4 k-anonymity graph

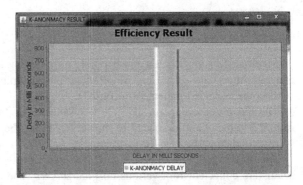

Fig. 5 Sw-sdf graph

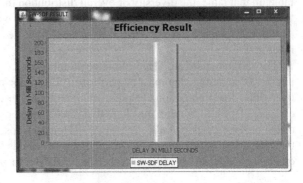

5 Conclusion and Future Work

Big data management system for personal privacy using sw-sdf uses less time for execution and data quality is preserved. Sensitivity of the record is indicated by using SW flag because whole record does not require privacy and the data utility is improved. SDF is another flag which also improves utility in the record containing sensitive weigh where a number of record owners are okay with revealing the identity some are not. So it can be decided by using the SDF flag.

QIDB anonymization is efficient in generalizing different quasi group independently. This method overcomes record linkage, probabilistic attacks and attribute linkage. This method works very well when concerned about frequency distribution of a particular sensitivity.

There are many ways in which this work can be extended. New future enhancement is that datasets can be used directly. The effect of multiple releases and sequential release of distributed records and its effect is also not considered. Different levels of weights can also be allocated and also can be extended to support unstructured data. Only five diseases are considered for the research purpose but it can be increased to any number the researchers want.

References

1. Kiran P and DrKavya N P "SW-SDF Based Personal Privacy with QIDB-Anonymization Method" (IJACSA) International Journal of Advanced Computer Science and Applications, Vol. 3, No. 8, 2012.
2. K. LeFevre, D. DeWitt, and R. Ramakrishnan. Incognito: Efficient full-domain k-anonymity. In ACM SIGMOD, 2005.
3. Ninghui Li, Tiancheng Li and Suresh Venkatasubramanian "t-Closeness: Privacy Beyond k-Anonymity and -Diversity" Department of Computer Science, Purdue University and AT&T Labs – Research.
4. Latanya Sweeney "K-Anonymity: A Model For Protecting Privacy" k-anonymity: a model for protecting privacy. International Journal on Uncertainty, Fuzziness and Knowledge-based Systems, 10 (5), 2002; 557–570.
5. Benjamin C.M. Fung, Ke Wang, and Philip S. Yu "Anonymizing Classification Data for Privacy Preservation" IEEE transactions on knowledge and data engineering, vol. 19, no. 5, may 2007.

Erratum to: An Amalgamated Strategy for Iris Recognition Employing Neural Network and Hamming Distance

Madhulika Pandey and Madhulika Bhatia

S.C. Satapathy et al. (eds.), *Information Systems Design and Intelligent Applications*, Advances in Intelligent Systems and Computing 434, DOI 10.1007/978-81-322-2752-6_73

Erratum DOI: 10.1007/978-81-322-2752-6_76

The name of second author was inadvertently removed by Volume Editor from Chapter 73 and now he would like to include the name of second author 'Madhulika Bhatia' through this erratum.

The online version of the original chapter can be found under
10.1007/978-81-322-2752-6_73

M. Pandey (✉)
Pune, India
e-mail: madhulika.pandey02@yahoo.in

M. Bhatia
Amity University, Noida, Uttar Pradesh, India
e-mail: mbhadauria@amity.edu

Erratum to: Emotion Recognition: A Step Ahead of Traditional Approaches

Surbhi Agarwal, Madhulika Bhatia and Madhurima Hooda

Erratum to:
Chapter "Emotion Recognition: A Step Ahead
of Traditional Approaches" in: S.C. Satapathy et al. (eds.),
Information Systems Design and Intelligent Applications,
Advances in Intelligent Systems and Computing 434,
https://doi.org/10.1007/978-81-322-2752-6_71

In the original version of the book, the incorrectly published author name "Madhulika Madhurima" has been now corrected to read as "Madhulika Bhatia" and "Madhurima Hooda" in the chapter "Emotion Recognition: A Step Ahead of Traditional Approaches".

The updated online version for this chapter can be found at
https://doi.org/10.1007/978-81-322-2752-6_71

Author Index

© Springer India 2016

S.C. Satapathy et al. (eds.), *Information Systems Design and Intelligent Applications*, Advances in Intelligent Systems and Computing 434, DOI 10.1007/978-81-322-2752-6